World Scientific Series on Emerging Technologies:
Avram Bar-Cohen Memorial Series – Volume 7

POLYMER SOFT MATTER SCIENCE & ENGINEERING

Volume 1

Advanced Polymer Life Science

World Scientific Series on Emerging Technologies: Avram Bar-Cohen Memorial Series

Print ISSN: 2737-5862
Online ISSN: 2737-5870

This compendium provides a comprehensive collection of the emergent applications of big data, machine learning, and artificial intelligence technologies to present day physical sciences ranging from materials theory and imaging to predictive synthesis and automated research. This area of research is among the most rapidly developing in the last several years in areas spanning materials science, chemistry, and condensed matter physics.

Written by world renowned researchers, the compilation of two authoritative volumes provides a distinct summary of the modern advances in instrument — driven data generation and analytics, establishing the links between the big data and predictive theories, and outlining the emerging field of data and physics-driven predictive and autonomous systems.

Published

Vol. 7 *Advanced Polymer Life Science*
edited by Jens Friedrichs, Uwe Freudenberg and Carsten Werner

Vol. 6 *Supercritical Fluids: Properties and Applications*
by Grazia Lamanna

Vol. 5 *Power Beaming: History, Theory, and Practice*
by Paul Jaffe, Tom Nugent, Bernd Strassner II and Mitchel Szazynski

Vol. 4 *System Sustainment: Acquisition and Engineering Processes for the Sustainment of Critical and Legacy Systems*
by Peter Sandborn and William Lucyshyn

More information on this series can also be found at
https://www.worldscientific.com/series/wset

World Scientific Series on Emerging Technologies:
Avram Bar-Cohen Memorial Series – Volume 7

POLYMER SOFT MATTER SCIENCE & ENGINEERING

Volume 1

Advanced Polymer Life Science

Editors

Jens Friedrichs
Leibniz Institute of Polymer Research Dresden, Germany

Uwe Freudenberg
Leibniz Institute of Polymer Research Dresden, Germany

Carsten Werner
Leibniz Institute of Polymer Research Dresden, Germany
TU Dresden, Germany

World Scientific

NEW JERSEY • LONDON • SINGAPORE • BEIJING • SHANGHAI • TAIPEI • CHENNAI

Published by

World Scientific Publishing Co. Pte. Ltd.
5 Toh Tuck Link, Singapore 596224
USA office: 27 Warren Street, Suite 401-402, Hackensack, NJ 07601
UK office: 57 Shelton Street, Covent Garden, London WC2H 9HE

Library of Congress Control Number: 2025001909

British Library Cataloguing-in-Publication Data
A catalogue record for this book is available from the British Library.

World Scientific Series on Emerging Technologies:
Avram Bar-Cohen Memorial Series — Vol. 7
ADVANCED POLYMER LIFE SCIENCE

ISBN 978-981-98-0677-5 (hardcover)
ISBN 978-981-98-0678-2 (ebook for institutions)
ISBN 978-981-98-0679-9 (ebook for individuals)

For any available supplementary material, please visit
https://www.worldscientific.com/worldscibooks/10.1142/14148#t=suppl

Desk Editor: Brittney Phitojo

Typeset by Stallion Press
Email: enquiries@stallionpress.com

https://doi.org/10.1142/9789819806782_fmatter

Foreword

The convergence of materials science and life sciences is unlocking unprecedented insights and opening new frontiers. We are now able to envision and create materials that mimic, interact with, and even enhance the complexity of living systems. As a result, the boundaries between living matter and engineered materials are gradually dissolving, paving the way for smarter, multifunctional systems.

At the heart of this transformation lies polymer science — a discipline that has evolved from the fundamental study of macromolecular structures into a vibrant, multidisciplinary nexus connecting biology, chemistry, physics, and engineering. Within this exciting landscape, the **Polymer Soft Matter Science and Engineering** series — part of the *World Scientific Series on Emerging Technologies* — showcases cutting-edge advances, illustrating the diverse and transformative potential of the field.

Polymers have long been the backbone of countless materials and technologies, driving innovations across industries ranging from healthcare and renewable energy to electronics and construction. As polymer science continues to mature, it increasingly bridges the gap between pure macromolecular science and highly specialized applications. This evolution reflects broader trends in science and technology, emphasizing the integration of biological principles, sophisticated computational tools, and sustainable practices.

However, the progression of polymer science is not merely technical — it also represents a paradigm shift in how we approach material innovation. By recognizing polymers as dynamic, adaptive systems that actively interact with their environments, we have unlocked the potential for materials that are not only functional but multifunctional, responsive, and self-sustaining. These advances are critical for addressing modern challenges and seizing emerging opportunities, demanding sustainable solutions, advanced functional materials, and the creative fusion of bioinspiration with artificial intelligence to guide the next generation of material design.

Carsten Werner
Scientific Director and Professor
Leibniz Institute of Polymer Research Dresden
and Dresden University of Technology

Jens Rieger
Independent Consultant
Formerly Senior Vice President, Advanced Materials & Systems
Research, Technology Scouting & Incubation, BASF SE

https://doi.org/10.1142/9789819806782_fmatter

Preface

Advanced Polymer Life Science, the first volume of the "Polymer Soft Matter Science and Engineering" series, provides a contemporary exploration of some of the most transformative areas of research in polymer science. It adopts a multidisciplinary perspective that reflects the field's rapid evolution and its growing integration with bioengineering, materials science, and life sciences. Contributions from leading experts demonstrate how polymers are being reimagined to address global challenges — with a particular focus on healthcare innovation.

At the core of this exploration lies a bioinspired approach, where researchers draw upon the principles of living systems to design materials that mimic nature's multifunctionality, adaptability, and efficiency. This philosophy is central to many of the topics discussed in this book, showcasing how the boundaries between biology and engineered materials are increasingly blurred.

For example, studies on *arthropod cuticle assembly* highlight how natural systems achieve exceptional strength, flexibility, and multifunctionality through intricate composite structures. By unraveling the principles of cuticle formation, researchers are pioneering innovative material designs that seamlessly integrate durability with adaptability.

Similarly, the dynamic complexity of *extracellular matrices (ECM)* — the biopolymeric meshworks that support cells in living organisms — serves as a blueprint for engineering biomaterials that replicate tissue environments. Combining *decellularized matrices* with polymers creates hybrid materials that promote cell growth, differentiation, and regeneration. This approach not only captures the multifaceted qualities of natural ECM but also holds significant promise for regenerative medicine and tissue engineering, advancing our ability to repair and replace damaged tissues.

Biomimetics extends beyond structural mimicry to embrace the dynamic, responsive properties of living systems. This is exemplified by the design of *supramolecular assemblies* and *hierarchically structured materials*, with the mussel byssus serving as a prime example. As a natural adhesive, the mussel byssus is a remarkable blend of strength and flexibility achieved through hierarchical design. By applying these principles, researchers are creating materials capable of adapting to environmental changes, self-healing, and performing exceptionally under demanding conditions.

At the cutting edge of polymer science is its seamless integration with molecular biology and nanotechnology — a synergy that has led to the development of *programmable DNA-based nanomaterials*. By leveraging the molecular recognition capabilities of DNA, these materials enable precise drug delivery, advanced molecular diagnostics, and even tissue repair. Similarly, *smart microgel and nanogel platforms* provide responsive systems that adapt to environmental changes, making them ideal for next-generation therapeutic applications.

Another exciting frontier is the development of *dynamic biomaterials* engineered through reversible covalent interactions. Their ability to respond to environmental stimuli — such as temperature, pH, or mechanical stress — unlocks transformative applications ranging from tissue scaffolds to self-healing devices. The development of *cell-compatible double-network hydrogels*, which mimic the mechanical and biochemical properties of natural tissues, exemplifies how dynamic biomaterials are advancing tissue engineering.

Beyond material design, polymers play a pivotal role in biomedical research, powering innovative systems that model complex biological processes and diseases. Examples include *tumor tissue engineering* and *neurodegeneration models*, where biomaterial-based platforms are revolutionizing our understanding of disease mechanisms. For instance, the use of polymeric materials to create *prostate cancer models* equips researchers with tools to study tumor growth, metastasis, and drug responses in a controlled environment. Similarly, polymeric biomaterials designed for studying neurodegeneration offer new insights into diseases such as Alzheimer's and Parkinson's, while simultaneously serving as platforms for testing cutting-edge therapeutic strategies.

Polymers are also central to advancing regenerative medicine, particularly through engineered *morphogen gradient systems* that precisely guide tissue development and repair. By mimicking the natural, concentration-dependent mechanisms of morphogens, these polymer-based platforms offer powerful tools for directing tissue regeneration and enhancing therapeutic outcomes.

For students and professionals in polymer science, bioengineering, or materials science, the chapters in this volume provide not only a comprehensive snapshot of current research but also a visionary roadmap for the future of the field. Ultimately, **Advanced Polymer Life Science** stands as a testament to the innovative merging of scientific disciplines and as a source of inspiration for the next generation of material scientists.

This volume marks the beginning of the "Polymer Soft Matter Science and Engineering" series, showcasing the breadth of exciting topics that emerge from the combination of polymer science and biology.

Jens Friedrichs, Uwe Freudenberg, Carsten Werner
Leibniz Institute of Polymer Research Dresden
and Dresden University of Technology

https://doi.org/10.1142/9789819806782_fmatter

About the Editors

Jens Friedrichs, a bioengineer by training, leads the research on biointerfacial phenomena at the Leibniz Institute of Polymer Research Dresden, Germany, and has extensive experience in surface-sensitive analytical techniques. His research interests include natural adhesion mechanisms and the control of cell adhesion on surfaces.

Uwe Freudenberg, a chemical engineer, directs research on cell-instructive polymer matrices at the Leibniz Institute of Polymer Research Dresden, Germany. He pioneered the development of modular glycosaminoglycan-based hydrogels as well as their customization for specific biomedical applications, including clinical translation.

Carsten Werner, a biochemist, is the Scientific Director of the Leibniz Institute of Polymer Research Dresden, Germany and holds a professorship for Biofunctional Polymer Materials at TU Dresden, Germany. He established a biomaterials program that combines exploring functional living matters from a materials science perspective with developing bioinspired polymer materials.

https://doi.org/10.1142/9789819806782_fmatter

Contents

https://doi.org/10.1142/9789819806782_0001

Chapter 1

Arthropod Cuticle Assembly: Bioinspired Insights for Materials Design

Aurimas Narkevicius[1], Michaela Wilsch-Bräuninger[2], Luca Bertinetti[1] and Yael Politi[1*]

[1]*Center of Biomolecular Engineering, B CUBE, TU Dresden, Tatzberg 41, Dresden, Germany*
[2]*Max Planck Institute of Molecular Cell Biology and Genetics, 01307 Dresden, Germany*
**Corresponding author: yael.politi@tu-dresden.de*

The arthropod cuticle is a versatile multifunctional material that captivates the observer with its iridescent colors and mechanical resilience. Despite being made of a limited set of building blocks — primarily chitin, proteins, and water — the arthropod cuticle demonstrates a diverse range of material properties. Like in many other biological materials, this resourcefulness emerges from hierarchical architecture and intricate compositional gradients that arise through the assembly of the building blocks in a cellular-controlled environment. Here, we first introduce the structure and composition of the arthropod cuticle followed by a discussion about the cuticle deposition. Given that the major steps in this process are not well known, beyond providing a summary of the key observations, we put forward questions and speculate upon cuticle deposition and, in particular, the control of its ultrastructure with the hope of inspiring new ideas in the field. We expect that a better understanding of the biological material formation mechanisms can provide new ideas for bioinspired materials design and processing irrespective of the actual building blocks.

1. Introduction

Nature accommodates a wide range of biological materials that, having gone through years of evolutionary adaptation within the boundaries of phylogenetic, ecological, and energetic constrains, show outstanding material properties and functions. Contrary to some simplifications, biological materials are not necessarily "optimally designed" to fulfill one particular function, given their evolutionary constraints, they rather function well enough to serve the organism at least during its lifespan. At the same time, this evolutionary legacy promotes impressive multifunctionality.[1,2]

Although many biological materials, like wood or wool, serve as high-quality raw resources, it is their design principles that are most inspiring (**Figure 1**). With the building blocks being generally restricted to the most abundant and accessible elements on Earth, the functionality of biological materials arises primarily from their hierarchical structuring and tailored compositional gradients. In contrast, the performance of engineered materials usually relies on a large variety of chemical elements and their compounds without the complexity over hierarchical levels.[3] Therefore, the performance

Figure 1. Arthropod cuticle serves as an archetype for bioinspired materials due to its interesting diverse designs and properties. In addition, the assembly processes can also promote the creation of new greener materials manufacturing strategies.

of engineered material could be improved by integrating the design principles seen in biology irrespective of the building blocks themselves. Moreover, biology can inspire more sustainable material fabrication methods because the processes take place in ambient conditions and in aqueous environments. In particular, self-assembly strategies within the biologically imposed boundary conditions give rise to structural and functional complexity, which contrasts with energy-intensive processes and organic solvents commonly used in human technologies. However, biological materials are formed bottom-up in a temporally and spatially controlled manner, and thus, their recreation is mostly beyond the capacity of current engineering technologies.

A number of excellent reviews on the composition, ultrastructure, and properties of biological and bioinspired materials, including aspects of sustainability, can be found in the literature.[4–7] However, relatively few discuss the processes involved in the assembly of biological materials, partly because it is the least understood aspect.[8,9] Here we discuss the deposition of a fiber-reinforced composite material, that is, the arthropod cuticle, without expanding into other systems for brevity. We first briefly describe the makeup of the arthropod cuticle and its relationship with property and function to provide sufficient context, followed by outlining what is known about the processes of how it is made and, finally, we pose several open questions that still remain to be answered. We believe that such a discussion is greatly needed to tackle the fundamental aspects of the making of biological materials as well as for designing bioinspired processes to produce new materials.

2. Arthropod cuticle

The phylum of *Arthropoda* is the most diverse within the animal kingdom and includes organisms such as crustaceans, insects, and arachnids. Despite showing high diversity in habitats, life cycles, shapes, and sizes, arthropods are defined by their segmented bodies and articulated appendages, the number and morphology of which are used for classification. A prominent characteristic of all arthropods, and most relevant to this chapter, is the presence of the cuticular exoskeleton — a fiber-reinforced composite material — primarily made of the polysaccharide chitin and a number of proteins.[10,11] The cuticle is typically hydrated and contains minor amounts of small cross-linking organic molecules and metal ions. In some specific insect clades and in most crustaceans, the cuticle also contains minerals, which are mainly calcium salts.[12,13] Although the arthropod cuticle serves primarily as a protective layer, it is inherently multifunctional and is used for vision, inter- and intraspecies communication, controlling hydration and gas exchange, and is used to form specialized structures such as mouthparts and claws.[4,7,11] In addition, cuticle or cuticle-like materials make up a large part of the respiratory organs of the arthropods, line the digestive tract, and serve for muscle attachment in tendon-like structures allowing locomotion. In this chapter, we concentrate on the cuticle that defines the integument of the arthropod.

2.1. *Cuticle organization*

The cuticle of the arthropod is made primarily of three principal layers: envelope, epicuticle, and procuticle.[14,15] Procuticle is often further subdivided into exocuticle, mesocuticle, and endocuticle.[11,16,17] A single layer of epidermal cells lines the cuticle from the inside. Encompassing the animal in a hard shell, the cuticle must be shed and replaced for the animal to grow in size. Although the cuticle does not contain living cells, it is highly dynamic as it undergoes remodeling, healing, and changes in hydration. We only briefly discuss the architecture of the cuticle here; deeper discussion on the topic can be found in many excellent reviews.[7,11,16,18]

The epicuticle and the outermost envelope layer are devoid of chitin and are made of cross-linked lipids and proteins.[19,20] Due to its composition, the envelope acts as a hydrophobic protective layer against dehydration and pathogen entry.[14] The envelope, which is tens of nanometers thick, is deposited first during new cuticle formation, acting as a protective layer between the forming cuticle and the old cuticle that is being digested in the process.[14,21] The thicker epicuticle has a varying thickness up to several micrometers and forms a stable protective layer, which also contributes to mechanical properties including wear and scratch resistance.[17,22,23] The envelope and epicuticle layer are often decorated by minute structures such as ridges, pillars, and nipples, which endow it with exquisite surface properties, including superhydrophobicity, superomniphobicity, antireflection, structural coloration, and more.[24,25]

The procuticle encompasses the bulk of the exoskeleton. It is situated directly under the epicuticle and above the epidermal cells and is defined as a chitin-containing layer.[11,14] The procuticle is usually further subdivided into the exocuticle and the endocuticle, based on the time of deposition. The exocuticle is laid down by the epidermal cells soon after the envelope and the epicuticle, but before the organism emerges from the previous molt or pupal casing, in the processes known as ecdysis or eclosion, respectively.[21,26] After the organism emerges, the exocuticle becomes chemically cross-linked, rendering it stiff and hard.[27,28] Notably, the exoskeleton of crustaceans also mineralizes at this stage.[29] The endocuticle, deposited exclusively after ecdysis, is usually thicker and typically less cross-linked than the exocuticle.[16] Another layer of procuticle, named mesocuticle, is often identified by histological staining in between the exocuticle and the endocuticle, which has transitory properties of the exocuticle and the endocuticle in terms of cross-linking.[16]

The procuticle has a general conserved architecture of chitin microfibrils organized in a protein matrix.[30,31] Although, such a description might be misleading, as a large part of the proteins is typically well ordered with respect to the chitin microfibril (**Figure 2**).[30,32] Within the procuticle, chitin microfibrils are generally aligned, giving an impression of the presence of stacked sheets. The chitin microfibril alignment direction continuously changes within the sheets across the stack, resulting in a helicoidal architecture.[33,34] This type of ordering is also found in many biological materials, such as bone and plant tissues, made of mineralized collagen and cellulose, respectively.[35–37]

The helicoidal architecture is described by the principal axis that defines the direction around which the sheets within the

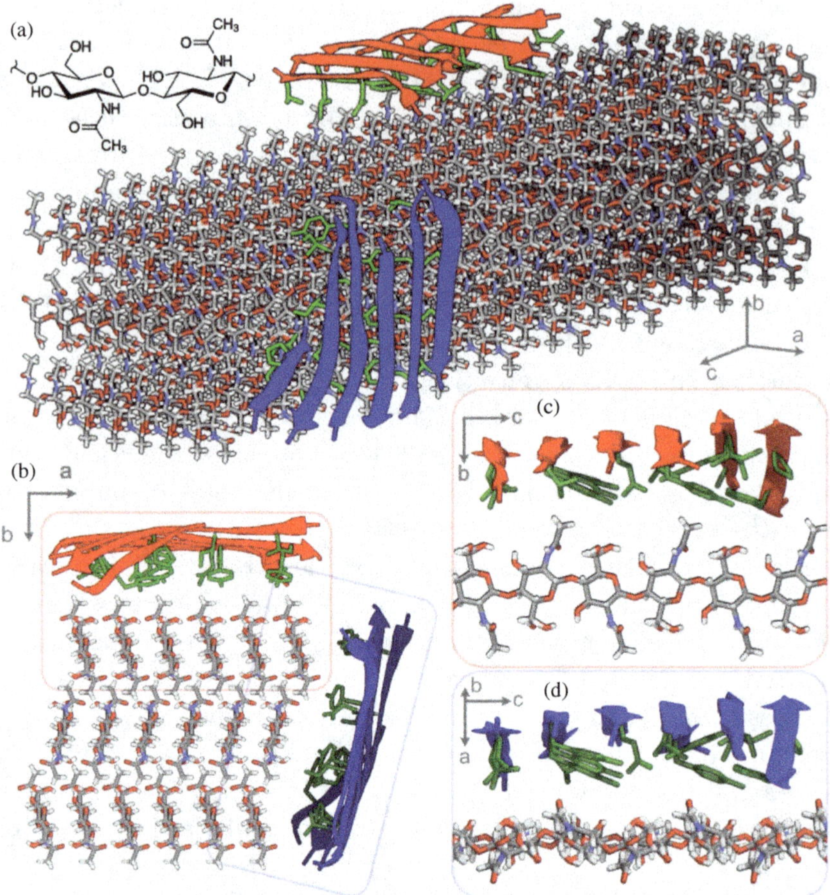

Figure 2. Chitin polymer chains are arranged in an antiparallel orientation in a crystal lattice and interact with chitin-binding domains (CBDs). (**a**) 3D rendering of an α-chitin crystal consisting of 18 units and terminated by (010) and (110) faces. The predicted tertiary structure of a CBD from *Drosophila melanogaster* cuticle (UniProt: A0A6J2SSP6, AlphaFold: A0A6J2SSP6) is docked on top of the two crystal surfaces, namely (010) and (110). The CBD β-sheet is displayed in red and blue when adsorbed to the (010) and the (110) surfaces respectively. (**b**) A view down the crystal c-axis. Note the agreement between the size of the crystal core and the protein β-sheet. (**c**, **d**) A zoom-in onto the CBD docking chitin crystal. The structural prediction suggests that Tyr moieties facing one face of the β-sheet orient to allow possible favorable interactions with chitin acetyl groups on (010) face (**c**) or stack with sugar rings on the (110) face.

stack revolve. Within such a structure, chitin microfibrils are aligned perpendicular to the principal axis and parallel to the surface of the integument. Interestingly, in the arthropod cuticle, the rotation direction of the helicoidal structure takes exclusively a left-handed sense.[11] In addition, the distance over which the helicoidal structure revolves by 360°, termed pitch, is typically submicrometric within the exocuticle and larger, up to hundreds of micrometers, in the endocuticle.[16,18] Notably, the helicoidal pitch usually shows variations and gradients even within a single cuticle type.[16,38] While the exocuticle is primarily found to be helicoidally arranged, the endocuticle shows a higher degree of ultrastructural variation; for example, unidirectional structures, where chitin microfibrils are well aligned to a single direction without rotation, as well as plywood-like organization with discrete rather than continuous changes in microfibril orientation.[11,39]

Ultrastructural organization has important implications on the mechanical properties of the material.[7,39,40] Chitin microfibers have anisotropic mechanical properties, with the Young's modulus along the long axis of the microfibril predicted to be around 100 GPa[41]; however, values obtained experimentally indicate lower values of around 60 GPa.[42,43] In addition, the mechanical properties of the protein matrix are estimated to be two orders of magnitude lower, most likely covering a large range depending on the hydration and cross-linking levels, which are in turn dependent on the protein sequence.[7,17] The helicoidal arrangement provides resistance to shear and bending deformations as well as improves resistance to crack propagation, whereas unidirectionally arranged chitin microfibrils within the cuticle maximize its tensile properties along the microfibril long axis, which are needed in structures such as tendons.[39] Overall, the helicoidal and the parallel fiber architectures are mechanically complementary, and in fact, organisms combine them within the cuticle in different proportions and relative organizations to attain the required material properties.[4,23] By combining structural and compositional modulations, the arthropod cuticle displays highly tunable mechanical properties; for example, the Young's modulus can span over four orders of magnitude.[7]

Helicoidal ultrastructure can also produce structural coloration.[11,44,45] This occurs when the helicoidal pitch is comparable to the wavelength of visible light, resulting in reflection of circularly polarized light, which has the same handedness as the helicoidal structure. While the wavelength of light reflected depends on the helicoidal pitch, the refractive indices of the material, and the angle of the illumination and observation, the reflectivity is determined by the overall thickness of the helicoidal structure as well as the birefringence of the material.[46] Nevertheless, the reflectance of helicoidal structures is limited to 50% of the light coming from usual light sources, such as the sun, which emit randomly polarized light.[46,47] These limitations, for example, in the case of the beetle *Plusiotis resplendens*, are overcome by incorporating additional optical components sandwiched between the helicoidal architectures.[45,47] Ultimately, any structure with periodicity comparable to the wavelength of visible light can produce structural coloration, where the symmetry and order or disorder level of the system determine the degree of angular reflectance and diffuse scattering.[44] In fact, arthropods use a wide variety of structures for structural coloration, including multilayers, diffraction gratings, photonic crystals, and more.[44,48] The fundamentals of structural coloration in biological materials are reviewed in detail elsewhere.[48]

The optical effects demonstrated by the cuticle are not limited to structural coloration. The reflectance of electromagnetic radiation in the near-infrared range from the cuticle is thought to contribute to passive thermoregulation.[49,50] Furthermore, the cuticle also makes up important parts of the cornea of many arthropod species; besides its antireflection properties, stemming from refractive index gradients associated with surface nanostructures, the cuticle may also serve as an efficient lens to collect light for vision by various mechanisms.[24,51]

2.2. *Chitin and proteins in arthropods*

Chitin is one of the most abundant biopolymers on the planet. In addition to arthropods, mollusks, sponges, annelids, cnidaria, nematodes, diatoms, and fungi also produce chitin.[52,53] Its primary role is to provide structural integrity and mechanical support. In nature, chitin is typically associated with other components such as proteins, other polysaccharides, and minerals.

The large biomass of the arthropods itself and their cuticle containing around 30% of chitin, makes them the most common source of chitin. In particular, crustacean exoskeletons (a waste by-product of the food industry) are a common source of raw material. Based on the annual crustacean harvest worldwide of around 20 million tonnes,[54] an estimated 3 million tonnes of chitin could be extracted. In contrast, the global cellulose pulp production is around 180 million tonnes originating from wood.[55] As such, chitin could find use in specialty products instead of resource-heavy industries and integrated in a circular economy.

However, chitin harvesting and extraction are not inherently sustainable. Animal rearing and chitin isolation have a notable environmental impact; in particular, crustacean cultivation is a highly energy- and water-intensive process.[54,56] Moreover, the isolation of chitin from crustacean sources requires demineralization, consuming large amounts of acids, and releasing carbon dioxide with additional environmentally impactful steps for deproteinization and depigmentation.[56,57] Other alternative resources of chitin include farmed insects and fungi, with an estimated lower environmental impact.[58,59] In any case, the extraction of most resources impacts the environment and so need to be carefully assessed and regulated to make them sustainable materials.

Chitin is made of N-acetyl-glucosamine units that are connected to form a linear polysaccharide. The degree of polymerization (DP) is expected to vary highly on the source of chitin as well as the purification processes.[52] The best estimates to date suggest that chitin in prawn exoskeleton has a DP of around 20,000, which is reduced by more than half upon conventional extraction methods.[60] This is especially striking given that the state-of-the-art synthetic polysaccharide synthesis methods allow polymerization only of up to several sugar units.[61]

Chitin is also characterized by its degree of acetylation (DA), which defines the level of acetylation with respect to the theoretical maximum. While it is difficult to determine with sufficient precision, DA values for purified chitin are usually reported to be in the range of 90–100%, which could partly stem from technical difficulties.[11,62–64] Deacetylation can have important consequences on the properties of chitin within the cuticle, by providing a pH-dependent positive charge, changing the levels of hydration propensity, and providing sites for regioselective chemical and physical interactions.[52,65] Importantly, not only the DA but also the pattern of acetylation can be important in defining the interactions with the proteins.[66,67] While chemical deacetylation is most likely providing a random pattern of deacetylation,[68] it is becoming clear that enzymatical deacetylation produces defined patterns.[69] Interestingly, the patterns of deacetylation of chitin oligomers can be important for intra- and interspecies signaling.[70,71] Multiple chitin deacetylases (CDAs) are found in insect cuticle although some deacetylase-like proteins appear to be catalytically inactive.[72] Many CDAs are expected to be active extracellularly on the nascent chitin, and have been demonstrated to be required for cuticle assembly.[73,74]

Chitin is largely insoluble in most solvents. While the insolubility of chitin is often cited as a limitation in chitin processing,[75] it can also be viewed as an advantage when low solubility of the final product is beneficial. The primary reason behind the insolubility of chitin is the extended network of intramolecular and intermolecular hydrogen bonds accompanied by van der Waals interactions, leading to orderly packing.[75,76] Chitin molecules are packed in one of the two different crystalline allomorphs, namely alpha and beta.[76–81] In α-chitin, the polymer chains are organized in an antiparallel manner, whereas in β-chitin, all the polymer chains are parallel. The antiparallel packing allows additional intermolecular hydrogen bonding, rendering the former more thermodynamically stable and less prone to swelling.[77] Not only is α-chitin more abundant in nature, it is also so far the only form to be crystallized in vitro.[78,82,83] β-chitin is found in the peritrophic matrix in arthropods, some mollusks, annelid worms, diatoms, and others.[84] Further in the text, we only refer to the α-chitin allomorph, unless stated otherwise.

The crystalline structure of chitin leads to its microfibrillar morphology. Chitin microfibrils (also referred to as nanofibrils at times) in arthropod cuticle have a cross-sectional diameter of 2–5 nm, depending on the organism or tissue studied.[85] The dimension of the cross-section combined with the known crystallographic unit cell[76] implicates 18–25 molecular chains aligned along their long axis per microfibril (**Figure 2**). Within the arthropod cuticle, the microfibrils are spaced by 5–7 nm center-to-center distance, resembling hexagonal packing.[85,86] Notably, due to its small size, the actual cross-sectional shape remains to be determined, although a polygon is well assumed.[11,85,87] In fact, in cases, in which the chitin microfibril cross-sections are considerably larger, as in the fibers synthesized by the algae *Phaeocystis*, a hexagonal cross-section has been determined.[87] The length of chitin microfibrils is not precisely known due to extraction challenges, but is expected to be on the order of micrometers. This is in accord with the 10-μm long chitin molecules, calculated based on the measured average molecular weight.[60] Notably, however, the fibril is most likely longer than the length of a homogenous crystal length, implicating defects or less ordered parts.

2.3. *Cuticular proteins*

Proteins within the arthropod cuticle constitute around 70%, and in some cases even more, of the dry weight, which signifies their relevance in determining the properties of the biological material as a whole.[27,88] The fact that proteins do not comprise a merely amorphous matrix has been postulated as early as 1950s, by showing that chitin crystals are covered by proteins with β-sheet structure in a regular manner.[30,32,89–91] Importantly, the removal of cuticular proteins leads to an irreversible chitin crystal coalescence.[32] The number of proteins and their amino acid composition vary significantly between species and within different tissues of the same organism.[88,92] Cuticular proteins studied to date are classified into 12 different protein groups based on their sequence.[93] Characteristically, these proteins have low composition complexity, with low abundance of cysteine, and high levels of glycine, alanine, histidine, and tyrosine amino acids.[93,94] In particular, tyrosine is implicated in chitin binding, tyrosine and histidine in enzymatically controlled cross-linking with catechol-based derivatives during sclerotization of the procuticle,[27,28,95,96] and histidine in forming stabilizing interactions via coordination with metal ions such as zinc.[97,98]

Of particular importance are cuticular proteins with a chitin-binding domain (CBD). These proteins mainly belong to one of the two classes, based on their CBD. The so-called Rebers and Riddiford (R&R) consensus is the most abundant CBD in the cuticle of all studied arthropods to date, and proteins with this binding domain belong to a protein family called cuticular proteins with R&R consensus, or CPR (**Figure 2**).[93,99,100] While various versions of the sequences are published, it is clear that the consensus pattern is highly conserved, containing a high proportion of glycine, tyrosine, and phenylalanine, where the aromatic residues are believed to directly participate in chitin binding.[100,101] This domain folds into a β-sheet composed of several antiparallel β-strands, for example, the cuticular protein of *Drosophila hydei* (UniProt: A0A6J2SSP6; AlphaFold A0A6J2SSP6) (**Figure 2**).[101,102]

Another prominent type of chitin-binding domain, CBD2, contains six conserved cysteine residues that form three disulfide bonds stabilizing a β-sheet, made of antiparallel β-strands, similar to the CBD mentioned above.[93] Unlike the R&R consensus, CBD2 is also abundant in both structural and catalytic proteins in other phyla. Cuticular proteins that contain CBD2 are grouped into a family called cuticular proteins akin to peritrophins (CPAP) and are present in the exoskeleton and the peritrophic membrane of many arthropods.[103–106]

Besides structured chitin-binding domains and N-terminus signaling sequence targeting the proteins into the extracellular matrix, the remaining parts of the cuticular structural proteins contain low sequence complexity and are predicted to be intrinsically disordered.[107] In this context, it is interesting to note that intrinsically disordered regions and high abundance of Tyr and His are characteristic of proteins that undergo liquid–liquid separation into condensed phases.[8] The consequences of variable amino acid composition, structure, or the presence of intrinsic disorder on the mechanical properties of the proteins, their interactions with chitin or each other via cross-linking, are still poorly understood.

Recent findings show that some cuticular proteins form condensates in vitro.[108–110] Specifically, two of the most abundant proteins within the mandibles of the migratory locust, implicated in directly promoting the exceptionally high mechanical properties of the material, undergo hierarchical self-assembly from condensed liquid phases to solid films.[109,110] The formation of protein-condensed liquid phases is particularly interesting in the context of protein delivery into the growing cuticle and the manner they infiltrate into the chitin-defined architecture.[8,111]

Besides structural proteins, there is a number of catalytic proteins present in the arthropod cuticle.[106] Many of them possess a chitin-binding domain, mainly CBD2, and have been shown to be essential for the correct assembly of the cuticle. The most prominent examples are chitinases and chitin deacetylases, which upon silencing in multiple organisms, lead to phenotypes with a highly disturbed cuticle ultrastructure and fatality in many cases.[103,105,112–114] In addition, catalytic proteins are also required for a well-timed cross-linking of the cuticle, which is especially prevalent within the exocuticle.[27,28] Thorough reviews of enzymatic cuticular proteins can be found elsewhere.[106,115,116]

Marine as well as terrestrial crustaceans often include a large amount of minerals in their cuticles. Calcium carbonate and calcium phosphates are the most abundant and are typically found in an amorphous form; however, crystalline polymorphs are also frequently found especially in crabs and in the epicuticle and the mandibles of some lobsters.[12] While biomineralization is widespread in crustaceans, it is much less abundant in insects. Some fly larvae (Diptera) use calcium carbonates or phosphates in both amorphous and crystalline forms to reinforce their pupal cases.[117–120] In addition, it is known that some beetles use biomineralization in their cuticles,[121] and yeast-farming ants reinforce their cuticles with an external layer of high magnesium calcite.[122] Despite these interesting observations, little is known about the mechanisms of biomineralization in these species. For more discussion on biomineralization in crustaceans, we refer the reader to several excellent reviews.[12,18,29]

In addition to the major components of the arthropod cuticle described above, a small amount of sclerotization agents has a strong effect on the mechanical properties of the material.[27,123] Sclerotization is the process of covalent enzymatic cross-linking of structural proteins and chitin. During sclerotization, N-acetyldopamine and N-β-alanyldopamine precursors are activated by phenoloxidases, quinone isomerase, and other enzymes found in the cuticle, leading to the formation of adducts and cross-links in the cuticle.[27,28,96,124,125] At least four different mechanisms are known to be involved in cuticle sclerotization, some of which endow the final cuticle with amber coloration. As such, the sclerotization process is also sometimes called tanning. Sclerotization not only hardens the material by introducing covalent cross-links, but is also correlated with drastic dehydration of the tanned cuticles.[17,27] However, it has also been shown that water loss in medfly pupa mutants that are unable to transport sclerotization precursors is similar to sclerotizing wild type.[126] The relationship between cuticle cross-linking, dehydration, and hardening is, however, not mechanistically clear.[28,126,127] Additionally, sclerotization and melanization mechanisms are also involved in wound healing and cuticle repair. A thorough review of cuticle sclerotization is provided elsewhere.[27,28]

3. Cuticle deposition

While a deeper understanding of the structure-property-function relationship of the arthropod cuticle allows translating the "design" principles of biological materials into engineered ones,[128,129] the recreation of the intricacies of the biological composites remains a challenge in its own right. Understanding the control over ultrastructure in the arthropod cuticle may lead to new approaches for materials engineering. In addition, applying the principles of biomimicry in the synthesis of engineered materials could provide sustainable strategies as biological material production occurs at ambient conditions and in an aqueous environment. However, we are limited by the lack of knowledge on the formation of many biological materials in general, and specifically on arthropod cuticle deposition. Here, it is our aim to not only summarize the current state of the art on arthropod cuticle deposition, but also to raise questions and start engaging in a much-needed discussion on the topic. We limit ourselves to Insecta, the most studied arthropod group in this regard. Even within this single class, some variations are found between species, different developmental stages, and the exact location investigated. Nevertheless, the general principles of the cuticle organization are mainly conserved and are expected to apply for the whole phylum.

3.1. *Cells that deposit the cuticle*

The arthropod cuticle is an extracellular matrix secreted by a single layer of epidermal cells, situated just beneath it.[113,114] To fulfill this role, the epidermal cells are highly polarized along and across the cell layer.[130] Planar cell polarity of the epidermal cells allows the cells to determine their position with respect to one another and within the whole body.[131] The basal side of the epidermal cells secretes a collagenous matrix, the so-called basal membrane, which is in contact with the hemolymph, which is functionally equivalent in many respects to mammalian blood.[130] The apical side of the epidermal cells directly faces the extracellular space into which the cuticular components are secreted.[11,132] The plasma membrane here is heavily modulated during cuticle deposition and takes shape akin to finger-like protrusions, termed microvilli, which are supported by actin filaments.[133] Other structures, which are ridge-like and named apical undulae, have been observed so far only in *Drosophila melanogaster* during the embryonic stage and are supported by microtubules running along the length of the ridges.[134,135]

Microvilli and apical undulae share many similarities. Both of these structures protrude out of the cell by a few hundred nanometers, with microvilli typically being longer.[133,134,136] At the tips of these structures, electron microscopy shows areas of high electron density, termed plaques, which suggests protein accumulation.[137] The plaques are tens of nanometers wide and are generally believed to be the primary location of chitin-synthesizing enzymes, namely chitin synthases (ChS); however, direct evidence is still missing. In addition, plaques appear concomitantly with microvilli or apical undulae long before the secretion of chitin, where their positions correlate with the sites of the initiation of the envelope deposition.[19] In addition, plaques are likely to have additional functions important for cuticle deposition. Microvilli are stabilized by high curvature proteins as well as actin-bundling proteins.[21,130] The microvilli are generally distributed on the plasma membrane in a pattern reminiscent of hexagonal packing.[130,138] However, in locusts, under particular rearing conditions, the microvilli become more ordered along a single direction, and fuse to form ridge-like structures, the importance of which is discussed later.[138] The spacing between the tips of such structures, assessed by the distribution of the plaques, indicates roughly hexagonal or isotropic packaging with a spacing of 120–160 nm.[130,138] Similar spacing, though more anisotropic, is observed in fused microvilli in locust, which appears smaller than the distance of 160–200 nm between the undulae ridges in *D. melanogaster*.[138] Such a distribution is potentially relevant in setting up the initial boundary condition for further assembly of the cuticle, and thus the control of its ultrastructure. In sample preparations, where the plaques remain attached to the deposited cuticle after removal of cells, the plaques appear periodically arranged along the chitin fibrils with similar spacing.[130] Nevertheless, the significance of the plaque distribution to the deposition of the arthropod cuticle is still to be determined.

While scientific literature may give an impression that microvilli are always present during cuticle deposition, it has recently been shown otherwise.[138] Furthermore, microvilli are typically portrayed pointing orthogonally away from the cell surface.[130] However, this is not a common observation in cuticle formation, where they are more commonly found extended at an angle to the surface of the cell.[130] During cuticle deposition in locusts, the apical side of the epidermis contains areas with and without microvilli (**Figure 3**).[138]

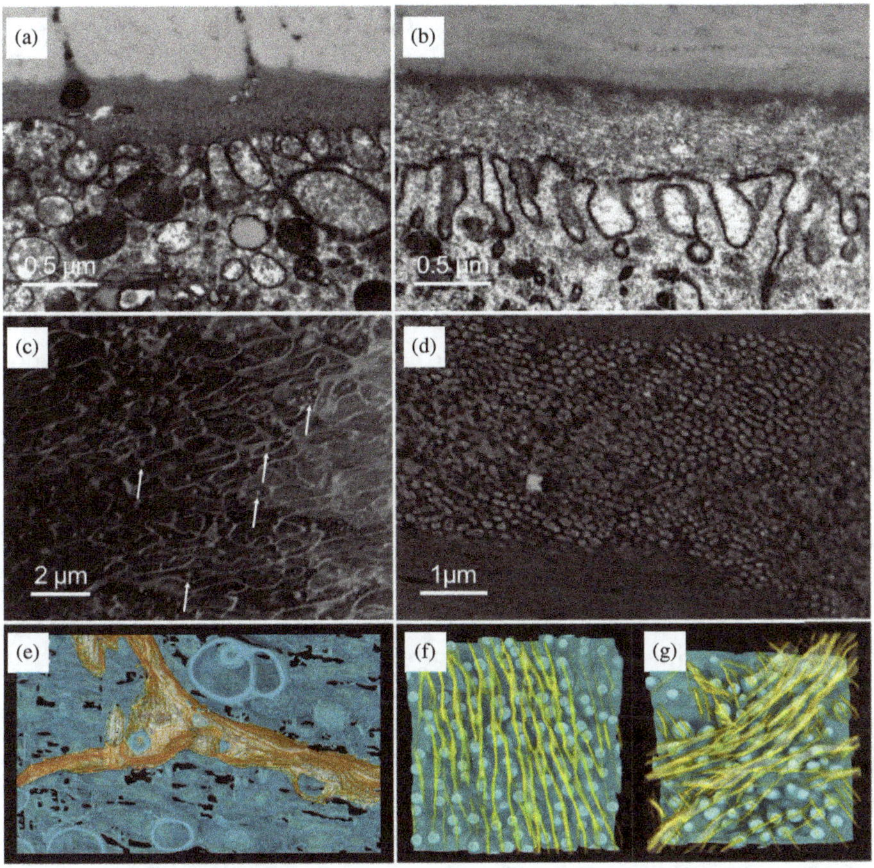

Figure 3. The apical cell surface during cuticle deposition in locust tibia shows regions with and without microvilli. (**a, b**) Transmission electron microscopy (TEM) micrograph of a growing locust cuticle. (**a**) Large vesicles are present in close proximity to the deposition zone, which itself appears densely stained. (**b**) Region showing microvilli on the same sample. The ChS plaques are seen as high electron-dense patches at the tips of microvilli. Fibrous texture is visible in the deposition zone below the denser mature cuticle. (**c–d**) Focused ion beam scanning electron microscopy (FIB/SEM) data showing regions without microvilli (view down the cell apical surface). Note the opposite contrast relative to TEM images. White circular objects resemble chitin synthase plaques, some of which are indicated by white arrows. (**d**) Region on the cuticle that contains microvilli. (**e**) 3D rendering of vesicle FIB/SEM stacks. View from within the cell toward the apical cell surface. The lateral cell membrane is depicted in brown, and the apical cell membrane and vesicles are labeled in cyan. (**f, g**) View of the apical cell surface showing microvilli (cyan) and fibers (yellow) in day (**f**) and night (**g**) deposition. (**a, b**) Unpublished data. (**c–g**) Adapted from Sviben et al.[138] under Creative Commons CC BY 4.0 license.

The regions void of microvilli show comparatively unorganized plasma membrane, including large folds and invaginations. Interestingly, these membrane structures also contain particles that resemble ChS plaques (**Figure 3c**). In addition, a high density of large vesicles appears to be fusing with the plasma membrane. It is speculated that this membrane topology is related to the temporal events of intensified release of proteins (discussed later), concomitant with plaque recycling.[138] Microvilli also disappear temporarily during molting when the epidermal cells disconnect from the old cuticle and reform to start to produce a new cuticle.[21] These observations, together with the acknowledgement of the better-understood microvilli dynamics in other biological systems,[139] lead to the hypothesis that the microvilli of the epidermal cell surface in arthropods are highly dynamic,[138] advocating an interesting direction for further investigations.

The valleys between extended microvilli are assumed to be the main location for the release of cuticular proteins via vesicle fusion.[136,140,141] Such fusion events are only observed occasionally in the electron micrographs, despite them having to account for the large amount of protein being secreted (cuticle contains 70% by weight or more of protein). Based on the volumetric imaging of cuticle deposition in locust tibia, we speculated about a potentially additional protein secretory event.[138] As mentioned before, regions without microvilli are often found at the apical cell surface, with an unusually large number of vesicles, including events of vesicles seemingly merging with the cell membrane.[138] It was suggested that at such a time point, the chitin secretion would be halted in that particular area and complemented by the secretion of cuticular proteins required to attain the final composition of the procuticle. However, these vesicles contain cuticular proteins, and the generality of this finding to other arthropod species are still to be determined. We also suggest that during this stage, the microvilli membrane is unfolded and partially recycled through endocytosis (**Figure 4**). Ultimately, a combination of these scenarios is very likely; some proteins involved in chitin processing and modification could be released by vesicular fusion at the microvilli valleys concomitantly with chitin synthesis and later complemented by an intensified secretion of cuticular proteins.

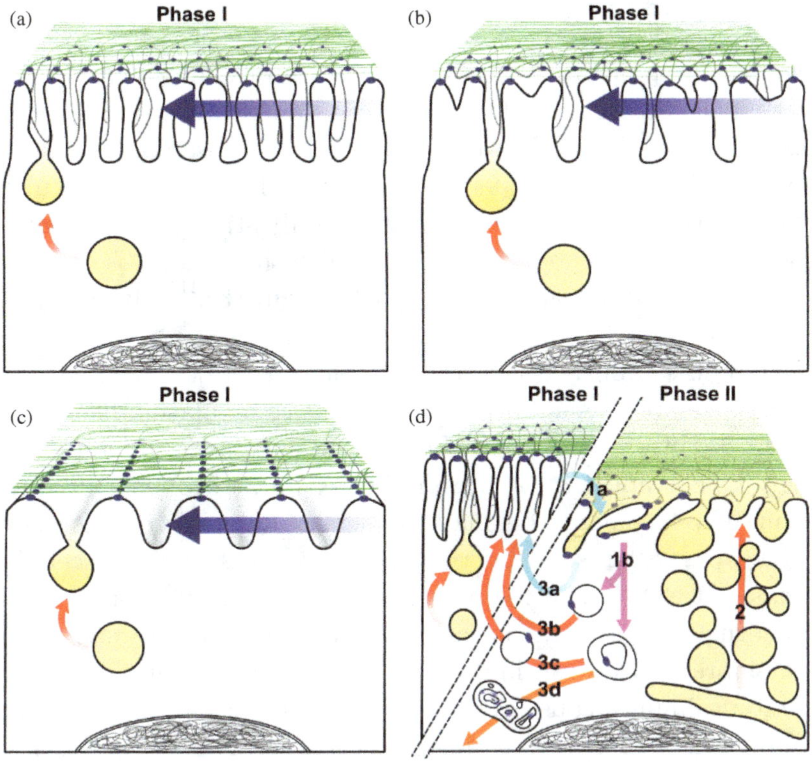

Figure 4. A hypothetical two-phase model for the arthropod cuticle formation. During *Phase I*, chitin is secreted by chitin synthases (ChS) situated as plaques at the tips of either (a) microvilli, (b) fused microvilli, or (c) undulea. These structures are dynamic and move synchronously as a wave in a single direction, depicted by a blue arrow. As the plaques are carried along while secreting chitin, chitin microfibrils are also oriented along the same direction. Since the movement direction changes in (**a**) and (**c**) over time, a plywood structure in the assembly zone is produced, whereas the movement direction remains constant in (**b**) producing a unidirectional stack. In *Phase I*, only some of the cuticular proteins, such as chitinases and deacetylases, are secreted by exocytosis at the valleys. Thereafter, the cuticle formation transitions from *Phase I* to *Phase II* (**d**), marked by the flattening of the microvilli via (1a) unfolding and/or (1b) endocytosis. As such, chitin secretion is halted. Concurrently, (2) cuticular protein secretion into the assembly zone intensifies, promoting the reorganization and maturation of the assembly zone. Then the transitions back to *Phase I* start again by (3) reforming microvilli with plaques by (3a) the refolding of the apical plasma membrane, or (3b) direct exocytosis or (3c) with additional processing, during which (3d) some chitin synthases and other membrane proteins are fully degraded in lysosomes.

3.2. *Assembly zone*

Cuticular components secreted into the extracellular space do not directly become a part of the procuticle, but rather reside first in a transitory layer (200–1000 nm thick) termed assembly zone, also known as the Schmidt layer.[132] The assembly zone is particularly visible in electron microscopy by its generally stronger staining with respect to the rest of the procuticle, indicating that it is less compacted in nature than the mature cuticle (**Figure 3**).[11,132] Interestingly, it also lacks lamellate order.[138]

It is well established that chitin molecules are extruded directly into the assembly zone during their synthesis by plasma membrane-embedded chitin synthases (ChS).[142] Mainly two different ChS, namely ChS-1 and ChS-2, are implicated in arthropod cuticle formation. These enzymes belong to a bigger family of proteins known as glycosyltransferases, together with enzymes such as cellulose synthases (CeS).[142] The enzymes catalyze the reaction of glycosidic bond formation and translocate the growing polymer chain across the membrane. CeS are known to form complexes of 18–36 units. Whereby, simultaneous extrusion of a single cellulose polymer per unit leads to cellulose fibrils with defined cross-sections.[143,144] A similar arrangement of chitin synthases is also expected, and is partly supported by observations such as in vitro catalytic activity and dimerization of purified yeast ChS-1.[145] Nevertheless, in natural cellulose, the molecules are arranged in parallel within a crystal,[146] whereas the molecules in α-chitin are antiparallel to each other, indicating that a direct analogy between chitin and cellulose synthases is not sufficient to explain chitin biosynthesis.[77] Indeed, there is still no convincing explanation for an antiparallel arrangement of chitin molecules in microfibrils to date.

Chitin molecules emerge from the plasma membrane perpendicular to it and immediately coalesce into a microfibril. The microfibrils are generally oriented within a plane perpendicular to the microvilli.[34,130,134,138] As such, there must be a bias for the planar ordering of chitin microfibrils, likely due to the confinement and density of the assembly zone. When epidermal cells are carefully removed, the most nascent cuticle layer, presumably the dehydrated assembly zone, shows that chitin microfibrils fibrils are well aligned along the plane of the cuticle spanning across the boundaries of neighboring cells.[130,140] Chitin microfibrils have an even higher hierarchical level of ordering within the assembly zone. Studies on cuticle formation in locusts show that when the endocuticle is destined to become helicoidally organized, the microfibrils within the assembly zone form discrete layers, where the fiber axes are rotated by around 50° between layers (**Figure 3g**).[138] Within such samples, the microvilli assumed a typical hexagonal organization. In contrast, when the endocuticle to be produced is unidirectionally oriented, the microfibrils within the assembly zone were similarly unidirectionally aligned, and the microvilli appeared fused into ridge-like structures in a direction parallel to the microfibril orientation (**Figure 3f**).[138] However, it is yet to be determined how these observations are reconciled with the apical undulae in *D. melanogaster* embryonic stages, where the chitin microfibrils in the assembly zone are oriented perpendicular to the undulae ridges, ultimately leading to a helicoidal ultrastructure of the procuticle (**Figure 4**).[134]

The material in the assembly zone is in a transitory state, and as such needs to mature to become a part of the procuticle. Fibril reorganization is expected during this process. As in the example above, where the chitin microfibril sheets arranged at 50° angles to each other in the assembly zone need to rearrange into a continuously rotating helicoidal structure as found in the mature cuticle.[138] In addition, the nascent cuticle needs to be both filled with more protein and dehydrated to transform into a compact procuticle. As such, we postulate that while the initial blueprint is set directly and actively by the apical modulations of epidermal cells, additional self-assembly of chitin (very likely tailored by co-assembly with cuticular proteins) takes place to produce the mature cuticle.

The involvement of proteins in determining the final cuticle architecture is evident from protein-silencing studies.[106,147,148] Many such modifications result in ultrastructural changes, which nevertheless vary based on the organism and the tissue investigated. Many such phenotypes are lethal, highlighting the importance of cuticle integrity for the fitness of the animal. The direct processing of chitin microfibrils in the assembly zone by enzymes that sever the polymer chains (chitinases) or deacetylate them (chitin deacetylases) appears to be vital for proper procuticle deposition.[106,148] For example, the pattern of deacetylation produced within the assembly zone could change the chitin–protein interactions. Importantly though, chitin-modifying enzymes need to first bind to nascent chitin polymer, and so other cuticular proteins could therefore be used to regulate how and to what extent chitin is processed by blocking their binding. Furthermore, by enveloping the chitin microfibrils, cuticular proteins can protect them from coalescence.[32] Lastly, protein binding to chitin microfibrils can directly alter their surface morphology and chemistry, thus potentially tailoring processes such as liquid crystalline self-assembly,[47,149] which is often implicated as the main driving force for the ultrastructural organization in the arthropod cuticle.[37,150–152]

3.3. *Self-assembly*

The idea that the cuticle is organized by self-assembly via liquid crystalline intermediate is particularly popular.[37,150–152] In particular, this hypothesis emerged around the 1960s, upon the realization that the helicoidal ultrastructure observed in the arthropod cuticle is reminiscent of the chiral nematic liquid crystalline phase and the unidirectional organization of the nematic liquid crystalline phase.[11,150–152] Of particular importance was also the observation that the helicoidal architecture found in the arthropod cuticle is always left-handed, inconsistent with the bilateral symmetry of the arthropods.[11] Similarly, chiral nematic liquid crystalline phases typically form only one-handedness structures dictated by the properties of the mesogen, the component that self-assembles. While the self-assembly theory has gained traction partly due to its explanatory simplicity of the process and the analogy to structures obtained via nematic or chiral nematic liquid crystalline phases, it has little to no direct evidence to date. It is important to note that the analogy based on the architecture of the liquid crystal and cuticle architecture does not immediately implicate the same mechanistic pathway.

For the nematic liquid crystalline phase to develop, at least two requirements have to be met.[36,47,153,154] First, the liquid crystalline phase needs to be fluid enough to allow the rearrangement of mesogens. The viscosity of the whole system may slow down or preclude self-assembly altogether. Second, mesogens themselves need to take an anisotropic shape that exhibits a certain degree of rigidity. For the liquid crystalline phase to become chiral, the mesogens themselves need to be chiral or the system needs to be doped by a chiral seed. Assuming the formation of such liquid crystalline phases in the case of the cuticle assembly, the most likely mesogen is chitin microfibrils, potentially coated by proteins, rather than individual chitin molecules. This is in contrast to other biological systems; for example, in bone during remodeling, individual collagen molecules (triple helices) organize into a cholesteric liquid crystalline phase prior to fibrogenesis.[36,155]

Within its limits, the most appropriate in vitro system to draw analogies to the cuticle assembly to date would be that of chitin nanocrystals (ChNCs). These nanoparticles are typically derived from purified chitin after its hydrolysis in hydrochloric acid, which results in ChNCs that are polydisperse and spindle-like shaped with the widths of tens of nanometers and lengths of a few hundred nanometers.[156–158] ChNCs are colloidally stable at least for days when dispersed in aqueous acidic media, which allows the particles to reorient freely to a preferred configuration. When the concentration of ChNCs in an aqueous medium reaches a critical concentration of about 3 wt% (2 vol%), droplets of tens of micrometers in diameter with higher density of ChNCs, termed tactoids, spontaneously form.[157–159] Within these droplets, ChNCs order themselves into a chiral nematic phase, while within the surrounding medium, ChNCs remain isotropically dispersed. Due to their higher density in comparison to the rest of the medium, tactoids sediment and merge slowly to form a continuous chiral nematic mesophase.

The volume fraction of the liquid crystalline phase increases with the ChNC concentration until another threshold concentration of around 12 wt% (9 vol%) is reached, when the whole suspension is self-assembled.[158,159] Upon further increase in ChNC concentration to roughly 20 wt% (14 vol%), the system gels, locking the chiral nematic structure in place, thus preventing further reorganization.[150] Ultimately, when water is continuously evaporated without disturbing the suspension, solid films are obtained with the preserved architecture, in which solid state is termed helicoidal, and appear analogous to the one found in the arthropod procuticle.[11,159,160] The analogy can be further extended in that the helicoidal solid-state structures made by self-assembly of ChNCs are always left-handed.

Despite these striking similarities, there are several key challenges with the analogy. First of all, ChNCs are significantly shorter than the chitin microfibrils in the arthropod cuticle. The increased length of the particles, while expected to lower the threshold concentrations for the self-assembly, also increases the viscosity of the medium, thus hindering the formation of the liquid crystalline phases. In particular, chitin nanoparticles that are a few micrometers in length do not exhibit self-assembly in vitro, potentially also due to their higher flexibility compared to ChNCs. Furthermore, ChNCs exhibit self-assembly at a limited range of pH (roughly from 2 to 6) and ionic strength (up to about 10 mM), the latter being incompatible with most biological systems.[158] In addition, the self-assembly of ChNCs follows a mechanism via tactoid formation and merging, which neither fits the dimensions of the assembly zone, nor has ever been observed within it, suggesting alternative self-assembly mechanisms. Lastly, the unidirectional alignment (such as found in some cuticles) has not been yet produced by self-assembly of ChNCs without the application of external aligning forces.[158]

Overall, despite the architectural similarity between chiral nematic liquid crystals and the helicoidal cuticle, liquid crystalline self-assembly behavior analogous to ChNC assembly is unlikely to occur in the assembly zone. Nevertheless, some aspects of ChNC self-assembly can be helpful in discussing the assembly events based on chitin–protein and protein–protein interactions. Of particular interest is the fact that chitin on its own already contains the structural information to promote exclusively left-handed chiral nematic structure. Ultimately, while self-assembly may indeed be the answer to arthropod cuticle organization, it is not a satisfactory answer in its own right unless we properly understand how the interactions between various cuticular components lead to and tailor the self-assembly.

3.4. *Boundary conditions*

The assembly of the cuticle is subject to the environment in the confined space of the assembly zone. Notwithstanding their importance to cuticle formation, the properties of the assembly zone are, however, largely unknown. As such, we outline some open questions regarding the boundary conditions defined by the apical plasma membrane of the cells and the assembly zone, which we believe are key to further our understanding of cuticle deposition and its ultrastructural control.

1. Do, and if so how, chitin synthases arrange into complexes within the plaques? What other proteins are present in the plaques and what are their functions?
2. How can directional chitin biosynthesis be reconciled with the antiparallel molecular chain orientation in chitin microfibrils? What are the processing steps required for the reorientation of polymer chains?
3. What is the physical state of the assembly zone? Is it a solid, a hydrogel, or a liquid? What are its mechanical properties? What are the pH, the key electrolytes, and their concentrations, and are they constant or do they vary spatially and temporally?
4. To what extent is chitin microfibril orientation in the deposition zone and the mature cuticle (e.g., helicoidal pitch) controlled by

microvilli organization, the rate of chitin secretion, and protein secretion?

5. What is the temporal and spatial control of protein secretion? Are there multiple modes of protein secretion and, if yes, which proteins are secreted by which modes?
6. How do proteins interact with chitin and with each other, and how do these interactions manifest toward higher-level organization of the cuticle and its material properties?
7. How does the assembly zone transform into the procuticle? Is it a continuous or a multistep process?

3.5. *A two-phase model on the procuticle deposition*

As a conclusion to this chapter, we suggest a speculative model for the control of the ultrastructure in the arthropod model, with the hope to motivate a much-needed discussion and stimulate further research in this direction.

In our model, the cells actively define the initial orientation of the chitin fibrils in the assembly zone. This is most likely done via the tailored organization of apical cell structures, as has been observed during cuticle deposition in locust and fruit fly embryo.[130,134,136,138] Chitin microfibrils emerging into a confined space from the plaques at the tips of the microvilli (or undulae) are aligned by coordinated migration of microvilli in one direction (**Figure 4a–c**). This movement is dictated by the cellular cytoskeleton, which allows coordination on a single and several cell level by the interactions between neighboring cells.[161] A change in the microvilli migration direction would also change the alignment direction of chitin microfibrils. In this way, sheets of chitin microfibrils aligned unidirectionally (**Figure 4b**) or at an angle with respect to each other (**Figure 4a, c**) can be deposited. Some cuticular proteins, for example, chitin-modifying enzymes and structural proteins, are secreted concomitantly with chitin synthesis via vesicular fusion with the plasma membrane at the valleys between the microvilli (**Figure 4a–c**). This allows to modify chitin surface chemistry to tailor for the interactions with proteins via the activity of chitinases and chitin deacetylases, as well as preventing chitin microfibril aggregation.

With the blueprint of the ultrastructure imposed by the epidermal cells, the structure further develops into a mature cuticle by self-assembly subject to the boundary conditions of the assembly zone. As such, the initial orientation of the microfibrils is adjusted toward the final arrangement, in particular, during the process of intensified protein secretion, during which chitin secretion is halted (**Figure 4d**). If the bulk of proteins are secreted sequentially to the chitin deposition, these cuticular proteins could form liquid condensates, which can facilitate the impregnation of a predeposited matrix, via shear thinning and capillary forces.[162] A similar mechanism has been suggested in the formation of the chitinous squid beak.[111] In this case, the plasma membrane, ChS, and plaques proteins are also recycled (**Figure 4d**).

Interactions between protein-decorated chitin microfibrils and other cuticular proteins dictate the rearrangement to obtain the final architecture that compacts and becomes locked in as a part of the procuticle. This could occur in areas with presumed intensified protein secretion, where the assembly zone appears tighter and more electron dense, in transmission electron micrographs, indicative of high protein content (**Figure 4d**). In particular, based on the binding patterns of the proteins to chitin microfibrils, the inherently chiral shape of the fibrils could be altered, thus tailoring the pitch of the helicoidal architecture or maintaining unidirectionally aligned fibers. The physicochemical conditions at the assembly zone, such as pH, ionic strength, and so on, could also be used to tailor chitin–protein and protein–protein interactions and, in turn the self-assembly, as seen in the in vitro system of ChNCs. Lastly, the microvilli with plaques at their tips would be reformed in such areas and the cycle starts over (**Figure 4d**).

4. Conclusion

The arthropod cuticle is a multifunctional material showing remarkable versatility in structure and properties. As such, it has captured the minds of researchers and inspired new material designs. Similarly, the process of the cuticle assembly itself could become a new source of inspiration for process engineering. However, summarizing the state-of-the-art knowledge on arthropod cuticle assembly, we emphasize that there are currently more questions than answers. As such, we outlined some of the major questions, which need to be answered to obtain a good explanation of how one of the defining features of arthropods, their cuticle, is produced. We also conjectured a hypothesis and a rough descriptive model on the process of the cuticle deposition, being fully aware of a needed thorough confirmation, but hoping to encourage a discussion and raise questions to help develop the field. We have no doubt that with the advancing technologies and with scientists becoming ever more interested in understanding cuticle deposition, the problem will be unraveled layer by layer via interdisciplinary research. We also anticipate that the findings will inspire the development of new manufacturing strategies for engineered materials, and as such contribute toward a more sustainable future.

References

1. Milo DS. *Good Enough: The Tolerance for Mediocrity in Nature and Society.* Cambridge, MA: Harvard University Press; 2019.
2. Van Nieuwenhoven RW, Drack M, Gebeshuber IC. Engineered materials: Bioinspired "good enough" versus maximized performance. *Adv Funct Mater.* 2023 Dec 14;34:2307127.
3. Eder M, Amini S, Fratzl P. Biological composites — complex structures for functional diversity. *Science.* 2018 Nov 2;362(6414): 543–547.
4. Politi Y, Bertinetti L, Fratzl P, Barth FG. The spider cuticle: A remarkable material toolbox for functional diversity. *Phil Trans R Soc A.* 2021 Sep 20;379(2206):20200332.
5. Harrington MJ, Fratzl P. Natural load-bearing protein materials. *Prog Mater Sci.* 2021 Jul 1;120:100767.
6. Meyers MA, Lin AYM, Seki Y, Chen PY, Kad BK, Bodde S. Structural biological composites: An overview. *JOM.* 2006 Jul 1;58(7):35–41.

7. Vincent JFV, Wegst UGK. Design and mechanical properties of insect cuticle. *Arthropod Struct Dev.* 2004 Jul 1;33(3):187–199.
8. Miserez A, Yu J, Mohammadi P. Protein-based biological materials: Molecular design and artificial production. *Chem Rev.* 2023 Mar 8;123(5):2049–2111.
9. Rising A, Harrington MJ. Biological materials processing: Time-tested tricks for sustainable fiber fabrication. *Chem Rev.* 2023 Mar 8;123(5):2155–2199.
10. Souza CP, Almeida BC, Colwell RR, Rivera ING. The importance of chitin in the marine environment. *Mar Biotechnol.* 2011 Oct 1;13(5):823–830.
11. Neville AC. *Biology of the Arthropod Cuticle.* 1st ed. Berlin: Springer; 1975. p. 460.
12. Lowenstam HA, Weiner S. *On Biomineralization.* New York: Oxford University Press; 1989. p. 335.
13. Romano P, Fabritius H, Raabe D. The exoskeleton of the lobster *Homarus americanus* as an example of a smart anisotropic biological material. *Acta Biomater.* 2007 May 1;3(3):301–319.
14. Locke M. The Wigglesworth Lecture: Insects for studying fundamental problems in biology. *J Insect Physiol.* 2001;47(4–5): 495–507.
15. Schwarz H, Moussian B. Electron-microscopic and genetic dissection of arthropod cuticle differentiation. In: Méndez-Vilas A, Díaz J, editors. *Modern Research and Educational Topics in Microscopy.* Badajoz: Formatex; 2007. p. 316–325.
16. Barth FG. Microfiber reinforcement of an arthropod cuticle. *ZZellforsch.* 1973 Sep 1;144(3):409–433.
17. Klocke D, Schmitz H. Water as a major modulator of the mechanical properties of insect cuticle. *Acta Biomater.* 2011 Jul 1;7(7): 2935–2942.
18. Fabritius H, Sachs C, Raabe D, Nikolov S, Friák M, Neugebauer J. Chitin in the exoskeletons of Arthropoda: From ancient design to novel materials science. In: Gupta NS, editor. *Chitin: Formation and Diagenesis [Internet].* Dordrecht: Springer Netherlands; 2011 [cited 2024 Aug 2. p. 35–60. doi:10.1007/978-90-481-9684-5-2.
19. Locke M. The structure and formation of the cuticulin layer in the epicuticle of an insect, *Calpodes ethlius* (Lepidoptera, Hesperiidae). *J Morphol.* 1966;118(4):461–494.

20. Locke M. Secretion of wax through the cuticle of insects. *Nature.* 1959 Dec;184(4703):1967–1967.
21. Locke M, Huie P. Apolysis and the turnover of plasma membrane plaques during cuticle formation in an insect. *Tissue Cell.* 1979; 11(2):277–291.
22. Blomquist GJ, Ginzel MD. Chemical ecology, biochemistry, and molecular biology of insect hydrocarbons. *Ann Rev Entomol.* 2021 Jan 7;66:45–60.
23. Tadayon M, Younes-Metzler O, Shelef Y, Zaslansky P, Rechels A, Berner A, et al. Adaptations for wear resistance and damage resilience: Micromechanics of spider cuticular "tools." *Adv Funct Mater.* 2020;30(32):2000400.
24. Watson GS, Watson JA, Cribb BW. Diversity of cuticular micro- and nanostructures on insects: Properties, functions, and potential applications. *Ann Rev Entomol.* 2017 Jan 31;62:185–205.
25. Helbig R, Nickerl J, Neinhuis C, Werner C. Smart skin patterns protect springtails. *PLOS One.* 2011 Sep 30;6(9):e25105.
26. Moussian B. Recent advances in understanding mechanisms of insect cuticle differentiation. *Insect Biochem Mol Biol.* 2010 May 1;40(5):363–375.
27. Andersen SO. Insect cuticular sclerotization: A review. *Insect Biochem Mol Biol.* 2010 Mar 1;40(3):166–178.
28. Sugumaran M. Cuticular sclerotization in insects: a critical review review. In: Sugumaran M, editor. *Advances in Insect Physiology [Internet].* London: Academic Press; 2022 [cited 2024 Aug 2]. p. 111–214. Available from: https://www.sciencedirect.com/science/article/pii/S0065280622000017.
29. Roer R, Dillaman R. The structure and calcification of the crustacean cuticle. *Am Zool.* 1984 Nov 1;24(4):893–909.
30. Blackwell J, Weih MA. Structure of chitin-protein complexes: Ovipositor of the ichneumon fly *Megarhyssa. J Mol Biol.* 1980 Feb 15;137(1):49–60.
31. Fraenkel G, Rudall KM. The structure of insect cuticles. *Proc R Soc London Ser B.* 1947;134(874):111–143.
32. Valverde Serrano C, Leemreize H, Bar-On B, Barth FG, Fratzl P, Zolotoyabko E, et al. Ordering of protein and water molecules at their interfaces with chitin nano-crystals. *J Struct Biol.* 2016 Feb 1;193(2):124–131.

33. Neville AC. Chitin orientation in cuticle and its control. In: Beament JWL, Treherne JE, Wigglesworth VB, editors. *Advances in Insect Physiology [Internet]*. London: Academic Press; 1967 [cited 2024 Aug 2]. p. 213–286. Available from: https://www.sciencedirect.com/science/article/pii/S006528060860209X.
34. Bouligand Y. Sur une architecture torsadée répandue dans de nombreuses cuticules d'Arthopodes. *C R Acad Sci (Paris)*. 1965;261: 3665–366.
35. Bouligand Y. Twisted fibrous arrangements in biological materials and cholesteric mesophases. *Tissue and Cell*. 1972 Jan 1;4(2):189–217.
36. Belamie E, Mosser G, Gobeaux F, Giraud-Guille MM. Possible transient liquid crystal phase during the laying out of connective tissues: α-chitin and collagen as models. *J Phys Condens Matter*. 2006 Mar;18(13):S115.
37. Mitov M. Cholesteric liquid crystals in living matter. *Soft Matter*. 2017 Jun 14;13(23):4176–4209.
38. Boßelmann F, Romano P, Fabritius H, Raabe D, Epple M. The composition of the exoskeleton of two crustacea: The American lobster *Homarus americanus* and the edible crab *Cancer pagurus*. *Thermochim Acta*. 2007 Oct 25;463(1):65–68.
39. Politi Y, Bar-On B, Fabritius HO. Mechanics of arthropod cuticle-versatility by structural and compositional variation. In: Estrin Y, Bréchet Y, Dunlop J, Fratzl P, editors. *Architectured Materials in Nature and Engineering: Archimats [Internet]*. Cham: Springer International Publishing; 2019 [cited 2024 Aug 2]. p. 287–327. doi:10.1007/978-3-030-11942-3-10.
40. Nikolov S, Fabritius H, Petrov M, Friák M, Lymperakis L, Sachs C, et al. Robustness and optimal use of design principles of arthropod exoskeletons studied by ab initio-based multiscale simulations. *J Mech Behav Biomed Mater*. 2011 Feb 1;4(2):129–145.
41. Chen P, Zhao C, Wang H, Li Y, Tan G, Shao Z, et al. Quantifying the contribution of the dispersion interaction and hydrogen bonding to the anisotropic elastic properties of chitin and chitosan. *Biomacromolecules*. 2022 Apr 11;23(4):1633–1642.
42. Ogawa Y, Hori R, Kim UJ, Wada M. Elastic modulus in the crystalline region and the thermal expansion coefficients of α-chitin determined using synchrotron radiated X-ray diffraction. *Carbohy Polym*. 2011 Jan 30;83(3):1213–1217.

43. Nishino T, Matsui R, Nakamae K. Elastic modulus of the crystalline regions of chitin and chitosan. *J Polym Sci Part B: Polym Phys.* 1999;37(11):1191–1196.
44. Kinoshita S, Yoshioka S. Structural colors in nature: The role of regularity and irregularity in the structure. *Chem Phys Chem.* 2005;6(8):1442–1459.
45. Caveney S. Cuticle reflectivity and optical activity in scarab beetles: The role of uric acid. *Proc R Soc London Ser B.* 1971; 178(1051):205–225.
46. Bay MM, Vignolini S, Vynck K. PyLlama: A stable and versatile Python toolkit for the electromagnetic modelling of multilayered anisotropic media. *Comput Phys Commun.* 2022 Apr 1;273:108256.
47. Frka-Petesic B, Parton TG, Honorato-Rios C, Narkevicius A, Ballu K, Shen Q, et al. Structural color from cellulose nanocrystals or chitin nanocrystals: Self-assembly, optics, and applications. *Chem Rev.* 2023 Dec 13;123(23):12595–12756.
48. Kinoshita S. *Structural Colors in the Realm of Nature Cover Structural Colors in the Realm of Nature.* Singapore: World Scientific; 2008. p. 368.
49. Allahverdyan K, Galstian T, Gevorgyan A, Hakobyan R. Could the cuticle of beetles serve also for their radiative thermoregulation? *Opt Photon J.* 2013 Nov 27;3(7):17–22.
50. Mitov M. Multifunctional reflectors in the carapace of scarab beetles. Proc SPIE. 2018 Sep 1;10735:107350B.
51. Spaeker O, Taylor GJ, Wilts BD, Slabý T, Abdel-Rahman MAK, Scoppola E, et al. Gradients of orientation, composition, and hydration of proteins for efficient light collection by the cornea of the horseshoe crab. *Adv Sci.* 2022;9(33):2203371.
52. Rinaudo M. Chitin and chitosan: Properties and applications. *Prog Polym Sci.* 2006 Jul;31(7):603–632.
53. Zakrzewski AC, Weigert A, Helm C, Adamski M, Adamska M, Bleidorn C, et al. Early divergence, broad distribution, and high diversity of animal chitin synthases. *Genome Biol Evol.* 2014 Feb 1;6(2):316–325.
54. Food and Agricultural Organization. The state of world fisheries and aquaculture 2022 [Internet]. FAO. 2022 [cited 2022 Aug 11]. Available from: http://www.fao.org/documents/card/en/c/cc0461en.

55. Małachowska E, Dubowik M, Lipkiewicz A, Przybysz K, Przybysz P. Analysis of cellulose pulp characteristics and processing parameters for efficient paper production. *Sustainability.* 2020 Jan;12(17):7219.
56. Boenish R, Kritzer JP, Kleisner K, Steneck RS, Werner KM, Zhu W, et al. The global rise of crustacean fisheries. *Front Ecol Environ.* 2022;20(2):102–110.
57. Berroci M, Vallejo C, Lizundia E. Environmental impact assessment of chitin nanofibril and nanocrystal isolation from fungi, shrimp shells, and crab shells. *ACS Sustainable Chem Eng.* 2022 Oct 31;10(43):14280–14293.
58. Pasquier E, Beaumont M, Mattos BD, Otoni CG, Winter A, Rosenau T, et al. Upcycling byproducts from insect (fly larvae and mealworm) farming into chitin nanofibers and films. *ACS Sustainable Chem Eng.* 2021 Oct 11;9(40):13618–13629.
59. Nawawi WMFBW, Jones M, Murphy RJ, Lee KY, Kontturi E, Bismarck A. Nanomaterials derived from fungal sources — is it the new hype? *Biomacromolecules.* 2020 Jan 13;21(1):30–55.
60. Wineinger HB, Shamshina JL, Kelly A, King C, Rogers RD. A method for determining the uniquely high molecular weight of chitin extracted from raw shrimp shells using ionic liquids. *Green Chem.* 2020 Jun 22;22(12):3734–3741.
61. Fittolani G, Tyrikos-Ergas T, Vargová D, Chaube MA, Delbianco M. Progress and challenges in the synthesis of sequence controlled polysaccharides. *Beilstein J Org Chem.* 2021 Aug 5;17(1): 1981–2025.
62. Heux L, Brugnerotto J, Desbrières J, Versali MF, Rinaudo M. Solid State NMR for determination of degree of acetylation of chitin and chitosan. *Biomacromolecules.* 2000 Dec 1;1(4):746–751.
63. Giles CH, Hassan ASA, Laidlaw M, Subramanian RVR. Adsorption at organic surfaces. *J Soc Dyers Colour.* 1958;74(9):647–654.
64. Waterhouse DF, Hackman RH, McKellar JW. An investigation of chitinase activity in cockroach and termite extracts. *J Insect Physiol.* 1961 Apr 1;6(2):96–112.
65. Li J, Revol JF, Marchessault RH. Effect of degree of deacetylation of chitin on the properties of chitin crystallites. *J Appl Polym Sci.* 1997;65(2):373–380.

66. Bonin M, Sreekumar S, Cord-Landwehr S, Moerschbacher BM. Preparation of defined chitosan oligosaccharides using chitin deacetylases. *Int J Mol Sci.* 2020 Jan;21(21):7835.
67. Arnold ND, Brück WM, Garbe D, Brück TB. Enzymatic modification of native chitin and conversion to specialty chemical products. *Mar Drugs.* 2020 Feb;18(2):93.
68. Weinhold MX, Sauvageau JCM, Kumirska J, Thöming J. Studies on acetylation patterns of different chitosan preparations. *Carbohydr Polym.* 2009 Nov 17;78(4):678–684.
69. Grifoll-Romero L, Pascual S, Aragunde H, Biarnés X, Planas A. Chitin deacetylases: Structures, specificities, and biotech applications. *Polymers.* 2018 Apr;10(4):352.
70. Gubaeva E, Gubaev A, Melcher RLJ, Cord-Landwehr S, Singh R, El Gueddari NE, et al. 'Slipped sandwich' model for chitin and chitosan perception in Arabidopsis. *MPMI.* 2018 Nov;31(11): 1145–1153.
71. Fuchs K, Cardona Gloria Y, Wolz O, Herster F, Sharma L, Dillen CA, et al. The fungal ligand chitin directly binds TLR2 and triggers inflammation dependent on oligomer size. *EMBO Rep.* 2018 Dec;19(12):e46065.
72. Arakane Y, Muthukrishnan S. Insect chitinase and chitinase-like proteins. *Cell Mol Life Sci.* 2010 Jan 1;67(2):201–216.
73. Luschnig S, Bätz T, Armbruster K, Krasnow MA. *Serpentine* and *vermiform* encode matrix proteins with chitin binding and deacetylation domains that limit tracheal tube length in *Drosophila. Curr Biol.* 2006 Jan 24;16(2):186–194.
74. Zhang T, Ma P, Zhou J, He Y, Liu W, Liu X, et al. Group I CDAs are responsible for a selective CHC-independent cuticular barrier in *Locusta migratoria. Pestic Biochem Physiol.* 2021 Jun 1;175:104854.
75. Roy JC, Salaün F, Giraud S, Ferri A, Roy JC, Salaün F, et al. Solubility of chitin: Solvents, solution behaviors and their related mechanisms. In: Xu Z, editor. *Solubility of Polysaccharides [Internet]*. London: IntechOpen; 2017. [cited 2024 Aug 2]. p. 109–127. Available from: https://www.intechopen.com/chapters/57402.
76. Sikorski P, Hori R, Wada M. Revisit of α-chitin crystal structure using high resolution X-ray diffraction data. *Biomacromolecules.* 2009 May 11;10(5):1100–1105.

77. Minke R, Blackwell J. The structure of α-chitin. *J Mol Biol.* 1978 Apr 5;120(2):167–181.
78. Rudall KM. The chitin/protein complexes of insect cuticles. In: Beament JWL, Treherne JE, Wigglesworth VB, editors. *Advances in Insect Physiology [Internet].* London: Academic Press; 1963 [cited 2024 Aug 2]. p. 257–313. Available from: https://www.sciencedirect.com/science/article/pii/S0065280608601770. Accessed 2024 Aug 2.
79. Blackwell J. Structure of β-chitin or parallel chain systems of poly-β-(1→4)-N-acetyl-D-glucosamine. *Biopolymers.* 1969;7(3):281–298.
80. Dweltz NE. The structure of β-chitin. *Biochimica et Biophysica Acta.* 1961 Aug 5;51(2):283–294.
81. Nishiyama Y, Noishiki Y, Wada M. X-ray structure of anhydrous β-chitin at 1 Å resolution. *Macromolecules.* 2011 Feb 22;44(4): 950–957.
82. Zhong C, Cooper A, Kapetanovic A, Fang Z, Zhang M, Rolandi M. A facile bottom-up route to self-assembled biogenic chitin nanofibers. *Soft Matter.* 2010 Nov 7;6(21):5298–5301.
83. Noishiki Y, Takami H, Nishiyama Y, Wada M, Okada S, Kuga S. Alkali-induced conversion of β-chitin to α-chitin. *Biomacromolecules.* 2003 Jul 1;4(4):896–899.
84. Karthick Rajan D, Mohan K, Rajarajeswaran J, Divya D, Kumar R, Kandasamy S, et al. β-chitin and chitosan from waste shells of edible mollusks as a functional ingredient. *Food Front.* 2024;5(1):46–72.
85. Giraud-Guille MM, Chanzy H, Vuong R. Chitin crystals in arthropod cuticles revealed by diffraction contrast transmission electron microscopy. *J Struct Biol.* 1990 May 1;103(3):232–240.
86. Erko M, Hartmann MA, Zlotnikov I, Valverde Serrano C, Fratzl P, Politi Y. Structural and mechanical properties of the arthropod cuticle: Comparison between the fang of the spider *Cupiennius salei* and the carapace of American lobster *Homarus americanus. J Struct Biol.* 2013 Aug 1;183(2):172–179.
87. Ogawa Y, Kimura S, Wada M, Kuga S. Crystal analysis and high-resolution imaging of microfibrillar α-chitin from *Phaeocystis. J Struct Biol.* 2010 Jul 1;171(1):111–116.
88. Willis JH. Cuticular proteins in insects and crustaceans. *Am Zool.* 1999 Jun 1;39(3):600–609.

89. Rudall KM. Conformation in chitin–protein complexes. In: Ramachandran GN, editor. *Conformation of Biopolymers [Internet]*. London: Academic Press; 1967 [cited 2024 Aug 2]. p. 751–765. Available from: https://www.sciencedirect.com/science/article/pii/B9781483228433500261.
90. Neville AC. *Biology of Fibrous Composites: Development beyond the Cell Membrane [Internet]*. Cambridge: Cambridge University Press; 1993 [cited 2024 Aug 2]. Available from: https://www.cambridge.org/core/books/biology-of-fibrous-composites/ED1EA61E5E7F1D3642DB4FB3BFE342C3.
91. Atkins E. Conformations in polysaccharides and complex carbohydrates. *J Biosci.* 1985 Aug 1;8(1):375–387.
92. Willis JH. The evolution and metamorphosis of arthropod proteomics and genomics. *Ann Rev Entomol.* 2018 Jan 7;63:1–13.
93. Willis JH. Structural cuticular proteins from arthropods: Annotation, nomenclature, and sequence characteristics in the genomics era. *Insect Biochem Mol Biol.* 2010 Mar 1;40(3):189–204.
94. Andersen SO, Hojrup P, Roepstorff P. Insect cuticular proteins. *Insect Biochem Mol Biol.* 1995 Feb 1;25(2):153–176.
95. Hopkins TL, Kramer KJ. Insect cuticle sclerotization. *Ann Rev Entomol.* 1992 Jan 1;37:273–302.
96. Mun S, Young Noh M, Dittmer NT, Muthukrishnan S, Kramer KJ, Kanost MR, et al. Cuticular protein with a low complexity sequence becomes cross-linked during insect cuticle sclerotization and is required for the adult molt. *Sci Rep.* 2015 May 21;5(1):10484.
97. Politi Y, Priewasser M, Pippel E, Zaslansky P, Hartmann J, Siegel S, et al. A spider's fang: How to design an injection needle using chitin-based composite material. *Adv Funct Mater.* 2012;22(12): 2519–2528.
98. Politi Y, Pippel E, Licuco-Massouh ACJ, Bertinetti L, Blumtritt H, Barth FG, et al. Nano-channels in the spider fang for the transport of Zn ions to cross-link His-rich proteins pre-deposited in the cuticle matrix. *Arthropod Struct Dev.* 2017 Jan 1;46(1):30–38.
99. Rebers JE, Riddiford LM. Structure and expression of a *Manduca sexta* larval cuticle gene homologous to *Drosophila* cuticle genes. *J Mol Biol.* 1988 Sep 20;203(2):411–423.
100. Rebers JE, Willis JH. A conserved domain in arthropod cuticular proteins binds chitin1. *Insect Biochem Mol Biol.* 2001 Oct 1;31(11): 1083–1193.

101. Heymann D, Mohanram H, Kumar A, Verma CS, Lescar J, Miserez A. Structure of a consensus chitin-binding domain revealed by solution NMR. *J Struct Biol.* 2021 Jun 1;213(2):107725.
102. Steinegger M, Söding J. Clustering huge protein sequence sets in linear time. *Nat Commun.* 2018 Jun 29;9(1):2542.
103. Pesch YY, Riedel D, Behr M. Obstructor A organizes matrix assembly at the apical cell surface to promote enzymatic cuticle maturation in *Drosophila. J Biol Chem.* 2015 Apr 17;290(16): 10071–10082.
104. Petkau G, Wingen C, Jussen LCA, Radtke T, Behr M. Obstructor-A is required for epithelial extracellular matrix dynamics, exoskeleton function, and tubulogenesis. *J Biol Chem.* 2012 Jun 15;287(25):21396–21405.
105. Abehsera S, Peles S, Tynyakov J, Bentov S, Aflalo ED, Li S, et al. MARS: A protein family involved in the formation of vertical skeletal elements. *J Struct Biol.* 2017 May 1;198(2):92–102.
106. Muthukrishnan S, Merzendorfer H, Arakane Y, Yang Q. Chitin organizing and modifying enzymes and proteins involved in remodeling of the insect cuticle. In: Yang Q, Fukamizo T, editors. *Targeting Chitin-Containing Organisms [Internet].* Singapore: Springer; 2019 [cited 2024 Aug 2]. p. 83–114. doi:10.1007/978-981-13-7318-3-5.
107. Andersen SO. Are structural proteins in insect cuticles dominated by intrinsically disordered regions? *Insect Biochem Mol Biol.* 2011 Aug 1;41(8):620–627.
108. Yamanaka M, Ishizaki Y, Nakagawa T, Taoka A, Fukumori Y. Purification and characterization of coacervate-forming cuticular proteins from *Papilio xuthus* pupae. *Zoolog Sci.* 2013 Jul;30(7): 534–542.
109. Qi H, Teng Y, Chen S, Yuan F, Chen L, Liang X, et al. Major structural protein in locust mandible capable of forming extraordinarily stiff materials via hierarchical self-assembly. *Matter.* 2024 Mar 6;7(3):1314–1329.
110. Qi H, Ding Y, Teng Y, Liang X, Chen L, Ma J, et al. A core structural protein that builds the locust mandible with a mechanical gradient. *ACS Nano.* 2023 Dec 26;17(24):25311–25321.
111. Tan Y, Hoon S, Guerette PA, Wei W, Ghadban A, Hao C, et al. Infiltration of chitin by protein coacervates defines the squid beak mechanical gradient. *Nat Chem Biol.* 2015 Jul;11(7):488–495.

112. Chaudhari SS, Arakane Y, Specht CA, Moussian B, Boyle DL, Park Y, et al. Knickkopf protein protects and organizes chitin in the newly synthesized insect exoskeleton. *Proc Natl Acad Sci.* 2011 Oct 11;108(41):17028–17033.
113. Moussian B. Molecular model of skeletal organization and differentiation. In: Cohen E, Moussian B, editors. *Extracellular Composite Matrices in Arthropods [Internet].* Cham: Springer International Publishing; 2016 [cited 2024 Aug 2]. p. 67–87. doi:10.1007/978-3-319-40740-1-3.
114. Moussian B, Letizia A, Martínez-Corrales G, Rotstein B, Casali A, Llimargas M. Deciphering the genetic programme triggering timely and spatially-regulated chitin deposition. *PLoS Genet.* 2015 Jan 24;11(1):e1004939.
115. Muthukrishnan S, Merzendorfer H, Arakane Y, Yang Q. Chitin metabolic pathways in insects and their regulation. In: Cohen E, Moussian B, editors. *Extracellular Composite Matrices in Arthropods [Internet].* Cham: Springer International Publishing; 2016 [cited 2024 Aug 2]. p. 31–65. doi:10.1007/978-3-319-40740-1-2.
116. Campli G, Volovych O, Kim K, Veldsman WP, Drage HB, Sheizaf I, et al. The moulting arthropod: A complete genetic toolkit review. *Biol Rev.* 2024 Jul 22;brv.13123.
117. Krueger RA, Broce AB, Hopkins TL, Kramer KJ. Calcium transport from Malpighian tubules to puparial cuticle of *Musca autumnalis. J Comp Physiol B.* 1988 Jul 1;158(4):413–419.
118. Darlington MV, Meyer HJ, Graf G, Freeman TP. The calcified puparium of the face fly, *Musca autumnalis* (Diptera: Muscidae). *J Insect Physiol.* 1983 Jan 1;29(2):157–162.
119. Fraenkel G, Hsiao C. Calcification, tanning, and the rôle of ecdyson in the formation of the puparium of the facefly, *Musca automnalis. J Insect Physiol.* 1967 Sep 1;13(9):1387–1394.
120. Roseland CR, Grodowitz MJ, Kramer KJ, Hopkins TL, Broce AB. Stabilization of mineralized and sclerotized puparial cuticle of muscid flies. *Insect Biochem.* 1985 Jan 1;15(4):521–528.
121. Leschen RAB, Cutler B. Cuticular calcium in beetles (Coleoptera: Tenebrionidae: Phrenapetinae). *Ann Entomol Soc Am.* 1994 Nov 1;87(6):918–921.
122. Li H, Sun CY, Fang Y, Carlson CM, Xu H, Ješovnik A, et al. Biomineral armor in leaf-cutter ants. *Nat Commun.* 2020 Nov 24;11(1):5792.

123. Moses DN, Pontin MG, Waite JH, Zok FW. Effects of hydration on mechanical properties of a highly sclerotized tissue. *Biophys J.* 2008 Apr 15;94(8):3266–3272.
124. Kerwin JL, Turecek F, Xu R, Kramer KJ, Hopkins TL, Gatlin CL, et al. Mass spectrometric analysis of catechol–histidine adducts from insect cuticle. *Anal Biochem.* 1999 Mar 15;268(2):229–237.
125. Kramer KJ, Hopkins TL, Schaefer J. Applications of solids NMR to the analysis of insect sclerotized structures. *Insect Biochem Mol Biol.* 1995 Dec 1;25(10):1067–1080.
126. Wappner P, Quesada-Allué LA. Water loss during cuticle sclerotization in the medfly *Ceratitis capitata* is independent of catecholamines. *J Insect Physiol.* 1996 Jul 1;42(7):705–709.
127. Miserez A, Schneberk T, Sun C, Zok FW, Waite JH. The transition from stiff to compliant materials in squid beaks. *Science.* 2008 Mar 28;319(5871):1816–1819.
128. Fratzl P. Biomimetic materials research: What can we really learn from nature's structural materials? *J R Soc Interface.* 2007 Mar 6;4(15):637–642.
129. Da Rosa RR, Fernandes SN, Mitov M, Godinho MH. Cellulose and chitin twisted structures: From nature to applications. *Adv Funct Mater.* 2023;2304286.
130. Locke M. Insect epidermal cells. In: Binnington K, Retnakaran A, editors. *Physiology of the Insect Epidermis.* East Melbourne: CSIRO, Australia; 1991. p. 1–22.
131. Eaton S. Cell biology of planar polarity transmission in the *Drosophila* wing. *Mech Dev.* 2003 Nov 1;120(11):1257–1264.
132. Schmidt EL. Observations on the subcuticular layer in the insect integument. *J Morphol.* 1956;99(1):211–231.
133. Locke M. The structure of an epidermal cell during the development of the protein epicuticle and the uptake of molting fluid in an insect. *J Morphol.* 1969;127(1):7–39.
134. Moussian B, Seifarth C, Müller U, Berger J, Schwarz H. Cuticle differentiation during *Drosophila* embryogenesis. *Arthropod Struct Dev.* 2006 Sep 1;35(3):137–152.
135. Moussian B. The arthropod cuticle. In: Minelli A, Boxshall G, Fusco G, editors. *Arthropod Biology and Evolution: Molecules, Development, Morphology [Internet].* Berlin: Springer; 2013 [cited 2024 Aug 2]. p. 171–196. doi:10.1007/978-3-642-36160-9-8.

136. Uv A, Moussian B. The apical plasma membrane of *Drosophila* embryonic epithelia. *Eur J Cell Biol.* 2010 Feb 1;89(2):208–211.
137. Locke M, Huie P. Bismuth staining for light and electron microscopy. *Tissue Cell.* 1977 Jan 1;9(2):347–371.
138. Sviben S, Spaeker O, Bennet M, Albéric M, Dirks JH, Moussian B, et al. Epidermal cell surface structure and chitin–protein co-assembly determine fiber architecture in the locust cuticle. *ACS Appl Mater Interfaces.* 2020 Jun 10;12(23):25581–25590.
139. Gorelik J, Shevchuk AI, Frolenkov GI, Diakonov IA, Lab MJ, Kros CJ, et al. Dynamic assembly of surface structures in living cells. *Proc Natl Acad Sci.* 2003 May 13;100(10):5819–5822.
140. Locke M. Epidermis. In: *Microscopic Anatomy of Invertebrates.* New York: Wiley-Liss; 1998. p. 75–138.
141. Locke M. Surface membranes, Golgi complexes, and vacuolar systems. *Ann Rev Entomol.* 2003 Jan 1;48:1–27.
142. Chen W, Cao P, Liu Y, Yu A, Wang D, Chen L, et al. Structural basis for directional chitin biosynthesis. *Nature.* 2022 Oct;610 (7931):402–428.
143. Perrin RM. Cellulose: How many cellulose synthases to make a plant? *Curr Biol.* 2001 Mar 20;11(6):R213–R216.
144. Haigler CH, Roberts AW. Structure/function relationships in the rosette cellulose synthesis complex illuminated by an evolutionary perspective. *Cellulose.* 2019 Jan 1;26(1):227–247.
145. Chen DD, Wang ZB, Wang LX, Zhao P, Yun CH, Bai L. Structure, catalysis, chitin transport, and selective inhibition of chitin synthase. *Nat Commun.* 2023 Aug 8;14(1):4776.
146. Peter Z. Order in cellulosics: Historical review of crystal structure research on cellulose. *Carbohydr Polym.* 2021 Feb 15;254:117417.
147. Behr M, Ganesan K. Improving polysaccharide-based chitin/chitosan-aerogel materials by learning from genetics and molecular biology. *Materials.* 2022 Jan;15(3):1041.
148. Cohen E, Moussian B, editors. *Extracellular Composite Matrices in Arthropods [Internet].* Cham: Springer International Publishing; 2016. [cited 2024 Mar 20]. Available from: http://link.springer.com/10.1007/978-3-319-40740-1.
149. Parker RM, Guidetti G, Williams CA, Zhao T, Narkevicius A, Vignolini S, et al. The self-assembly of cellulose nanocrystals: Hierarchical design of visual appearance. *Adv Mater.* 2018;30(19): 1704477.

150. Neville AC, Luke BM. A two-system model for chitin-protein complexes in insect cuticles. *Tissue Cell.* 1969 Jan 1;1(4):689–707.
151. Bouligand Y. Liquid crystals and biological morphogenesis: Ancient and new questions. *CR Chim.* 2008;11(3):281–296.
152. Giraud-Guille MM. Plywood structures in nature. *Curr Opin Solid State Mater Sci.* 1998 Jun 1;3(3):221–227.
153. Schütz C, Bruckner JR, Honorato-Rios C, Tosheva Z, Anyfantakis M, Lagerwall JPF. From equilibrium liquid crystal formation and kinetic arrest to photonic bandgap films using suspensions of cellulose nanocrystals. *Crystals.* 2020 Mar;10(3):199.
154. Onsager L. The effects of shape on the interaction of colloidal particles. *Ann N Y Acad Sci.* 1949;51(4):627–659.
155. Robin M, Djediat C, Bardouil A, Baccile N, Chareyron C, Zizak I, et al. Acidic osteoid templates the plywood structure of bone tissue. *Adv Sci.* 2024;11(9):2304454.
156. Marchessault RH, Morehead FF, Walter NM. Liquid crystal systems from fibrillar polysaccharides. *Nature.* 1959 Aug; 184(4686):632–633.
157. Revol JF, Marchessault RH. *In vitro* chiral nematic ordering of chitin crystallites. *Int J Biol Macromol.* 1993 Dec 1;15(6): 329–335.
158. Belamie E, Davidson P, Giraud-Guille MM. Structure and chirality of the nematic phase in α-chitin suspensions. *J Phys Chem B.* 2004 Sep 1;108(39):14991–15000.
159. Narkevicius A, Steiner LM, Parker RM, Ogawa Y, Frka-Petesic B, Vignolini S. Controlling the self-assembly behavior of aqueous chitin nanocrystal suspensions. *Biomacromolecules.* 2019 Jul 8;20(7):2830–2838.
160. Narkevicius A, Parker RM, Ferrer-Orri J, Parton TG, Lu Z, van de Kerkhof GT, et al. Revealing the structural coloration of self-assembled chitin nanocrystal films. *Adv Mater.* 2022;34(31):2203300.
161. Diaz de la Loza MC, Thompson BJ. Forces shaping the *Drosophila* wing. *Mech Dev.* 2017 Apr 1;144:23–32.
162. Sun Y, Lim ZW, Guo Q, Yu J, Miserez A. Liquid–liquid phase separation of proteins and peptides derived from biological materials: Discovery, protein engineering, and emerging applications. *MRS Bull.* 2020 Dec 1;45(12):1039–1047.

https://doi.org/10.1142/9789819806782_0002

Chapter 2

Biomimicking Tissue Microenvironments via Decellularized Matrix-Polymer Combinations

Hugo Edgar-Vilar, Francisca P. Correia, Mariana-Tomás de Carvalho, Margarida Henriques-Pereira, Vítor M. Gaspar* and João F. Mano*

Department of Chemistry,
CICECO — Aveiro Institute of Materials, University of Aveiro,
Campus Universitário de Santiago, 3810-193 Aveiro, Portugal
**Corresponding authors: vm.gaspar@ua.pt and jmano@ua.pt*

In nature, the extracellular matrix (ECM) provides dynamic support and biophysical signals that guide cellular behavior, tissue morphogenesis, and repair. Decellularized ECM (dECM) formulations have been rapidly emerging as valuable biomaterials to mimic tissue microenvironments owing to their tissue-matched bioactivity and cell bioinstructive features. Leveraging on the dECM's inherent chemical reactivity to create new classes of combinatorial materials with improved features and biophysical functionalities is highly relevant for numerous biomedical applications. This chapter delves into the combination of tissue-derived dECMs with diverse classes of synthetic or naturally available polymers. Different chemical toolboxes that enable the combination of dECM and polymeric biomaterials are highlighted, considering their ease of implementation and costs. Recent advances exploring the fabrication of dECM–polymer hybrid materials exhibiting tunable mechanical properties, controlled degradation rates, and increased biomimicry are also showcased. Such advanced combinations have the potential to continue contributing to improving tissue regeneration approaches and unlocking the bioengineering of increasingly meaningful in vitro disease models.

1. Introduction

The extracellular matrix (ECM) — a noncellular component ubiquitous across all tissues and organs — is not merely a structural support upon which the cells reside, but a complex hydrated meshwork of biomacromolecules that ultimately provides biochemical and biomechanical cues that are essential for tissue morphogenesis, regeneration, and homeostasis.[1–3] The ECM composition, comprising several types of collagens, glycoproteins, glycosaminoglycans (GAGs), growth factors, cytokines, and ECM-modifying enzymes, as well as its topological features, is naturally tailored to meet the living tissue's requirements.[4–6] This biomolecular diversity is instrumental in mediating tissue-specific responses under varying physiological conditions across our lifetime, underscoring the ECM's central role both in homeostasis and diseased states.[4,7] Disease progression often marks a dysregulation in ECM composition and mechanics, where the overexpression and cross-linking of some ECM components (i.e., collagen) can lead to fibrosis and a consequential increase in tissue stiffness, as seen in cancer or cardiovascular diseases. Conversely, pathologies like osteoarthritis exhibit excessive ECM degradation.[4,8–11]

Recapitulating and exploring ECM biomolecular diversity in the context of regenerative medicine and disease modeling have received significant interest in the last decade with numerous biomaterials being synthesized to mimic different ECM features and components. Despite numerous advances and relevant results, most of these biomaterials are unable to provide the multitude of bioinstructive cues that native ECMs impart in adhered cells.

Among the approaches explored to date, the processing and utilization of decellularized ECM (dECM) obtained from animal or human tissues is particularly noteworthy, as it recapitulates the native tissue microenvironments' biochemical and biophysical cues.[2,3,12] dECM-based biomaterials from various tissues including breast,[13] liver,[14] or lung[15] have already been investigated for tissue repair and in vitro disease modeling. Notably, skin-derived dECM biomaterials have already received FDA approval for clinical use in medical products — Skin patch AlloDerm Select™ RTM (BioHorizons).[3,16]

Methodologically, ECM decellularization involves the removal of cellular and genetic components from tissues while upkeeping most ECM components (i.e., proteins and glycosaminoglycans). However, standardization of this process remains elusive due to tissue- and application-specific variations with influence from factors like tissue source, surgical processing of tissue pieces and their dimensions, as well as the lipid content. The latter is particularly relevant as it varies among the different tissues and donors, being a source of variability.[2,3] To address these challenges, researchers have developed a wide variety of decellularization protocols that rely on immersion or perfusion with chemical or biological agents, or physical stresses that disrupt the cell membrane to induce cell death and that can also break specific cell–ECM interactions.[1,3] Gentler physical methods (e.g., freeze-thawing) better preserve ECM components but may leave residual cellular debris, potentially leading to the activation of the host immune system and ultimately to immune rejection of implantable dECM biomaterials. Conversely, chemical methods (e.g., Triton X) and enzymes (e.g., trypsin) efficiently remove cellular components but can alter ECM composition and structure. Each method has its trade-offs, but usually combining these methods is essential to maximize cellular debris removal.[1,16,17] Decellularization protocols often necessitate additional preparatory steps such as delipidation or demineralization for tissues with high fat or mineral content.[2,17] Postprocessing steps include sterilization (e.g., peracetic acid), aimed at eliminating toxic components to improve the biocompatibility of the dECM,[3,17] and preservation (e.g., lyophilization), which extends the shelf life, but may compromise biomechanical properties, posing challenges to clinical applicability.[3,18]

Decellularization yields two main products depending on the desired application: intact whole-organ scaffolds retaining the original 3D structure, or loose dECM biomolecular formulations (that can be further processed into surface coatings) for creating bioinks for 3D bioprinting, hydrogels for 3D cell culture, or other platforms.[3,19] For applications requiring structure and mechanical integrity, intact decellularized tissue is preferred. Whole organs are generally decellularized through perfusion-based techniques, preserving vascular and neural networks for potential recellularization with autologous or stem cells, having the potential for future creation of biofunctional tissue-engineered organs that can be implanted to address donor organ shortage.[3,6,17]

Innovative cell-free strategies are being explored to direct the body's endogenous stem cells to damaged areas, leveraging the ECM's influence on stem cell behavior to achieve desired tissue repair outcomes in situ.[1,19,20] On the other hand, soluble dECM formulations offer versatility, enabling the formation of bioactive and cytocompatible hydrogels under physiological conditions, which can provide a conducive microenvironment to promote cellular adhesion, infiltration, migration, and tissue-specific support.[2,18,21]

Despite their potential, the decellularization and solubilization processes crucially alter ECM biomechanical properties including porosity, stiffness, and fiber alignment.[19] This is mainly due to the denaturation and/or fragmentation induced in key structural ECM proteins, like collagen and fibronectin, during the decellularization process.[2,3] Such alterations can impact its ability to self-gel or to support cellular infiltration and migration, both in vitro and in vivo. Besides this, decellularization and downstream processing techniques may restrict dECM processing via additive manufacturing techniques such as extrusion 3D bioprinting due to the suboptimal rheological properties of these formulations.[2,18] To enhance the mechanical properties and ensure a stable 3D network, dECM-based composites are being explored, where dECM is combined with other synthetic or natural materials and polymers (**Figure 1**).

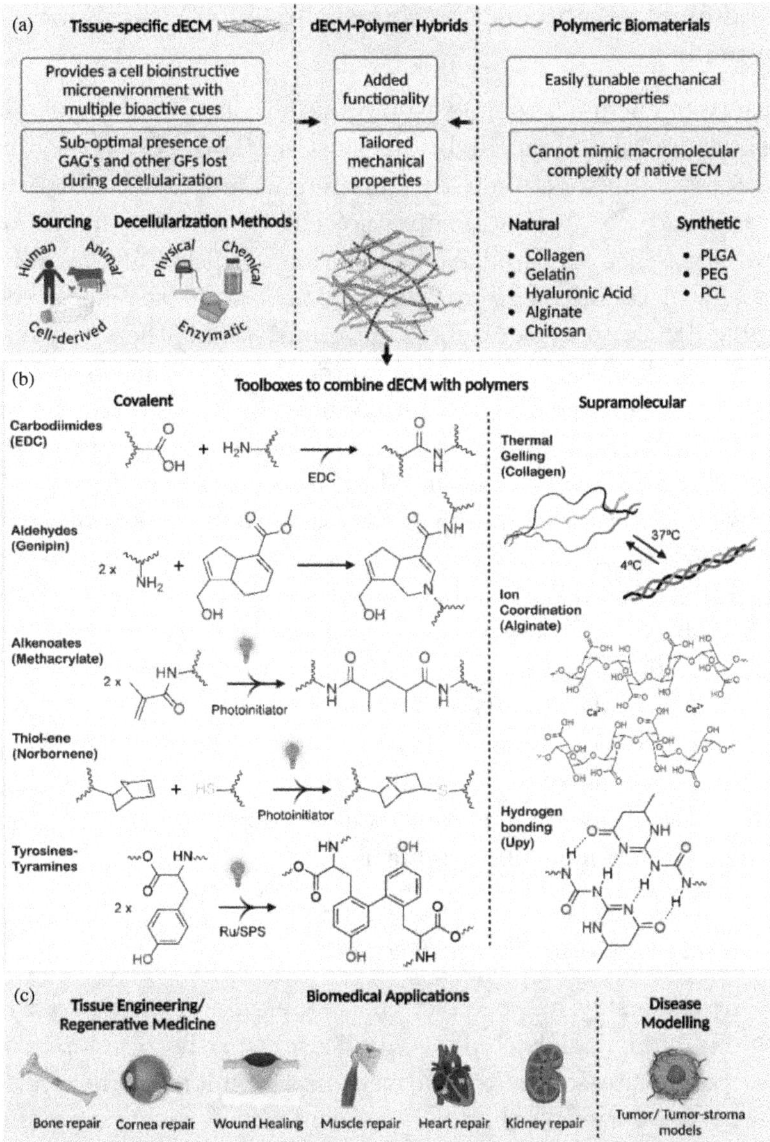

Figure 1. Decellularized matrix-polymers combinations. (**a**) Properties, sourcing, and decellularization methods of decellularized extracellular matrix biomaterials (left); Properties and common polymeric biomaterials; Properties of dECM-polymer hybrids (middle); (**b**) Chemical toolboxes typically employed to combine dECM with polymers; (**c**) Published applications of dECM hybrid materials. Created in BioRender.

2. Chemical toolboxes for generating dECM–polymer hybrids

A variety of chemical toolboxes and methodologies are available to generate dECM–polymer hybrids. Most methods rely on the creation of cross-links between the different components and can be separated into covalent and supramolecular methods, though often several strategies are used in conjunction. Some strategies rely on the usage of chemical groups already present in the materials to assemble dECM–polymer hybrid materials, while others rely on the chemical modification of pristine materials by adding user-defined moieties that enable the combination through different chemical linkages. This last approach is generally more complex but provides increased control over the final hybrid material formulations. While effective, some of these approaches, especially those that require the chemical modification of the dECM, can alter its native structure or bioactivity, as well as reduce its cytocompatibility due to potentially cytotoxic effects from some reagents that can remain in formulations after chemical modification.[1] In this subsection, the different chemical toolboxes that have been explored to date for generating dECM–polymer hybrid materials will be showcased and classified in terms of the chemical linkages that are established between the different macromolecular components: (i) covalent and/or (ii) supramolecular (**Table 1**).

2.1. *Covalent coupling toolboxes*

Covalent chemistry has been extensively employed in the synthesis of various biomaterials, but its application in the development of dECM composites remains relatively underexplored. The available toolboxes are very well characterized and offer a wide range of tunability. Furthermore, covalent approaches generally impart increased mechanical strength and stability to the resulting materials.

2.1.1. *Carbodiimides*

One of the most common cross-linking methods is carbodiimide cross-linking in which carbodiimide-containing compounds react

Table 1. Summary of conventionally employed chemical toolboxes for the production of dECM–polymer hybrid biomaterials.

Type of coupling toolbox	Type of cross-linking	Composited materials	dECM origin	Application	Main conclusions	Ref.
Covalent	Carbodiimides (EDC/NHS, DTMM, DCC, CMC, etc.)	Amine and carboxylic acid containing polymers, DCB, gelatin	Porcine knee meniscus	Knee meniscus repair	Regenerated rabbit knee meniscus with dECM and bovine bone composite, enhancing strength, and supporting fibrochondrocyte proliferation.	22
			Human amniotic membrane	Peripheral nerve repair	EDC-NHS cross-linked gelatin nanofiber and decellularized amniotic membrane, creating a stable composite for improved peripheral nerve repair.	23
			Porcine corneal stroma	Cornea modeling	Dual cross-linked hydrogel using porcine decellularized corneal stroma matrix and HAMA enables suture-free corneal repair, promoting rapid healing and minimal scarring.	24

(*Continued*)

Table 1. (*Continued*)

Type of coupling toolbox	Type of cross-linking	Composited materials	dECM origin	Application	Main conclusions	Ref.
	Aldehydes (e.g., Genipin)	Primary amine-containing polymers, chitosan, collagen-modified PEG	Porcine heart	Cardiac patches	Genipin-cross-linked cardiac ECM–chitosan patches, enhanced MSC culture, showed biocompatibility, and effectively healed acute and chronic myocardial infarction in mice.	25
Covalent			**Porcine urinary bladder, liver,** pancreas, and small intestine	**Tissue glue**	**Genipin-4 arm-PEG creates tunable, adhesive dECM** materials but shows limited biocompatibility and tissue integration in subdermal implants.	**26**
	Alkenoates (methacrylate, acrylate, etc.)	Materials modified with alkenoates like methacrylate, GelMA, HAMA	Porcine kidney	Kidney modeling	Kidney methacrylated dECM, HA, gelatin, and glycerol bioink were developed supporting bioprinted kidney cells and mimic native tissue.	27
	Thiol-ene	Materials modified with thiol and ene groups (or only one if a linker is used)	Calf knee	Cartilage modeling	Development of PVA-Nb and cartilage dECM bioink.	28

Covalent			Porcine knee cartilage	Cartilage modeling	Comparison of printability and mechanical properties of combinations of methacrylated or pentanoated cartilage dECM and HA.	29
	Tyrosines-Tyramines	Compounds containing tyramine, fibroin, collagen, HA–Tyr	Porcine pancreas	Tumor modeling	Development of a method to add HA to dECM hydrogels using light-activated HA–Tyr, preserving dECM's integrity.	30
Supramolecular	Thermal Gelling	Collagen	Brain	Spinal cord injury	Porcine brain dECM hydrogels promoted M2 macrophage polarization, reduced inflammation, and enhanced recovery, showing potential for acute SCI treatment.	31
	Ion Coordination	Alginate	Bovine aorta	Aorta graft	Alginate, tyraminated hyaluronan, and decellularized aorta were combined into an ink leveraged to produce tough and elastic aorta grafts.	32

(*Continued*)

Table 1. (*Continued*)

Type of coupling toolbox	Type of cross-linking	Composited materials	dECM origin	Application	Main conclusions	Ref.
			Mouse lung	Lung modeling	A bioactive 3D printable ink of lung dECM and alginate created strong scaffolds, with models showing tissue maturation.	33
Supramolecular	Hydrogen Bonding	Freeze-dried materials and polymers modified with UPy	Porcine femoral condyles	Cartilage repair	A waterborne polyurethane and cartilage dECM ink was cryoprinted and leveraged to regenerate auricular cartilage in rabbit models.	34
			Porcine urinary bladder	Abdominal wall defects	Enhanced dECM patches outperformed synthetic meshes, promoting collagen deposition and blood vessel formation, and reducing macrophage accumulation for abdominal wall repair.	35

CMC: carbonyldiimidazole; DCC: N,N′-dicyclohexylcarbodiimide; dECM: decellularized extracellular matrix; DTMM: N-dimethyltryptamine; EDC: 1-ethyl-3-(3-dimethylaminopropyl) carbodiimide; GelMA: gelatin methacrylate; HA: hyaluronic acid; HAMA: hyaluronic acid methacrylate; HA–Tyr: hyaluronic acid–tyramine; NHS: N-hydroxysuccinimide; PEG: polyethylene glycol; PVA-Nb: cis-5-norbornene-endo-2,3-dicarboxylic anhydride; SCI: spinal cord injury; UPy: ureido-4[1H]-pyrimidinone.

with carboxylic acids generating an o-acylisourea ester that can then bond to primary amines forming amide cross-links.[36] Note that most proteins are rich in carboxylic acid groups (i.e., such as those found on the side chains of the amino acids aspartate and glutamate) and primary amine groups (i.e., such as those found on the side chains of lysine and hydroxylysine); thus, carbodiimide-containing compounds such as 1-ethyl-3-(3-dimethylaminopropyl)-carbodiimide (EDC) and N-cyclohexyl-N′-(2-morpholinoethyl) carbodiimide metho-p-toluenesulfonate (CMC) are frequently used to cross-link scaffolds comprised of proteins like gelatin, collagen, or more relevantly, dECM.[22] More commonly, this kind of reaction is performed with catalyzing agents that serve to increase the extent of the reaction, and thus providing higher cross-linking degrees; N-hydroxysuccinimide (NHS) is the most commonly used. It reacts with intermediate and unstable o-acylisourea esters, generating instead more stable ester–NHS intermediates that can readily react with amines forming amide cross-links.[23,24,36] Because the ester–NHS complex is so stable, it is considered that NHS activates the esters and some teams exploit this to cross-link a non-dECM membrane with an excess of NHS, leaving unreacted activated esters. This membrane can then be compressed into a dECM membrane, forming a composite with cross-links between the two parts but not between the dECM, maximizing its native bioactivity.[23]

2.1.2. *Aldehydes*

Glutaraldehyde is another compound with a lot of usage for cross-linking biomaterials containing aldehydes or primary amines, namely proteins, dECM, and polysaccharides such as chitosan. It contains two aldehyde groups that can readily engage in aldol condensation reactions (aldehyde + aldehyde) or formation of Schiff bases (aldehyde + amine). However, glutaraldehyde is relatively cytotoxic, limiting its applications. Genipin is a compound extracted from gardenia fruits that is increasingly replacing glutaraldehyde, as it can react with two separate primary amines (i) through a nucleophilic addition-elimination reaction of the heterocyclic chain, forming an intermediate glutaraldehyde group under acidic and neutral pH and (ii) through a nucleophilic substitution of the ether to a secondary amide, forming cross-links between the two original amines. Genipin can also polymerize, allowing for longer cross-links via short genipin copolymers. Genipin cross-linking exhibits excellent cytocompatibility, though this comes at the cost of slower reaction kinetics, generally requiring a few hours to a few days of reaction time.[25,37] Genipin can also be conjugated to synthetic polymers, providing slow but spontaneous cross-linking once the polymer is mixed with dECM.[26]

2.1.3. *Photocrosslinking: alkenoates, thiol-ene, tyrosines*

Photocrosslinking methods rely on spontaneous, light-induced cross-linking and can offer improved printability, enabling the creation of self-standing and well-defined 3D structures. These methods depend on the usage of photoinitiator compounds and usually the modification of polymers with suitable light-responsive chemical moieties. Different studies have attempted to develop quickly polymerizable dECMs containing photoreactive groups. One of the most used groups is alkenoates with carbon–carbon double bonds at the opposite end of the carboxylic acid such as acrylate or, more commonly used, methacrylate.[27,29] In the presence of blue or ultraviolet (UV) light, photoinitiators such as lithium phenyl-2,4,6-trimethybenzoylphosphinate (LAP) or Irgacure break down forming free radicals capable of reacting with double bonds present in the alkenoates, leading to the formation of covalent bonds between the alkenoate groups. In this context, dECM has been functionalized with methacrylic groups from different tissue sources, such as tendon, cartilage, and kidney, to create hydrogels suitable for bioprinting.[27] Like the already discussed alkenoate approach, thiol-ene photopolymerization is also a radical-mediated reaction, requiring the use of photoinitiators, but unlike the former approach, this reaction forms covalent bonds between two different chemical groups, thiols and carbon–carbon double bonds (i.e., "alkenes"). Thiol-ene photopolymerization tends to have fast kinetics and is less susceptible to oxygen inhibition, so fewer radicals are required to initiate the polymerization process, reducing radical-induced damage to cells.[28] On the other hand, this chemistry requires the use of polymers or linkers modified with both thiols and -enes (usually norbornenes, and to a lesser extent, pentanoate), being more difficult to implement. Despite all the advantages of these methods, the incorporation of photosensitive groups often requires a significant modification of the dECM with synthetic chemical compounds. This is a laborious process and can reduce dECM bioactivity. A different approach is being explored, aiming to minimize the need for chemical modification of dECM formulations. This approach relies on the usage of tris(bipyridine) ruthenium (II) and sodium persulfate, or Ru/SPS, as a cross-linking agent, establishing tyrosine–tyrosine bonds. Under blue light (i.e., 400–450 nm), the ruthenium complex is excited and transfers an electron to SPS, generating a sulfate radical and oxidizing the ruthenium atom to Ru (III). The generated radical and the Ru (III) then oxidizes aromatic groups such as those already present in amino acid residues (i.e., tyrosine, histidine, etc.) as well as, to a lesser extent, lysine.[38] This oxidation transfers to these groups the free radicals that can react with each other, forming covalent cross-links.[30] This approach can be used to photocrosslink dECM with no prior modification, to cross-link dECM composites with unmodified proteins, or with polysaccharides modified with aromatic groups such as tyramine. This last approach is particularly promising as despite requiring modification of the second polymer, many polysaccharides such as hyaluronic acid (HA) are relatively simple to modify and are already present in the natural ECM, though the decellularization process often removes them.[30]

2.1.4. *Alternative and emerging covalent coupling approaches*

Apart from the aforementioned approaches, other covalent strategies based on light irradiation are rapidly emerging. Riboflavin — a photoinitiator that can cross-link proteins under UV light — has been recently demonstrated to provide fast and efficient photocrosslinking of dECM and collagen hydrogels, being a relevant alternative for dECM hybrid materials assembly.[39] Other interesting approaches are those based on enzyme-based cross-linking techniques such as transglutaminase cross-linking. This enzyme is already widely used in the food industry and to a lesser degree in tissue engineering research to cross-link proteins. It induces the formation of an isopeptide bond between glutamine and lysine residues, and it is sometimes used to stiffen dECM hydrogels, but its use for hybrid materials remains largely unexplored. Another such case is the combined use of horseradish peroxidase and hydrogen peroxide. Together these two compounds oxidize the phenolic ring of tyrosine residues originating in dityrosine cross-links. This approach is also relatively common in the tissue engineering field to cross-link proteins and has the potential to be used for dECM composites.[32]

2.2. *Supramolecular coupling toolboxes*

Unlike covalent toolboxes, whose cross-links tend to be permanent in physiological conditions, supramolecular alternatives are generally reversible and dynamic, often leading to materials with intricate mechanical properties and self-healing. This dynamic nature is increasingly sought after, as it can better recapitulate the dynamic and viscoelastic mechanical properties of tissues ECM, providing a more relevant and bioinstructive environment for cells.

2.2.1. *Thermal gelling*

Collagens are the most abundant protein in our bodies as they have a fundamental structural role. At physiological conditions (neutral pH, 37°C), most collagens self-assemble into rigid, rod-like, triple α-helix monomers stabilized by a central hydrophobic core along the entire length of the helix. This hydrophobic core is mostly composed of side chains of hydrophobic amino acids such as valine, lysine, and phenylalanine. The outside of the fibers is stabilized by an assortment of polar interactions between the external amide groups and polar residues. These polar groups also contribute, along with some external hydrophobic moieties, to the self-assembly of the triple helix monomers into interconnected collagen fibers. In our bodies, these fibers are covalently cross-linked into a very strong structure, but in vitro, this usually isn't done, resulting in a supramolecular dynamic 3D scaffold. However, at low temperatures such as 4°C, hydrophobic interactions become weaker, reversibly disassembling the connections between the collagen monomers and disrupting the structure of the triple helix resulting in the dissolution of the interconnected network. This temperature-dependent behavior allows for easy and reversible cross-linking of collagen-based materials, and as most dECMs are rich in collagens, they tend to form soft hydrogels at 37°C. Building on this natural cross-linking ability, unmodified dECM has been combined with collagen, forming a supramolecular composite.[31]

2.2.2. *Ion coordination*

Another very simple approach to generate hybrid materials is based on the combination of dECM with alginate, an algae polysaccharide that can be easily cross-linked by the presence of divalent cations such as calcium ions (Ca^{2+}) and others. These ions coordinate with two alginate chains, forming dynamic bonds. dECM–alginate hybrid materials have particular relevance for embedded 3D bioprinting applications because the ink can be printed into a support bath containing Ca^{2+}, thus instantaneously cross-linking upon contact, allowing for very good print fidelities.[32,33]

2.2.3. *Hydrogen bonding*

Many polymers spontaneously form hydrogen bonds, yet, in most cases, this by itself is not enough to form a cohesive material. However, there are ways to increase the amount of hydrogen bonds a polymer forms. Freeze-drying is a method where a material is frozen and placed under a low-temperature vacuum, sublimating the ice. When performed on a polymeric hydrogel, this method creates a porous scaffold with the polymers concentrated into discrete lamellas.[40] This is beneficial on three fronts — for one, as the material is dehydrated, it can easily be stored until use; second, porosity is known to improve cell function; and lastly, the localized increased polymer concentration allows for an increased number of polymer–polymer entanglements and other noncovalent interactions such as hydrogen bonding. Freeze-drying is thus often used to supplement the stiffness and mechanical strength of many composite materials.[22,34]

Another more complex way to introduce hydrogen bonds in a composite material is by modifying it with groups that readily form hydrogen bonds. This idea started being explored very recently for dECM hybrid materials with ureido-4[1H]-pyrimidinone (UPy), a compound known for its self-healing properties. UPy readily forms an orthogonal quadruple hydrogen bond with another UPy moiety, akin to the nucleotide bases of nucleic acids. A polymer can be modified with UPy moieties and mixed with dECM, originating a tailored supramolecular composite.[35]

3. Combinations of dECM with polymers: A new class of hybrid materials

The underlying concept under the integration of dECM biomaterials with polymers is that a potential increase in the overall bioactive, mechanical, physicochemical, or processing properties of the resulting materials can be achieved. This is grounded on the dECM's natural bioactive features and on the mechanical properties, viscosity, bioactivity, biodegradability, and/or electric conduction potential that can be provided by different synthetic or natural polymers. Such hybrid materials have already been demonstrated to exhibit improved mechanical properties, tunable degradation rates, and compatibility with sophisticated fabrication techniques, such as extrusion 3D bioprinting.[34,41] dECM–polymer composite scaffolds also allow the incorporation of peptides and growth factors, further enhancing the bioactivity and performance of dECM.[3,17] Two main categories of polymers can be used in combination with dECM to augment its features: (i) natural polymers and (ii) synthetic polymers, as summarized in **Table 2**.

Table 2. Examples of dECM–polymer hybrid materials.

Polymer origin	Composite scaffold	Advantages	Disadvantages	dECM origin	Ref.
Natural	dECM–Collagen	Bioactivity, optical transparency, support for cell adhesion and migration	Sourcing challenges (especially for human collagen) Complex processability due to temperature requirements Suboptimal mechanical properties for some load-bearing applications	Porcine brain	31
				Porcine cartilage and skin	42
	dECM–Gelatin	Bioactivity, cost-effectiveness, reduced immunogenicity, support for cell adhesion and migration	Sourcing challenges (especially for human origin gelatin) Complex processability due to temperature requirements Suboptimal mechanical properties for some load-bearing applications	Rabbit bone	43
				Human amniotic membrane	23
				Human dental pulp	44
	dECM–Alginate	Enhanced 3D printability, long-term stability, bioactivity	Limited control over degradation Potential variability associated with biopolymers Relatively limited control over biodegradation	Porcine bone	45
				Human cartilage	46
				Porcine kidney	47
	dECM–Hyaluronic **Acid**	Increased biomimicry, bioactivity, enhanced **cell–matrix interactions,**	Potential variability associated with biopolymers **Cost**	Bovine cartilage	48
				Porcine cornea	24
				Porcine pancreas	**30**

(*Continued*)

Table 2. (*Continued*)

Polymer origin	Composite scaffold	Advantages	Disadvantages	dECM origin	Ref.
	dECM–Chitosan	Bioactivity, promotion of angiogenesis for tissue repair	Reduced water solubility of biopolymer component Relatively limited control over biodegradation	Porcine nerve	49
				Human and rat cartilage	46,50
Synthetic	dECM–PLGA	Mechanical strength and stability, controlled drug delivery features, tunable degradation rate of synthetic block	Relatively limited cell–ECM interactions Difficulty in achieving homogeneous formulations	hUCMSC cells	51
				Cartilage	52
				Porcine skeletal muscle	53
				Porcine dermis	54
	dECM–PEG	Hydration and hydrophilicity, versatility, 3D printability	Limited cell–ECM interactions Challenges associated with PEG immunogenicity	Porcine heart	55
				Rat single liver lobe	56
				Human placenta	57
	dECM–PCL	Biodegradability, mechanical strength and stability, controlled drug delivery features, versatility	Relatively limited cell–ECM interactions Difficulty in achieving homogeneous formulations	Osteoblast hFOB cells, human dermal fibroblast cells, rat MSC cells, chondrocyte cells, EA.hy926 endothelial cells, and rabbit primary chondrocytes	58,60
				Pig auricular cartilage	61

3.1. *Composites of dECM with natural polymers*

Natural polymers have been widely implemented in the fabrication of scaffolds for tissue engineering. Such polymers have been shown to promote tissue repair and regeneration with suitable biocompatibility profiles and good biodegradability.[17] In this framework, this section will cover the latest advancements in dECM composite scaffolds, which are created by combining dECM with naturally occurring proteins (e.g., collagen) and with natural polysaccharides (e.g., alginate and HA).

3.1.1. *Gelatin*

Gelatin, a collagen derivative, has been widely used in combination with dECM-based materials since it maintains the original RGD (Arg–Gly–Asp) sequence that is crucial for cell adhesion and has lower immunogenicity. In a recent study, a composite membrane comprising the combination of gelatin with human decellularized amniotic membrane (dAM) through interfacial covalent bonding was described.[26] The dAM has unique features for promoting peripheral nerve injury repair due to its pro-regenerative ECM structure and anti-inflammatory properties. In this sense, the combination of gelatin with dAM was used in this study with the goal of maintaining dECM biological activity and reducing its degradation rate. The viability of Schwann cells, a predominant type of glial cells that have a crucial role in the formation of the myelin sheath of axons, was evaluated in dAM–gelatin membranes. The results revealed enhanced Schwann cell adhesion and proliferation, and also macrophage polarization, in vivo.[26]

To increase the processability of gelatin and the tunability of its mechanical properties under stimuli, some chemical modifications have been implemented, namely the modification with methacryloyl groups (gelatin methacrylate [GelMA]) to enable the assembly of tunable hydrogels. This modification does not affect the RGD sequence of gelatin and the metalloprotein ratio in the matrix. To date, the combination of dECM–GelMA composites in tissue regeneration studies has demonstrated very good results in the regeneration of several tissues. Indeed, in a recent study, an injectable granular hydrogel made from decellularized dental pulp matrix and GelMA was developed with the purpose of bone reconstruction (**Figure 2a**).[44] To address that, GelMA microspheres were first

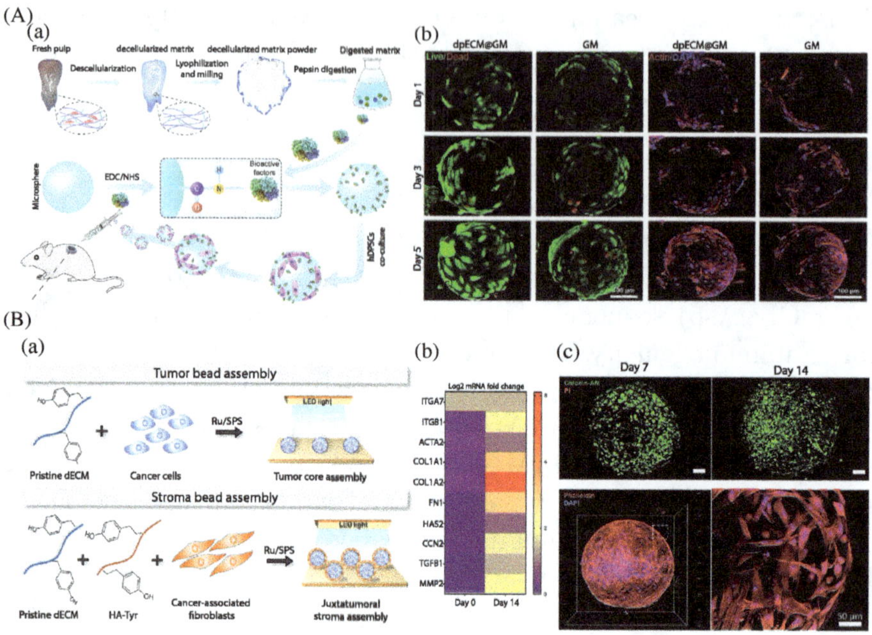

Figure 2. dECM hybrid materials formed with natural polymers. (A) (a) Schematic illustration demonstrating the processes involved in the fabrication of hydrogel microspheres functionalized with decellularized tooth pulp matrix and their application in regenerative endodontic therapy. (b) Live/dead cell staining of hDPSCs on the dECM composite hydrogel and staining of β-actin filaments (stained with phalloidin) for accessing cell adhesion and spreading phenotype. Reproduced with permission from Zheng et al.[44] Copyright 2023, Elsevier. **(B)** (a) Schematic representation of dECM–hyaluronan hybrid materials used for assembling core–shell tumor–stroma in vitro models using superhydrophobic surfaces. (b) Results from the evaluation of mRNA expression levels of important genes related to the development of pancreatic adenocarcinoma. (c) Results of live/dead 3D viability assay of core–shell microtumors, at days 7 and 14 of culture (upper panels) and the results of the staining with phalloidin at day 14 of culture (lower panels). Reproduced with permission from Monteiro et al.[30] Copyright 2023, Wiley.

generated by a standard protocol of electrospinning, and then they were combined with the dECM from human dental pulp. For this combination, N-ethyl-N′-(3-dimethylaminopropyl) carbodiimide/N-hydroxysuccinimide was utilized to promote the reaction of the carboxyl groups and the amino residue on both the microsphere surface and the dECM. Finally, these dECM–GelMA microspheres were tested in terms of biocompatibility and ability to promote bone regeneration. In fact, the results of the culture of human dental pulp stem cells (hDPSC) on these microspheres demonstrate not only good cell proliferation, but also the promotion of odontogenic, angiogenic, and neurogenic differentiation of hDPSCs in vitro (**Figure 2a**). An in vivo study was also conducted to evaluate the ability to promote regeneration. In brief, the microspheres were transferred into tooth slices and then implanted into the subcutaneous of immunodeficient nude mice. The results revealed the presence of more formed blood vessels confirmed by the presence of $CD31^+$, compared to the other control groups, and also the presence of collagen fibers and new dentin, crucial to recapitulate the in vivo scenario. The most important aspect observed was the indication of dental pulp–dentin complex regeneration.

3.1.2. *Hyaluronic acid*

HA is a natural polysaccharide present in tissues ECM, with an important regeneration role in the human body. HA chemical moieties render it especially prone to chemical modifications with various types of functional moieties such as methacryloyl groups (HA methacrylate [HAMA]) that enable the assembly of HA hydrogels using light stimuli.[24] Owing to its biological relevance, several other HA derivatives functionalized with valuable chemical moieties (i.e., cyclodextrins or cucurbiturils, norbornenes, maleimides, etc.) that add further functionality have been formulated to date. The combination of dECM-HA to form hydrogels is particularly interesting to be explored not only due to its natural presence, but also due to the major loss of HA during most decellularization processes. This is a major drawback considering that HA plays an important role in several diseases and in tissue homeostasis. A common chemical modification for the establishment of dECM–HA hydrogels is the addition of tyramine groups to the HA backbone, to promote the formation of tyramine–tyrosine covalent bonds with dECM under exposure to visible light and in the presence of a photoinitiator.[30] In a recent study, such methodology was followed with the aim of assembling 3D pancreatic cancer models that better recapitulate the tumor microenvironment composition and cell–matrix interactions (**Figure 2b**). In order to accomplish this, HA was first chemically functionalized to include tyramine groups (HA-Tyr). This was performed to promote the formation of tyramine–tyrosine covalent interactions with dECM when exposed to visible light (i.e., 400–450 nm) and in the presence of the photoinitiator ruthenium/sodium persulfate (Ru/SPS). The cancer models were designed to comprise a core of tumor cells and a shell of stroma cells. The dECM–HATyr hydrogel was used to include stroma cells and to form the outer layer of the generated 3D spherical structures, recapitulating the mechanical properties of the ECM sense by these cells in vivo and providing important biofactors.

In another recent study, a porcine decellularized corneal stroma matrix was combined with light-responsive HAMA.[24] This approach yielded the fabrication of a double-network hydrogel through a noncompetitive double cross-linking process using CMC/NHS and LAP-induced photocrosslinking, with the aim of creating a hydrogel that can be used to repair corneal defects. The results obtained revealed that the dECM–HAMA hybrid hydrogel supported the adhesion, viability, and proliferation of the corneal stromal cells and that after eight weeks of observation, the experimental rabbit eyes filled with this hydrogel remained transparent and exhibited accelerated corneal re-epithelialization and wound healing.

3.2. *Composites of dECM with synthetic polymers*

Synthetic polymers have been widely explored in the last decade for advancing tissue engineering and disease modeling, as most of them confer the possibility of enhancing various properties of proteinaceous hydrogels (i.e., mechanical, electrically conductive, viscoelastic properties, etc.).[21] To date, different synthetic polymers have been explored to prepare dECM-based composites with excellent biomimetic properties, strong stability in vivo, and tunable mechanical properties (**Figure 3**). This section will therefore, discuss the most recent developments in dECM–synthetic polymer hybrid scaffolds.

3.2.1. *Polylactic acid-co-glycolic acid*

Polylactic acid-co-glycolic acid (PLGA) is an FDA-approved synthetic polymer with extensive application in various biomedical industries. PLGA possesses tunable molecular weight, biodegradation rate, and chemical versatility, rendering it highly attractive for combination with other biomaterials. The combination of dECM with PLGA has been recently implemented, demonstrating improved mechanical properties, strength, and degradation rate, and even good cell growth and adhesion in these composites. To achieve the desired mechanical properties of different organs and tissues, the ratios of dECM–PLGA must be investigated.[49] In an elegant study, an ice particle leaching method was used for assembling a kidney scaffold with different ratios of kidney dECM and PLGA (**Figure 3a**).[59] The results indicate that a concentration of 10% dECM demonstrated decreased compressive moduli and compressive stress, essential for kidney regeneration. These tunable mechanical properties conferred by this combination have also been shown in other studies[51,59] to promote the activation of several biochemical pathways that are crucial for the initiation of tissue regeneration, namely by enhancing the expression of matrix metalloproteinases and collagens. In a more recent study, the authors combined porcine skeletal muscle dECM with microscale fibrillated PLGA to fabricate a nanofibrous structure that recapitulates the skeletal muscle tissue.[53] To achieve this goal, the dECM was first chemically modified with methacrylate (MA) groups that improve the structural stability for the next step, which consisted of the formation of nanofiber of dECM-methacryloyl (dECM-MA) using electrospinning and photocrossinking (UV light, 400 mW cm^{-2}, 60 min) techniques. After this

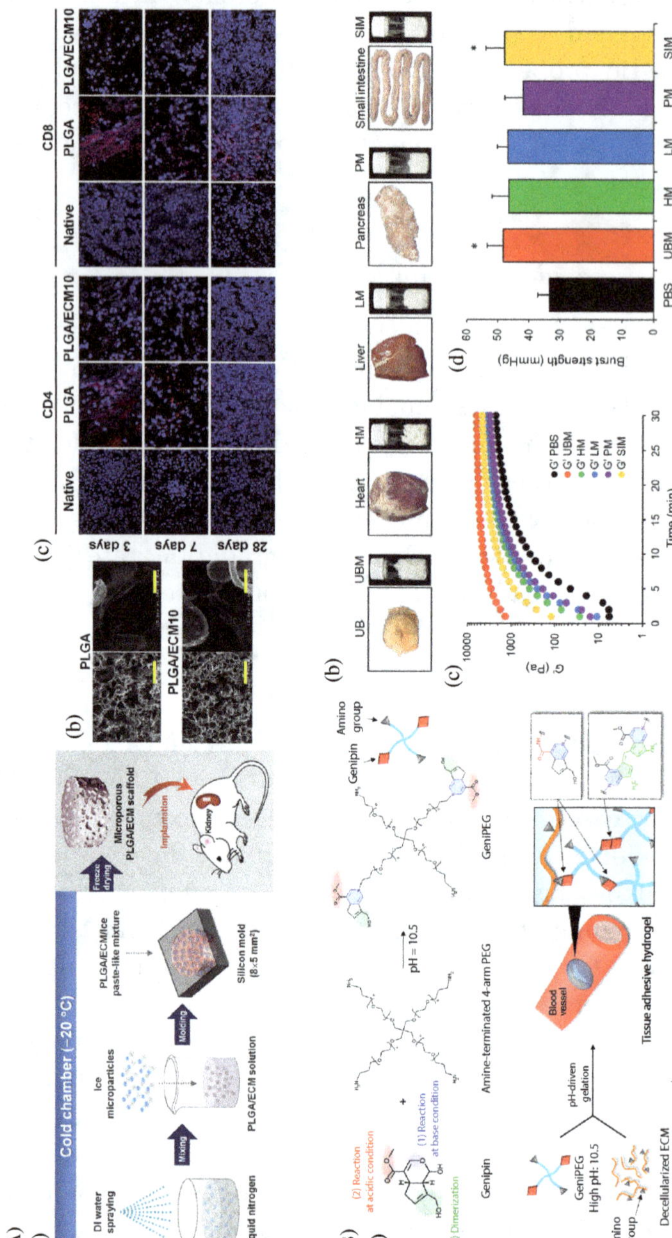
(A)
(a)
Cold chamber (−20 °C)
DI water spraying
Liquid nitrogen
Mixing
Ice microparticles
PLGA/ECM solution
Molding
PLGA/ECM/ice paste-like mixture
Silicon mold (8×5 mm²)
Freeze drying
Microporous PLGA/ECM scaffold
Implantation
Kidney
(b)
PLGA
PLGA/ECM10
(c)
CD4
CD8
Native
PLGA
PLGA/ECM10
3 days
7 days
28 days
(B)
(a)
(2) Reaction at acidic condition
(1) Reaction at base condition
(3) Dimerization
Genipin
Amine-terminated 4-arm PEG
pH = 10.5
GeniPEG
Genipin
Amino group
GeniPEG High pH: 10.5
Decellularized ECM Low pH: 2.5
pH-driven gelation
Blood vessel
Tissue adhesive hydrogel
(b)
UB
UBM
Heart
HM
Liver
LM
Pancreas
PM
Small intestine
SIM
(c)
G' (Pa)
Time (min)
G' PBS
G' UBM
G' HM
G' LM
G' PM
G' SIM
(d)
Burst strength (mmHg)
PBS
UBM
HM
LM
PM
SIM

Figure 3. dECM hybrid materials formed with synthetic polymers. (**A**) (a) Schematics illustrating the process of creating porous PLGA/ECM scaffolds using an ice microparticle leaching technique. (b) SEM images reveal the differences in the structure of PLGA and PLGA/ECM hydrogels. (c) Immunohistochemical analysis of the implanted region with T cell markers. Reprinted with permission.[59] Copyright 2016, American Chemical Society. (**B**) (a) Schematic representation of the synthesis of GeniPEG. (b) Macroscopic photos of the decellularized tissues used: porcine-derived UB, heart, liver, pancreas, and small intestine. (c) Rheological measurement of GeniPEG-cross-linked dECM composites hydrogels. (d) Burst strength of GeniPEG-cross-linked hydrogels using collagen casing. Reproduced with permission from Nishiguchi and Taguchi.[26] Copyright 2021, Elsevier.

step, uniaxially aligned nanofibers of dECM-methacryloyl were deposited onto PLGA-printed constructs by electrostatic alignment. This is especially important to ensure the correct topographical cues needed for muscle regeneration. Through the implementation of this multiscale composite scaffold, myotubes were produced and sustained in the absence of myogenic medium conditions, and hMPCs were aligned parallel to the topographical cue.

3.2.2. *Polyethylene glycol*

Polyethylene glycol (PEG) is a common synthetic polymer widely combined with other materials. In fact, in the last years, it has been conjugated with dECM once it allows a low amount of interaction with cells enabling control over cell responses by conjugating other bioactive molecules, such as cell-adhesive ligands or cell-cleavable moieties. This conjugation and its degree can be modulated in a way of controlling, for example, cell adhesion and the stiffness of the materials. For instance, a dECM-based tissue adhesive was developed based on the pH-dependent reactivity of genipin (gelator).[26] The combination reported in this study between dECM from different tissues (urinary bladder, heart, liver, pancreas, and small intestine) with genipin-terminated four-arm PEG, demonstrated the nucleophilic attack to methyl ester in GeniPEG by amino groups in the urinary bladder, that is determinant for the gelation speed (**Figure 3b**). The results also elucidated that at pH values less than 2.5, the gelation speed was fast and suitable for use as a sprayable tissue adhesive. In this study, the investigators also explored the capacity of this dECM–GeniPEG combination to be implemented for the leakage of body fluid and blood from wounds after surgeries. The findings showed that GeniPEG reacted more quickly at low pH, which is associated with improved hydrogel mechanical characteristics and interfacial adhesion strength due to tissue-resident ECM protein reactions. Considering the findings, it was determined that this tissue-adhesive hydrogel based on dextran may be used as a supportive material to lessen postoperative blood loss and expedite surgery.

4. Conclusions and outlook

In recent years, the rapidly emerging exploitation of advanced dECM-based biomaterials has marked a significant evolution from the initial decellularization of simple tissue sections to the complex and successful preparation of whole decellularized organ scaffolds. In this framework, the unique advantages of soluble dECM biomaterials, such as their inherent biological degradability, cell-supporting properties, and multiple bioinstructive features, underscore their prominence over other processable biomaterials. These features have also been recently enhanced through successful combination with different natural and/or synthetic polymers as herein overviewed. The combination of cell or tissue-sourced dECMs with other polymers has already been shown to improve its processing via state-of-the-art 3D printing technologies, as well as to impart relevant characteristics such as porosity, tunable mechanical properties (including improved stiffness and viscoelastic responses), as well as programmable biodegradation. Alongside dECM–polymer combinations, particularly those resorting to the incorporation of naturally present hyaluronan, have already opened new opportunities to establish new in vitro models of healthy and diseased tissues that more faithfully recapitulate native in vivo microenvironments. Significant advances in this field are expected in the future, with the combination of dECMs with other key GAGs that are removed during decellularization processes. Moreover, the combination of dECMs with other classes of synthetic polymers, including those exhibiting conductive features, is envisioned to generate innovative dECM-based piezoelectric materials in the foreseeable future.

Despite the significant potential of dECM–polymer hybrid materials, several challenges remain to be addressed for their widespread adoption. The inherent dECM batch-to-batch variability associated with dECM and dECM–polymer mixing, their laborious processing and manufacture, the material sourcing requirements (i.e., animal vs. human species), and standardization are key parameters that must be addressed for broader applicability. Continuous efforts are also needed to further optimize matrix decellularization processes and to assure dECM biocompatibility with minimal loss of naturally present components (i.e., soluble growth factors, extracellular vesicles, GAGs), that are key for its bioinstructive properties. In addition, assuring cost-effective and scalable manufacturing processes will be a major target for future clinical translation and widespread application of dECM scaffolds. These foreseen developments will undoubtedly contribute for establishing dECM–polymer materials as relevant toolboxes for multiple biomedical applications or for the development of organic electronics and soft robotics.

References

1. Noro J, Vilaça-Faria H, Reis RL, Pirraco RP. Extracellular matrix-derived materials for tissue engineering and regenerative medicine: A journey from isolation to characterization and application. *Bioact Mater.* 2024;34:494–519.
2. Kort-Mascort J, Flores-Torres S, Peza-Chavez O, Jang JH, Pardo LA, Tran SD, et al. Decellularized ECM hydrogels: Prior use considerations, applications, and opportunities in tissue engineering and biofabrication. *Biomater Sci.* 2023;11(2):400–431.
3. Golebiowska AA, Intravaia JT, Sathe VM, Kumbar SG, Nukavarapu SP. Decellularized extracellular matrix biomaterials for regenerative therapies: Advances, challenges, and clinical prospects. *Bioact Mater.* 2024;32:98–123.
4. Cox TR. The matrix in cancer. *Nat Rev Cancer.* 2021;21:217–238.
5. Pati F, Jang J, Ha DH, Won Kim S, Rhie JW, Shim JH, et al. Printing three-dimensional tissue analogues with decellularized extracellular matrix bioink. *Nat Commun.* 2014;5(1):3935.
6. Hussey GS, Dziki JL, Badylak SF. Extracellular matrix-based materials for regenerative medicine. *Nat Rev Mater.* 2018;3:59–173.
7. Bonnans C, Chou J, Werb Z. Remodelling the extracellular matrix in development and disease. *Nat Rev Mol Cell Biol.* 2014;15:786–801.
8. Lampi MC, Reinhart-King CA. Targeting extracellular matrix stiffness to attenuate disease: From molecular mechanisms to clinical trials. *Sci Transl Med.* 2018;10:422.

9. Ferreira LP, Gaspar VM, Mano JF. Decellularized extracellular matrix for bioengineering physiomimetic 3D in vitro tumor models. *Trends Biotechnol.* 2020;30:1397–1414.
10. Iozzo RV, Gubbiotti MA. Extracellular matrix: The driving force of mammalian diseases. *Matrix Biol.* 2018;71–72:1–9.
11. Park JH, Jo SB, Lee JH, Lee HH, Knowles JC, Kim HW. Materials and extracellular matrix rigidity highlighted in tissue damages and diseases: Implication for biomaterials design and therapeutic targets. *Bioact Mater.* 2023;20:381–403.
12. Barthold JE, St. Martin BM, Sridhar SL, Vernerey F, Schneider SE, Wacquez A, et al. Recellularization and integration of dense extracellular matrix by percolation of tissue microparticles. *Adv Funct Mater.* 2021;31:2103355.
13. Jin Q, Liu G, Li S, Yuan H, Yun Z, Zhang W, et al. Decellularized breast matrix as bioactive microenvironment for in vitro three-dimensional cancer culture. *J Cell Physiol.* 2019;234:3425–3435.
14. Shaheen MF, Joo DJ, Ross JJ, Anderson BD, Chen HS, Huebert RC, et al. Sustained perfusion of revascularized bioengineered livers heterotopically transplanted into immunosuppressed pigs. *Nat Biomed Eng.* 2020;4:437–445.
15. Ott HC, Clippinger B, Conrad C, Schuetz C, Pomerantseva I, Ikonomou L, et al. Regeneration and orthotopic transplantation of a bioartificial lung. *Nat Med.* 2010;16(8):927–933.
16. Zhang X, Chen X, Hong H, Hu R, Liu J, Liu C. Decellularized extracellular matrix scaffolds: Recent trends and emerging strategies in tissue engineering. *Bioact Mater.* 2022;10:15–31.
17. Mendibil U, Ruiz-Hernandez R, Retegi-Carrion S, Garcia-Urquia N, Olalde-Graells B, Abarrategi A. Molecular sciences tissue-specific decellularization methods: Rationale and strategies to achieve regenerative compounds. *Int J Mol Sci.* 2020;21(15):5447.
18. Zhang H, Wang Y, Zheng Z, Wei X, Chen L, Wu Y, et al. Strategies for improving the 3D printability of decellularized extracellular matrix bioink. *Theranostics.* 2023;13(8):2562–2587.
19. Liao J, Xu B, Zhang R, Fan Y, Xie H, Li X. Applications of decellularized materials in tissue engineering: Advantages, drawbacks and current improvements, and future perspectives. *J Mater Chem B.* 2020;8:10023–10049.

20. Zhu M, Li W, Dong X, Yuan X, Midgley AC, Chang H, et al. In vivo engineered extracellular matrix scaffolds with instructive niches for oriented tissue regeneration. *Nat Commun.* 2019;10:4620.
21. Xu P, Kankala RK, Wang S, Chen A. Decellularized extracellular matrix-based composite scaffolds for tissue engineering and regenerative medicine. *Regen Biomater.* 2023;11:rbad107.
22. Yuan Z, Liu S, Hao C, Guo W, Gao S, Wang M, et al. AMECM/DCB scaffold prompts successful total meniscus reconstruction in a rabbit total meniscectomy model. *Biomaterials.* 2016;11:13–26.
23. Chen L, Song X, Yao Z, Zhou C, Yang J, Yang Q, et al. Gelatin nanofiber-reinforced decellularized amniotic membrane promotes axon regeneration and functional recovery in the surgical treatment of peripheral nerve injury. *Biomaterials.* 2023;300:122207.
24. Shen X, Li S, Zhao X, Han J, Chen J, Rao Z, et al. Dual-crosslinked regenerative hydrogel for sutureless long-term repair of corneal defect. *Bioact Mater.* 2023;20:434–448.
25. Efraim Y, Sarig H, Cohen Anavy N, Sarig U, de Berardinis E, Chaw SY, et al. Biohybrid cardiac ECM-based hydrogels improve long term cardiac function post myocardial infarction. *Acta Biomater.* 2017;50:220–233.
26. Nishiguchi A, Taguchi T. A pH-driven genipin gelator to engineer decellularized extracellular matrix-based tissue adhesives. *Acta Biomater.* 2021;131:211–221.
27. Ali M, Pr K, Yoo JJ, Zahran F, Atala A, Lee SJ. A photo-crosslinkable kidney ECM-derived bioink accelerates renal tissue formation. *Adv Healthc Mater.* 2019;8(7):e1800992.
28. Setayeshmehr M, Hafeez S, van Blitterswijk C, Moroni L, Mota C, Baker MB. Bioprinting via a dual-gel bioink based on poly(vinyl alcohol) and solubilized extracellular matrix towards cartilage engineering. *Int J Mol Sci.* 2021;22(8):3901.
29. Kiyotake EA, Thomas EE, Iribagiza C, Detamore MS. High-stiffness, fast-crosslinking, cartilage matrix bioinks. *J Biomech.* 2023;148:111471.
30. Monteiro MV, Henriques-Pereira M, Neves BM, Duarte ID, Gaspar VM, Mano JF. Photo-compartmentalized decellularized matrix-hyaluronan hybrid units for pancreatic tumor-stroma modeling. *Adv Funct Mater.* 2024;34(6):2305473.
31. Hong JY, Seo Y, Davaa G, Kim HW, Kim SH, Hyun JK. Decellularized brain matrix enhances macrophage polarization and

functional improvements in rat spinal cord injury. *Acta Biomater.* 2020 101:357–371.

32. Isik M, Karakaya E, Arslan TS, Atila D, Erdogan YK, Arslan YE, et al. 3D Printing of extracellular matrix-based multicomponent, all-natural, highly elastic, and functional materials toward vascular tissue engineering. *Adv Healthc Mater.* 2023;12:2203044.
33. De Santis MM, Alsafadi HN, Tas S, Bölükbas DA, Prithiviraj S, Da Silva IAN, et al. Extracellular-matrix-reinforced bioinks for 3D bioprinting human tissue. *Adv Mater.* 2021;33:2005476.
34. Chen M, Li YY, Liu S, Feng Z, Wang H, Yang D, et al. Hierarchical macro-microporous WPU-ECM scaffolds combined with microfracture promote in situ articular cartilage regeneration in rabbits. *Bioact Mater.* 2021;6(7):1932–1944.
35. Nishiguchi A, Ito S, Nagasaka K, Taguchi T. Tissue-adhesive decellularized extracellular matrix patches reinforced by a supramolecular gelator to repair abdominal wall defects. *Biomacromolecules.* 2023;24:1545–1554.
36. Jayachandran B, Parvin TN, Alam MM, Chanda K, MM B. Insights on chemical crosslinking strategies for proteins. *Molecules.* 2022;27(23):8124.
37. Gao M, Wang Y, He Y, Li Y, Wu Q, Yang G, et al. Comparative evaluation of decellularized porcine liver matrices crosslinked with different chemical and natural crosslinking agents. *Xenotransplantation.* 2019;26(1):e12470.
38. Gulzar A, Yıldız E, Kaleli HN, Nazeer MA, Zibandeh N, Malik AN, et al. Ruthenium-induced corneal collagen crosslinking under visible light. *Acta Biomater.* 2022;147:198–208.
39. Tirella A, Liberto T, Ahluwalia A. Riboflavin and collagen: New crosslinking methods to tailor the stiffness of hydrogels. *Mater Lett.* 2012;74:58–61.
40. Sukul M, Sahariah P, Lauzon HL, Borges J, Másson M, Mano JF, et al. In vitro biological response of human osteoblasts in 3D chitosan sponges with controlled degree of deacetylation and molecular weight. *Carbohydr Polym.* 2021;254:117434.
41. Elmashhady HH, Kraemer BA, Patel KH, Sell SA, Garg K. Decellularized extracellular matrices for tissue engineering applications. *Electrospinning.* 2017;1(1):87–99.
42. Hua Y, Huo Y, Bai B, Hao J, Hu G, Ci Z, et al. Fabrication of biphasic cartilage-bone integrated scaffolds based on tissue-specific

photo-crosslinkable acellular matrix hydrogels. *Mater Today Bio.* 2022;17:100489.

43. Kara A, Distler T, Polley C, Schneidereit D, Seitz H, Friedrich O, et al. 3D printed gelatin/decellularized bone composite scaffolds for bone tissue engineering: Fabrication, characterization and cytocompatibility study. *Mater Today Bio.* 2022;15:100309.
44. Zheng L, Liu Y, Jiang L, Wang X, Chen Y, Li L, et al. Injectable decellularized dental pulp matrix-functionalized hydrogel microspheres for endodontic regeneration. *Acta Biomater.* 2023;156: 37–48.
45. Lee J, Hong J, Kim WJ, Kim GH. Bone-derived dECM/alginate bioink for fabricating a 3D cell-laden mesh structure for bone tissue engineering. *Carbohydr Polym.* 2020;250:116914.
46. Upadhyay U, Kolla S, Maredupaka S, Priya S, Srinivasulu K, Chelluri LK. Development of an alginate–chitosan biopolymer composite with dECM bioink additive for organ-on-a-chip articular cartilage. *Sci Rep.* 2024;14:11765.
47. Chu TL, Tripathi G, Park M, Bae SH, Lee BT. In-vitro and in-vivo biocompatibility of dECM-alginate as a promising candidate in cell delivery for kidney regeneration. *Int J Biol Macromol.* 2022;211: 616–625.
48. Guo P, Jiang N, Mini C, Miklosic G, Zhu S, Vernengo AJ, et al. Decellularized extracellular matrix particle-based biomaterials for cartilage repair applications. *J Mater Sci Technol.* 2023;160:194–203.
49. Kong Y, Wang D, Wei Q, Yang Y. Nerve decellularized matrix composite scaffold with high antibacterial activity for nerve regeneration. *Front Bioeng Biotechnol.* 2022;9:840421.
50. Chen D, Zhang Y, Lin Q, Chen D, Li X, Dai J, et al. The effect of cartilage decellularized extracellular matrix-chitosan compound on treating knee osteoarthritis in rats. *Peer J.* 2021;9:e12188.
51. Deng M, Tan J, Hu C, Hou T, Peng W, Liu J, et al. Modification of PLGA scaffold by MSC-derived extracellular matrix combats macrophage inflammation to initiate bone regeneration via TGF-β-induced protein. *Adv Healthc Mater.* 2020;9(13):2000353.
52. Chen Y, Chen LF, Wang Y, Duan YY, Luo SC, Zhang JT, et al. Modeling dECM-based inflammatory cartilage microtissues in vitro for drug screening. *Compos B Eng.* 2023;250:110437.
53. Lee H, Kim WJ, Lee JU, Yoo JJ, Kim GH, Lee SJ. Effect of hierarchical scaffold consisting of aligned dECM nanofibers and

poly(lactide- co-glycolide) struts on the orientation and maturation of human muscle progenitor cells. *ACS Appl Mater Interfaces.* 2019;11(43):39449–39458.
54. Fang Y, Han Y, Wang S, Chen J, Dai K, Xiong Y, et al. Three-dimensional printing bilayer membranous nanofiber scaffold for inhibiting scar hyperplasia of skin. *Biomater Adv.* 2022;138:212951.
55. Gómez-Cid L, López-Donaire ML, Velasco D, Marín V, González MI, Salinas B, et al. Cardiac extracellular matrix hydrogel enriched with polyethylene glycol presents improved gelation time and increased on-target site retention of extracellular vesicles. *Int J Mol Sci.* 2021;26;22(17):9226.
56. Zhang JS, Wang ZB, Lai ZZ, Yang JW, Song WJ, Wei YB, et al. Polyethylene glycol crosslinked decellularized single liver lobe scaffolds with vascular endothelial growth factor promotes angiogenesis in vivo. *HBPD Int.* 2023;22(6):622–631.
57. Fan Y, Lüchow M, Badria A, Hutchinson DJ, Malkoch M. Placenta powder-infused thiol-ene PEG hydrogels as potential tissue engineering scaffolds. *Biomacromolecules.* 2023;24(4):1617–1626.
58. Wang W, Zhou X, Yin Z, Yu X. Fabrication and evaluation of porous dECM/PCL scaffolds for bone tissue engineering. *J Funct Biomater.* 2023;14(7):343.
59. Lih E, Park KW, Chun SY, Kim H, Kwon TG, Joung YK, et al. Biomimetic porous PLGA scaffolds incorporating decellularized extracellular matrix for kidney tissue regeneration. *ACS Appl Mater Interfaces.* 2016;8(33):21145–21154.
60. Xu J, Fang Q, Liu Y, Zhou Y, Ye Z , Tan WS. In situ ornamenting poly(ε-caprolactone) electrospun fibers with different fiber diameters using chondrocyte-derived extracellular matrix for chondrogenesis of mesenchymal stem cells. *Colloids Surf B Biointerfaces.* 2021;197:111374.
61. Feng B, Ji T, Wang X, Fu W, Ye L, Zhang H, et al. Engineering cartilage tissue based on cartilage-derived extracellular matrix cECM/PCL hybrid nanofibrous scaffold. *Mater Des.* 2020;193:108773.

https://doi.org/10.1142/9789819806782_0003

Chapter 3

Engineering the Extracellular Matrix and its Interactions with Biomolecules

Priscilla S. Briquez

Department of General and Visceral Surgery, Medical Center — University of Freiburg, Breisacher Strasse, 79106 Freiburg, Germany
priscilla.briquez@uniklinik-freiburg.de

The extracellular matrix (ECM) can be seen as a physiological 3D material that supports the cell's activities in tissues and organs. It is an insoluble scaffold made of structural fibrillar proteins, further decorated with glycoproteins and proteoglycans, which functionalize it. The high diversity of ECM components and their assembly precisely define the mechanical, chemical, and biological properties of the scaffold. In recent years, the use of ECM-mimetic natural and synthetic biomaterials has rapidly developed for regenerative medicine purposes to replace defective tissues and organs. However, polymeric materials often fail to fully recapitulate the ECM properties, requiring further engineering to optimize their interactions with cells and control cell behavior. Polymeric materials have indeed been functionalized via chemical or enzymatic conjugation and via protein engineering approaches to enhance their biocompatibility, degradability, and bioactivity. In this chapter, we first recapitulate the key properties of the physiological ECM, before presenting some natural and synthetic polymeric materials commonly used in regenerative medicine to mimic the ECM. We then highlight strategies that have been developed to optimize these materials and provide examples of their applications. Lastly, we discuss ongoing challenges in the field and future avenues to overcome them.

1. Introduction

Although overlooked, the extracellular matrix (ECM) plays a fundamental role in tissues and organs. The ECM is the noncellular part of tissues that supports their structure and functions. While it has been long regarded as a passive fibrous network providing mechanical support to cells, we now know that the ECM constitutes a highly dynamic physiological biomaterial that actively regulates the behavior of cells, in addition to its mechanical function. The role of the ECM has been extensively explored in the context of tissue repair and engineering, to understand how to recreate tissues and organs for application in regenerative medicine. Indeed, scientists and engineers have developed biomaterials that mimic the ECM in terms of biocompatibility and biological function, to provide long-term replacement solutions for tissues that are difficult to heal because of trauma or underlying diseases.[1] To date, biomaterials are widely used in clinics and provide important therapeutic options, serving as prosthesis, implants, and wound dressings, to cite a few examples. The strong ongoing research in regenerative medicine is additionally fueled by the field of artificial organs, which has emerged in clinics and is actively growing. In fact, one unmet goal of regenerative medicine is to create fully functional replacement tissues or whole organs that are ready for transplantation into patients, reducing reliance on organ donation.[2]

Starting from the latter half of the 20th century, we have witnessed a revolution in the development of ECM-mimetic biomaterials. These materials are made of advanced polymeric materials, engineered for cell–materials interactions, biocompatibility, and biodegradability.[3,4] In addition, they have been widely functionalized with biomolecules to control cell morphogenetic behaviors, including cell survival, proliferation, migration, and differentiation.[5,6] Many studies have detailed the mechanisms by which the native ECM remodels itself and stores bioactive compounds to instruct cell behaviors with specific kinetics and patterns. Our increasing knowledge of the ECM continues to inspire the development of novel materials with enhanced performance. In addition, the fast development of proteomics, DNA sequencing and gene expression quantification, molecular biology, and engineering methods allows the identification of particularly potent biomolecules or protein domains with strong regulatory effects on cells, as well as their incorporation into polymeric materials. Thus, the fundamental understanding of the ECM structure and functions guides the design and engineering of innovative advanced polymeric biomaterials.

In this chapter, we will first provide an overview of the key ECM components, structure, and functions. While both the composition and structure of the ECM define its functions, it is the functions that are crucial to recapitulate when designing materials for therapeutic applications. Therefore, we will discuss how advanced polymeric materials can be functionalized to mimic the ECM using molecular engineering techniques.

2. ECM components, structure, and functions

The ECM is a complex and dynamic network of proteins and glycosaminoglycans (GAGs) that constitutes the noncellular portion of tissues. The different ECM proteins can be categorized into two main types: the fibrous structural proteins, which provide a core 3D fibrillar structure to the ECM, and the adhesive glycoproteins, which "decorate" the fibrous network and functionalize it with cell adhesion sites, soluble molecule-binding sites, and protease cleavage sites, among others.[5,7,8] In addition, the ECM contains proteoglycans and GAGs, which are highly hydrophilic and importantly regulate the water content in tissues, among other functions.

2.1. *Structural role of the ECM*

The fibrous structural proteins of the ECM provide mechanical support and cell adhesion sites, serving as a scaffold to hold cells within a structured 3D microenvironment. Therefore at a macroscopic scale, the ECM defines the tissue shape and resistance to deformation, whereas at a microscopic scale, it provides the cells with multiple biophysical cues, including stiffness, degree of cross-linking, elasticity, topology, and porosity of the microenvironment.[9,10] All these parameters regulate cell morphogenetic behaviors (e.g., survival, adhesion, migration, proliferation, and differentiation).

The main fibrous structural proteins in the interstitial ECM are collagens and elastin. Collagens are widely known to provide the stability and tensile strength to tissues. In the interstitial ECM (**Figure 1a, b**), the most abundant collagen is type I collagen; it represents about 30% of the whole-body protein content and 80% of all collagens. Type I collagen forms triple helix fibers that assemble into a stable 3D network. It is highly present in connective tissues throughout the body and is particularly enriched in bone, skin, tendons, and ligaments. Another relatively abundant collagen is type III collagen, representing about 10% of total collagen. Type III collagen has a similar triple helix structure as type I collagen, yet forms slightly thinner fibrils. It is found in high concentration in the skin and is also produced as a temporary scaffold during wound healing before being gradually replaced by collagen I.[5] Type II collagen is also similar in structure to type I and III collagens, and is abundant in cartilage. In basement membranes, which are specialized ECM sheets underlying endothelial or epithelial cells in tissues, type IV collagen is one of the main structural components although it is a nonfibrillar collagen (**Figure 1a, c**).

Intermingled into the collagen network is a network of elastin fibers, which provides elasticity to tissues. Elastin is a stretchy protein that can be extended to more than twice its length and can recoil to restore the original size and shape of the tissue.

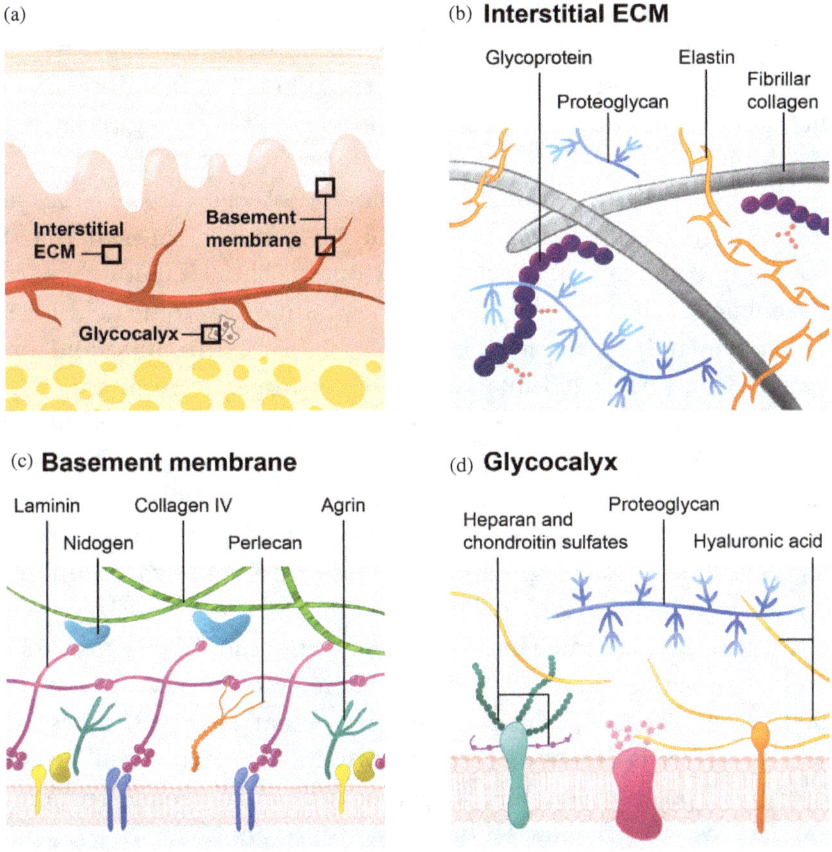

Figure 1. Different types of ECMs in the body. (**a**) Different ECM compositions in healthy skin tissue. (**b**) Composition of the interstitial ECM. (**c**) Key components and structure of the basement membrane. Depicted cell membrane receptors are integrins (blue) and dystroglycan (yellow and dark yellow). (**d**) Composition of the pericellular matrix, also called the glycocalyx. Proteins in the cell membrane are syndecan (green), an integral glycoprotein (pink), and CD44 receptor (orange).

2.2. *Biochemical and signaling roles of the ECM*

In addition to fibrillar proteins, various adhesive glycoproteins are incorporated into the 3D core structure of the ECM, such as fibronectin, tenascin, osteopontin, vitronectin, and laminin. While these glycoproteins also modulate the mechanical properties of the ECM, they importantly functionalize it with integrin-binding sites to further improve cell adhesion, growth factor-binding domains to control cell behavior, and protease cleavage sites to regulate ECM degradability.[3]

The ECM also contains GAGs and proteoglycans. GAGs are linear polysaccharides that are highly negatively charged due to their high composition in sulfate groups. They include heparan sulfate, chondroitin sulfate, dermatan sulfate, hyaluronic acid (HA), and heparin polysaccharide. They are highly hydrophilic and provide the tissue with gel-like compressive properties. On the other hand, proteoglycans are composed of a core protein covalently linked to GAGs. Some examples include decorin, biglycan, lumican, aggrecan, and perlecan. Both GAGs and proteoglycans are involved in tissue hydration, filling, and regulation of the space between cells, as well as in cell signaling, although through slightly different mechanisms. In addition to being present in the interstitial ECM and in basement membranes, proteoglycans and GAGs are abundant in the pericellular ECM directly connected to the cell surface, called the glycocalyx, which means the "sugar coat"[11] (**Figure 1d**). The glycocalyx is essential to ensure hydration at the cell surface, protection from physical and mechanical damages, sequestration of biomolecules, and regulation of cell–cell interactions.

2.3. *ECM remodeling*

The ECM undergoes constant turnover regulated by proteases and protease inhibitors, which further controls the cell microenvironment and its processes. For example, degradation of the ECM allows cells to create a path that facilitates migration. ECM breakdown can also generate the release of matrikines, which are bioactive signaling peptides derived from cleaved ECM proteins. Furthermore, ECM proteolysis can make accessible cryptic sites that are hidden to the cells in an intact ECM, yet have signaling functions upon exposure.[12] Therefore, ECM remodeling in tissues strongly modulates cell fate.

Recently, the increased interest in studying the ECM has led scientists to gather all ECM components under the name "matrisome." This term, which originally appeared in the publication by Hynes and Naba in 2012,[13] refers to all ECM components, including biomolecules that regulate and remodel the ECM, such as the ECM-degrading proteases and inhibitors (e.g., matrix metalloproteinases [MMPs], tissue inhibitors of matrix metalloproteinases [TIMPs]) and ECM-bound bioactive molecules (e.g., growth factors or cytokines).

3. ECM-mimetic polymeric biomaterials

Despite some similarities in composition, the ECM is highly specialized to the different organs and optimized to best serve their functions. For example, collagen fibers are dense and packed in the skin, dense and aligned in a specific direction in tendons and ligaments, and relatively loose and compliant in the brain and in adipose tissues. This important ECM diversity is central to consider when developing ECM-mimetic materials to tailor their design to the specific envisioned applications.

3.1. *Natural and synthetic polymeric materials used for ECM mimicry*

Due to their 3D fibrous and polymeric nature, the ECM has been widely mimicked using polymeric biomaterials. Polymers are typically large molecules made up of repeated subunits, from natural or synthetic origin. Many of these polymeric materials have been developed as ECM substitutes to treat wounds or repair tissues after diseases. Indeed, they have been specifically designed to provide a supportive and instructive cell-friendly microenvironment for tissues to regrow. To optimally do so, they have been modified to promote cell survival, migration, proliferation, or differentiation in a controlled manner.[14,15]

Natural polymers are commonly derived from living organisms. For example, polymers can be extracted from human or animal tissues or blood, such as collagen and gelatin (a form of denatured collagen), HA, and fibrin.[3,16,17] Rather than extracting the polymers, another approach consists in fully removing the cellular content of tissues while keeping the matrix intact, to generate decellularized ECM as a biomaterial.[18] Other animal-derived polymers include silk, made from silkworms and spiders, or chitosan, extracted from the shell of crustaceans. Lastly, natural polymers can be derived from plants, such as cellulose and alginate, respectively extracted from the cell walls of plants and brown algae. Due to their origins, natural biomaterials often provide a biocompatible and cell-friendly environment, for example, with natural cell adhesion sites and proteolytic sites, which facilitate their use for biomedical applications. However, they require highly optimized purification and quality control to limit batch-to-batch variability and pathogen transmission from the natural material to the patient.[19] In addition, while many of their properties can be tuned to some degree (e.g., their concentration, stiffness), tailored modifications of their physicochemical properties remain relatively restricted.

On the contrary, polymers can be of synthetic origin, that is, synthesized from chemical compounds, and therefore be manufactured in a very standardized and reproducible way. In addition, their chemical origin makes them highly tunable to meet the specific properties required for a biomedical application. Examples of synthetic polymers include polyethylene glycol (PEG), polyethylene oxide (PEO), poly(lactic acid) (PLA), poly(glycolic acid) (PGA) and the copolymer poly(lactic-co-glycolic) acid (PLGA), polyurethane (PU), polyvinyl alcohol (PVA), poly(ethylene terephthalate) (PET), to cite a few.[20] These polymers have been used in a wide range of tissue engineering applications: heart valve replacement, bone substitutes, soft tissue engineering, sutures, and drug delivery systems. However, synthetic polymers do not inherently provide a cell-friendly and biodegradable microenvironment and often require additional engineering to be optimized for biomedical applications.[1]

3.2. *Methods to design ECM-mimetic biomaterials*

In many biomedical applications, especially in tissue engineering, natural and synthetic biomaterials have been utilized in the form of hydrogels, in which the cross-linked network of polymers retains a large amount of water.[21] High water content is essential for the cells and is an important feature of the physiological ECM. In addition, the polymeric 3D network in hydrogels provides additional biomechanical and physical support (porosity, stiffness, elasticity), mimicking the fibrous network of the ECM.

Apart from hydrogels, many techniques have been developed to mimic the structure of the ECM. For example, fibrous biomaterials can be manufactured via electrospinning, a technique in which a high voltage is applied onto a syringe loaded with a polymeric solution, which allows the polymer to form a charged stream of liquid that solidifies in flight and gets elongated to nanometer scale before collection.[22] In addition, stereolithography and 3D printing of polymers have been optimized to recapitulate ECM micro- and macrostructures.[23] Other methods have been developed to make porous materials from polymers, such as gas foaming, freeze-drying, and solvent casting and particulate leaching (SCPL).[24] Similar to the latter, sacrificial matrices have been used to create complex structures into polymer materials. For example, a polymer scaffold can be created around or within a temporary sacrificial matrix made of salts that will be dissolved at the end of the process to generate interconnected pores or channels in the polymeric scaffold.[25]

While ECM-mimetic biomaterials have been extensively realized in 3D to recapitulate the physiological ECM, material surface modifications and coatings have also been applied to enhance cell adhesion, signaling, and tissue control.[26] Mimicry of the ECM in 2D has been achieved using multiple techniques,[27] one of the simplest being the direct adsorption of ECM-derived peptides or proteins, such as collagen, fibronectin, or laminin, on material surfaces. Surfaces can also be functionalized to allow covalent immobilization of peptides and proteins, to enhance protein retention and coating stability. ECM-mimetic surfaces have also been created using layer-by-layer assembly technology, in which alternating layers of oppositely charged biomolecules are deposited, enabling the formation of nanostructured coatings with multiple ECM components.[28] Moreover, 2D micropatterned ECMs can be generated via microcontact printing, for example, using soft lithography or photolitography,[29] offering highly precise spatial control of the deposited ECM.

4. Engineering advanced polymeric biomaterials to recapitulate the native ECM

4.1. *Engineering direct cell–material interactions*

Despite the multitude of polymeric biomaterials that have been developed to mimic the ECM, biomaterials often need optimization to properly recapitulate the ECM complexity and functions. For example, biomaterials may lack adhesion sites for cells, particularly in the case of synthetic polymers. Cells indeed need to sense their microenvironment via specific interactions between their mechanoreceptors, such as integrins, and the fibrous network to survive and determine their fate. Upon interaction, integrins firmly attach to the ECM components, inducing cytoskeleton rearrangement and triggering intracellular signaling.

Therefore, materials that lack cell adhesion sites are poorly cell friendly and their use for biomedical applications is limited. To overcome this limitation, researchers have incorporated integrin-binding sites into polymeric biomaterials, such as the tri-amino acid peptide arginine-glycine-aspartic acid (RGD), to substantially increase cell adhesion and make the material suitable for cells (**Figure 2a**). For example, the incorporation of RGD into PEG hydrogels promotes the spreading of human fibroblasts,[30] survival of mesenchymal stem cells,[31] and mineralization of bone marrow stromal cells,[32] as compared to nonfunctionalized PEG. RGD is a well-known motif found in many ECM components, which promiscuously binds multiple integrins, including $\alpha_{IIb}\beta_3$, $\alpha_5\beta_1$, $\alpha_5\beta_6$, $\alpha_8\beta_1$, $\alpha_V\beta_1$,

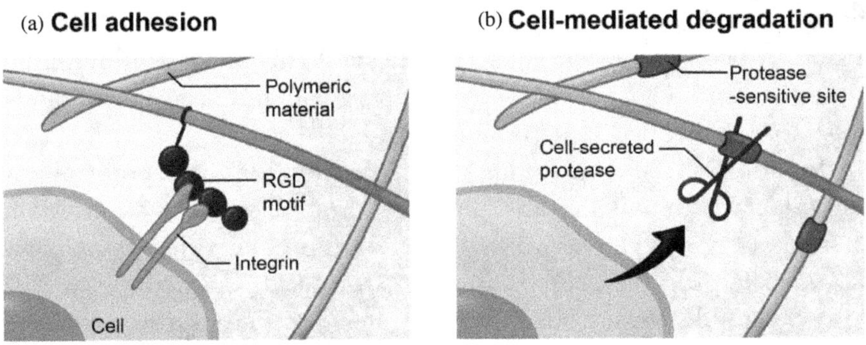

Figure 2. Engineering of ECM-mimetic polymeric materials enhanced for cell adhesion. Using (**a**) integrin-binding RGD motif or (**b**) degradability by cell-secreted proteases.

$\alpha_V\beta_3$, $\alpha_V\beta_5$, and $\alpha_V\beta_6$.[33] Other peptides with enhanced integrin-binding specificity have been also used to more precisely target relevant integrins, enabling better control of biological processes in specific applications. Indeed, the laminin-derived sequences IKVAV and YIGSR have been used to target the integrins $\alpha_3\beta_1$, $\alpha_4\beta_1$, and $\alpha_6\beta_1$[34] and the collagen-derived sequence DGEA to target the integrin $\alpha_2\beta_1$. The incorporation of the IKVAV peptide in self-assembling nanofibers, for example, has been shown to improve neurite outgrowth and reduce glial scar after injury.[35] Similarly, hydrogels that specifically contain a cell-adhesion peptide for the $\alpha_6\beta_1$ integrin, but not for $\alpha_V\beta_3$, have recently been developed to highlight the different roles of these integrins during angiogenesis.[36]

Instead of peptides, materials can incorporate or be coated with full-length ECM proteins. Interestingly, it has been highlighted that the surface chemistry of the material determines the conformation of the coated ECM proteins, and therefore the accessibility to specific integrin-binding sites. For example, fibronectin adopts a fibrillar shape when adsorbed on poly(ethyl acrylate) material and a globular shape when adsorbed on a poly(methyl acrylate) material, resulting in differences in cell adhesion.[37,38]

One important limitation of cell-adhesion peptides and proteins is their susceptibility to enzymatic digestion upon implantation into tissues. To overcome this, the use of robust synthetic alternatives has been explored, such as cationic-hydrophobic amphiphilic α- or β-amino acid polymers, to improve cell–material interactions in a more controlled and tunable manner.[39,40]

In addition to controlling cell adhesion via incorporation of cell-adhesive domains, some material properties such as stiffness and elasticity can directly affect cell adhesion and subsequent morphogenetic processes. For example, fibroblasts seeded onto stiff polyacrylamide gels coated with collagen differentiate into profibrotic myofibroblasts via the activation of mechanosignaling pathways, in contrast to fibroblasts seeded on soft polyacrylamide gels.[41] In addition, the topography of the material, that is, its micro- and nanostructures (e.g., porosity, surface roughness), can be engineered with microfabrication or nanoimprinting techniques to modulate cell–material interactions and cell behaviors.[42]

4.2. *Engineering material remodeling*

Physiologically, the cells populating the ECM produce proteases to degrade it, protease inhibitors to protect it, and new ECM components to replace it. Cell-mediated remodeling of polymeric materials is therefore an important feature to consider to increase material biocompatibility.

Similar to the incorporation of cell adhesion sites, natural and synthetic biomaterials have been engineered to incorporate protease cleavage sites, when not originally present (**Figure 2b**). For example, PEG hydrogels have been modified with MMP-sensitive sequences to allow cell invasion and tissue growth.[43–45] MMPs are a family of more than 20 matrix proteases that degrade specific ECM components, such as collagen, elastin, glycoproteins (e.g., fibronectin), and proteoglycans. Other protease-sensitive sequences have also been used to engineer biomaterials, for example, the plasmin-cleavage sequence, which is naturally present in fibrin clots.[46] The choice of the cleavage sites to incorporate in a biomaterial must be made according to the specific enzymes secreted by the invading cells as well as the degradation time required for the specific targeted biomedical application.[47]

In addition to modulating cell invasion into materials, it has been shown that the degradation of some natural polymers by proteases exposes or releases matricryptic sites and matrikines that signal cells.[12] For example, degradation of the basement membrane, including collagen IV, has been shown to release matrikines that display antitumor effects by modulating the angiogenic response.[48] However, to date, engineering of polymeric materials using matricryptic sites or matrikines remains relatively unexplored.

Lastly, biomaterials can be engineered with protease inhibitors to slow down their degradation time. For instance, fibrin is widely used in the clinic as a tissue glue, yet its rapid degradation has hindered its full potential for tissue anastomosis. To extend its half-life, fibrin has been supplemented with aprotinin, a bovine-derived broad-spectrum serine protease inhibitor,[49] or with the plasmin inhibitor α_2-antiplasmin, which can covalently cross-link into the fibrin polymer network.[50]

4.3. *Engineering GAG–material interactions*

GAGs in the ECM are essential to keep high hydration of the tissue while displaying bioactive properties on cells. Indeed, GAGs modulate cell adhesion via specific receptors (e.g., via the hyaluronan receptor CD44) and cell signaling via the sequestration of bioactive molecules (see more details in Section 4.4). Consequently, GAGs have been commonly used in tissue engineering either as core polymers to develop biomaterials or to functionalize other polymeric materials. GAGs-based hydrogels, notably based on HA, have been particularly used for skin, cartilage, vasculature, bone, or neural tissues.[51,52] To date, multiple strategies have been developed to cross-link HA hydrogels via photoreactions, thermoreactions, enzymatic reactions, synthetic polymer cross-linkers, or via specific interactions of protein domains.[53] Using similar cross-linking methodology, HA-based hydrogels can be functionalized with peptides to acquire diverse functions, including incorporation of the RGD motif to enhance cell adhesion or of protease-sensitive domains to increase biodegradability.[54]

In many instances, however, GAGs have been copolymerized with other core fibrous materials, such as PEG, collagen, or fibrinogen, rather than being used as a stand-alone material. For example, a hybrid HA and PEG hydrogel was developed by functionalizing the HA and PEG chains with complementary transglutaminase substrate sequences that allow their covalent bonding by the coagulation factor XIII (FXIII).[55] Furthermore, the HA–PEG hydrogel was engineered with multiple stiffness, with the RGD motif and with MMP-sensitive sequences. Ultimately, the soft, RGD-containing MMP-sensitive hydrogels significantly enhance myoblast spreading and modulate cellular phenotypes. This highlights how advanced polymeric materials can be precisely tailored when designed for specific applications. In addition, HA has been used as a copolymer with collagen or fibrinogen-based scaffolds for cell encapsulation and cartilage repair.[56,57] Thus, all these advanced polymeric materials resemble the native ECM by mimicking its structure, biocompatibility, and cell-mediated biodegradability.

4.4. *Engineering bioactive protein–material interactions*

Naturally, the ECM contains a multitude of components that interact with bioactive molecules, such as growth factors. Growth factors are a group of potent signaling proteins that control key morphogenetic cellular processes. Therefore, they are involved in a variety of mechanisms during embryogenesis and tissue regeneration, including tissue growth, angiogenesis, and inflammation. However, the delivery of growth factors as bulk soluble proteins in regenerative medicine has overall been disappointing from a therapeutic perspective. Indeed, preclinical and clinical studies have highlighted that the interactions between growth factors and the ECM are essential to control the spatiotemporal release of the growth factors in the tissue, which strongly modulate their bioactivity. Thus, the ECM acts as a reservoir of growth factors and releases them in a controlled, slow, and sustainable manner.

To mimic this ECM function, a multitude of polymeric materials have been engineered as drug carriers, especially to deliver growth factors. Growth factor retention has been achieved using physical or chemical interactions with the polymeric material (**Figure 3a**). For example, the vascular endothelial growth factor-A (VEGF-A) has been physically encapsulated into degradable hydrophobic PLGA microspheres, which are then entrapped into a PLGA hydrogel, to substantially delay and sustain the release of VEGF-A.[58] In addition, some growth factors directly attach to the polymeric materials via electrostatic or hydrophobic interactions, which can be used to control the growth factor release (**Figure 3b**). For instance, chemically modifying gelatin by anionic succinylation enhanced the retention of the stromal cell-derived factor 1α (SDF1α), improving its pro-angiogenic effects.[59]

Another approach that has been widely explored to improve growth factor retention into polymeric materials is via covalent cross-linking. Growth factor cross-linking can be achieved via chemical conjugation or enzymatic reactions (**Figure 3c**). For example, the insulin-like growth factor-1 (IGF-1) has been modified with a transglutaminase substrate sequence of FXIII to allow covalent bonding into fibrin–collagen hybrid hydrogels.[60] However, because covalent attachment of growth factors may occur at the receptor-binding site or may be too strong to allow their release from the material, an MMP-cleavage sequence has been added between the FXIII substrate sequence and IGF-1. This system is therefore tailored to release IGF-1 only at the vicinity of protease-expressing cells, in a "cell-demand" fashion.

Apart from covalent immobilization, growth factors have been incorporated into ECM-mimetic materials via noncovalent interactions. Particularly, some growth factors — often referred as heparin-binding — are known to display specific affinity to heparin and other GAGs. This encouraged scientists to functionalize polymeric matrices with heparin to increase the retention of these factors (**Figure 3d**). Incorporation of heparin in star-shaped PEG

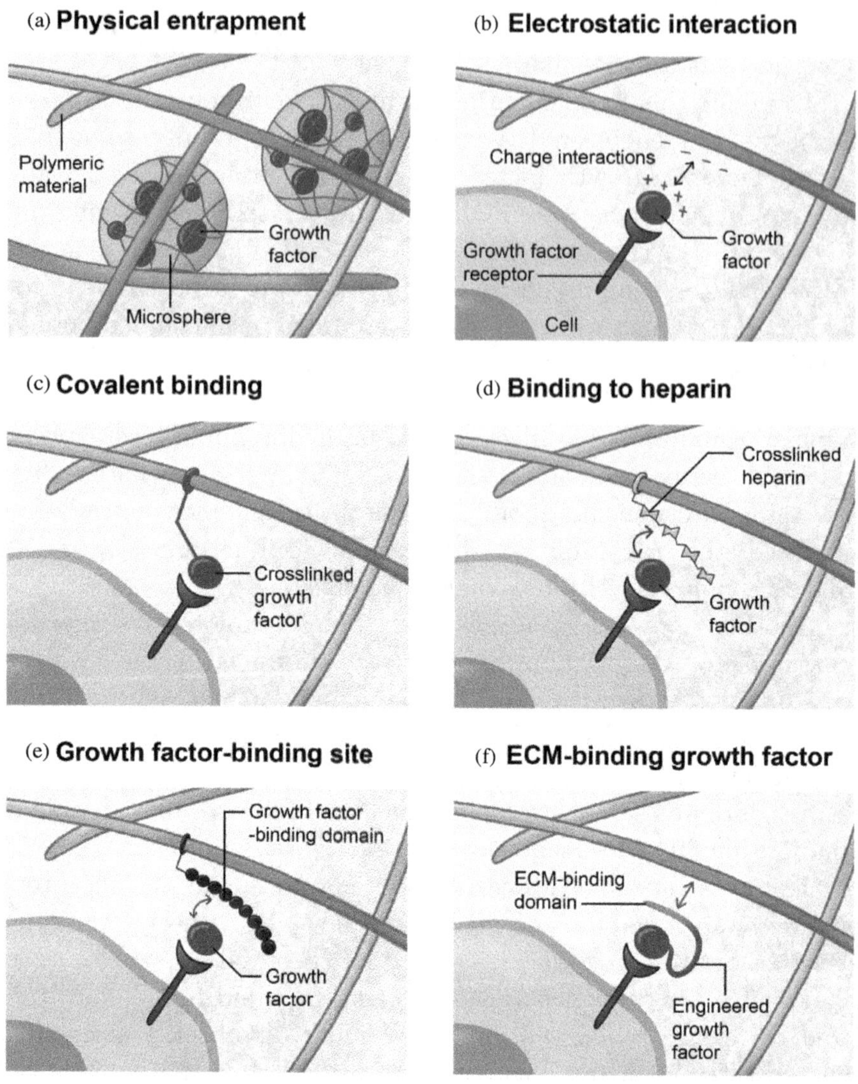

Figure 3. Multiple strategies controlling growth factor release from polymeric materials. Growth factor release can be slowed by (**a**) physical encapsulation into microspheres, (**b**) via electrostatic interactions, (**c**) via chemical or enzymatic cross-linking into the polymeric material, (**d**) via incorporation of heparin in the material, (**e**) via incorporation of specific growth factor-binding domain in the material, or (**f**) via engineering of the growth factor for enhanced binding to the polymeric material.

hydrogels, for example, successfully controlled the release of a multitude of growth factors, including fibroblast growth factor-2 (FGF-2), bone morphogenetic protein-2 (BMP-2), transforming growth factor β (TGF-β), nerve growth factor (NGF), glial cell-line neurotrophic factor (GDNF), VEGF-A, and SDF1α.[61] Other GAGs with specific affinity for certain growth factors have been isolated and used to engineer delivery systems; this is the case of the heparan sulfate HS7 that binds and increases the activity of VEGF-A, inducing prolonged signaling in endothelial cells and increased angiogenesis.[62]

Similarly, polymeric materials have been supplemented with ECM proteins or domains that display specific affinity to multiple growth factors, such as domains of fibronectin, vitronectin, fibrinogen, and laminin[63–65] (**Figure 3e**). Interestingly, it has been shown that the incorporation of the growth factor-binding domain of fibrinogen into synthetic PEG material, further engineered for integrin binding and protease sensitivity, permits effective control of growth factor release and fully recapitulates the effect of the natural fibrin material in supporting wound healing.[63] This further highlights that key ECM functions can be successfully recapitulated to develop therapeutic synthetic matrices. In another example, a recombinant fragment of fibronectin has been designed to fuse in close proximity the integrin-binding site and the growth factor-binding site of fibronectin. This fibronectin fragment has been used to engineer fibrin matrices that subsequently permits the sustained release of VEGF-A and platelet-derived growth factor (PDGF-BB). Furthermore, the fibronectin fragment enhanced both growth factor receptor signaling and integrin signaling in a synergistic manner by inducing molecular clustering at the cell surface.[64] Besides ECM proteins, fragments of anti-growth factor antibodies or growth factor receptors can be utilized to sequester growth factors with high affinity and high specificity into biomaterials.[66] The incorporation of these fragments themselves into core polymeric materials can be achieved via covalent immobilization or noncovalent but high-affinity binding. For example, in a recent study, a dual-affinity bridge protein made of the growth factor-binding domain of laminin fused to an anti-collagen antibody fragment has been successfully developed to retain BMP-2 into collagen sponges and promote bone regeneration for spine fusion.[67]

In contrast to engineering the matrix for enhanced affinity to growth factors, one can engineer growth factors to display tailored affinity to polymeric materials (**Figure 3f**). For example, growth factors that do not naturally bind to heparin have been added with a heparin-binding domain.[68] In addition to enhancing the binding of growth factors to heparin and GAGs, growth factors have also been fused to the domain that specifically binds to fibrous ECM proteins such as type I collagen.[69] Finally, peptides that display high yet promiscuous affinity for multiple ECM proteins have been discovered and fused to growth factors, to confer them the ability to interact with a multitude of ECM components.[70,71] Such ECM-binding domains have been discovered in placental growth factor-2 (PlGF-2) or amphiregulin (AREG), for example. Their fusion to VEGF-A, PDGF-BB, BMP-2, or interleukin-1 receptor antagonist (IL-1Ra) has substantially improved tissue healing in rodent models of skin, bone, skeletal muscle, and myocardium regeneration. One strong advantage of engineering the growth factor rather than the polymeric matrix is that these factors could be delivered from exogenous implantable materials as well as from the physiological ECM upon direct injection into the tissue as a soluble factor.[70]

Overall, multiple strategies have been validated to engineer biomaterials with active molecules, leading to the development of therapeutic advanced polymeric materials. While many examples have been here illustrated with growth factors, these approaches could be applied to other families of bioactive molecules, such as cytokines, hormones, and neuromodulators, to expand their applications in biomedicine.

5. Conclusion

In this chapter, we have discussed the composition, structure, and functions of the different physiological ECMs and seen that this has been a source of inspiration for biomaterial scientists, to develop advanced polymeric ECM-mimetic biomaterials for therapeutic purposes in tissue regeneration. Our understanding of the role of ECM has provided central design criteria to engineer materials that are biocompatible and instructive to cells. Importantly, multiple strategies have been evaluated to fine-tune materials' biomechanical, cell adhesive, and biodegradability properties for specific biomedical applications. In addition, polymeric materials have been extensively functionalized with bioactive molecules to further control cell morphogenetic behaviors and enhance therapeutic outcomes.

One important challenge that still hinders the development of effective therapeutic materials is our inability to precisely define the ideal properties of a material in order to cure a specific disease. Indeed, diseases are often very complex and evolve over time. It is therefore challenging to predict what therapeutic microenvironment, cell behaviors, and sequence of events would be necessary to rescue a diseased tissue. A second challenge to consider when designing therapeutic materials is the heavy regulatory path toward clinical translation. While the complexity of the physiological ECM is undoubtedly important to fulfill its biological functions, the clinical translation of complex multicomponent materials has been very limited due to the high costs and duration of its safety and efficacy assessments, required for marketed clinical use. Therefore, novel innovative strategies are needed to overcome these challenges, potentially via the design of "even smarter" biomaterials.

Besides, many aspects of ECM biology remain to be elucidated, and it is very likely that new ECM functions will be highlighted in the future. This would excitingly lead to novel designs of integrative ECM-mimetic materials. For example, the role of the ECM in regulating immune responses has started to be explored and next-generation immunomodulatory biomaterials are being generated to better control tissue regeneration and prevent wound chronicity or scar formation.[72] Apart from tissue regeneration, advanced polymeric materials can be used in a variety of applications, including in other biomedical research fields (e.g., cancer, neurological diseases) for therapeutic purposes or as tools to develop diagnosis platforms or to deepen our understanding of biological processes. Therefore, while a multitude of advanced polymeric materials have been already developed, ongoing challenges and new discoveries continue to drive this highly active field of research.

References

1. Lutolf MP, Hubbell JA. Synthetic biomaterials as instructive extracellular microenvironments for morphogenesis in tissue engineering. *Nat Biotechnol.* 2005;23:47–55.
2. Mandrycky C, Phong K, Zheng Y. Tissue engineering toward organ-specific regeneration and disease modeling. *MRS Commun.* 2005;7;332–347.
3. Rice JJ, Martino MM, Laporte LD, Tortelli F, Briquez PS, Hubbell JA. Engineering the regenerative microenvironment with biomaterials. *Adv Healthc Mater.* 2013;2:57–71.
4. Martino MM, Briquez PS, Maruyama K, Hubbell JA. Extracellular matrix-inspired growth factor delivery systems for bone regeneration. *Adv Drug Deliv Rev.* 2015;94:1–12.
5. Briquez PS, Hubbell JA, Martino MM. Extracellular matrix-inspired growth factor delivery systems for skin wound healing. *Adv Wound Care.* 2015;4:479–489.
6. Rodrigo-Navarro A, Sankaran S, Dalby MJ, Campo AD, Salmeron-Sanchez M. Engineered living biomaterials. *Nat Rev Mater.* 2021;6:1175–1190.
7. Kular JK, Basu S, Sharma RI. The extracellular matrix: Structure, composition, age-related differences, tools for analysis and applications for tissue engineering. *J Tissue Eng.* 2014;5:2041731414557112.
8. Theocharis AD, Skandalis SS, Gialeli C, Karamanos NK. Extracellular matrix structure. *Adv Drug Deliv Rev.* 2016;97:4–27.
9. Young JL, Holle AW, Spatz JP. Nanoscale and mechanical properties of the physiological cell–ECM microenvironment. *Exp Cell Res.* 2016;343:3–6.
10. Saraswathibhatla A, Indana D, Chaudhuri O. Cell–extracellular matrix mechanotransduction in 3D. *Nat Rev Mol Cell Biol.* 2023; 24:495–516.

11. Gaudette S, Hughes D, Boller M. The endothelial glycocalyx: Structure and function in health and critical illness. *J Vet Emerg Crit Care.* 2020;30:117–134.
12. Ricard∆Blum S, Salza R. Matricryptins and matrikines: Biologically active fragments of the extracellular matrix. *Exp Dermatol.* 2014;23:457–463.
13. Hynes RO, Naba A. Overview of the matrisome — an inventory of extracellular matrix constituents and functions. *Cold Spring Harb Perspect Biol.* 2012;4:a004903.
14. Place ES, Evans ND, Stevens MM. Complexity in biomaterials for tissue engineering. *Nat Mater.* 2009;8:457–470.
15. Hussey GS, Dziki JL, Badylak SF. Extracellular matrix-based materials for regenerative medicine. *Nat Rev Mater.* 2018;3:159–173.
16. Sell SA, Wolfe PS, Garg K, McCool JM, Rodriguez IA, Bowlin GL. The use of natural polymers in tissue engineering: A focus on electrospun extracellular matrix analogues. *Polymers.* 2010;2: 522–553.
17. Fan J, Abedi-Dorcheh K, Vaziri AS, Kazemi-Aghdam F, Rafieyan S, Sohrabinejad M, et al. A review of recent advances in natural polymer-based scaffolds for musculoskeletal tissue engineering. *Polymers.* 2022;14:2097.
18. Zhang X, Chen X, Hong H, Hu R, Liu J, Liu C. Decellularized extracellular matrix scaffolds: Recent trends and emerging strategies in tissue engineering. *Bioact Mater.* 2022;10:15–31.
19. Lei B, Guo B, Rambhia KJ, Ma PX. Hybrid polymer biomaterials for bone tissue regeneration. *Front Med.* 2019;13:189–201.
20. Guo B, Ma PX. Synthetic biodegradable functional polymers for tissue engineering: A brief review. *Sci China Chem.* 2014;57:490–500.
21. Lee KY, Mooney DJ. Hydrogels for tissue engineering. *Chem Rev.* 2001;101:1869–1880.
22. Xue J, Wu T, Dai Y, Xia Y. Electrospinning and electrospun nanofibers: Methods, materials, and applications. *Chem Rev.* 2019;119:5298–5415.
23. Chia HN, Wu BM. Recent advances in 3D printing of biomaterials. *J Biol Eng.* 2015;9:4.
24. Babaie E, Bhaduri SB. Fabrication aspects of porous biomaterials in orthopedic applications: A review. *ACS Biomater Sci Eng.* 2018;4: 1–39.

25. Liu S, Wang T, Li S, Wang X. Application status of sacrificial biomaterials in 3D bioprinting. *Polymers.* 2022;14:2182..
26. Shekaran A, Garcia AJ. Nanoscale engineering of extracellular matrix-mimetic bioadhesive surfaces and implants for tissue engineering. *Biochim Biophys Acta (BBA) — Gen Subj.* 2011;1810:350–360.
27. Aazmi A, Zhang D, Mazzaglia C, Yu M, Wang Z, Yang H, et al. Biofabrication methods for reconstructing extracellular matrix mimetics. *Bioact Mater.* 2024;31:475–496.
28. Ren K, Hu M, Zhang H, Li B, Lei W, Chen J, et al. Layer-by-layer assembly as a robust method to construct extracellular matrix mimic surfaces to modulate cell behavior. *Prog Polym Sci.* 2019;92: 1–34.
29. Sgarbi N, Pisignano D, Benedetto FD, Gigli G, Cingolani R, Rinaldi R. Self-assembled extracellular matrix protein networks by microcontact printing. *Biomaterials.* 2004;25:1349–1353.
30. Hern DL, Hubbell JA. Incorporation of adhesion peptides into nonadhesive hydrogels useful for tissue resurfacing. *J Biomed Mater Res.* 1998;39:266–276.
31. Salinas CN, Anseth KS. The influence of the RGD peptide motif and its contextual presentation in PEG gels on human mesenchymal stem cell viability. *J Tissue Eng Regen Med.* 2008;2:296–304.
32. Yang F, Williams CG, Wang D, Lee H, Manson PN, Elisseeff J. The effect of incorporating RGD adhesive peptide in polyethylene glycol diacrylate hydrogel on osteogenesis of bone marrow stromal cells. *Biomaterials.* 2005;26:5991–5998.
33. Humphries JD. Integrin ligands at a glance. *J Cell Sci.* 2006;119: 3901–3903.
34. Frith JE, Mills RJ, Hudson JE, Cooper-White JJ. Tailored integrin–extracellular matrix interactions to direct human mesenchymal stem cell differentiation. *Stem Cells Dev.* 2012;21:2442–2456.
35. Tysseling-Mattiace VM, Sahni V, Niece KL, Birch D, Czeisler C, Fehlings, MG, et al. Self-assembling nanofibers inhibit glial scar formation and promote axon elongation after spinal cord injury. *J Neurosci.* 2008;28:3814–3823.
36. Moreira HR, Rodrigues DB, Freitas-Ribeiro S, da Silva LP, da S Morais A, Jarnalo M, et al. Integrin-specific hydrogels for growth factor-free vasculogenesis. *NPJ Regen Med.* 2022;7:57.

37. Grigoriou E, Cantini M, Dalby MJ, Petersen A, Salmeron-Sanchez M. Cell migration on material-driven fibronectin microenvironments. *Biomater Sci.* 2017;5:1326–1333.
38. Mao Y, Schwarzbauer JE. Fibronectin fibrillogenesis, a cell-mediated matrix assembly process. *Matrix Biol.* 2005;24:389–399.
39. Sun H, Meng F, Dias AA, Hendriks M, Feijen J, Zhong Z. α-Amino acid containing degradable polymers as functional biomaterials: Rational design, synthetic pathway, and biomedical applications. *Biomacromolecules.* 2011;12:1937–1955.
40. Chen Q, Zhang D, Zhang W, Zhang H, Zou J, Chen M, et al. Dual mechanism β-amino acid polymers promoting cell adhesion. *Nat Commun.* 2021;12:562.
41. Huang X, Yang N, Fiore VF, Barker TH, Sun Y, Morris SW, et al. Matrix stiffness–induced myofibroblast differentiation is mediated by intrinsic mechanotransduction. *Am J Respir Cell Mol Biol.* 2012;47:340–348.
42. Charras G, Sahai E. Physical influences of the extracellular environment on cell migration. *Nat Rev Mol Cell Biol.* 2014;15:813–824.
43. Lutolf MP, Lauer-Fields JL, Schmoekel HG, Metters AT, Weber FE, Fields GB, Hubbell JA. Synthetic matrix metalloproteinase-sensitive hydrogels for the conduction of tissue regeneration: Engineering cell-invasion characteristics. *Proc Natl Acad Sci U S A.* 2003;100:5413–5418.
44. Patterson J, Hubbell JA. SPARC-derived protease substrates to enhance the plasmin sensitivity of molecularly engineered PEG hydrogels. *Biomaterials.* 2011;32:1301–1310.
45. Oliver-Cervelló L, Martin-Gómez H, Gonzalez-Garcia C, Salmeron-Sanchez M, Ginebra M-P, Mas-Moruno C. Protease-degradable hydrogels with multifunctional biomimetic peptides for bone tissue engineering. *Front Bioeng Biotechnol.* 2023;11:1192436.
46. Patterson J, Hubbell JA. Enhanced proteolytic degradation of molecularly engineered PEG hydrogels in response to MMP-1 and MMP-2. *Biomaterials.* 2010;31:7836–7845.
47. Beamish JA, Juliar BA, Cleveland DS, Busch ME, Nimmagadda L, Putnam AJ. Deciphering the relative roles of matrix metalloproteinase- and plasmin-mediated matrix degradation during capillary morphogenesis using engineered hydrogels. *J Biomed Mater Res Part B: Appl Biomater.* 2019;107:2507–2516.

48. Monboisse JC, Oudart JB, Ramont L, Brassart-Pasco S, Maquart FX. Matrikines from basement membrane collagens: A new anti-cancer strategy. *Biochim Biophys Acta — Gen Subj.* 2014;1840: 2589–2598.
49. Lorentz KM, Kontos S, Frey P, Hubbell JA. Engineered aprotinin for improved stability of fibrin biomaterials. *Biomaterials.* 2010;31(1):1–9.
50. Liu J, Solanki A, White MJV, Hubbell JA, Briquez PS. Therapeutic use of α2-antiplasmin as an antifibrinolytic and hemostatic agent in surgery and regenerative medicine. *NPJ Reg Med.* 2022;7:34.
51. Tsanaktsidou E, Kammona O, Kiparissides C. Recent developments in hyaluronic acid-based hydrogels for cartilage tissue engineering applications. *Polymers.* 2022;14:839.
52. Ding YW, Zhang XW, Mi CH, Qi XY, Zhou J, Wei DX. Recent advances in hyaluronic acid-based hydrogels for 3D bioprinting in tissue engineering applications. *Smart Mater Med.* 2023;4:59–68.
53. Neves MI, Araújo M, Moroni L, Silva RMPd, Barrias CC. Glycosaminoglycan-inspired biomaterials for the development of bioactive hydrogel networks. *Molecules.* 2020;25:978.
54. Kim J, Kim IS, Cho TH, Kim HC, Yoon SJ, Choi J, et al. In vivo evaluation of MMP sensitive high-molecular weight HA-based hydrogels for bone tissue engineering. *J Biomed Mater Res Part A.* 2010;95A:673–681.
55. Ranga A, Lutolf MP, Hilborn J, Ossipov DA. Hyaluronic acid hydrogels formed in situ by transglutaminase-catalyzed reaction. *Biomacromolecules.* 2016;17:1553–1560.
56. Huang S, Wang C, Xu J, Ma L, Gao C. In situ assembly of fibrinogen/hyaluronic acid hydrogel via knob-hole interaction for 3D cellular engineering. *Bioact Mater.* 2017;2:253–259.
57. Gao Y, Liu Q, Kong W, Wang J, He L, Guo L, et al. Activated hyaluronic acid/collagen composite hydrogel with tunable physical properties and improved biological properties. *Int J Biol Macromol.* 2020;164:2186–2196.
58. Ennett AB, Kaigler D, Mooney DJ. Temporally regulated delivery of VEGF in vitro and in vivo. *J Biomed Mater Res Part A.* 2006;79A:176–184.
59. Kimura Y, Tabata Y. Controlled release of stromal-cell-derived factor-1 from gelatin hydrogels enhances angiogenesis. *J Biomater Sci Polym Ed.* 2010;21:37–51.

60. Vardar E, Larsson HM, Engelhardt EM, Pinnagoda K, Briquez PS, Hubbell JA, et al. IGF-1-containing multi-layered collagen-fibrin hybrid scaffolds for bladder tissue engineering. *Acta Biomater.* 2016;41:75–85.
61. Freudenberg U, Liang Y, Kiick KL, Werner C. Glycosaminoglycan-based biohybrid hydrogels: A sweet and smart choice for multifunctional biomaterials. *Adv Mater.* 2016;28:8861–8891.
62. Wang C, Poon S, Murali S, Koo CY, Bell TJ, Hinkley SF, et al. Engineering a vascular endothelial growth factor 165-binding heparan sulfate for vascular therapy. *Biomaterials.* 2014;35:6776–6786.
63. Martino MM, Briquez PS, Ranga A, Lutolf MP, Hubbell JA. Heparin-binding domain of fibrin(ogen) binds growth factors and promotes tissue repair when incorporated within a synthetic matrix. *Proc Natl Acad Sci U S A.* 2013;110:4563–4568.
64. Martino MM, Tortelli F, Mochizuki M, Traub S, Ben-David D, Kuhn GA, et al. Engineering the growth factor microenvironment with fibronectin domains to promote wound and bone tissue healing. *Sci Transl Med.* 2011;3:100ra89.
65. Ishihara J, Ishihara A, Fukunaga K, Sasaki K, White MJV, Briquez PS, et al. Laminin heparin-binding peptides bind to several growth factors and enhance diabetic wound healing. *Nat Commun.* 2018;9:2163.
66. Impellitteri NA, Toepke MW, Levengood SKL, Murphy WL. Specific VEGF sequestering and release using peptide-functionalized hydrogel microspheres. *Biomaterials.* 2012;33:3475–3484.
67. Briquez PS, Tsai HM, Watkins EA, Hubbell JA. Engineered bridge protein with dual affinity for bone morphogenetic protein-2 and collagen enhances bone regeneration for spinal fusion. *Sci Adv.* 2021;7:eabh4302.
68. Florine EM, Miller RE, Liebesny PH, Mroszczyk KA, Lee RT, Patwari P, et al. Delivering heparin-binding insulin-like growth factor 1 with self-assembling peptide hydrogels. *Tissue Eng. Part A.* 2015;21:637–646.
69. Zhang J, Ding L, Zhao Y, Sun W, Chen B, Lin H, et al. Collagen-targeting vascular endothelial growth factor improves cardiac performance after myocardial infarction. *Circulation.* 2009;119:1776–1784.
70. Martino MM, Briquez PS, Güç E, Tortelli F, Kilarski WW, Metzger S, et al. Growth factors engineered for super-affinity to the extracellular matrix enhance tissue healing. *Science.* 2014;343:885–888.

71. Alshoubaki YK, Lu YZ, Legrand JMD, Karami R, Fossat M, Salimova E, et al. A superior extracellular matrix binding motif to enhance the regenerative activity and safety of therapeutic proteins. *NPJ Reg Med.* 2023;8:25.
72. Julier Z, Park AJ, Briquez PS, Martino MM. Promoting tissue regeneration by modulating the immune system. *Acta Biomater.* 2017;53:13–28.

https://doi.org/10.1142/9789819806782_0004

Chapter 4

Designed Multicomponent Glycan-based Biopolymer Condensates

Simran Dewan[1] and Ayala Lampel[1–5,*]

[1]*Shmunis School of Biomedicine and Cancer Research, George S. Wise Faculty of Life Sciences, Tel Aviv University, Tel Aviv, 69978, Israel*

[2]*Center for Nanoscience and Nanotechnology, Tel Aviv University, Tel Aviv, 69978, Israel*

[3]*Sagol Center for Regenerative Biotechnology, Tel Aviv University, Tel Aviv, 69978, Israel*

[4]*Center for the Physics and Chemistry of Living Systems, Tel Aviv University, Tel Aviv, 69978, Israel*

[5]*Leibniz Institute of Polymer Research Dresden, Max Bergmann Center of Biomaterials, 01060 Dresden, Germany*

**Corresponding author: ayalalampel@tauex.tau.ac.il*

The design of synthetic biomolecular condensates that are formed by liquid–liquid phase separation of various building blocks — including those derived from amino acids, nucleic acids, and synthetic polymers — underscores the wide range of interactions that influence the properties and functions of biomolecular condensates. Of particular interest are naturally occurring glycan-based polymers, such as chitosan, hyaluronic acid, heparin, dextran sulfate (DS), pectin, carrageenan, and alginate. These glycan polymers possess distinctive structural and biochemical properties, making them promising candidates for the construction of synthetic condensates. Thus, the versatile biological roles of glycans and their biocompatibility and biodegradability open tremendous opportunities to utilize glycan-based condensates for a wide range of technologies, from tissue engineering and regenerative medicine to drug delivery and food applications. However, understanding the physicochemical properties of glycans is essential for determining how these condensates form, as well as their stability and functionality. A detailed examination of the features and interactions of these building blocks is critical for designing complex, multicomponent glycan-based condensates with customized functionalities. In this chapter, we focus on the emerging field of glycan-based condensates. We introduce the chemical composition, biological roles, and functionalities of naturally occurring glycan-based polymers and discuss the significance of glycans as building blocks of synthetic condensates. Additionally, we review recent work on the regulation of the properties of glycan condensates and their applications and discuss the remaining challenges and trends in this field.

1. Introduction

1.1. *Background on biomolecular condensates*

Most intracellular biomolecular processes, such as metabolic pathways and gene control, are compartmentalized within membrane-bound and membraneless organelles. Conventionally, the first category of organelles is intracellular compartments surrounded by lipid membranes, for example, the nucleus, the mitochondria, and the endoplasmic reticulum (ER), whereas the organelles from the second category, also termed biomolecular condensates, are dense liquid-like phases that lack membranous boundaries.[1] Biomolecular condensates include organelles that are found in distinct parts of the cell, that is, the cytoplasm, nucleus, or the mitochondria, and have various functionalities. Examples include the cytoplasmic P granules and stress granules,[2] the nucleoli and Cajal bodies, which are found in the nucleus and are involved in the assembly of the ribosome,[3] the spliceosomes,[4] and the mitochondrial ribonucleic acid (RNA) granules, which play a role in RNA metabolism.[5]

Overall, these organelles contribute significantly to the spatial organization within the cell[6] and to enhancing the efficiency and specificity of cellular functions.[7,8] Additionally, the availability and activity of the constituent molecules of biomolecular condensates can be regulated,[9] and by sequestering or releasing proteins and RNA, condensates can regulate processes like gene expression, signal transduction, and metabolism.[10] Under stress conditions, cells form condensates, for example, stress granules that protect and regulate mRNAs, facilitating cell adaptation to adverse environments.[11] Several condensates are involved in signaling pathways, where they act as hubs that bring together signaling molecules to amplify and coordinate cellular responses.[12]

The suggested mechanism of formation of intracellular condensates involves liquid–liquid phase separation (LLPS) of intrinsically disordered proteins (IDPs), which contain intrinsically disordered regions (IDRs).[13,14] Low-complexity motifs within IDPs promote LLPS; these motifs are composed of repeating amino acids such as glycine (G), which provides flexibility, serine (S), glutamine (Q), and asparagine (N), which promote protein–protein or protein–solvent hydrogen bonding, and the aromatic amino acids phenylalanine (F) and tyrosine (Y), which facilitate π-mediated intermolecular interactions.[8,15,16] Overall, the multitude of functionalities of biomolecular condensates and their involvement in diverse physiological and pathological cellular processes highlight the multiplexity of cellular organization and is of prime research interest in recent years.[17]

1.2. *Liquid–liquid phase separation*

LLPS is characterized by the formation of a distinct, dense phase within the surrounding medium (**Figure 1a**) formed by weak multivalent protein–protein and protein–RNA interactions.[18] Due to these weak multivalent contacts, the LLPS process is sensitive to changes in the cellular environment and can be triggered by a variety of factors, including temperature, pH, posttranslational modifications (PMTs), and critical concentrations of protein, RNA,

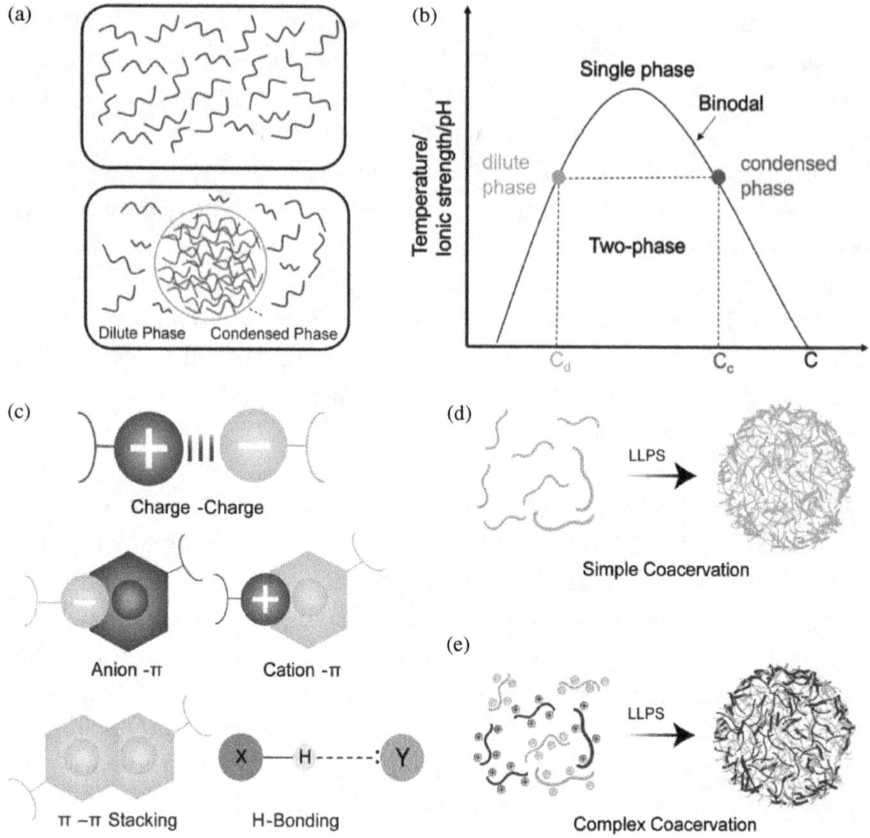

Figure 1. (**a**) Schematic illustration of demixing of two molecules, which results in the formation of dense and dilute phases. (**b**) Illustration of a simple phase diagram showing the dependency of LLPS on temperature, ionic strength, and pH. (**c**) Types of attractive forces that are involved in LLPS. Schematics of simple (**d**) and complex (**e**) coacervation.

and ions.[9] The coacervation process involves demixing of biopolymers to form a polymer-rich liquid phase, that is, the coacervate phase, and a polymer-depleted dilute phase[13] (**Figure 1b**). Coacervation is driven by different types of interactions including hydrogen bonding,[19] electrostatic, hydrophobic,[20] cation-π, and anion-π interactions[21] (**Figure 1c**).

LLPS can be divided into simple and complex coacervation, both considered as associative phase separation.[22] In simple coacervation, a single macromolecule, typically a protein, polyprotein, peptide, or nucleic acid,[17] exhibits attractive interactions making it insoluble at specific solution temperatures, pH levels, and salt concentrations among other chemical or physical triggers (**Figure 1d**). In complex coacervation, electrostatic interactions between multiple macromolecule species, including two or more oppositely charged biopolymers, result in LLPS (**Figure 1e**). The process is considered to be mainly driven by the entropy gained during the association between oppositely charged polyelectrolytes due to the release of small counterions.[11,23] Complex coacervation strongly depends on the chemical nature of the polyelectrolytes and the solution environments, including the ionic strength, ion composition, pH, and charge ratio.[24,25] Examples of biological condensates that are formed by simple coacervation include the mussel foot protein (mfp) condensates,[26] tau protein condensates, which interact with tubulin to control microtubule polymerization and disassembly,[27,28] and cytoplasmic structures like stress granules that are formed by environmental stresses and govern mRNA metabolism.[29] Examples of condensates from the second category that are formed by complex coacervation include the nuclear organelles Cajal bodies, which are formed by coacervation of small nuclear ribonucleoproteins (snRNPs) and related proteins.[30]

1.3. *Synthetic condensates*

With an increasing understanding of the biophysical mechanisms that drive LLPS of cellular condensates, that is, stress granules, germ granules,[31] and P bodies,[32] increasing efforts are being made to design synthetic condensates.[33] These synthetic compartments,[33] also termed "artificial cells," are developed as engineered analogs of naturally occurring membraneless organelles for a wide range of applications[34] including compartmentalization of enzymatic reactions,[35–37] controlled partitioning and release[38] of therapeutic molecules,[39] and downregulation or silencing pathways by the intake of specific target molecules.[40]

Bioengineered synthetic condensates[33] localize chemical species,[41] biochemical gradients, and functional building blocks to control biological or chemical processes. Their ability to mimic the crowded interiors of cells and facilitate molecular interaction with their surroundings makes them open reactors that can selectively sequester biomolecules and enhance the kinetics of catalytic processes.[41,42] A relatively large number of studies have shown that the enzymatic reaction rate can be increased by compartmentalization in condensates including the carboxylation of ribulose-1,5-bisphosphate,[43] RNA polymerization,[44] production of 3′,5′-cyclic guanosine monophosphate-adenosine monophosphate (GMP-AMP),[45] and glucose oxidation/peroxidation.[46] In addition, LLPS can enhance the kinetics of nucleation and the assembly of natural cellular assemblies, that is, actin filaments and microtubules.[47,48] Moreover, the selective sequestration and exclusion of specific molecules in synthetic condensates can facilitate the protection of the payload molecules by preventing their degradation or inhibition by excluded molecules[33,49] (**Figure 2a**).

Cellular biomolecular condensates might change their physical properties in response to environmental cues, that is, pH, ionic strength, temperature, or mechanical force, owing to the weak localized molecular interactions that underlie their formation.[50] Temperature-sensitive biomolecular condensates formed by heterogeneous nuclear ribonucleoprotein A1 (hnRNPA1) were recently reported, showing either phase separation or enhanced mixing above a critical temperature threshold or in response to heat stress.[29] Moreover, changes to the physical properties of condensates upon environmental cues may lead to arrested droplet dynamics and protein aggregation, which has been related to neurodegenerative diseases.[51,52] For instance, the condensates that are formed by tau protein are sensitive to environmental changes, that is, to temperature or ionic strength, where the latter can result in phase transition of condensates into pathological aggregation.[53] Such transitions can lead to the formation of neurofibrillary tangles, which are characteristic of Alzheimer's pathology.

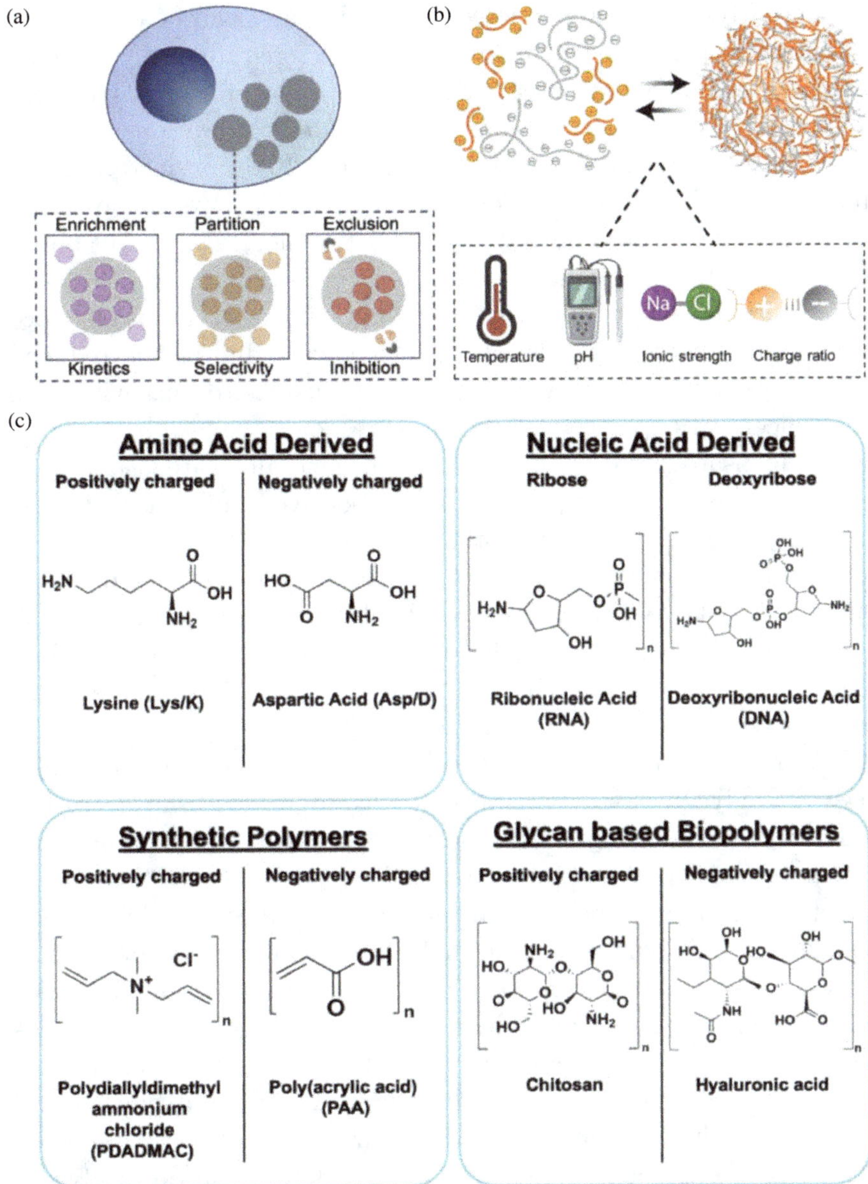

Figure 2. Schematic illustration of (**a**) the functionality of biomolecular condensates and (**b**) their reversibility and responsiveness toward environmental cues including temperature, pH, ionic strength, and charge ratio. (**c**) The four main categories of building blocks of designed biomolecular condensates.

Similarly, synthetic biomolecular condensates can be engineered as stimuli-responsive species that can phase separate or disassemble upon a specific (bio)chemical or physical trigger, for example, irradiation,[38] enzymatic activation,[54,55] oxidation,[56] alterations in pH,[57] solution temperature,[58] and time[59] (**Figure 2b**). For instance, pH fluctuations affect the complex coacervation of polyelectrolytes by altering their protonation states.[60] Furthermore, the addition of inorganic salts has frequently been employed as a means to disassemble droplets that are formed by complex coacervation of oppositely charged polyelectrolytes.[61,62] This occurs due to the reduced entropy gain associated with the release of counterions when the salt concentration exceeds a critical threshold.[50]

Condensate building blocks can be generally categorized into several groups, including (i) amino acid-derived molecules, that is, engineered proteins,[33] polypeptides, or peptides,[63] where in multi-component systems, which rely on complex coacervation, these building blocks typically include the basic amino acids arginine (Arg or R) or lysine (Lys or K)[64]; (ii) nucleic acid-derived building blocks, that is, RNA or deoxyribonucleic acid (DNA)[65]; (iii) synthetic polymers, for example, poly(diallyldimethylammonium chloride) (PDADMAC)[66] or poly(acrylic acid) (PAA)[67]; and (iv) naturally occurring glycan-based polymers, for example, chitosan or hyaluronic acid (HA)[68] (**Figure 2c**), where the latter group of building blocks is the focus of this chapter. The study of LLPS across a variety of building blocks — including amino acid-, nucleic acids-, and synthetic polymers-derived molecules — highlights how diverse interactions shape the properties and functionalities of biomolecular condensates. Among these, naturally occurring glycan-based polymers, such as chitosan, HA, heparin, dextran sulfate, pectin, carrageenan, and alginate, exhibit unique structural and biochemical properties, which make them particularly promising for the construction of synthetic condensates. Thus, the utilization of glycan-based building blocks opens opportunities in biomedical and biotechnological applications, such as enzymatic compartments, delivery systems, biosensors, and diagnostics. Yet, understanding the physicochemical properties of glycans is crucial for determining the formation, stability, and functionality of the resulting condensates. Thus, a thorough examination of the characteristics and interactions of these building blocks can facilitate the design of complex, multicomponent glycan-based condensates with tailored functionalities.

The focus of this chapter is the emerging field of glycan-based condensates, which offers insights into how glycans, with their unique biophysical properties — such as extensive hydration and branched structures — contribute to distinctive condensation mechanisms. In Section 2, we introduce the chemical composition, biological roles, and functionalities of naturally occurring glycan-based polymers. Subsequently, in Section 3, we discuss the potential and emerging significance of glycans as building blocks of designed condensates. In Sections 4 and 5, we review work on the regulation of properties of glycan-based condensates and their application, respectively. Lastly, in Section 6, we discuss the remaining challenges and future trends of glycan-based condensates.

2. Glycan-based building blocks of designed condensates

2.1. *Biological functions of glycans*

Glycans are complex carbohydrates that are increasingly recognized for their structural diversity and versatile functionality, as they serve as structural components in various biological systems. In plants and fungi, glycans such as cellulose and pectin are integral to the cell wall, providing mechanical strength and rigidity.[69] In arthropods such as insects, glycans are the key component of the exoskeleton, and in multicellular organisms, glycans are constituents of matrix molecules, such as proteoglycans. These glycans maintain tissue structure, porosity, and integrity, often containing binding sites for other specific glycans that aid in the overall matrix organization.[70] Additionally, structural modifications induced by glycans can offer benefits such as enhanced protein solubility and protection against degradation, thereby preserving structural integrity and physiological functions.[71] In the ER, glycans attached to nascent proteins assist in their proper folding, where misfolded glycoproteins are often translocated back to the cytoplasm for degradation. Glycans are also involved in cell recognition and signaling. For instance, the adhesive selectin proteins mediate interactions between blood and vascular cells by recognizing glycans on various ligands.[72]

Glycans play crucial roles in antigen recognition and modulation of immune responses. Glycan-binding proteins like galectin act as pathogen-associated molecular patterns recognized by pattern recognition receptors on immune cells, facilitating the detection and clearance of pathogens.[73] Changes in glycosylation can modulate immune cell functions, as seen in tumor-associated glycan alterations that contribute to immunoevasion and cancer progression.[74] Glycans also play a role in energy metabolism, serving as a reservoir of energy. Polysaccharides like starch and glycogen act as energy stores in plants and animals, respectively, with their breakdown providing glucose for cellular metabolism.[75]

2.2. *Chemical structure and properties of glycans*

Glycans are carbohydrate structures composed of multiple monosaccharides linked by glycosidic bonds. They are classified into different types, including N-linked glycans, O-linked glycans, and glycosaminoglycans (GAGs). N-linked glycans, attached to asparagine residues during protein synthesis in the ER, are crucial for proper protein folding and function. O-linked glycans, typically linked to serine or threonine residues, play significant roles in various cellular processes, that is, cell–cell communication, immune responses, and cell differentiation.[76] GAGs, including the glycans heparin or heparan sulfate (HS-GAGs), chondroitin sulfate/dermatan sulfate (CS-GAGs), HA, and keratan sulfate, are characterized by repeating disaccharide units and high negative charge density due to the presence of sulfate and carboxyl groups. They are essential for maintaining structural integrity in cells and tissues.[70]

GAGs such as HS interact with growth factors and cytokines, modulating cellular responses to environmental stresses such as inflammation or injury.[77]

In this section, we will first briefly introduce the chemical structure and properties of glycans that are utilized as LLPS building blocks.

Chitosan is a cationic polysaccharide derived from the deacetylation of chitin. It is composed of 2-acetamido-2-deoxy-D-glucopyranose and 2-amino-2-deoxy-D-glucopyranose units, with varying degrees of acetylation.[78,79] The cationic nature of chitosan facilitates multivalent interactions, similar to proteins with IDRs. The amino and hydroxyl groups of chitosan promote electrostatic interactions and hydrogen bonding, respectively. These interactions can facilitate the formation of polymer networks with anionic counterparts, which can enhance the mechanical stability of biomaterials.[80] As discussed in Sections 3–5, these electrostatic interactions are key to the formation of synthetic condensates.

HA is an anionic GAG composed of repeating D-glucuronic acid and N-acetyl-D-glucosamine units. HA is one of the major components of the extracellular matrix (ECM), where it maintains hydration, elasticity, and structural integrity.[81] The high propensity of HA to bind to water molecules is essential for the regulation of cell proliferation and migration.[82,83] HA facilitates wound healing, tissue repair, and medical applications like osteoarthritis treatment and skin hydration.[84] By interacting with receptors such as CD44 and receptor for HA-mediated motility (RHAMM), HA triggers signaling cascades that affect cell adhesion and migration.[85] The hydroxyl, carboxyl, and acetamido groups in HA facilitate hydrogen bonding and electrostatic interactions, which can promote LLPS and formation of biomolecular condensates.[86]

Heparin is a highly sulfated GAG that consists of repeating disaccharide units of iduronic acid and glucosamine with varying sulfation. Heparin interacts with numerous proteins, including antithrombin, to prevent excessive blood clotting, and binds to growth factors, cytokines, and chemokines, influencing cell signaling and immune responses.[87–89] By interacting with positively charged proteins, including those with arginine-rich motifs, heparin can induce or stabilize biomolecular condensates.[90]

Dextran is a microorganism-derived polysaccharide, composed of glucose units linked by α-1,6-glycosidic bonds with α-1,3-branching.[91] *Dextran sulfate (DS)* is a highly sulfated derivative, which has ~2.3 sulfate groups per glucosyl residue, giving it strong polyanionic properties.[92] In cell culture, DS acts as an anticoagulant, preventing cell aggregation and enhancing cell growth and protein production. DS is also utilized in biomolecule separation and purification, particularly through interactions with plasma lipoproteins, demonstrating its broad utility in biochemical research and pharmaceuticals.[93] The sulfate groups of DS facilitate electrostatic and hydrogen bonding interactions with protein amine groups, leading to phase separation and formation of synthetic condensates.[94]

Alginate, a naturally occurring polysaccharide derived from algae, is renowned for its biocompatibility and versatility in biomedical applications.[95] Composed of guluronic and mannuronic acid units, alginate's structure affects its physical properties and gelation behavior. Its ability to form hydrogels via ionic cross-linking with divalent cations like Ca^{2+} allows the creation of ECM-like 3D materials.[96] Alginate's acidic groups, which facilitate interactions with extracellular proteins and other cationic biomolecules through ionic and hydrogen bonding interactions,[97] can facilitate the formation of synthetic condensates for biomedical applications, that is, in controlled drug delivery.[98]

Pectin, a complex polysaccharide found in plant cell walls, is essential for cell adhesion and structural integrity, particularly in fruits and vegetables.[99] It is composed primarily of galacturonic acid and varies in esterification. This typical food-grade anionic polysaccharide has been widely used to interact with proteins to improve their functionality.[100] Pectin–polysaccharide condensates were reported to serve as drug delivery platforms for colon-targeted drug release.[98]

Carrageenan, a natural linear sulfated polysaccharide derived from seaweeds, is valued for its gel-forming, thickening, and stabilizing properties, which are utilized in the food, pharmaceutical, and cosmetics industries.[101] Structurally, carrageenan features alternating 3-linked and 4-linked D-galactose residues, modified by 3,6-anhydrogalactose and various sulfation patterns. These modifications contribute to its ability to form strong gels and films.[102] Carrageenan can interact with basic functional groups of biomolecules through electrostatic interactions, which enhances its immunomodulatory, anticoagulant, and antiviral activities.[103] *Iota carrageenan* (ι-carrageenan), a linear sulfated polysaccharide extracted from red seaweed, can undergo phase separation as a factor of pH and concentration.[104]

3. Formation of glycan-based biomolecular condensates

While there are no known examples of biological condensates that are formed by glycans, several reports point at the effect of carbohydrates on intracellular phase separation. Thus, O-linked glycans, particularly O-GlcNAcylation, are known to modify protein interactions and behavior within cellular compartments and can directly regulate LLPS.[105] Specific studies have shown that O-GlcNAcylation reduces phase separation propensity, modulates the fluidity of condensates, and affects the long-term stability of phase-separated structures.[106] GAGs, including HS, heparin, and HA, interact with a variety of proteins, mainly through attractive electrostatic interactions, influencing both protein self-assembly and LLPS.[24,107–109] For instance, the LLPS of heparin with tau protein strongly correlates with tau aggregation associated with Alzheimer's disease.[110]

Synthetic glycan-based condensates include phase-separated complexes of glycan polymers and charged proteins, including heparin and lysozyme condensates, which are highly sensitive to ionic strength,[107] and HS and basic fibroblast growth factor (bFGF) condensates, which are formed on the cell surface.[109] Tau protein, which aggregates in pathological states into amyloid fibrils associated with neurodegenerative diseases, phase separates with heparin and HA along with other various polyanions, for example, poly(A) RNA and poly(dA) DNA. The phase separation of tau with either heparin or HA occurs under specific conditions and is thought to be driven by electrostatic and hydrophobic interactions. The fact that these two chemically distinctive glycan polymers undergo LLPS with tau suggests that LLPS can occur independently of the polyanion's chemical identity, as a factor of other conditions, that is, concentration of the polymers, ionic strength, and so on.[110] At high ionic strength, LLPS is considered to be mediated by nonelectrostatic attractive forces, including hydrophobic interactions. Furthermore, various desulfated heparins affect LLPS and tau aggregation differently, highlighting the complex interplay between different factors in regulating tau condensates.[110]

4. Regulation of LLPS and properties of glycan condensates

Most glycan-based condensates are formed through complex coacervation of anionic or cationic glycans with oppositely charged polyelectrolytes,[5] where the phase separation process depends on various factors such as environmental conditions, that is, pH and ionic strength, and molecular determinants, that is, chain length, charge stoichiometry, and polymer concentration.[50,111]

The complexation and phase separation of glycans depend on solution pH, which regulates the degree of ionization. For instance, the pKa of the carboxyl group of HA is around 3–4, and thus, it is negatively charged in slightly acidic to neutral conditions, whereas the pKa of the amine group of chitosan is around 6.5, and thus, it is positively charged in slightly acidic conditions. Therefore, HA–chitosan LLPS is favored at a pH range between these two pKa values.[24] Additionally, the ionic strength of the solution impacts the ionization, and in turn, the complexation of glycans with oppositely charged polyelectrolytes. The effect of ionic strength depends on the chemical composition and concentration of the building blocks (**Figure 3**). Lower ionic strength can promote phase separation by promoting the solubility of the polymers. This allows the oppositely charged polymers to interact, facilitating coacervation. However, higher ionic strength can inhibit phase separation due to reduced repulsive electrostatic interactions, which is known as charge screening.[112] For instance, the phase separation propensity of chitosan and ovalbumin is influenced by the addition of NaCl. Lower ionic strength (≤ 200 mM NaCl) promotes the formation of chitosan–ovalbumin coacervates, while higher ionic strength (>200 mM NaCl) inhibits their formation due to charge screening by salt ions.[112] Moreover, Delair et al. show that the formation and stability of polyelectrolyte complexes formed by chitosan and DS are influenced by ionic strength and temperature, where the addition of NaCl controls the assembly or disassembly of the complexes, with a minimum of 1 mol/L NaCl required to prevent phase separation through charge screening. This assembly process is reversible, and condensates can be reformed by the removal of salt through dialysis.[113]

Other factors that affect glycan LLPS are charge density, which can be regulated, for instance, by varying the degree of glycan deacetylation,[24] and total polymer concentration (C_p), which directly correlates with turbidity increase and complex formation.[100] Notably, $C_p >> C_{sat}$ may lead to increased turbidity as a result of aggregation and precipitation rather than coacervation. Kayitmazer et al. systematically studied how factors like ionic strength, degree of ionization, molecular weight, and charge stoichiometry influence complex coacervation between HA and chitosan.[24] They show that high ionic strength regulates phase separation versus precipitation by charge screening. The pH-dependent degree of ionization, along with the molecular weight and charge density of the polymers, significantly affect phase separation, with an optimal balance needed to prevent precipitation.

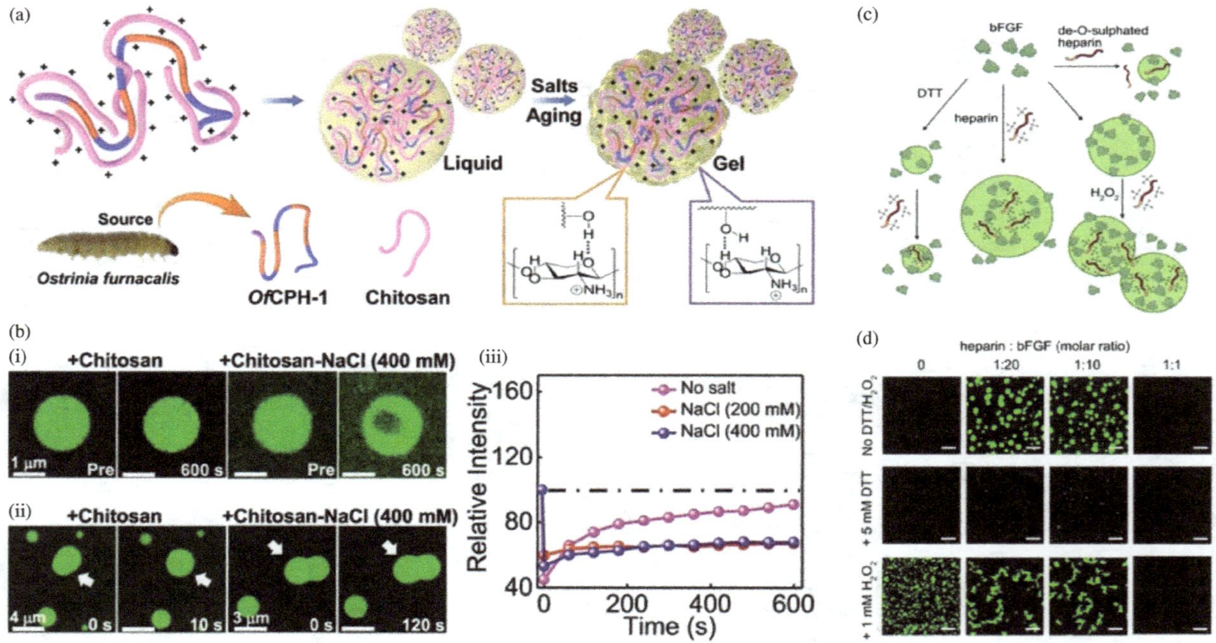

Figure 3. Regulation of LLPS by chemical and molecular determinants. (**a**) Coacervation between chitosan and insect cuticular protein from the larva head of *Ostrinia furnacalis.* (**b**) (i, ii) FRAP and fusion TEM images of OfCPH-1-chitosan coacervates, which were taken after adding 400 mM NaCl for 10 minutes, indicating that the microdroplets were liquid-like. Scale bars = 50 μm. (iii) Representation of relative fluorescence intensities versus time of the corresponding FRAP images. (**c**) Schematic illustration of bFGF assembly under various oxidative status with DTT, H_2O_2, heparin, and de-O-sulfated heparin. (**d**) Confocal imaging of different molar ratios of heparin–bFGF with H_2O_2 and DTT. (a, b) Reproduced with permission from Gong et al.[116] Copyright 2022 American Chemical Society. (**c, d**), Reproduced with permission from Springer Nature, Copyright 2022 Nature Communications.[109]

Two recent studies by Kamperman and coworkers[108] and Perry et al.[114] focus on the effect of ionic strength and pH on the viscoelastic properties of HA–chitosan condenses. They show that both factors modulate the material properties of glycan condensates, which transition from elastic solids to fluid-like materials, depending on the environmental conditions. Kamperman et al. also show that cosolvents like methanol or ethanol can influence condensate stability by altering the dielectric constant, thereby enhancing electrostatic interactions and increasing stability. The presence of alcohols increases the modulus of the condensates, suggesting dielectric-induced stiffening and slower relaxation dynamics.[108] This work suggests that cosolvents can be used to fine-tune the material properties of phase-separated materials for applications like electrospinning of nanofibers, which could be used to create scaffolds that support tissue growth and regeneration as well as improved wound-healing matrices.

Similarly, a study on chitosan–DS and chitosan–heparin complexes[115] shows that controlled assembly and disassembly of the complexes can be regulated by tuning the salt concentrations, where higher salt leads to more compact and stable particles. Interactions of chitosan with various proteins, such as insect cuticular protein, soybean whey protein, soybean protein isolate, pea protein isolate, canola protein isolate, ovalbumin, and gelatin, demonstrate its ability to form versatile multicomponent condensates. The LLPS propensities of these systems are regulated by pH, ionic strength, and protein-to-polysaccharide ratios.[112,116–121] Cuticular proteins interact with chitosan through hydrogen bonding where the size and formation of microdroplets depend on the concentrations of protein and chitosan.[116] The fluidity of these condensates is influenced by salt concentration, where in the absence of NaCl, the droplets are liquid-like and show high fluorescence recovery in fluorescence recovery after photobleaching (FRAP) analysis.[122] However, at high 0.2–0.4 M NaCl, the droplets transition to a gel-like state, showing almost no fluorescence recovery. Similarly, the phase behavior of soybean whey protein–chitosan condensates is regulated by pH and protein-to-polysaccharide ratios.[117,118] Zeta potential analysis shows that chitosan is positively charged below pH 8.3 and that the protein is negatively charged above pH 4.3. Thus, the phase separation of chitosan and soybean whey protein occurs optimally between pH 6.0 and 6.5, influenced by protein-to-polysaccharide ratio, temperature, and NaCl concentration. Higher temperatures initially increase the turbidity but decrease the coacervate yield, while NaCl inhibits coacervation by disrupting electrostatic interactions. Pea protein isolate–chitosan[119] and canola protein isolate–chitosan[120] coacervates are similarly regulated by pH and electrostatic interactions, with the strongest binding around pH 5.5 for pea protein isolate–chitosan and pH 6.0 for canola protein isolate–chitosan. Ovalbumin and gelatin also form coacervates with chitosan, influenced by ionic strength and pH. For ovalbumin–chitosan, pH 4.0 and 5.5 are critical for coacervate formation, where low ionic strength promotes phase separation.[112] In gelatin–chitosan complexes, salt ions are essential to achieve charge neutralization and form stable complexes.[121]

Several studies report on multicomponent condensates that are formed by LLPS of heparin and bFGF,[109] the chemokine CCL5,[123] tau protein,[124,110,123,125,126] and lysozyme.[107] LLPS of heparin and bFGF is driven by electrostatic interactions and results in fluid-like condensates, characterized by their ability to merge and show high FRAP. The oxidation status of bFGF affects this behavior, as reducing conditions inhibit phase separation, while oxidizing conditions lead to gel-like aggregates. Higher concentrations of heparin can prevent droplet formation, demonstrating reentrant behavior similar to that observed for RNA phase separation.[109] Heparin was reported as the main building block of biohybrid hydrogels,[127,128] which can effectively modulate chemokine gradients and reduce inflammation in chronic wounds by sequestering pro-inflammatory chemokines like MCP-1 and IL-8.[129] At specific conditions, chemokines like CCL5 undergo LLPS with heparin to form liquid droplets.[123] High salt concentrations weakens CCL5–heparin LLPS, which are driven by weak electrostatic interactions, and results in the diffusion of chemokines in solution to form a gradient. These results can facilitate the development of tools to modulate immune cell migration and inflammatory responses.[123]

Tau protein, associated with Alzheimer's disease, undergoes LLPS driven by electrostatic interactions. Pathogenic mutations in tau accelerate the transition from liquid to solid phase within droplets, leading to rapid fibrillation.[125] In the presence of heparin, certain tau variants show rapid solidification, with slow FRAP. Yet, the condensates that are formed by complex coacervation of heparin and truncated tau variants are reversible and do not promote amyloid aggregation.[125] Frokjaer et al.[107] showed that lysozyme–heparin interactions result in the formation of insoluble complexes that can be dissolved by NaCl, which suggests that an increase in the ionic strength of the solution results in charge screening and a decrease in the number of ionized groups on heparin that are available for interaction with the protein. However, it is important to note that these complexes do not dissociate completely, as some soluble complexes might also be present.[107] Complex coacervation between the protein and heparin may lead to protein denaturation and aggregation over time, which highlights the need for careful evaluation of heparin or similar highly charged polymer building blocks for applications in drug delivery and protein separation processes.

HA can form multicomponent complex coacervates with recombinant mussel adhesives[130–132] and tau protein,[110] where various factors regulate the interactions between HA and each of the proteins. Tirrell et al. demonstrated the potential of HA and recombinant mussel adhesive protein coacervates to serve as coating material for titanium surfaces, to promote osteoblast proliferation.[130] Optimal phase separation occurs at a 3:7 HA-to-fp-151-RGD ratio, with shear-thinning behavior observed at optimal mixing ratios. HA coacervation with tau protein was also shown to not promote tau aggregation, indicating that amyloid aggregation is independent of HA–tau complex coacervation.[110]

Lastly, plant glycans including pectin, alginate, and carrageenan also participate in LLPS to form multicomponent condensates with various proteins. Several studies report on the complex coacervation of pea protein isolate[100,133,134] and gelatin[98] with pectin and alginate, and that of canola protein isolate[104] with alginate and carrageenan. In addition, lysozyme[135] and soybean whey protein[136] also form multicomponent condensates with alginate and carrageenan, respectively, all of which are regulated by various external triggers such as pH, temperature, ionic strength, and total biopolymer concentration. Pectin–pea protein isolate coacervates[100,133] are significantly influenced by pH. The formation of soluble complexes and complex coacervates shifts toward higher pH as the pea protein isolate–pectin mixing ratio increases. The combination of higher concentration of protein with lower concentration of pectin (20:1) promotes the formation of complex coacervates. Generally, the charge density of pectin is determined by the degree of esterification of the (-COOH) groups in galacturonan and the degree of blockiness, which is the distribution pattern of unesterified blocks of galacturonic acid of pectin backbone, which in turn influences phase behavior. Low methoxyl pectin with higher charge density shows a broader pH range for complex coacervation compared to high methoxyl pectin. Pectin and alginate behave differently upon complexation with gelatin.[98] Alginate is more suitable than pectin for complex coacervation with gelatin due to its lower propensity of aggregation, smaller and more uniform particle size, easier dispersion in water, and faster drug release. Nickerson et al. reported that the complex coacervation of canola protein isolate and polysaccharides (alginate and Δ-carrageenan)[104] depends on the biopolymer mixing ratio, particularly at ratios below 20:1 canola protein isolate: anionic polysaccharide. Primarily electrostatic interactions drive this complexation followed by hydrogen bonding and limited involvement of hydrophobic interactions. The addition of NaCl (0.1 M) significantly inhibits complex formation, indicating the importance of electrostatic interactions, whereas the presence of urea (0.1 M) shifted critical pH values to lower pH values, suggesting a secondary role of hydrogen bonding in complex formation. The solubility of canola protein isolate–alginate complexes is significantly enhanced compared to the protein alone or in complexation with ι-carrageenan complexes, which could have practical implications for food applications.[104] Finally, lysozyme–alginate[135] and soybean whey protein–carrageenan[136] complex coacervates are both regulated by ionic strength, leading to different phase separation behaviors. Strong interactions result in aggregation, while weaker interactions favor coacervation.

5. Applications of glycan-based condensates

Biomolecular glycan-based condensates have significant potential for diverse applications in biotechnology and medicine owing to their biocompatibility, biodegradability, and nontoxicity.[137] These membraneless dynamic structures, which compartmentalize and concentrate biomolecules, can be harnessed for targeted drug delivery, gene therapy, as food stabilizers, and for the development of novel biomaterials.[68,98,109,113,136]

HA–chitosan complex coacervates serve as promising scaffolds for cartilage tissue engineering due to their ability to encapsulate bone marrow stem cells, maintain cell viability, and support cell–material interaction.[68] These complex coacervates are used to produce biopolymer nanofibers, which have high porosity and a large surface-to-volume ratio, which make them suitable for applications in tissue repair and as wound healing matrices.[115] Complex coacervates formed by HA and recombinant mussel adhesive protein can be used as a coating material on titanium surfaces to enhance osteoblast proliferation, which is beneficial for orthopedic implants and can ultimately help to improve the integration and longevity of implants in the body. Condensates that are formed by HA and mussel adhesive protein have low interfacial energy, which can be exploited for encapsulation and delivery of ECM components, cells, and nutrients. This can be particularly useful in tissue engineering and regenerative medicine.[130]

GAGs-based materials including biohybrid hydrogels have tremendous potential in tissue regeneration applications due to their multifaceted cell-instructive functions.[138] These biohybrid materials demonstrate significant potential for use in cell replacement therapies for neurodegenerative diseases, such as Parkinson's and Huntington's disease.[128] In the context of biomolecular condensates, heparin promotes tau fibrillation and forms complex coacervates with tau protein by LLPS. These in vitro models could be used for high-throughput screening of compounds that can modulate tau LLPS and fibrillation, aiding in the discovery of new drugs for neurodegenerative diseases.[125] Luo et al. show that HS acts as a platform to induce the phase separation of bFGF on the cell surface, facilitating its recruitment and concentration at the cell surface. The phase separation of bFGF with HS enhances the formation of the bFGF-HSPG-FGFR ternary complex, which is crucial for effective signal transduction, specifically activating the Ras-ERK pathway.[109] This process also stabilizes bFGF, protecting it from proteolytic degradation and promoting effective signal transduction, which is crucial for tissue regeneration and wound healing (**Figure 4**).

Glycan-based condensates possess diverse functionality, particularly in the food and pharmaceutical industries involving plant-derived polysaccharides. Complex coacervates formed by pea protein isolate and sugar beet pectin can be used to encapsulate bioactive compounds like polyphenols or vitamins, which can protect sensitive ingredients and control their release.[133] Similarly, pectin–gelatin and alginate–gelatin condensates are useful for controlled drug delivery.[98] Alginate–gelatin coacervates can encapsulate various drugs, for example, the antibiotic drug metronidazole hydrochloride and the anti-inflammatory drugs diclofenac sodium and indomethacin, which can improve the drug absorption rates and reduce stomach irritation, indicating a potential application in improving the gastrointestinal tolerability of certain medications.[98] Overall, the versatile biological roles of glycans and their biocompatibility and biodegradability open tremendous opportunities to utilize glycan-based condensates for a wide range of technologies, from tissue engineering and regenerative medicine to drug delivery and food applications.

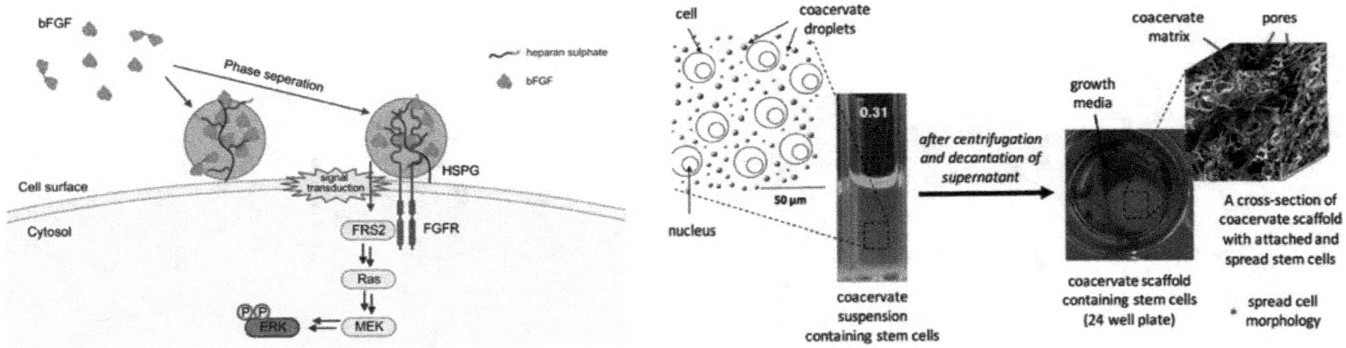

(a) (b)

Figure 4. Applications of glycan-based coacervates. (**a**) Schematic model of bFGF–HS phase separation on the cell surface triggering signaling transduction of Ras-ERK pathway. (**b**) HA–chitosan coacervates provide a favorable microenvironment for the cultivation and proliferation of rat bone marrow stem cells (rBMSCs), demonstrating high cell viability and well-spread cell morphology as well as high water content, interconnected porous structure, and noncytotoxicity, making them suitable as scaffolding materials for cartilage tissue engineering. (a) Reproduced with permission from Springer Nature, Copyright 2022 Nature Communications.[109] (b) Reproduced with permission from Acar et al.[68] Copyright 2018 American Chemical Society.

6. Challenges and future directions in the design of glycan-based condensates

Given the immense significance of glycans in structural integrity, immune modulation, pathogen interaction, and molecular mimicry, utilizing glycan-based building blocks in the design of synthetic condensates offers numerous advantages. Specifically, their ability to form hydrated, bioactive structures that interact seamlessly with biological systems makes glycans ideal building blocks for the construction of synthetic condensates. Yet, despite these benefits, current model systems (**Table 1**) also present certain limitations that must be carefully considered, including poor control of condensate size, lack of stability in high ionic strength, and limited long-term stability.[94]

Table 1. Summary of glycan-based building blocks of synthetic condensates, the trigger for LLPS, and the potential application for each condensate system.

Glycan building block–counterpart building block	Trigger for LLPS	Potential applications	References
Chitosan–hyaluronic acid	Ionic strength, temperature, pH, cosolvent addition, humidity	Tissue scaffolds, wound dressings, drug carriers	115, 108
Chitosan–heparin	Ionic strength, temperature, pH	Drug nanodelivery, cancer therapy, vaccine adjuvantation	94
Chitosan–dextran sulfate	Ionic strength, temperature, pH	Antiretroviral drug delivery, protein sorption for targeting	94, 113
Chitosan–insect cuticular protein	Ionic strength, time-dependent aging	Insect cuticle-inspired biomaterials, composite biomimetic materials	116

Table 1. (*Continued*)

Glycan building block–counterpart building block	Trigger for LLPS	Potential applications	References
Chitosan–soybean whey protein	Ionic strength, pH, protein-to-polysaccharide ratio ($R_{\mathrm{SWP/CHI}}$)	Food formulations, microreactors	117
Chitosan–soybean pea isolate	Ionic strength, pH, temperature, protein-to-polysaccharide ratio ($R_{\mathrm{SPi/CHI}}$)	Enzyme immobilization, flavor microencapsulation, cosmetics	118
Chitosan–pea protein isolate	pH, protein-to-polysaccharide ratio ($R_{\mathrm{PEA\ PROTEIN\ ISOLATE/CHI}}$)	Enzymatic immobilization, tissue engineering, controlled drug delivery vehicle	119
Chitosan–canola protein isolate	Ionic strength, pH, temperature, protein-to-polysaccharide ratio ($R_{\mathrm{CPI/CHI}}$)	Food and pharmaceutical microencapsulation, emulsifiers, and stabilizers in food applications	120
Chitosan–ovalbumin	Ionic strength, pH, temperature, protein-to-polysaccharide ratio	Food texture and stability, thickening and coating agent in the food applications	112
Chitosan–gelatin	Ionic strength, temperature, pH, shear rate	Enzymatic immobilization, tissue engineering	121
Chitosan–tsao-gum	pH, gum concentration	Microencapsulation, fat substitutes	139
Hyaluronic acid–tau protein	Ionic strength, polyanion chemistry,	Developing diagnostic tools for dementia, Pick's disease, and chronic traumatic encephalopathy	110

(*Continued*)

Table 1. (*Continued*)

Glycan building block–counterpart building block	Trigger for LLPS	Potential applications	References
Hyaluronic acid–mussel adhesive protein	Ionic strength, pH, shear rate	Bioreactors, bioadhesives, bone graft binders	130–132
Heparin–lysozyme	Ionic strength	Microencapsulation, protein purification process	107
Heparin–tau protein	Ionic strength, amphiphilic molecules, that is, 1,6-hexanediol	Drug development, treat or prevent tauopathies	110, 124, 125
Heparin–fibroblast growth factor	Ionic strength	Tissue repair, wound healing, embryonic development	109
Pectin–pea protein isolate	Ionic strength, pH, charge density, protein-to-polysaccharide ratio	Food formulations, microencapsulation of vitamins, probiotics, antioxidants	100
Pectin–gelatin	pH, temperature	Encapsulation of proteins and human cells, sustained drug release	98
Alginate–pea protein isolate	pH, biopolymer mixing ratios	Nanoparticles and microcapsules for protecting and delivering bioactive compounds to targeted sites	134
Alginate–gelatin	pH, temperature	Colon-targeted drug release, encapsulation of biological and flavoring agents	98

Table 1. (*Continued*)

Glycan building block–counterpart building block	Trigger for LLPS	Potential applications	References
Alginate–lysozyme	Ionic strength, temperature, charge distribution, polymer flexibility, mixing ratios, and total concentration	Protein purification, antibacterial films, drug delivery vehicles, microencapsulation.	135
Alginate–canola protein isolate	pH, temperature, biopolymer mixing ratio	Emulsification properties, microencapsulation, biofuels, and bioplastics	104
Carrageenan–soybean whey protein	pH, protein–polysaccharide mixing ratio, total biopolymer concentration	Cancer-preventive agents, microreactors, high-purity protein isolation	136
Carrageenan–canola protein isolate	pH, temperature, biopolymer mixing ratio	Emulsification properties, microencapsulation	104

The electrospinning of HA–chitosan coacervates highlights the role of alcohol cosolvents in modulating phase behavior and material properties, offering a scalable alternative to toxic solvents for nanofiber production.[115] These electrospun biopolymer nanofibers show promise for biomedical uses such as wound dressings, tissue scaffolds, and drug delivery. Exploring electrospinning with other biopolymers could broaden applications and enhance fiber properties. In addition, many of the reported complex coacervate systems such as HA–mussel adhesive proteins,[132] soybean protein isolate–chitosan,[118] pea protein isolate–beet pectin,[133] alginate–gelatin,[98] and canola protein isolate–chitosan[120] have the potential to be utilized in a range of biomedical applications within the food and pharmaceutical industries. An improved understanding of the coacervation mechanisms and environmental factors, along with the development of standardized protocols for reproducible formulations, will be crucial for the translation of glycan condensates.

In conclusion, this chapter highlights the emerging field of glycan-based condensates, offering significant insights into their unique condensation mechanisms. The distinctive biophysical properties of glycans, including extensive hydration and complex branching structures, contribute to their phase separation and condensate formation. While the complexation between glycan building blocks and biomolecules has gained theoretical and practical interest, the field remains underexplored, particularly in terms of self-coacervation and interactions with polysaccharides or proteins. These glycan-based condensates are governed by multiple interactions and are highly responsive to chemical and external triggers, which modulate their dynamics and material properties. Deeper investigation into their molecular mechanisms of glycan condensates will pave the way for the development of new technologies in fields such as food applications, targeted drug delivery, and tissue engineering, where the precision control of condensate formation can unlock new biotechnological potentials.

References

1. Banani SF, Lee HO, Hyman AA, Rosen MK. Biomolecular condensates: Organizers of cellular biochemistry. *Nat Rev Mol Cell Biol.* 2017;18(5):285–298.
2. Liu J, Zhorabek F, Chau Y. Biomaterial design inspired by membraneless organelles. *Matter.* 2022;5(9):2787–2812.
3. Bratek-Skicki A, Van Nerom M, Maes D, Tompa P. Biological colloids: Unique properties of membraneless organelles in the cell. *Adv Colloid Interface Sci.* 2022;310:102777.
4. Sridharan S, Hernandez-Armendariz A, Kurzawa N, Potel CM, Memon D, Beltrao P, et al. Systematic discovery of biomolecular condensate-specific protein phosphorylation. *Nat Chem Biol.* 2022;18(10):1104–1114.
5. Gomes E, Shorter J. The molecular language of membraneless organelles. *J Biol Chem.* 2019;294(18):7115–7127.

6. Zhang H, Ji X, Li P, Liu C, Lou J, Wang Z, et al. Liquid-liquid phase separation in biology: Mechanisms, physiological functions and human diseases. *Sci China Life Sci.* 2020;63(7):953–985.
7. Boisvert FM, van Koningsbruggen S, Navascués J, Lamond AI. The multifunctional nucleolus. *Nat Rev Mol Cell Biol.* 2007;8(7):574–585.
8. Lyon AS, Peeples WB, Rosen MK. A framework for understanding the functions of biomolecular condensates across scales. *Nat Rev Mol Cell Biol.* 2021;22(3):215–235.
9. Bratek-Skicki A, Pancsa R, Meszaros B, Van Lindt J, Tompa P. A guide to regulation of the formation of biomolecular condensates. *FEBS J.* 2020;287(10):1924–1935.
10. Roden C, Gladfelter AS. RNA contributions to the form and function of biomolecular condensates. *Nat Rev Mol Cell Biol.* 2021;22(3):183–195.
11. Kayitmazer AB. Thermodynamics of complex coacervation. *Adv Colloid Interface Sci.* 2017;239:169–177.
12. Decker CJ, Parker R. P-bodies and stress granules: Possible roles in the control of translation and mRNA degradation. *Cold Spring Harb Perspect Biol.* 2012;4(9):a012286.
13. Shin Y, Brangwynne CP. Liquid phase condensation in cell physiology and disease. *Science.* 2017;357(6357):eaaf4382.
14. Rekhi S, Garcia CG, Barai M, Rizuan A, Schuster BS, Kiick KL, et al. Expanding the molecular language of protein liquid-liquid phase separation. *Nat Chem.* 2024 Mar 29;16:1113–1124.
15. Leshem AB, Sloan-Dennison S, Massarano T, Ben-David S, Graham D, Faulds K, et al. Biomolecular condensates formed by designer minimalistic peptides. *Nat Commun.* 2023;14(1):421.
16. Netzer A, Leshem AB, Veretnik S, Edelstein I, Lampel A. Regulation of peptide liquid-liquid phase separation by aromatic amino acid composition. *Small.* 2024;20(38):e2401665.
17. Hyman AA, Weber CA, Jülicher F. Liquid-liquid phase separation in biology. *Annu Rev Cell Dev Biol.* 2014;30(1):39–58.
18. Boeynaems S, Alberti S, Fawzi NL, Mittag T, Polymenidou M, Rousseau F, et al. Protein phase separation: A new phase in cell biology. *Trends Cell Biol.* 2018;28(6):420–435.
19. Gabryelczyk B, Cai H, Shi X, Sun Y, Swinkels PJ, Salentinig S, et al. Hydrogen bond guidance and aromatic stacking drive

liquid-liquid phase separation of intrinsically disordered histidine-rich peptides. *Nat Commun.* 2019;10(1):5465.

20. Wei W, Tan Y, Martinez Rodriguez NR, Yu J, Israelachvili JN, Waite JH. A mussel-derived one component adhesive coacervate. *Acta Biomater.* 2014;10(4):1663–1670.
21. Kim S, Yoo HY, Huang J, Lee Y, Park S, Park Y, et al. Salt triggers the simple coacervation of an underwater adhesive when cations meet aromatic π electrons in seawater. *ACS Nano.* 2017; 11(7):6764–6772.
22. Astoricchio E, Alfano C, Rajendran L, Temussi PA, Pastore A. The wide world of coacervates: From the sea to neurodegeneration. *Trends Biochem Sci.* 2020;45(8):706–717.
23. Mitrea DM, Mittasch M, Gomes BF, Klein IA, Murcko MA. Modulating biomolecular condensates: A novel approach to drug discovery. *Nat Rev Drug Discov.* 2022;21(11):841–862.
24. Kayitmazer AB, Koksal AF, Kilic Iyilik E. Complex coacervation of hyaluronic acid and chitosan: Effects of pH, ionic strength, charge density, chain length and the charge ratio. *Soft Matter.* 2015;11(44):8605–8612.
25. Lindhoud S, Vries R de, Norde W, Stuart MAC. Structure and stability of complex coacervate core micelles with lysozyme. *Biomacromolecules.* 2007;8(7):2219–2227.
26. Chen X, Gao Y, Wang Y, Pan G. Mussel-inspired peptide mimicking: An emerging strategy for surface bioengineering of medical implants. *Smart Mater Med.* 2021;2:26–37.
27. Wegmann S, Eftekharzadeh B, Tepper K, Zoltowska KM, Bennett RE, Dujardin S, et al. Tau protein liquid-liquid phase separation can initiate tau aggregation. *EMBO J.* 2018;37(7):e98049.
28. Najafi S, Lin Y, Longhini AP, Zhang X, Delaney KT, Kosik KS, et al. Liquid-liquid phase separation of Tau by self and complex coacervation. *Protein Sci.* 2021;30(7):1393–1407.
29. Molliex A, Temirov J, Lee J, Coughlin M, Kanagaraj AP, Kim HJ, et al. Phase separation by low complexity domains promotes stress granule assembly and drives pathological fibrillization. *Cell.* 2015;163(1):123–133.
30. Machyna M, Heyn P, Neugebauer KM. Cajal bodies: Where form meets function. *Wiley Interdiscip Rev RNA.* 2013;4(1):17–34.

31. Protter DSW, Parker R. Principles and properties of stress granules. *Trends Cell Biol.* 2016;26(9):668–679.
32. Parker R, Sheth U. P bodies and the control of mRNA translation and degradation. *Mol Cell.* 2007;25(5):635–646.
33. Dai Y, You L, Chilkoti A. Engineering synthetic biomolecular condensates. *Nat Rev Bioeng.* 2023 Apr 17;1:466–480.
34. Cook AB, Novosedlik S, van Hest JCM. Complex coacervate materials as artificial cells. *Acc Mater Res.* 2023;4(3):287–298.
35. Faltova L, Küffner AM, Hondele M, Weis K, Arosio P. Multifunctional protein materials and microreactors using low complexity domains as molecular adhesives. *ACS Nano.* 2018;12(10):9991–9999.
36. Capasso Palmiero U, Küffner AM, Krumeich F, Faltova L, Arosio P. Adaptive chemoenzymatic microreactors composed of inorganic nanoparticles and bioinspired intrinsically disordered proteins. *Angew Chem Int Ed.* 2020;59(21):8138–8142.
37. Gao N, Mann S. Membranized coacervate microdroplets: From versatile protocell models to cytomimetic materials. *Acc Chem Res.* 2023;56(3):297–307.
38. Katzir I, Haimov E, Lampel A. Tuning the dynamics of viral-factories-inspired compartments formed by peptide-RNA liquid-liquid phase separation. *Adv Mater.* 2022;34(47):e2206371.
39. Liu J, Spruijt E, Miserez A, Langer R. Peptide-based liquid droplets as emerging delivery vehicles. *Nat Rev Mater.* 2023 Jan 9;8:139–141.
40. O'Flynn BG, Mittag T. The role of liquid-liquid phase separation in regulating enzyme activity. *Curr Opin Cell Biol.* 2021;69: 70–79.
41. Smokers IBA, Visser BS, Slootbeek AD, Huck WTS, Spruijt E. How droplets can accelerate reactions — coacervate protocells as catalytic microcompartments. *Acc Chem Res.* 2024;57(14):1885–1895.
42. Peters RJRW, Louzao I, van Hest JCM. From polymeric nanoreactors to artificial organelles. *Chem Sci.* 2012;3(2):335–342.
43. Rosenzweig ESF, Xu B, Kuhn Cuellar L, Martinez-Sanchez A, Schaffer M, Strauss M. The eukaryotic CO2-concentrating organelle is liquid-like and exhibits dynamic reorganization. *Cell.* 2017;171(1):148–162. e19.

44. Poudyal RR, Guth-Metzler RM, Veenis AJ, Frankel EA, Keating CD, Bevilacqua PC. Template-directed RNA polymerization and enhanced ribozyme catalysis inside membraneless compartments formed by coacervates. *Nat Commun.* 2019;10(1):490.
45. Du M, Chen ZJ. DNA-induced liquid phase condensation of cGAS activates innate immune signaling. *Science.* 2018;361(6403): 704–709.
46. Kojima T, Takayama S. Membraneless compartmentalization facilitates enzymatic cascade reactions and reduces substrate inhibition. *ACS Appl Mater Interfaces.* 2018;10(38):32782–32791.
47. Case LB, Zhang X, Ditlev JA, Rosen MK. Stoichiometry controls activity of phase-separated clusters of actin signaling proteins. *Science.* 2019;363(6431):1093–1097.
48. Woodruff JB, Ferreira Gomes B, Widlund PO, Mahamid J, Honigmann A, Hyman AA. The centrosome is a selective condensate that nucleates microtubules by concentrating tubulin. *Cell.* 2017;169(6):1066–1077.e10.
49. Garabedian MV, Wang W, Dabdoub JB, Tong M, Caldwell RM, Benman W, et al. Designer membraneless organelles sequester native factors for control of cell behavior. *Nat Chem Biol.* 2021;17(9):998–1007.
50. Priftis D, Tirrell M. Phase behaviour and complex coacervation of aqueous polypeptide solutions. *Soft Matter.* 2012;8(36):9396–9405.
51. Linsenmeier M, Faltova L, Morelli C, Capasso Palmiero U, Seiffert C, Küffner AM, et al. The interface of condensates of the hnRNPA1 low-complexity domain promotes formation of amyloid fibrils. *Nat Chem.* 2023;15(10):1340–1349.
52. Lipiński WP, Visser BS, Robu I, Fakhree MA, Lindhoud S, Claessens MM, et al. Biomolecular condensates can both accelerate and suppress aggregation of α-synuclein. *Sci Adv.* 2022; 8(48):eabq6495.
53. Liu Y-Q, Liang C-Q, Chen Z-W, Hu J, Hu JJ, Luo YY, et al. 14-3-3ζ participates in the phase separation of phosphorylated and glycated tau and modulates the physiological and pathological functions of tau. *ACS Chem Neurosci.* 2023;14(7):1220–1225.
54. Nakashima KK, Baaij JF, Spruijt E. Reversible generation of coacervate droplets in an enzymatic network. *Soft Matter.* 2018;14(3): 361–367.

55. Aumiller WM, Keating CD. Phosphorylation-mediated RNA/peptide complex coacervation as a model for intracellular liquid organelles. *Nat Chem.* 2016;8(2):129–137.
56. Abbas M, Law JO, Grellscheid SN, Huck WTS, Spruijt E. Peptide-based coacervate-core vesicles with semipermeable membranes. *Adv Mater.* 2022;34(34):e2202913.
57. Saini B, Singh RR, Nayak D, Mukherjee TK. Biocompatible pH-Responsive luminescent coacervate nanodroplets from carbon dots and poly(diallyldimethylammonium chloride) toward theranostic applications. *ACS Appl Nano Mater.* 2020;3(6):5826–5837.
58. Sahoo S, Bera S, Maiti S, Dhara D. Temperature- and composition-dependent DNA condensation by thermosensitive block copolymers. *ACS Omega.* 2017;2(11):7946–7958.
59. Gaash D, Dewan S, Leshem AB, Jaiswal KS, Jelinek R, Lampel A. Modulating the optical properties of carbon dots by peptide condensates. *Chem Commun.* 2023;59(82):12298–12301.
60. Lin K, Jing B, Zhu Y. pH-Dependent complexation and polyelectrolyte chain conformation of polyzwitterion-polycation coacervates in salted water. *Soft Matter.* 2021;17(39):8937–8949.
61. Perry S, Li Y, Priftis D, Leon L, Tirrell M. The effect of salt on the complex coacervation of vinyl polyelectrolytes. *Polymers (Basel).* 2014;6(6):1756–1772.
62. Lawrence PG, Lapitsky Y. Ionically cross-linked poly(allylamine) as a stimulus-responsive underwater adhesive: Ionic strength and pH effects. *Langmuir.* 2015;31(4):1564–1574.
63. Abbas M, Lipiński WP, Wang J, Spruijt E. Peptide-based coacervates as biomimetic protocells. *Chem Soc Rev.* 2021;50(6):3690–3705.
64. Hong Y, Najafi S, Casey T, Shea J-E, Han S-I, Hwang DS. Hydrophobicity of arginine leads to reentrant liquid-liquid phase separation behaviors of arginine-rich proteins. *Nat Commun.* 2022;13(1):7326.
65. Dai Z, Yang X. The regulation of liquid-liquid phase separated condensates containing nucleic acids. *FEBS J.* 2024;291(11):2320–2331.
66. Biswas S, Hecht AL, Noble SA, Huang Q, Gillilan RE, Xu AY. Understanding the impacts of molecular and macromolecular

crowding agents on protein-polymer complex Coacervates. *Biomacromolecules.* 2023;24(11):4771–4782.

67. Spruijt E, Westphal AH, Borst JW, Cohen Stuart MA, van der Gucht J. Binodal compositions of polyelectrolyte complexes. *Macromolecules.* 2010;43(15):6476–6484.
68. Acar OK, Kayitmazer AB, Torun Kose G. Hyaluronic acid/chitosan coacervate-based scaffolds. *Biomacromolecules.* 2018;19(4):1198–1211.
69. Strasser R. Plant protein glycosylation. *Glycobiology.* 2016;26(9): 926–939.
70. Merry CLR, Lindahl U, Couchman J, Esko JD. Proteoglycans and sulfated glycosaminoglycans. In: Varki A, Cummings RD, Esko JD, editors. *Essentials of Glycobiology.* 4th ed. New York: Cold Spring Harbor Laboratory Press; 2022. p. 487–504.
71. Kuribara T, Totani K. Structural insights into N-linked glycan-mediated protein folding from chemical and biological perspectives. *Curr Opin Struct Biol.* 2021;68:41–47.
72. Tedder TF, Steeber DA, Chen A, Engel P. The selectins: Vascular adhesion molecules. *FASEB J.* 1995;9(10):866–873.
73. Cerliani JP, Stowell SR, Mascanfroni ID, Arthur CM, Cummings RD, Rabinovich GA. Expanding the universe of cytokines and pattern recognition receptors: Galectins and glycans in innate immunity. *J Clin Immunol.* 2011;31(1):10–21.
74. Nardy AFFR, Freire-de-Lima L, Freire-de-Lima CG, Morrot A. The sweet side of immune evasion: Role of glycans in the mechanisms of cancer progression. *Front Oncol.* 2016;6:54.
75. Brewer CF, Dam TK. Essentials of Glycobiology. In: Varki A, Cummings R, Esko J, Freeze H, Hart G, Marth J, editors. New York: Cold Spring Harbor Laboratory Press; 1999. p. 653.
76. van Kooyk Y, Rabinovich GA. Protein-glycan interactions in the control of innate and adaptive immune responses. *Nat Immunol.* 2008;9(6):593–601.
77. Stringer SE, Gallagher JT. Heparan sulphate. *Int J Biochem Cell Biol.* 1997;29(5):709–714.
78. Pillai CKS, Paul W, Sharma CP. Chitin and chitosan polymers: Chemistry, solubility and fiber formation. *Prog Polym Sci.* 2009;34(7):641–678.
79. Aranaz I, Alcántara AR, Civera MC, Arias C, Elorza B, Heras Caballero A, et al. Chitosan: An overview of its properties and applications. *Polymers (Basel).* 2021;13(19).

80. Li B, Elango J, Wu W. Recent advancement of molecular structure and biomaterial function of chitosan from marine organisms for pharmaceutical and nutraceutical application. *Appl Sci.* 2020; 10(14):4719.
81. Garantziotis S, Savani RC. Hyaluronan biology: A complex balancing act of structure, function, location and context. *Matrix Biol.* 2019;78–79:1–10.
82. Dovedytis M, Liu ZJ, Bartlett S. Hyaluronic acid and its biomedical applications: A review. *Eng Regen.* 2020;1:102–113.
83. Itano N, Atsumi F, Sawai T, Yamada Y, Miyaishi O, Senga T. Abnormal accumulation of hyaluronan matrix diminishes contact inhibition of cell growth and promotes cell migration. *Proc Natl Acad Sci U S A.* 2002;99(6):3609–3614.
84. Yewdall NA, André AAM, Lu T, Spruijt E. Coacervates as models of membraneless organelles. *Curr Opin Colloid Interface Sci.* 2021;52:101416.
85. Misra S, Hascall VC, Markwald RR, Ghatak S. Interactions between hyaluronan and its receptors (CD44, RHAMM) regulate the activities of inflammation and cancer. *Front Immunol.* 2015;6:201.
86. Tang JD, Caliari SR, Lampe KJ. Temperature-dependent complex coacervation of engineered elastin-like polypeptide and hyaluronic acid polyelectrolytes. *Biomacromolecules.* 2018;19(10):3925–3935.
87. Casu B. Heparin structure. *Haemostasis.* 1990;20(Suppl 1):62–73.
88. Lee MS, Kong J. Heparin: Physiology, pharmacology, and clinical application. *Rev Cardiovasc Med.* 2015;16(3):189–199.
89. Alquwaizani M, Buckley L, Adams C, Fanikos J. Anticoagulants: A review of the pharmacology, dosing, and complications. *Curr Emerg Hosp Med Rep.* 2013;1(2):83–97.
90. Garg DK, Bhat R. Modulation of assembly of TDP-43 low-complexity domain by heparin: From droplets to amyloid fibrils. *Biophys J.* 2022;121(13):2568–2582.
91. Sidebotham RL. Dextrans. *Adv Carbohydr Chem Biochem.* 1974;30:371–444.
92. Rankin JC, Jeanes A. Evaluation of the periodate oxidation method for structural analysis of dextrans. *J Am Chem Soc.* 1954;76(17): 4435–4441.
93. Zaslavsky BY, Ferreira LA, Uversky VN, editors. Biophysical principles of liquid–liquid phase separation. In: *Droplets of Life.* Amsterdam: Elsevier; 2023. p. 3–82.

94. Costalat M, Alcouffe P, David L, Delair T. Controlling the complexation of polysaccharides into multi-functional colloidal assemblies for nanomedicine. *J Colloid Interface Sci.* 2014;430:147–156.
95. Mollah MZI, Zahid HM, Mahal Z, Faruque MRI, Khandaker MU. The usages and potential uses of alginate for healthcare applications. *Front Mol Biosci.* 2021;8:719972.
96. Neves MI, Moroni L, Barrias CC. Modulating alginate hydrogels for improved biological performance as cellular 3D microenvironments. *Front Bioeng Biotechnol.* 2020;8:665.
97. Wingender J. Interactions of alginate with exoenzymes. In: Gacesa P, Russell NJ, editors. *Pseudomonas Infection and Alginates: Biochemistry, Genetics and Pathology.* Berlin: Springer; 1990. p. 160–180.
98. Saravanan M, Rao KP. Pectin–gelatin and alginate–gelatin complex coacervation for controlled drug delivery: Influence of anionic polysaccharides and drugs being encapsulated on physicochemical properties of microcapsules. *Carbohydr Polym.* 2010;80(3): 808–816.
99. Chandrayan P. Biological function(s) and application (s) of pectin and pectin degrading enzymes. *Biosci BiotechRes.* 2018;15(1): 87–100.
100. Lan Y, Ohm J-B, Chen B, Rao J. Phase behavior, thermodynamic and microstructure of concentrated pea protein isolate-pectin mixture: Effect of pH, biopolymer ratio and pectin charge density. *Food Hydrocoll.* 2020;101:105556.
101. Rupert R, Rodrigues KF, Thien VY, Yong WTL. Carrageenan from *Kappaphycus alvarezii* (Rhodophyta, Solieriaceae): Metabolism, structure, production, and application. *Front Plant Sci.* 2022;13:859635.
102. Campo VL, Kawano DF, Silva DB da, Carvalho I. Carrageenans: Biological properties, chemical modifications and structural analysis — A review. *Carbohydr Polym.* 2009;77(2):167–180.
103. Lopes AH, Silva RL, Fonseca MD, Gomes FI, Maganin AG, Ribeiro LS, et al. Molecular basis of carrageenan-induced cytokines production in macrophages. *Cell Commun Signal.* 2020;18(1):141.
104. Klassen DR, Elmer CM, Nickerson MT. Associative phase separation involving canola protein isolate with both sulphated and carboxylated polysaccharides. *Food Chem.* 2011;126(3):1094–1101.

105. Ohn T, Kedersha N, Hickman T, Tisdale S, Anderson P. A functional RNAi screen links O-GlcNAc modification of ribosomal proteins to stress granule and processing body assembly. *Nat Cell Biol.* 2008;10(10):1224–1231.
106. Nosella ML, Tereshchenko M, Pritišanac I, Chong PA, Toretsky JA, Lee HO, et al. O-GlcNAcylation reduces phase separation and aggregation of the EWS N-terminal low complexity region. *BioRxiv.* 2021 May 11;11:2021–2025.
107. van de Weert M, Andersen MB, Frokjaer S. Complex coacervation of lysozyme and heparin: Complex characterization and protein stability. *Pharm Res.* 2004;21(12):2354–2359.
108. Es Sayed J, Caïto C, Arunachalam A, Amirsadeghi A, van Westerveld L, Maret D, et al. Effect of dynamically arrested domains on the phase behavior, linear viscoelasticity and microstructure of hyaluronic acid — chitosan complex coacervates. *Macromolecules.* 2023;56(15):5891–5904.
109. Xue S, Zhou F, Zhao T, Zhao H, Wang X, Chen L, et al. Phase separation on cell surface facilitates bFGF signal transduction with heparan sulphate. *Nat Commun.* 2022;13(1):1112.
110. Lin Y, Fichou Y, Zeng Z, Hu NY, Han S. Electrostatically driven complex coacervation and amyloid aggregation of tau are independent processes with overlapping conditions. *ACS Chem Neurosci.* 2020;11(4):615–627.
111. Villegas JA, Heidenreich M, Levy ED. Molecular and environmental determinants of biomolecular condensate formation. *Nat Chem Biol.* 2022;18(12):1319–1329.
112. Xiong W, Ren C, Jin W, Tian J, Wang Y, Shah BR, et al. Ovalbumin-chitosan complex coacervation: Phase behavior, thermodynamic and rheological properties. *Food Hydrocoll.* 2016;61: 895–902.
113. Costalat M, David L, Delair T. Reversible controlled assembly of chitosan and dextran sulfate: A new method for nanoparticle elaboration. *Carbohydr Polym.* 2014;102:717–726.
114. Sun J, Schiffman JD, Perry SL. Linear viscoelasticity and time–alcohol superposition of chitosan/hyaluronic acid complex coacervates. *ACS Appl Polym Mater.* 2022;4(3):1617–1625.
115. Sun J, Perry SL, Schiffman JD. Electrospinning nanofibers from chitosan/hyaluronic acid complex coacervates. *Biomacromolecules.* 2019;20(11):4191–4198.

116. Gong Q, Chen L, Wang J, Yuan F, Ma Z, Chen G, et al. Coassembly of a new insect cuticular protein and chitosan via liquid-liquid phase separation. *Biomacromolecules.* 2022;23(6): 2562–2571.
117. Li X, Dong D, Hua Y, Chen Y, Kong X, Zhang C. Soybean whey protein/chitosan complex behavior and selective recovery of Kunitz trypsin inhibitor. *J Agric Food Chem.* 2014;62(29): 7279–7286.
118. Huang G-Q, Sun Y-T, Xiao J-X, Yang J. Complex coacervation of soybean protein isolate and chitosan. *Food Chem.* 2012;135(2): 534–539.
119. Elmer C, Karaca AC, Low NH, Nickerson MT. Complex coacervation in pea protein isolate–chitosan mixtures. *Food Res Int.* 2011; 44(5):1441–1446.
120. Chang PG, Gupta R, Timilsena YP, Adhikari B. Optimisation of the complex coacervation between canola protein isolate and chitosan. *J Food Eng.* 2016;191:58–66.
121. Gupta AN, Bohidar HB, Aswal VK. Surface patch binding induced intermolecular complexation and phase separation in aqueous solutions of similarly charged gelatin-chitosan molecules. *J Phys Chem B.* 2007;111(34):10137–10145.
122. Sprague BL, McNally JG. FRAP analysis of binding: Proper and fitting. *Trends Cell Biol.* 2005;15(2):84–91.
123. Yu X, Duan G, Pei P, Chen L, Gu R, Hu W, et al. Heparan sulfate-dependent phase separation of CCL5 and its chemotactic activity. *eLife.* 2024;13: RP93871.
124. Lin Y, Fichou Y, Longhini AP, Llanes LC, Yin P, Bazan GC, et al. Liquid-liquid phase separation of tau driven by hydrophobic interaction facilitates fibrillization of tau. *J Mol Biol.* 2021; 433(2):166731.
125. Boyko S, Surewicz K, Surewicz WK. Regulatory mechanisms of tau protein fibrillation under the conditions of liquid-liquid phase separation. *Proc Natl Acad Sci U S A.* 2020;117(50):31882–31890.
126. Zhang X, Lin Y, Eschmann NA, Zhou H, Rauch JN, Hernandez I, et al. RNA stores tau reversibly in complex coacervates. *PLoS Biol.* 2017;15(7):e2002183.
127. Zieris A, Prokoph S, Levental KR, Welzel PB, Grimmer M, Freudenberg U, et al. FGF-2 and VEGF functionalization of

starPEG-heparin hydrogels to modulate biomolecular and physical cues of angiogenesis. *Biomaterials.* 2010;31(31):7985–7994.

128. Freudenberg U, Hermann A, Welzel PB, Stirl K, Schwarz SC, Grimmer M, et al. A star-PEG-heparin hydrogel platform to aid cell replacement therapies for neurodegenerative diseases. *Biomaterials.* 2009;30(28):5049–5060.
129. Lohmann N, Schirmer L, Atallah P, Wandel E, Ferrer RA, Werner C, et al. Glycosaminoglycan-based hydrogels capture inflammatory chemokines and rescue defective wound healing in mice. *Sci Transl Med.* 2017;9(386):eaai9044.
130. Hwang DS, Waite JH, Tirrell M. Promotion of osteoblast proliferation on complex coacervation-based hyaluronic acid — recombinant mussel adhesive protein coatings on titanium. *Biomaterials.* 2010;31(6):1080–1084.
131. Hwang DS, Zeng H, Srivastava A, Krogstad DV, Tirrell M, Israelachvili JN, et al. Viscosity and interfacial properties in a mussel-inspired adhesive coacervate. *Soft Matter.* 2010;6(14): 3232–3236.
132. Kim HJ, Yang B, Park TY, Lim S, Cha HJ. Complex coacervates based on recombinant mussel adhesive proteins: Their characterization and applications. *Soft Matter.* 2017;13(42):7704–7716.
133. Lan Y, Ohm J-B, Chen B, Rao J. Phase behavior and complex coacervation of concentrated pea protein isolate-beet pectin solution. *Food Chem.* 2020;307:125536.
134. Klemmer KJ, Waldner L, Stone A, Low NH, Nickerson MT. Complex coacervation of pea protein isolate and alginate polysaccharides. *Food Chem.* 2012;130(3):710–715.
135. Vakeri A, Boire A, Davy J, Hamon P, Bouchoux A, Bouhallab S, et al. Coacervation and aggregation in lysozyme/alginate mixtures. *Food Hydrocoll.* 2024;156:110359.
136. Li X, Hua Y, Chen Y, Kong X, Zhang C. The selective complex behavior between soybean whey proteins and ι-carrageenan and isolation of the major proteins of the soybean whey. *Food Hydrocoll.* 2016;56:207–217.
137. Kumar MNVR, Muzzarelli RAA, Muzzarelli C, Sashiwa H, Domb AJ. Chitosan chemistry and pharmaceutical perspectives. *Chem Rev.* 2004;104(12):6017–6084.

138. Freudenberg U, Liang Y, Kiick KL, Werner C. Glycosaminoglycan-based biohybrid hydrogels: A sweet and smart choice for multi-functional biomaterials. *Adv Mater.* 2016;28(40):8861–8891.
139. You G, Liu XL, Zhao MM. Preparation and characterization of hsian-tsao gum and chitosan complex coacervates. *Food Hydrocoll.* 2018;74:255–266.

https://doi.org/10.1142/9789819806782_0005

Chapter 5

Fabricating Hierarchically Structured Materials from Fluid Condensates: Lessons from the Mussel Byssus

Max J. Renner-Rao and Matthew J. Harrington*

Department of Chemistry, McGill University, 801 Sherbrooke Street West, Montreal, Quebec H3A 0B8, Canada
**Corresponding author: matt.harrington@mcgill.ca*

The field of bioinspired materials aims to reproduce the excellent properties of biological materials. However, replicating their structural hierarchies and subtle chemistry is challenging using existing material fabrication platforms. Thus, we must first elucidate the principles underlying bottom-up self-assembly of biological materials. Here, we focus on mussel byssus fabrication. The byssus is an external proteinaceous anchor fashioned by mussels, which comprises three functionally distinct yet structurally integrated components — (1) a tough, self-healing fibrous core; (2) a hard yet extensible coating; and (3) a porous underwater glue. Each byssus part consists of distinct proteins that organize into complex hierarchical structures with different cross-linking strategies, ranging from strong covalent bonds to transient metal-coordination interactions. Understanding how this complex structure self-assembles in minutes is challenging, but recent cross-disciplinary studies have significantly advanced our understanding. Most notably, mussels utilize fluid protein condensates — coacervates and liquid crystals (LCs) — as concentrated, yet malleable precursor phases that rapidly transition into a hierarchically structured solid via bio-friendly physicochemical triggers (e.g., ionic strength, pH, and mechanical shear). Studies reveal that the stimuli-responsive proteins comprising condensates undergo conformation and cross-linking transitions that trigger rapid bottom-up self-assembly. This new understanding is already influencing the improved design of mussel-inspired polymeric materials.

1. Introduction: Fluid condensates in nature

Biological fluid condensates consist of polymeric biomolecules (e.g., proteins and polynucleotides), which, under specific physicochemical conditions, phase separate out of the bulk solution to form dense fluid phases characterized by high concentration (relative to the depleted bulk phase), low surface tension, tendency to coalesce with other condensates, and viscoelastic mechanical response.[1–3] First identified in living organisms in the context of tropoelastin[4] — the precursor for the elastic fibers in vertebrate connective tissues — biological fluid condensates have gained enormous attention in the past decade following their discovery in intracellular environments.[3] Transient intracellular condensates (often dubbed "membraneless organelles") are proposed to play important roles in cellular functions — however, this extremely active topic is reviewed at length elsewhere.[3] Notably, the discovery of tropoelastin condensates in the 1950s clearly indicates that these condensed fluid protein phases also have an extracellular function in material assembly.[4] In fact, in recent years, researchers have identified protein fluid condensates — in the form of both coacervates and LCs — which are involved in the formation of numerous materials, including spider silk,[5] squid beaks,[6] velvet worm slime,[7] sandcastle worm cement,[8] and the focus of the current chapter, mussel byssus (**Figure 1**).[9] Understanding the principles underlying the formation of these materials provides a remarkable opportunity for humans to learn how to develop new polymeric materials that exhibit excellent performance and sustainable life cycles.[10,11] In this chapter, we provide

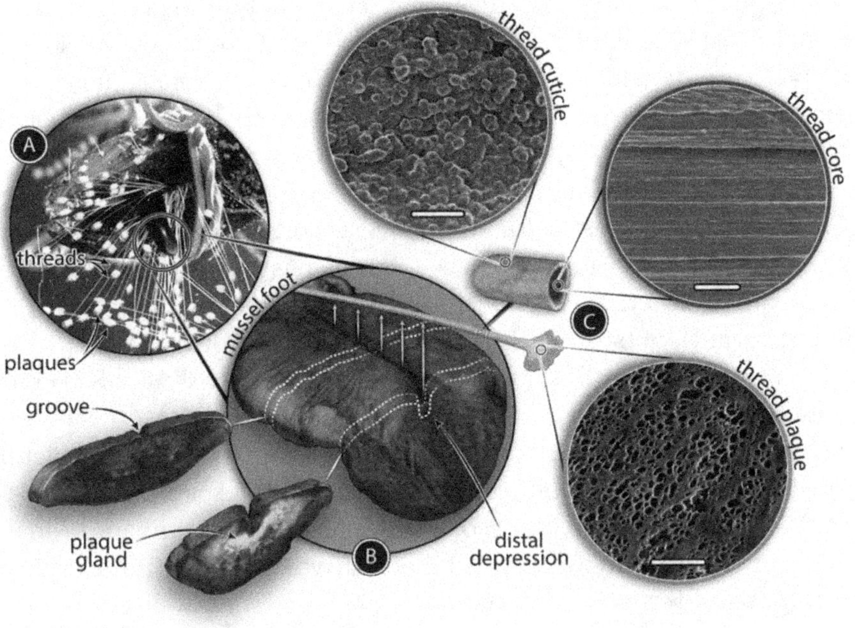

Figure 1. Overview of mussel byssus and foot. Mussels attach to surfaces with a byssus that comprises numerous individual fibrous threads covered with a protective cuticle and glued to the surface with a porous adhesive plaque. Each thread is fabricated individually by an organ known as the mussel foot, which has specialized secretory glands in which the protein precursors of the byssus are synthesized, stored, and eventually secreted to form threads. Reproduced from Priemel et al.[28] under CC BY 4.0 license.

a review of the current understanding of how the mussel byssus is fabricated from the bottom-up self-assembly of fluid protein condensates.

2. Nature as a model for materials science

In 1907, Leo Hendrik Baekeland developed a phenol–formaldehyde resin, one of humanity's first fully synthetic materials.[12,13] Trademarked under the name Bakelite, this resin was strong, lightweight, insulating, and water-impermeable and could easily be molded and cast into an endless variety of shapes, leading to quick adoption by industry.[12,13] Baekeland correctly deemed that this temperature-dependent pliability would be revolutionary and coined a general term for his phenolic polymers: plastics. Although Bakelite became largely obsolete following the technological advances of World War II,[13] its legacy can be seen in the innumerable varieties of plastic found in every corner of human society. Bakelite lives on in a much more literal sense, however — jewelry cast from synthetic resin in the 1930s can be found to this day, intact, in antique stores; assorted Bakelite components occupy the lower strata of landfills; even more lie at the bottom of oceans, lakes, and rivers.[14,15] The monomers that made these objects will likely never be recovered, and fresh phenol (sourced from petroleum feedstocks) is running out. A century after Baekeland's industry-transforming discovery, the ecological and economic impact of plastics and other unrecyclable objects drives scientific research of sustainable alternatives. Although displacing such well-established products might seem impossible, an even older class of polymeric materials has emerged as a potential solution: biological materials.

Humanity's first forays into toolmaking naturally employed the materials nearest at hand. Wood, stone, sinew, and bone were joined to more refined technologies as societies developed but were then gradually phased out following the Industrial Revolution, primarily due to the ever-growing demand for larger, stronger building materials.[16] While metal alloys reigned supreme in strength, the rapid depletion of near-surface ores and the rising environmental cost of mining operations encouraged the use of lighter alternatives. Wood, although strong for its weight, is limited to the shape of the tree it was taken from, with old-growth trees representing another dwindling resource.[16,17] Conversely, polymers (e.g., plastics), composites (concrete), glasses, and ceramics could all be molded or cast and thus began to figure prominently in every aspect of human life.[16] Synthetic fibers are used in a variety of applications outside textiles. Kevlar became a household name, is stronger than steel, and is used in bulletproof vests, parachute cords, and high-performance sports equipment, among other things.[18,19] However, one year before Stephanie Kwolek's invention of polyaramid textiles (one of which was later branded as Kevlar), another assembly of ultra-strong fibers was "discovered" in backyards and basements: spider webs.

In 1964, Lucas[20] was the first to investigate the mechanical properties of spider silk, finding the dragline fibers produced by European garden spiders (*Araneus diadematus*) not only more extensible than steel or nylon but also tougher.[20,21] While subsequent research identified spider species with superior silks,[22,23] this initial research highlighted an important advantage of biological materials over their artificial counterparts; they combined several desirable properties into a single construct, whereas synthetic materials typically excelled in only one domain. For example, modern Kevlar is stiffer than any known silk but can barely stretch due to a para-crystalline internal structure and thus exhibits a toughness more than threefold lower than dragline silk.[18,19,21] However, if spider silks were stronger, lighter, and made of naturally abundant amino acids, why didn't they displace synthetic polymers?

The most familiar biological materials are familiar because they can be produced and used at a scale relevant to human demand. While silkworm cocoons can be harvested and spun into textiles, the cannibalistic tendencies of spiders proved to be incompatible with large-scale farming efforts, thereby driving research into alternative production methods.[22] One successful approach has been *bioinspiration*, a methodology that aims to establish a deep understanding of the physicochemical interactions at play in biological structures in order to extract design concepts that can be used to create high-performance, sustainable synthetic materials. After decades of investigation into the composition and assembly of arachnid silks, modern science can bacterially express large quantities of optimized silk-like proteins and spin them into fibers with toughness equal to native dragline silk.[11,22,24,25] Thus, a material evolved to catch flies may soon catch bullets. Many of the material challenges faced by society have already been overcome by natural selection, albeit at too small a scale for direct application, so the ongoing success of silk research continues to prompt interest in other types of super-tough protein fibers that may benefit from the bioinspired treatment.[26]

While an amazing achievement, the story of artificial spider silk also offers a cautionary tale that it is not sufficient to simply understand structure–function relationships and be able to express copious amounts of protein precursors (including in the milk of transgenic goats and mice).[27] Rather, a deep understanding of the physicochemical parameters underlying material assembly and processing was necessary to realize synthetic silks.[11] Indeed, compared to what we currently understand about the structure–function relationships defining the properties of countless biological materials, we know very little about how most of these materials are fabricated. This realization has led a handful of researchers to take a step back, focusing their studies on understanding the physical and chemical parameters at play in the self-assembly of biological materials. The rest of the chapter focuses on the mussel byssus as a model system for bioinspired material fabrication.

3. Mussels: A marine model system for bioinspiration

While there are innumerable examples of natural structural adaptations, special consideration must be given to the potential ease with which these materials can be translated into modern technology. For example, sinew (i.e., tendons and ligaments) has a rich history of human use and is in fact incredibly tough proteinaceous fiber,[21] but it is the one that is generated and maintained within organisms by cellular activity,[29] thereby presenting clear challenges for biomimicry. Fortunately, one type of organism can produce stable tendon-like fibers outside its body in a way that is potentially much more amenable to human manufacturing techniques. Marine mussels anchor themselves by producing a series of thin fibers that connect their shell-enclosed bodies to the coastal rocks where they live and feed; the collection of fibers is known as a *byssus*, with individual fibers referred to as byssal threads (**Figures 1–4**).[9] Only a few minutes are required for a mussel to generate a byssal thread, and it does so in its native seawater milieu,[28,30,31] a habitat whose

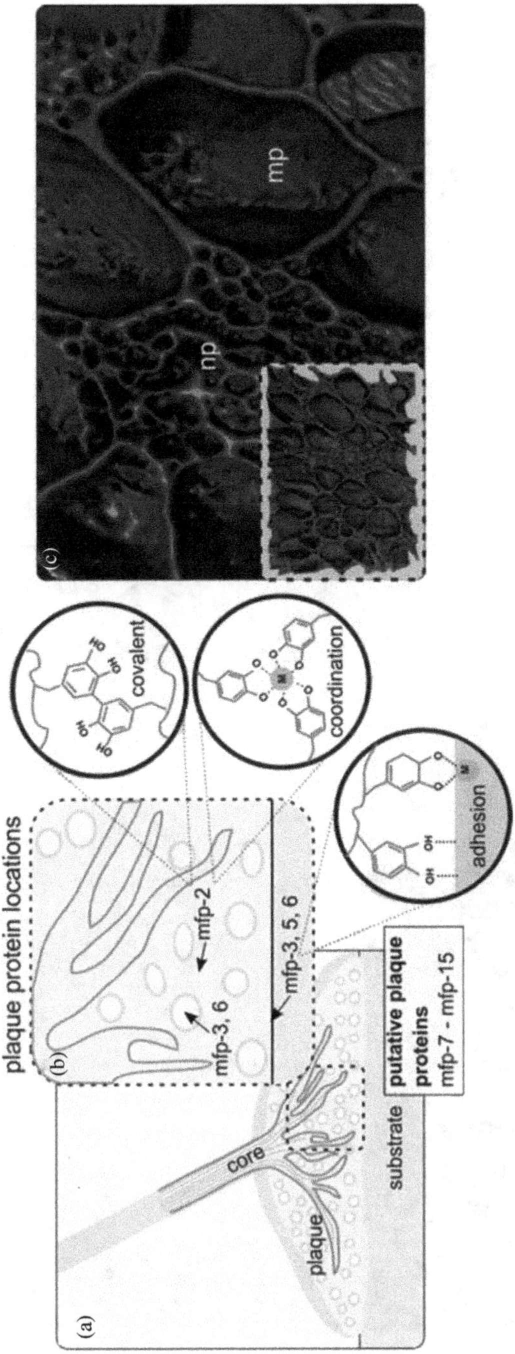

Figure 2. Mussel byssus plaque hierarchical structure. (a and b) Schematic drawing of the plaque showing the localization of different plaque precursor proteins and the molecular interactions within. (c) Focused ion beam scanning electron microscopy (FIB-SEM) reconstruction of a plaque showing the microporous (mp) and nanoporous (np) structure. Adapted with permission from Renner-Rao et al.[32]

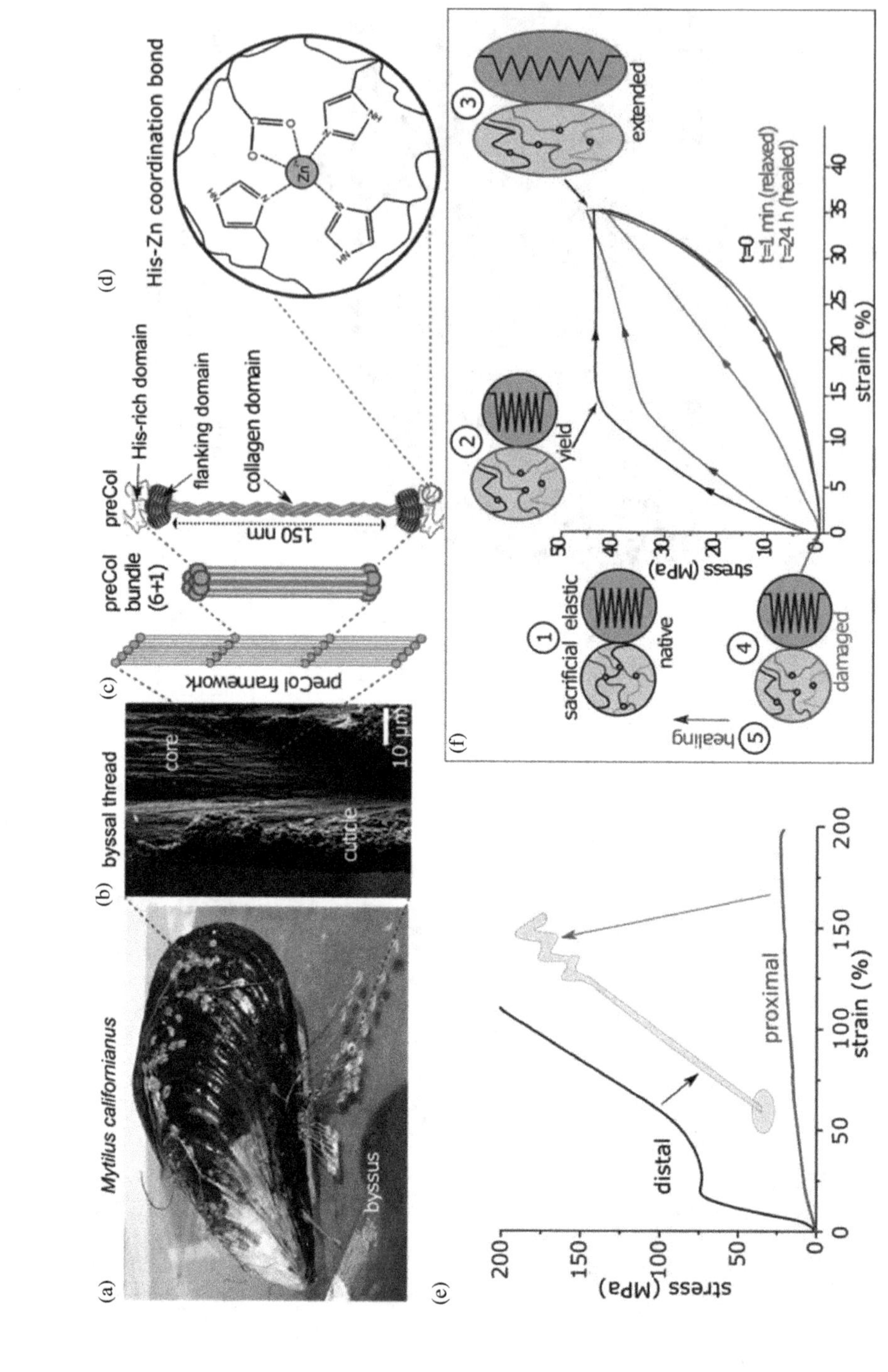
(a)
Mytilus californianus
byssus
(b)
byssal thread
cuticle
core
10 μm
(c)
preCol framework
preCol bundle (6+1)
preCol
His-rich domain
flanking domain
collagen domain
150 nm
(d)
His-Zn coordination bond
(e)
stress (MPa)
strain (%)
distal
proximal
(f)
sacrificial
elastic
native
healing
damaged
yield
extended
stress (MPa)
strain (%)
t=0
t=1 min (relaxed)
t=24 h (healed)

Figure 3. Mussel byssus core structure–function relationships. (**a**) Photo of the California mussel *Mytilus californianus* attached to a surface with its byssus. (**b**) SEM image of a byssal thread cut open to expose the fibrous core. (**c**) Hierarchical organization of preCols within the thread core. (**d**) Schematic of His-Zn^{2+} metal-coordination bonds present in the terminal domains of the preCols. (**e**) Tensile pull-to-break stress–strain curves of the distal and proximal regions of byssal threads. (**f**) Tensile cyclic stress–strain curves of a distal byssal thread cycled three times to 35% strain, highlighting thread damage after the first cycle and the self-healing response following a 24-hour resting period. Schematic images show the role that sacrificial bonds (His-Zn cross-links) and hidden length (beta-sheet structure) play in this process. Adapted with permission from Reinecke et al.[41]

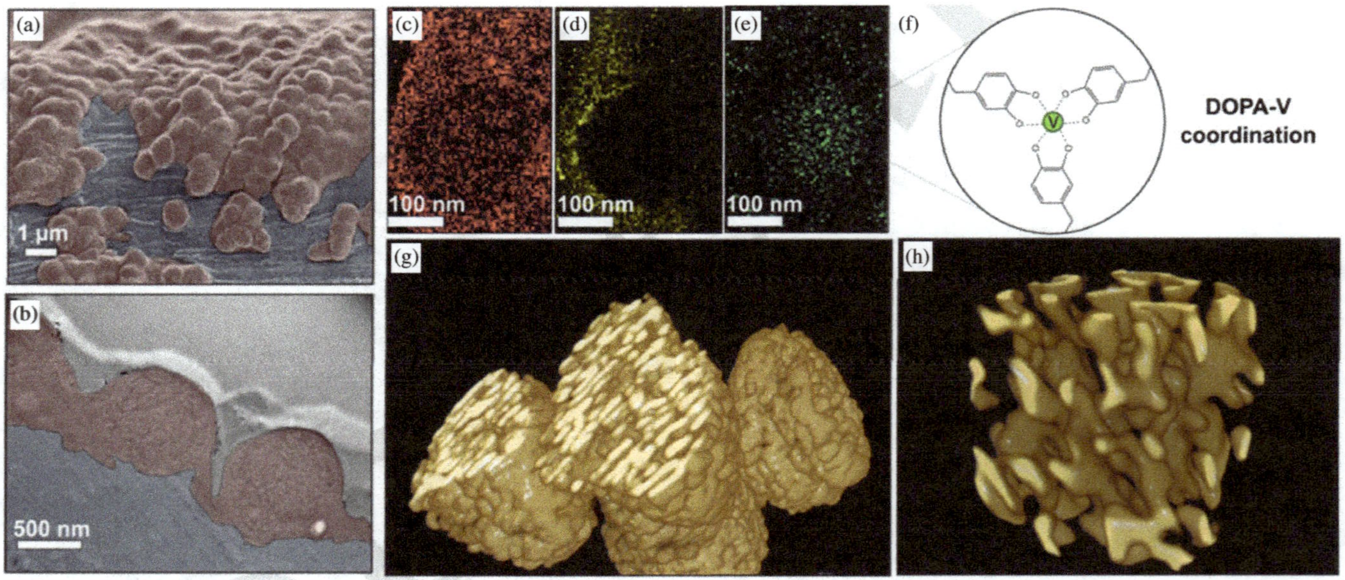

Figure 4. Mussel byssus cuticle hierarchical structure. (**a**) SEM and (**b**) transmission electron microscopy (TEM) images of the byssus cuticle (red) covering the core (blue), highlighting its granular structure. (**c–e**) TEM–EDX images of the cuticle showing the distribution of (**c**) sulfur, (**d**) iron, and (**e**) vanadium. (**f**) Schematic of tris-DOPA–V metal-coordination complex present in the granules of the cuticle. (**g**) FIB–SEM reconstruction of the bicontinuous brainy morphology present in the granules. (**h**) Higher magnification image of the inside of one of the granules in (**g**). Adapted from Jehle et al.[44] under CC BY 4.0 license.

peculiar environmental stresses have endowed the byssus with unique mechanical properties.

Byssal threads must dissipate the kinetic energy of wave impacts, a function that requires a specific combination of material properties including excellent toughness, extensibility, hysteresis, self-healing, and underwater adhesion,[9] all of which have been honed by millions of years of evolution. Unlike spiders, which produce and combine several types of silk within a web,[22,25] mussels integrate every functionality within a single thread, denoted by various zones that are compositionally and structurally distinct from one another.[28,33] Starting at the furthest point from the mussel (i.e., the thread-to-substrate interface), we find an adhesive plaque that resembles a flattened cone with a porous interior structure[34] (**Figures 1** and **2**); this plaque can be attached to nearly any type of surface (including Teflon) and remains strongly adhered in air and water alike[35] — an important adaptation for intertidal life and of considerable interest to the design of synthetic surgical adhesives.[36] The plaque tapers down to the fibrous portion of the thread, the porous interior giving way to a tightly packed, semi-crystalline fiber core.[37] While the toughness of the overall fiber is of the same order as Kevlar,[21] its material properties change along its length.[38,39] In the genus *Mytilus*, the fiber's distal portion (i.e., furthest from the mussel body) has an elastic stiffness of roughly 500–800 MPa, while the proximal portion is an order of magnitude less stiff but can extend significantly further before breaking (**Figure 3**).[21,38–40] This combination of properties means that whole fibers have an elastic stiffness of ~300 MPa but can extend to nearly twice their original length or ~80% past their yield point, a massive regime of pseudoplastic deformation wherein the bulk of impact energy is dissipated.[9,21] Notably, the mechanical mismatch between the distal and proximal regions of the thread is mitigated through a mechanically graded transition of several hundred microns, which serves to avoid stress concentrations at their interface.[39]

Given the chaotic nature of its environment, the mussel also produces a thin, abrasion-resistant coating that covers the entire byssal thread, shielding it from wave-borne debris (**Figure 4**).[42–44]

This structure, known as the cuticle, must not only be quite stiff and hard to serve a protecting function but also conform, flex, and extend with the underlying fiber core, a combination of properties that are once again challenging to replicate in synthetic materials.[42,45] Although the various regions of the cuticle, core, and plaque have very different morphologies and mechanical properties, they must connect to each other in a structurally sound manner; otherwise, failure in any one portion or interface could mean failure of the entire thread. Moreover, the entire byssus must anchor into the soft living tissue of the mussel, which occurs through a structure known as the stem root, which is in the form of ~50 biopolymer sheets that interdigitate with the living extracellular matrix of the mussel, creating an incredibly strong yet releasable biointerface unlike anything previously reported in nature.[46,47] This capacity for the mussel to join extremely hard rocky substrates to soft tissue in a saline environment via an intricate mechanical gradient is of extreme interest not only to the biomedical field (e.g., implant design, surgical glues) but also for the design of high-performance technical polymers.[11,36,48,49] Yet, in addition to these remarkable properties, each thread is produced in just minutes under bio-friendly conditions.[28] How does this work? Here, we take a closer look at the protein components of the byssus and the mechanisms by which they achieve such excellent and diverse mechanical properties.

3.1. *Byssal proteins: Compositions, structures, and functions*

Many members of class *Bivalvia* produce byssal threads, but their compositions, structures, and functions are quite varied.[50] Therefore, we will confine ourselves to the description of the genus *Mytilus*, with special attention given to the common North Atlantic blue mussel (*M. edulis*), although close cousins *M. californianus* and *M. galloprovincialis* are also well studied and will serve as important comparators. Over a dozen protein families compose the byssus,[31,51] some of which can be lumped together according to their function and location within the byssal thread. The mussel foot proteins (mfps) found in the plaque (**Figure 2**) play various roles in substrate adhesion (mfp-3 and 5),[52–54] structural cohesion (mfp-2 and 4),[55,56] and oxidative protection (mfp-6).[57] This last function requires an explanation of the mechanisms by which the plaque precursors are thought to stick to the substrate and each other. Every byssal protein is adorned with the post-translationally modified amino acid 3,4-dihydroxyphenylalanine (DOPA) in quantities ranging from a few residues to a stunning 30 mol% in the adhesive mfp-5.[31] Although DOPA and the tyrosinase enzymes that catalyze its production are found ubiquitously throughout nature (e.g., in the production of dopamine), its incorporation into proteins is rare. And yet, DOPA residues are conserved throughout the genus *Mytilus* because they play a critical multifunctional role.[9,31]

At seawater pH (~8.2), the catechol form of DOPA is oxidized to DOPA–quinone, which can undergo radical-mediated coupling with other quinones or form Michael adducts with histidine, lysine, or cysteine.[58,59] Although these strong covalent interprotein bonds enhance structural cohesion, some DOPA must remain in the reduced catechol form to perform adhesive functions, forming either hydrogen or metal-coordination bonds with the substrate[31,59] and forming cohesive metal-coordination bonds with Fe and V ions (**Figure 2**).[60,61] Since the spontaneous oxidation of DOPA in seawater would quickly diminish the plaque's adhesive capabilities, the cysteine-rich mfp-6 has been proposed to serve as a built-in reducing agent, donating electrons to quinones via the formation of disulfide bridges.[57] A similar redox system is thought to exist in the cuticle, which consists of DOPA-rich mfp-1 granules embedded in a cysteine-rich matrix.[44,62] In addition, the well-studied mfps are accompanied by a slew of putative proteins (mfp-7 to mfp-19), which have only been recently identified through transcriptomic investigation of *M. californianus* foot glands.[51] Although some of these putative proteins have been detected in native plaque secretions, their functions remain largely unknown. Some have sequence motifs that resemble well-understood adhesive proteins, but most only contain a few potential DOPA residues.[51] In particular, unusual are those proteins enriched in histidine (mfp-7, 9, 11, and 12), whose role in the plaque has never been established but might bear similarities to the mechanisms found within the fiber core.[41]

Breaking the mfp naming convention, the fiber core components can be grouped as collagenous structural proteins (preCols), thread matrix protein-1 (TMP-1) and proximal thread matrix protein-1 PTMP-1) (**Figure 3**). There are three known types of preCols making up most of the core mass: preCol-D, found in the stiff distal portion of the thread; preCol-P, found in the compliant proximal portion; and preCol-NG, which is evenly distributed throughout the entire fiber (NG standing for "non-gradient").[40,63] While mfp-1 to 6 lack any intrinsic secondary structure,[9,64] the preCols are split into five well-defined domains.[63] Like tendon type-I collagen, the preCols' central domains consist of a long series of Gly-X-Y repeats, with Y occasionally being triple helix-stabilizing hydroxyprolines.[63] Although tendon is slightly stiffer and much more resilient than byssal threads, the sequences found at either end of the preCol collagen domain impart higher extensibility and higher toughness (i.e., better energy dissipation) to the fiber core.[21,41] These flanking domains are silk-like β-sheets in preCol-D, while those in preCol-P resemble spring-like elastin.[63] At the C- and N-termini of the preCols are short histidine- and DOPA-enriched sequences that can form DOPA-based covalent bonds,[58] form histidine-based metal-coordination bonds with Zn,[65] and undergo pH-triggered folding into β-sheets[66] (**Figure 3**). A rupture of histidine–Zn-coordination bonds dissipates energy upon fiber loading, but their reformation upon resting can provide a cell-free self-healing capability, thereby drastically increasing functional lifetime (**Figure 3**).[41,65] While this self-healing strategy is also employed in the plaque and cuticle and has become a major focus in mussel-inspired design,[48,67] the properties of laboratory-made materials that rely on protein–metal-coordination rarely match up to the native byssus.

3.2. *Establishing structural hierarchy through a biological fiber factory*

Nature's real gift for material design lies in turning weak molecular chains into strong structures via careful organization at multiple length scales, a phenomenon referred to as *structural hierarchy.* The preCols, for example, are arranged with nanometer precision within the fiber core. The collagen domains not only form a triple helix, but these helices also further organize into 6+1 hexagonal bundles, which then pack together into a semi-crystalline array (**Figure 3**).[37] In fact, the composition, structural organization, and ensuing mechanical properties of the byssus are so similar to tendons that they may serve as a tendon analog, a cell scaffold that could be used to study biomechanical phenomena or grow replacement tissue.[68] The porous plaque and granular cuticle both possess their own unique hierarchical organizations (**Figures 2 and 4**), which require the correct spatial localization of several different proteins,[31,34,44] but the means of achieving this level of complexity in a minutes-long fabrication process remains unclear.

A closer look at the bivalve thread-building machinery reveals a system analogous to industrial plastic injection molding (**Figure 5**), a toxic and energy-intensive process wherein solid thermoplastic resin pellets (typically based on petroleum by-products, e.g., polyethylene) are melted and injected into a mold, where they harden into the desired shape. While superficially similar to injection molding, byssus fabrication by mussels instead proceeds as a biologically friendly process occurring under ambient conditions in which the precursor phase comprises encapsulated fluid protein condensates and the mold is found in the byssus-generating organ known as the *foot* (**Figure 5**), so named due to the fact that mussels can extend them past the confines of their shells for rudimentary locomotion.[28,30,69] On one side of this foot is a furrowed channel (the *ventral groove*) connected to a roughly hemispherical indent (the *distal depression*), both of which are open to the surrounding seawater when not in

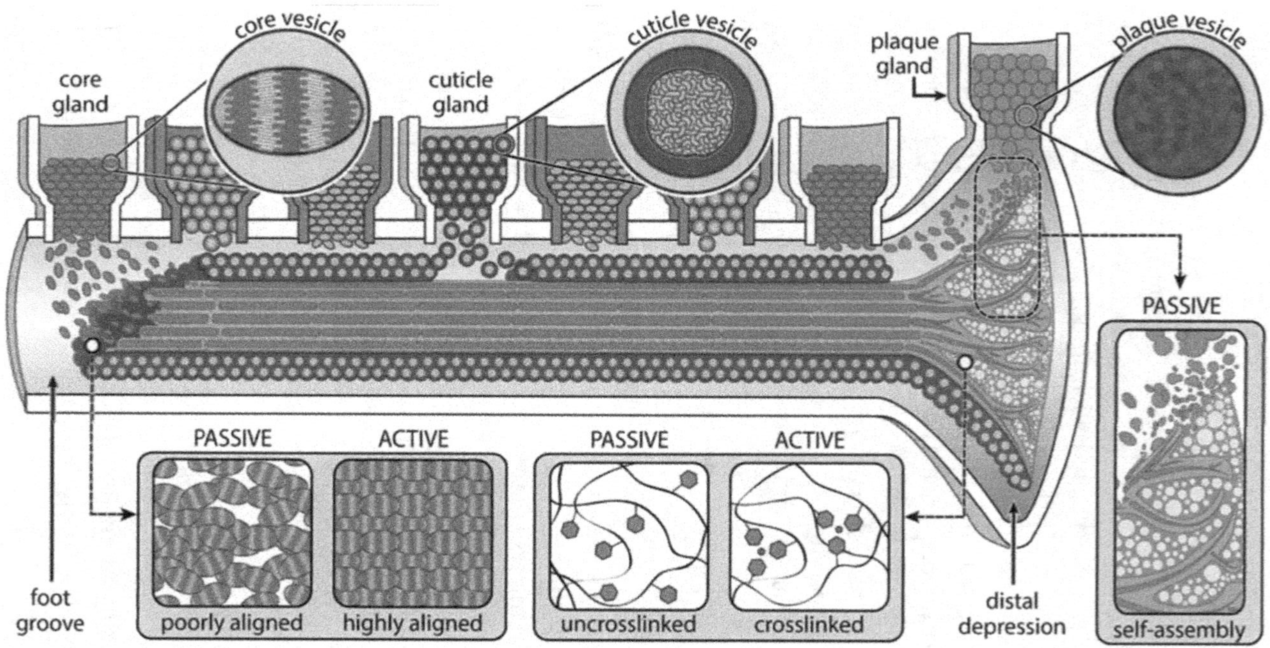

Figure 5. Schematic overview of mussel byssus fabrication process. Schematic representation of the byssal thread fabrication process, highlighting similarities to a polymer injection molding process. Liquid crystalline vesicles from the core gland are secreted into the ventral groove and actively aligned by the mussel foot before hardening. At the same time, secretory vesicles from the plaque gland are secreted into the distal depression, where they undergo phase separation, forming a porous structure and mixing with the core vesicles to form root-like structures in the plaque. Finally, the secretory vesicles from the cuticle gland are secreted, covering both the core and the plaque. The cuticle and plaque are cross-linked through the addition of metal ions that form bonds with DOPA residues. Reproduced from Priemel et al.[28] under CC BY 4.0 license.

use (**Figure 1**).[28,30] When conditions are favorable (i.e., submerged underwater with no nearby predators), the foot will probe nearby surfaces, searching for a stiff, unmoving substrate.[69] Once satisfied with its choice, it will press the distal depression to this substrate and close the ventral groove, forming a sealed cavity (the "mold") into which protein precursors are secreted (the "injection")[31,61,70]; the plaque is assembled within the distal depression and the fiber is formed along the ventral groove, both of which open up and release the completed byssal thread once it has solidified, freeing up the foot for more thread-building (**Figure 5**).

During byssus assembly, the precursor phase is in the form of extracellular vesicles that are produced and stored within specialized glands in the foot.[28,71–73] These micron-scale phospholipid membranes enclose condensed fluid protein phases, with TEM investigations of sectioned foot tissue revealing three distinct vesicle morphologies corresponding to plaque, core, and cuticle precursors (**Figure 5**).[28,44,61,72–74] Given that the byssal proteins are enriched in pH-responsive residues (DOPA, His, Asp, and Glu)[9,51] and the distal depression is acidified (~pH 3)[70] before vesicle secretion, it was proposed that the liquid-to-solid phase transition is triggered by an increase in pH and the addition of metals.[28,58,61,66] However, the source of either chemical stimulus was unclear. There is no evidence of metal ions within the vesicles,[28,58] and while exposure to seawater would indeed raise the pH, solidification would need to be near-instantaneous to prevent material diffusing away upon thread release. Diffusion presents another challenge even while the foot is sealed; how do the fluid contents of multiple vesicle types segregate to form distinct hierarchical substructures when entropy should drive the mixing of all the precursors into one homogeneous mass?

4. Condensed fluid phases I: Coacervates

If one were trying to make an injection-molded material with compositionally distinct areas, two obvious solutions present themselves. The first is to inject one precursor, wait for it to set, inject the next precursor, and so on. However, this is a time-consuming process that offers little in terms of microscale architecture. The second faster approach would be to inject every component simultaneously *as a phase-separated mixture*. Liquid–liquid phase separation (LLPS) is familiar to most in the context of nonpolar–polar mixtures (e.g., oil and water), but nature and polymer science have leveraged all-aqueous LLPS phenomena for a variety of functions.[3] Proteins phase separating from aqueous solutions are termed *coacervates* and constitute a growing area of research, particularly since coacervates were found to form intracellular membraneless compartments[75] (e.g., the nucleolus), to help transcribe genes,[76] and to rearrange cytoskeletal elements in eukaryotic cells.[77] In the context of byssus, coacervation presents itself as a solution to the diffusion problem. Numerous purified or recombinantly expressed byssal precursors have been shown to form coacervates in vitro via the aid of various triggers, such as pH, ionic strength, or the addition of polyelectrolytes.[78–80] While the exact mechanism varies for each protein, a general trend emerges for all coacervating systems: (1) the summed strength of non-covalent interprotein interactions must overcome the entropy of mixing, and (2) the coacervate components must be intrinsically disordered, that is, flexible and mobile enough for individual amino acids to sample a large volume and interact with one another (**Figure 6**).[1]

Although the final product of biological material assembly is often intricate hierarchical structures, the driving forces for biological phase separation are relatively simple. After all, some of the earliest instances of life may have required spontaneous phase separation to concentrate enough material for survival and replication.[1] At the heart of coacervation phenomena is the balance between interpolymer attraction and thermal motion. For an intuitive example, let us examine the association of two oppositely charged polyelectrolytes into a complex coacervate, which typically occurs when the polymers approach a 1:1 charge ratio (note that they do not need to be of similar length or charge density for this to occur).[1,81] At one extreme, the interaction between the two components is too weak to overcome random motion, and they remain free to diffuse

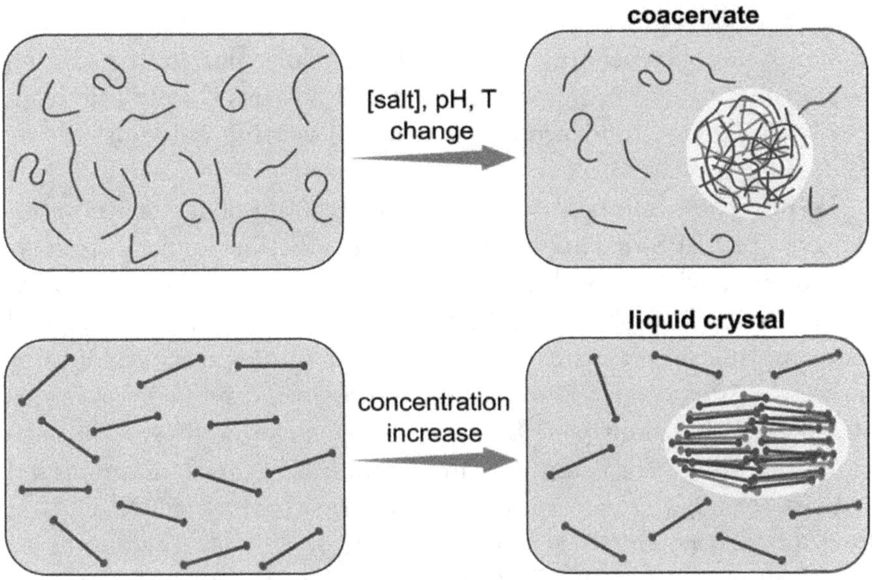

Figure 6. Schematic of biomolecular coacervate and liquid crystal formation. Fluid condensates of biomolecules include both coacervates and liquid crystals. In both cases, the biomolecules become locally concentrated but remain fluid and malleable, providing an ideal precursor phase for forming materials. In coacervates, favorable non-covalent interactions will lead to the phase separation of the condensed phase under certain physicochemical conditions. In lyotropic liquid crystals, high-aspect-ratio biomolecules are driven to self-organize into anisotropic fluid condensates when a critical concentration is passed, which is primarily driven by entropic factors arising from excluded volume considerations. Reproduced with permission from Rising and Harrington.[11]

throughout a single-phase system. This may occur due to charge screening, pH-dependent neutralization of charged groups, solution temperature (**Figure 6**) or simply because the polymers are too short or too dilute to effectively associate.[1,81] Conversely, an exceedingly strong attraction will stop all Brownian motion, trapping the components in a solid phase (e.g., a gel or precipitate).[1,81] Between these two extremes is a regime where the entropic cost of forming a network is slightly outweighed by the enthalpic contribution of charge–charge binding, leading to energetically favorable LLPS.

This thermodynamic balance dictates the relative amounts of water and polymer in each phase, yet all the molecules in the system remain free to move, imparting the dense, viscous coacervate droplets with distinct fluid properties, such as wetting onto surfaces or coalescing together.[1,3,82]

While the same rules apply for single component coacervates, the type of LLPS-driving interaction has a distinct impact on polymer design, with peptides depending on hydrogen bonding, π–π, π-cation, hydrophobic, or other self-interacting motifs typically requiring non-interacting spacers between the "sticky" residues to prevent aggregation.[1] The variety of possible driving interactions also means that multiple LLPS mechanisms can be present in individual peptides. For example, mfp-3 contains regularly interspersed positive charges (e.g., arginine), aromatic residues (DOPA), and hydrophobic residues (tryptophan), which can all contribute to coacervation depending on the pH as well as concentration and charge of ions in the environment.[1,83] Although high-chain flexibility is required in coacervating motifs, these relatively short regions can remain attached to highly ordered sections (e.g., enzymatic or fluorescent domains) and still impart LLPS behavior, allowing coacervate droplets to act as membraneless bioreactors.[75,84] However, the structurally rigid preCols do not contain any coacervating motifs, and thus, another type of condensed liquid phase must be introduced.

4.1. *Condensed fluid phases II: Liquid crystals*

While not unheard of in biological materials, LC mesophases are a somewhat rarer phenomenon than coacervates,[85] and layered arrangements (i.e., smectic LCs) are extremely uncommon for proteins. Like coacervates, certain general molecular features are critical for the formation of LC phases, notably long persistence lengths (rigidity) and geometrically anisotropic shapes (rods, discs, and bowls).[10,86,87] LC phase transitions can even be likened to those observed in coacervating systems: a freely diffusing isotropic system at one extreme, a crystalline solid at the other, and an intermediate regime wherein molecular motion is only partially arrested.[81] However, the driving forces responsible for the orientational ordering of LC mesogens are very different from enthalpically favored coacervate systems. For the sake of simplicity and biological relevance, we will limit discussion to lyotropic (i.e., concentration-dependent) LC systems, although many parallels can be drawn to the widely familiar thermotropic LCs. In the case of a single-component lyotropic LC, phase separation and orientational ordering minimize the total excluded volume of the system, a purely entropic effect best described by Onsager's theory for rigid colloids (**Figure 6**).[86–88] While this effect is dependent only on shape and concentration of the mesogen, biological LCs can also arise in complex mixtures, environments where the composition and enthalpic interactions of the components can have a distinct effect on the LC phase behavior. Both segregative and associative phase separation have been observed in LC systems, typically driven by charge–charge interactions between the rigid mesogens and flexible polymers.[86,87,89] These dynamic and condensed systems are particularly interesting due to their capacity to locally accumulate mesogens past their critical LC threshold and allowing the formation of LC droplets at significantly lower concentrations compared to single-component systems.[86,87] Type-I collagen, amyloid fibrils, chitin/cellulose nanocrystals (CNCs), and a variety of other rod-like biogenic mesogens have been shown to form nematic and cholesteric phases whose presence is essential for the formation of solid biological structures (e.g., arthropod cuticle).[85,90,91] Oddly, the only other protein-based smectic LC ever observed was composed of tobacco mosaic virus (TMV), whose formation into LC mesophases occurs *in vitro* only under artificially high concentrations and plays no obvious biological role.[92] Since Onsager's theory only accounts for isotropic–nematic phase transitions, the transverse ordering observed in core vesicles might arise from the high monodispersity and penta-block copolymer structure of preCol mesogens,[74] although this hypothesis remains unconfirmed.

5. Materials processing à la Moule: Tracking the transition from fluid to solid

While it is clear that mussels harness protein fluid condensates — both coacervates and LCs — to fabricate the byssus, the process by which this occurs in real time in the native mussel has remained shrouded in mystery. This is largely an experimental challenge considering that byssus assembly occurs in just minutes and remains hidden within the confines of the foot, which completely obscures the secretion of the precursor vesicles and their subsequent solidification. What are the driving forces for this rapid transition? How do the complex hierarchical structures, which characterize the native thread, and which are so critical for the material's properties, suddenly "materialize" in the most literal sense? And most importantly, for translating this biological process toward sustainable fabrication of synthetic polymeric materials — what are the physical and chemical triggers that are guiding this process versus those that are biologically regulated?

In recent years, researchers have begun slowly chipping away at these questions through a cross-disciplinary approach combining traditional methods from molecular biology and biochemistry with advanced imaging and spectroscopic methods from chemistry and materials science.[28,44,58,61,74,93] Importantly, following a method developed by Tamarin in the 1970s, byssus formation can be artificially induced via injection of potassium chloride into the pedal ganglion of the foot — a dense nerve bundle at the base of the mussel foot that serves as a control center for foot-related activities, including byssus formation (**Figure 7**).[94] This results in the muscular contraction of the foot and the sudden release of the vesicular contents of the core, cuticle, and plaque secretory glands into the ventral groove and distal depression. The resulting "induced thread" is a far cry from the native thread — much weaker, less fibrillar, and structurally disorganized.[28] Yet, considering the paralysis of the mussel foot during the induction process, comparing the morphology of the induced thread with the native thread provides a means of deducing which aspects of thread fabrication result from

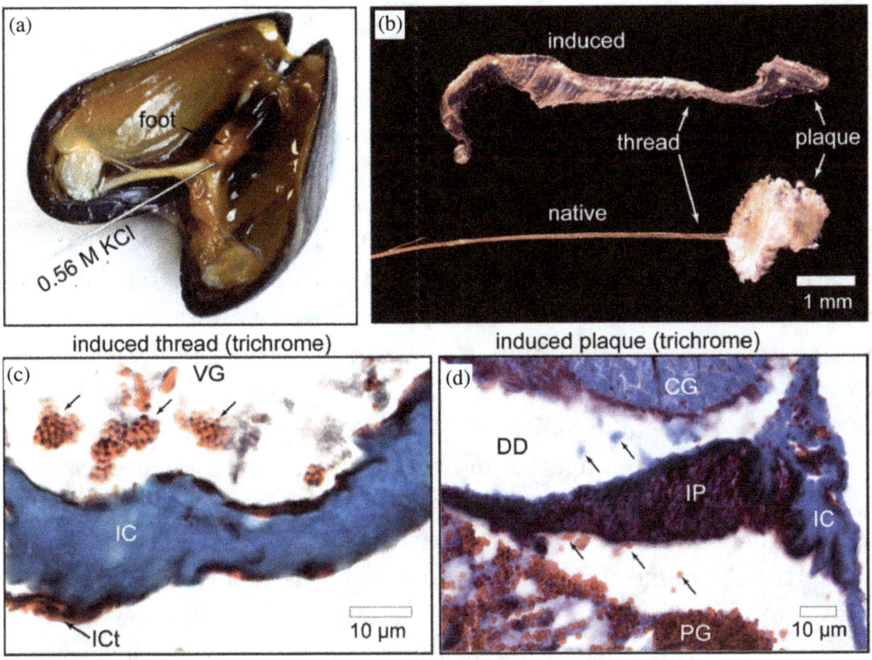

Figure 7. Tracking byssus thread fabrication with induced threads. (**a**) Mussels can be induced to secrete the byssus precursor proteins by the injection of potassium chloride into the pedal ganglion of the foot. (**b**) Image comparing the morphologies of a native and an induced byssal thread. (**c** and **d**) Histological sections of the induced thread fabrication process stained with Masson's trichrome showing the formation of (**c**) the core and cuticle and (**d**) the core and plaque (VG: ventral groove; IC: induced core; ICt: induced cuticle; DD: distal depression; CG: core gland; PG: plaque gland; IP: induced plaque). Reproduced from Priemel et al.[28] under CC BY 4.0 license.

spontaneous self-assembly and which require biologically controlled intervention.[28] Moreover, by stopping the induction process at various timepoints, it has been possible to follow the different temporal stages of the fabrication process and deduce the key sequential steps of assembly (**Figure 7**). By analyzing the different timepoints of assembly with a combination of histological staining, confocal Raman spectroscopy, and advanced electron microscopy, a new model of the native assembly process has emerged over the past few years.[28,44,58,61] We summarize the key findings — organized into separate sections on core, cuticle, and plaque formation.

5.1. *Induced core formation*

As mentioned earlier, the core of the byssal threads is a remarkable biopolymeric fiber with properties that outperform most artificial plastics — it has a toughness comparable to Kevlar, it is self-healing, it is functionally graded along its length (spanning an order of magnitude in stiffness), and it is biosourced and biodegradable.[9] How is such a high degree of structural order achieved in just minutes under ambient conditions? Core vesicles look like 2-μm-long rugby balls. This elongated morphology, properly termed *prolate spheroid*, is extremely unusual given that the vast majority of biological and manmade vesicles adopt a spherical shape which maximizes their volume-to-area ratio and minimizes surface tension.[74,95] The nanostructure of the vesicle interior provides clues to this mystery; in TEM-imaged longitudinal sections, a light–dark banding pattern repeats every ~180 nm (**Figure 8**).[72,74] At higher magnification, an even finer periodic pattern appears perpendicular to the banding lines (**Figure 8**). Conversely, transverse sections have no lines whatsoever, only a series of dark-staining dots.[74] These features are consistent with the predicted size of the preCol 6 + 1 bundles and demonstrate that the proteins are likely stored as an anisotropic fluid phase: an LC (**Figure 8**).[74,95]

Analysis of induced thread secretions with polarized light microscopy and polarized Raman spectroscopy reveals that the aligned LC phases within the micron-scale core secretory vesicles enable the formation of well-aligned domains ranging up to hundreds of microns in length, but not the centimeter length scale observed in native threads (**Figure 8**).[28,95] It was hypothesized that this higher degree of alignment involves a biological input that is lost during the paralysis associated with thread induction. During native byssus formation, the mussel foot appears to "pull" on the substrate before releasing the assembled thread; these distinctive

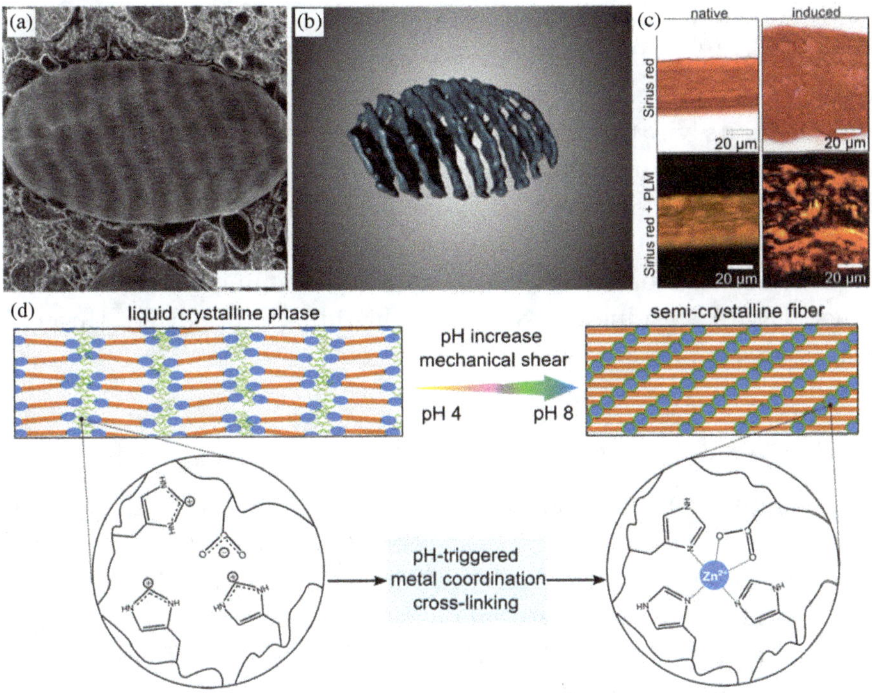

Figure 8. **Byssus core formation.** (**a**) TEM image of core secretory vesicles showing banding pattern characteristic of the smectic liquid crystalline arrangement of the preCol precursors. (**b**) 3D FIB–SEM reconstruction of a single core secretory vesicle showing the smectic layered structure. (**c**) Sirius red-stained histological sections of a native versus induced thread core in light and polarized light microscopy showing poor alignment of induced threads. (**d**) Schematic representation of proposed thread assembly process by which smectic liquid crystal phase of preCols transitions to a highly aligned tough fiber through mechanical alignment and a pH trigger, resulting in deprotonation of histidine residues and metal cross-linking in the terminal preCols domains. Adapted with permission from Rising and Harrington,[11] Priemel et al.,[28] and Jehle et al.[74]

muscular contractions may produce shear forces that help align the micron-size LC droplets of preCols (**Figure 8**).[28] Their solidification was proposed to occur rapidly through the formation of transient metal-coordination bonds triggered by the pH-dependent deprotonation of histidine residues in the terminal domains of the preCols, which was supported through several *in vitro* studies using peptides (**Figure 8**).[44,66,96] Further studies on induced thread formation revealed that permanent covalent curing of the preCols occurs over several hours via the formation of catechol-based oxidative cross-linking mediated by highly conserved DOPA residues at the preCol termini.[58] In this case, the spontaneous oxidation of DOPA at the basic pH of seawater led to oxidation to the quinone form and subsequent spontaneous cross-linking with other amino acids, including other DOPA residues, lysine, and histidine, supported through peptide-based studies.[58]

5.2. *Induced cuticle formation*

The highly unusual combination of hardness and extensibility exhibited by the cuticle arises from a characteristic composite-like structure in which hard granules enriched in DOPA–metal-coordination cross-links are embedded in a continuous matrix of cysteine-rich proteins.[42,44,45,60] However, how do the fluid protein precursors quickly adopt this complex, microscale heterogeneity in just minutes? Studies have revealed that the key to the assembly of the composite cuticle structure arises from the biphasic organization of the cuticle proteins within the secretory vesicles (**Figure 9**).[44,73] These two immiscible condensates consist of cysteine-rich cuticle proteins comprising the outer layer (named the protomatrix [PM]) and DOPA-rich mfp-1 proteins comprising the inner core (named the protogranule [PG]), which exhibit a characteristic brainy structure that is proposed to arise from the block copolymer structure of mfp-1.[44] Examining induced threads at the very earliest stages of cuticle formation reveals that the PM of the adjacent cuticle vesicles fuse during secretion while the PG remains separated and intact.[44] The fused vesicles then spread over the already formed core, melding with the PM layers of other vesicles already layered on the threads, creating a continuous cysteine-rich matrix phase with embedded granules (**Figure 9**).[28,44] It was observed that the initial stages of cuticle formation occur in the absence of metals and that metals are apparently added later, with vanadium localizing in the granule and iron localizing in the matrix with the sulfur.[28,44,58] The mechanism by which the two metals segregate is still not clear, but it has been proposed that the different binding kinetics of Fe and V for DOPA result in V concentrated in the granules.[44,97,98] Unlike the core, evidence suggests that the DOPA in the cuticle does not oxidize and covalently cross-link immediately, likely due to the presence of reducing cysteine residues, which can reduce the oxidized DOPA back to the catechol form (**Figure 9**).[58]

5.3. *Induced plaque formation*

The adhesive plaque has been the subject of intense interest for more than half a century due to its ability to achieve robust adhesion on the wet, fouled surfaces of intertidal zones, indicating a potential to inspire novel synthetic wet adhesives for biomedical and technical applications.[9,49] While much has been learned of the chemical mechanisms of adhesion (focused primarily on the key role of DOPA-rich proteins in mediating adhesion and curing through various intermolecular interactions), much less is understood about the process by which these micro- and nanoporous underwater glues are fabricated. Induced thread studies revealed that the contents of the plaque vesicles are secreted as a condensed fluid phase, carried through a network of micro-channels (known as the longitudinal ducts [LDs]) in which the condensates fuse together and apparently undergo LLPS before solidifying into a porous structure (**Figure 9**).[31,32,61,71] In addition, the protein condensate phase is mixed together with Fe, V, and DOPA-enriched metal storage particles (MSPs).[61] Within the low pH environment of the fluid plaque condensate, the metals contained within MSPs are slowly released, diffusing into the plaque precursors and leading to the formation of DOPA–metal-coordination cross-links once the pH is increased in the seawater environments (**Figure 9**).[61] This approach resembles a microfluidic-like mixing that leads to a more subtle and slow cross-linking, compared to current methods

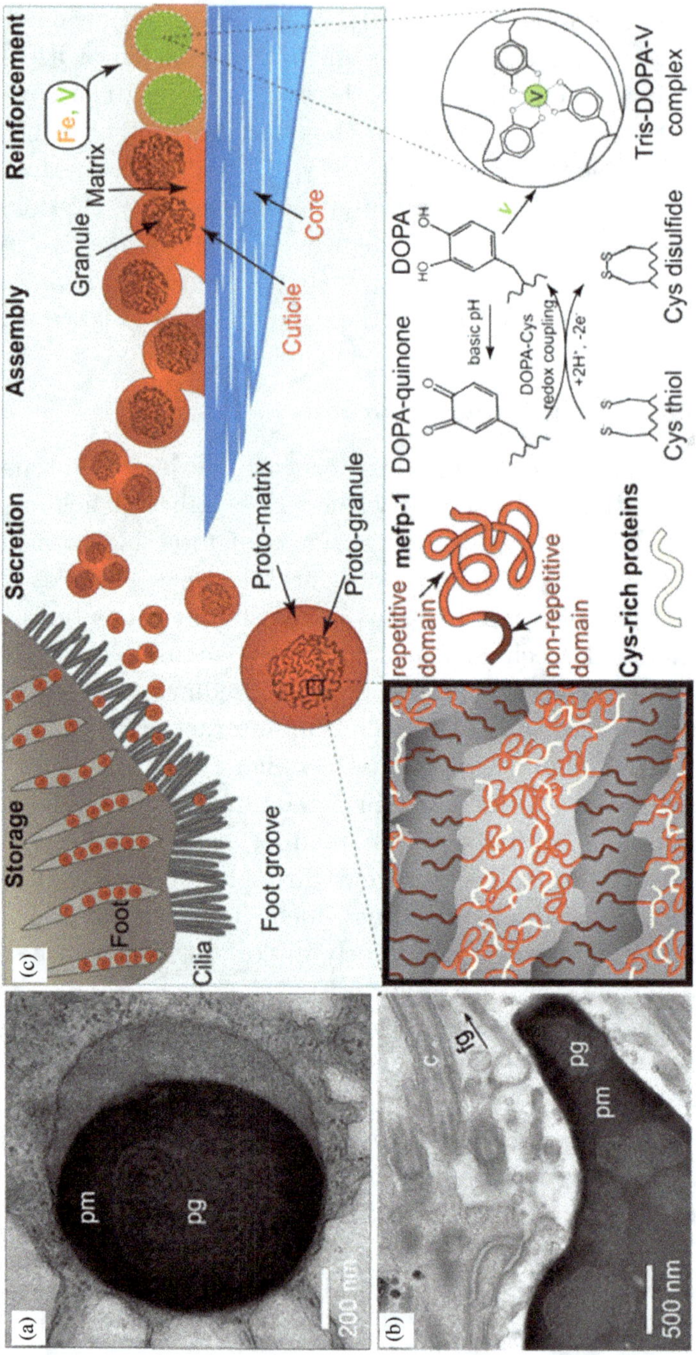
(a)
pm
pg
200 nm
(b)
c
fg
pg
pm
500 nm
(c)
Storage
Secretion
Assembly
Reinforcement
Foot
Cilia
Foot groove
Proto-matrix
Proto-granule
Granule
Matrix
Cuticle
Core
Fe, V
repetitive domain
mefp-1
non-repetitive domain
Cys-rich proteins
DOPA-quinone
DOPA
basic pH
DOPA-Cys redox coupling
$+2H^+, -2e^-$
Cys thiol
Cys disulfide
V
Tris-DOPA-V complex

Figure 9. Byssus cuticle formation. (**a**) TEM image of a cuticle secretory vesicle showing the biphasic condensate structure consisting of the DOPA-rich protogranule (pg) and the cysteine-rich protomatrix (pm). (**b**) TEM image of the cuticle secretion process showing the contents of multiple cuticle vesicles fusing as they pass through the cilia (c) to the foot groove (fg). (**c**) Schematic of the cuticle formation process showing the secretion of cuticle secretory vesicles, which fuse and then spread over the surface of the already formed core. Notably, the protomatrix coalesces, while the protogranules stay distinct. Metal ions are added at some point with vanadium localizing in the DOPA-rich granule and iron in the matrix. The brain-like structure of the granule is thought to arise from the mixing of the block-copolymer-like mfp-1 with the cysteine-rich proteins, while the cysteines are proposed to function in keeping DOPA in a reduced form. Reproduced from Jehle et al.[44] under CC BY 4.0 license.

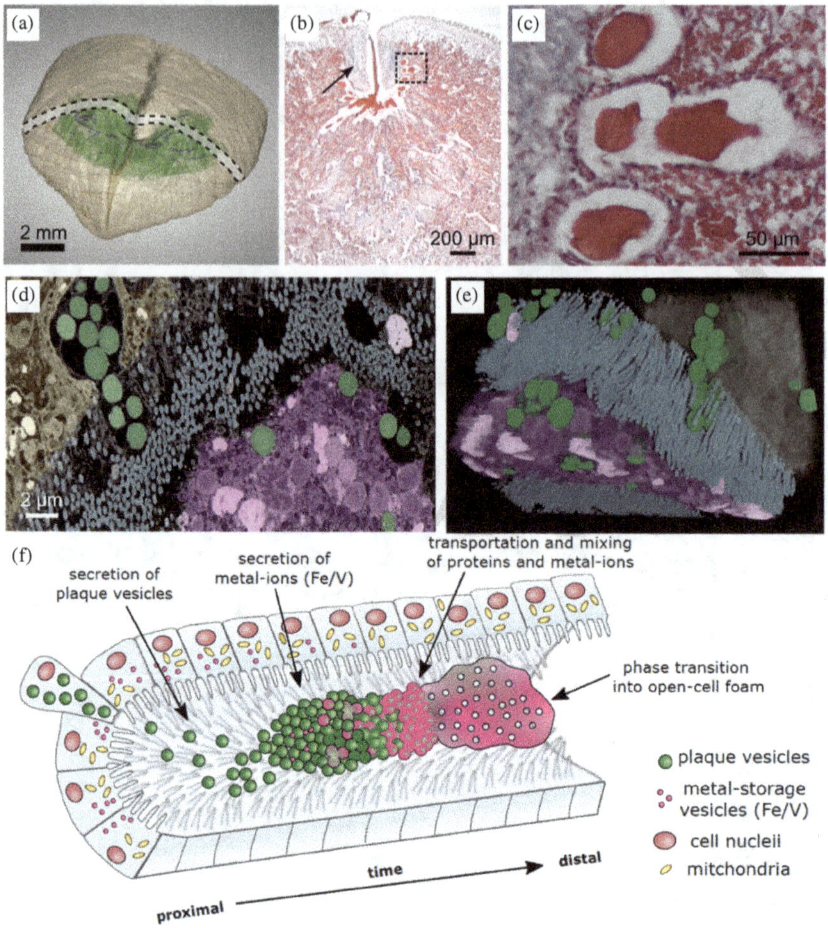

Figure 10. Byssus plaque formation. (**a**) Reconstructed micro computed tomography (μCT) image highlighting the plaque gland in green. (**b**) Histological section of the plaque gland tissue showing secretion of plaque secretory vesicle contents into the network of longitudinal ducts leading to the distal depression. (**c**) Higher magnification image from (**b**) showing the red-staining plaque condensates within the longitudinal ducts. (**d**) SEM image in false color showing the secretory vesicles (green) being secreted past the cilia (blue) into a longitudinal duct. The vesicles coalesce to form the nascent plaque secretion (purple). (**e**) FIB–SEM reconstruction of the dataset from which (**d**) arises. (**f**) Schematic illustration of the plaque formation process within a longitudinal duct. The fluid condensates within the plaque vesicles fuse and coalesce and are mixed with metal storage particles (MSPs), which help cure the adhesive through the formation of DOPA–metal-coordination cross-links. Adapted with permission from Tobias et al.[61]

of mussel-inspired metallopolymeric materials.[48,99,100] Similar to the cuticle (and unlike the core), the DOPA in the plaque does not rapidly oxidize, even at basic seawater conditions.[58] This redox protection is attributed to the presence of reducing Cys-rich protein mfp-6 and metal coordination.[57,101] It is worth noting that the plaque and the core are formed concurrently during thread fabrication, with the core and the plaque vesicles secreted in the same region at the same time.[28] Yet, the fluid condensates do not mix with one another. Rather, they form finger-like structures that anchor the core fibers into the porous plaque (like the roots of a tree into soil), producing a mechanically stable interface with increased surface area (**Figure 7**). The process by which this occurs is still not understood; it has been suggested that differences in interfacial energy lead to the complex root-like structure through fingering instabilities,[28] but this hypothesis remains to be tested.

6. Byssal precursor secretory vesicles: Stockpiling biomolecular building blocks

Induced assembly studies have indicated that although some of the structural hierarchy in byssal threads can be achieved by physicochemical self-assembly, a certain amount of active processing is required to form a fully functional material.[28,44,58,61] Yet, given the nature of these experiments, it is extremely challenging to test the hypothesis that mechanical or chemical changes regulated by the organism influence the structure and cross-linking of the final product. There have been numerous studies focused on individual byssal proteins, but it is obvious at this point that the core, cuticle, and plaque each consist of multiple proteins that work together to create the final structure and properties. Thus, studies focused on single proteins fall short of the mark. What is needed are studies that examine the native mixture of proteins in their appropriate proportions under the native conditions required to form their characteristic condensate phases. To this end, researchers have recently turned to the investigation of intact secretory vesicles extracted and purified from the mussel foot tissue.[32,39,95]

Asked simply, which proteins are in which vesicles and where do they end up in the byssus? Complicating this question, there are only three vesicle morphologies that need to encompass the nearly 20 separate protein families involved, many of them known solely through transcriptomic analysis.[9,31,51] While we can confidently assign mfp-1 to the cuticle, mfp-2 through mfp-6 to the plaque, and preCols to the core,[9] the matter becomes less definite when considering the newly identified putative proteins mfp-7 to mfp-19[51] and when localizing the proteins that make up the interface between various regions (e.g., mfp-4).[56] Furthermore, the morphological similarity of vesicles does not guarantee that they contain the same protein contents. For example, there could potentially be several subpopulations of each vesicle type containing slightly different contents or proportions of proteins — a hypothesis that may explain the compositional heterogeneity observed in different zones of the threads (e.g., the proximal-to-distal distribution of preCol-P and preCol-D).[38]

Although the advantages of using condensed liquid phases for building hierarchical structures have now become apparent, several unknown steps still lie between vesicle-bound precursors and mature byssal threads. Electron microscopy offers unparalleled spatial resolution of the core mesogens,[72] but comparatively little information about the exact proteins is present. Histological staining of thread and tissue sections provides insight into composition[28] but yields no clues as to what triggered the incredible morphological transformation of the precursors during assembly. *In vitro* studies of purified or recombinant byssal proteins permit tight control of the chemical environment[79,83,102,103] but potentially exclude multiprotein interactions of the native system and ignore those precursors not yet identified. Thus, we require a means by which we can complement these previous methodologies and fill some of the knowledge gaps surrounding byssus biofabrication. Considering that the byssus vesicles stored in the mussel foot secretory are thought to contain every byssal precursor, known and unknown, in the correct proportions and physical state required for byssus formation,[28] they provide an ideal model system for *in vitro* investigation of the physicochemical parameters that govern assembly. Until recently, a 1976 publication by Bdolah and Keller was the only report of successful isolation of core vesicles from the foot of *M. californianus* using a discontinuous density gradient.[104] Following in the footsteps of this study, researchers recently adapted the methodology from this earlier study to successfully purify both core and plaque vesicles in large enough quantities to initiate focused *in vitro* studies that were able to specifically test some of the hypotheses generated in the studies on induced thread formation.[32,39,95] Here, we highlight the key findings of these studies.

6.1. *Core vesicles*

Extracted core vesicles were found in the lowest density layer of the discontinuous gradient used for purification.[95] The vesicles remained stable for at least a week under the slightly acidic, hyperosmotic, and reducing conditions of the extraction buffer, which are ostensibly matched to the conditions of the proteins within the vesicles. Consistent with the ultrastructural investigations suggesting LC organization of the preCols, core vesicles exhibited a notable birefringence with polarized light microscopy (**Figure 11**), while Raman imaging of the vesicles showed polarization dependence, indicating protein anisotropy.[95] Hypoosmotic conditions (e.g., distilled water) caused rapid swelling of the vesicles, followed by a spheroid-to-spherical shape transition — this change in morphology, and the concomitant loss of birefringence, support the hypothesis of a lyotropic (concentration-dependent) LC phase within the core vesicles.[95] Rough estimates place the concentration of preCols in the vesicles at approximately 200 mg/ml, which is in line with the critical concentrations for LC phase transition for other rod-like biomolecules, including type I collagen and cellulose.[85] The molecular crowding at play due to the preCols being confined within a membrane likely plays a key role in the longitudinal alignment of the lyotropic phase;[105] however, there is good reason to believe that

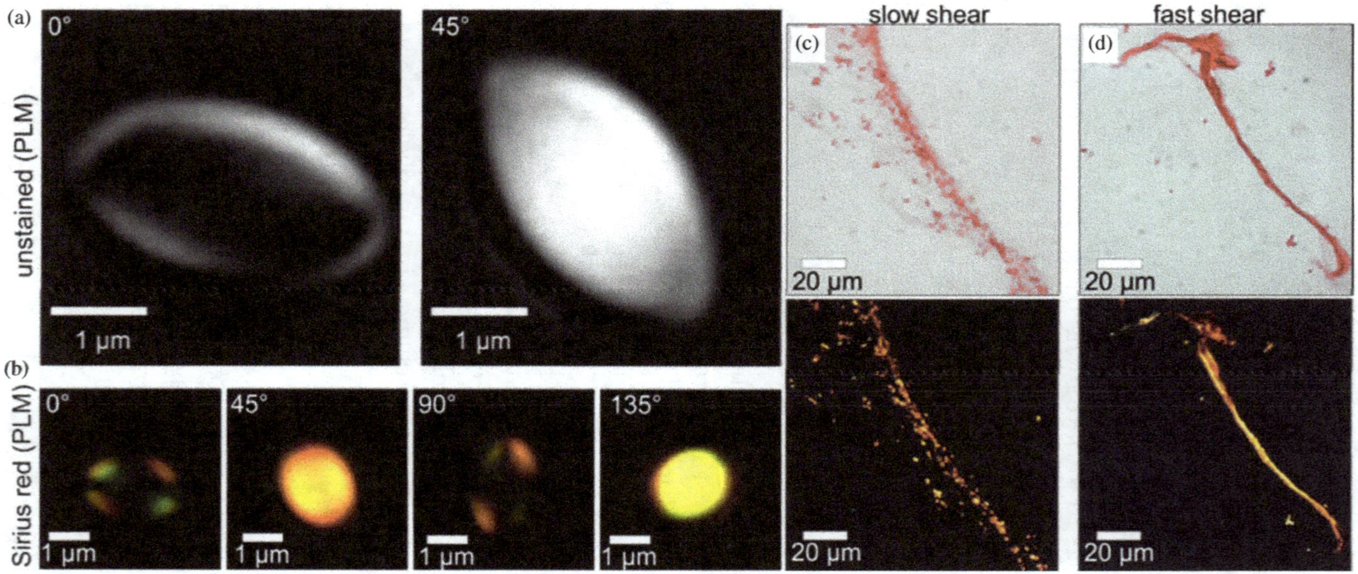

Figure 11. Core secretory vesicles. (**a**) Polarized light microscopy images of isolated core secretory vesicles showing birefringence due to the liquid crystalline structure. (**b**) Polarized light microscopy images of a core secretory vesicle stained with Sirius red, which enhances the birefringence. (**c** and **d**) Light microscopy (above) and polarized light microscopy (below) images of core secretory vesicles to which slow (**c**) and fast (**d**) shear forces were applied showing alignment of vesicles and fiber formation, respectively. Strong birefringence in the lower images confirms the high alignment of preCols. Adapted with permission from Renner-Rao et al.[95]

it is the pentablock copolymer-like distribution of charged/uncharged and polar/nonpolar amino acids along the length of the preCols that likely drives the characteristic and distinctive layering of the smectic LC phase.[63,74] Eventually, the influx of water into the vesicles under hypoosmotic conditions leads to the bursting of the vesicles, leaving a tangled, fibrous mess of preCols on the surface, which completely lacks the hierarchical organization characteristic of the native thread.[95] However, the mechanical application of shear forces led to the alignment of vesicles, and their subsequent rupture produced coherent, highly birefringent fibers (**Figure 11**). This directed assembly offers support for the hypothesis that mechanical forces play a crucial role in the attainment of higher-order structural organization during the natural byssus formation process.

A characteristic feature of the byssal thread core that has long fascinated researchers is its division into two structurally and functionally distinct regions — the soft and stretchy proximal end and the tough and stiff distal end.[38] It was long suspected that the interface between distal and proximal comprised a functional gradient arising from a graded distribution of two preCol variants (preCol-P and preCol-D),[38,103] but until recently, this hypothesis had not been directly tested, and it was completely unclear how the mussel might fabricate such a gradient during thread formation. During the vesicle extraction procedure, it was noted that the core vesicles had an unusually high range of density values, consistently distributing themselves across three different density cushions in the discontinuous density gradient.[39] Notably, the different vesicle fractions present in these layers exhibited different spectral signatures with Raman spectroscopy, matching spectra acquired from the proximal and distal regions of both the thread and the foot secretory glands.[39] Moreover, mass spectrometric investigation of the vesicle contents indicated different proportions of preCol-P and preCol-D/NG in the different vesicle populations. Formation of fibers from the two morphologically identical, but compositionally distinct vesicle extracts leads to very different micro-morphologies — fibers pulled from preCol-D-rich extracts exhibited a highly linear, distal-like structure, while fibers pulled from preCol-P-rich extracts exhibited a corrugated, proximal-like structure.[39] Importantly, fibers formed from chromatographically purified preCol-P did not exhibit proximal-like structure,[103] suggesting that another component in the proximal core vesicles — likely PTMP-1 — influences the final structure of the fiber.[106,107] These data provide strong evidence that the functional thread gradient and the characteristic morphologies self-assemble through mixing of various core vesicle subpopulations with distinct contents (preCol-D vs. preCol-P/PTMP-1).[39]

6.2. *Plaque vesicles*

During the development of the core vesicle purification protocol, it was observed that higher density layers of the discontinuous gradient contained spherical vesicles with diameters in the range of 2 μm (**Figure 12**), morphologically similar to the plaque secretory vesicles observed in the foot tissue — a conjecture that was confirmed spectroscopically.[28,32,61] This fortuitous observation opened up the previously inaccessible possibility of exploring plaque formation in its native context through *in vitro* assembly studies. First is the key question of how many proteins are in the plaque and whether all these proteins are stored together in individual vesicles or whether, similar to the core vesicles, there might be more than one subvariant of plaque vesicles. Biochemical analysis of the plaque vesicle proteins revealed that vesicles contain up to 10 different proteins, enriched in DOPA and histidine (based on staining of SDS-PAGE gels).[32] Mass spectrometric investigation of the plaque vesicle contents revealed that these proteins comprise at least mfp-2, 3, 5, and 6 and likely also putative His-rich proteins mfp-11 and -12, consistent with the SDS-PAGE analysis. Notably, single-vesicle spectroscopic analysis of numerous individual vesicles showed that they all give a nearly identical spectral fingerprint, suggesting that they all contain highly similar protein contents.[32] This was an unexpected finding given that different plaque proteins are thought to localize to specific regions,[31,80,93] suggesting that there must be an undiscovered mechanism by which the different proteins self-organize into the micro- and nanoporous structure of the plaque. The first

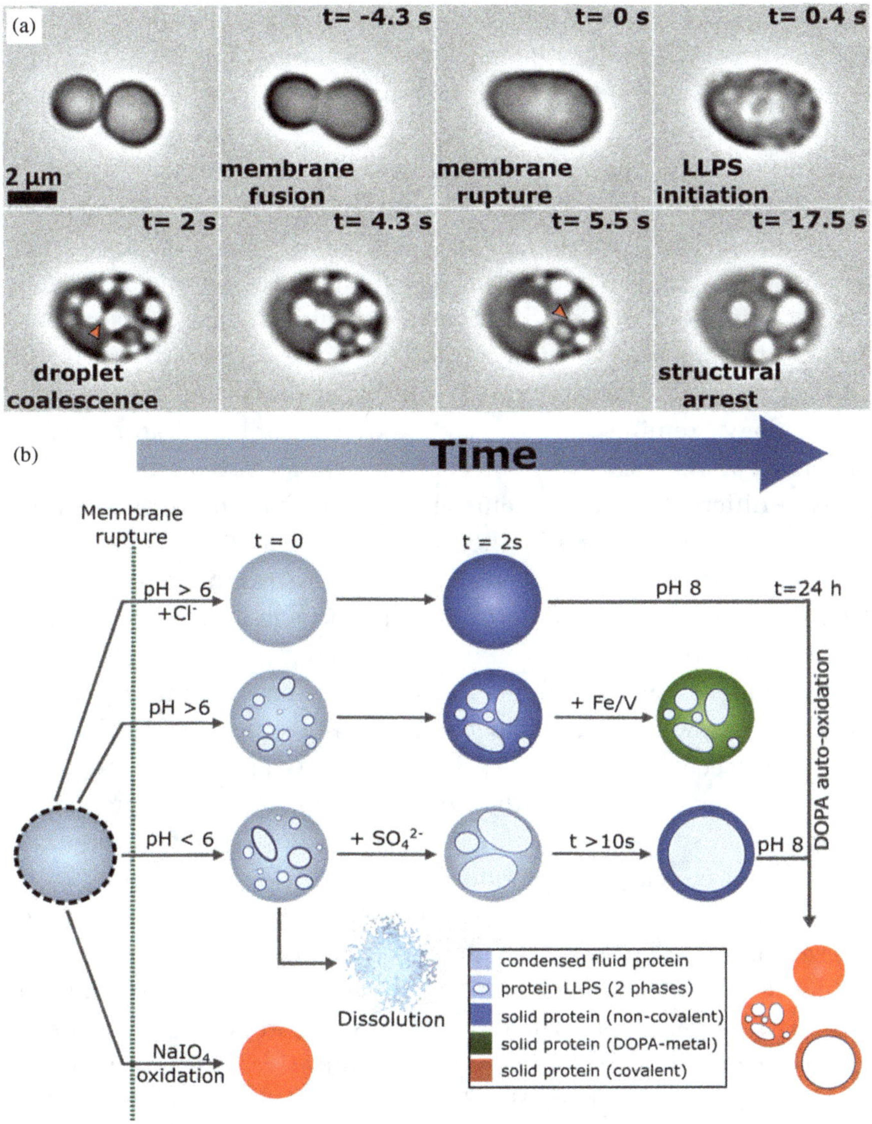

Figure 12. Plaque secretory vesicle. (**a**) Still images taken from a video of isolated plaque secretory vesicles being ruptured in hypoosmotic conditions and undergoing subsequent phase separation to yield a microporous solid. (**b**) Schematic overview of chemical control over plaque secretory vesicle phase separation and cross-linking. Adapted with permission from Renner-Rao et al.[32]

insights into this formation process were deduced through *in vitro* investigations in which plaque vesicles were ruptured through osmotic pressure in hypoosmotic buffers with controlled pH and salt content.[32]

Rupturing plaque secretory vesicles in pure water resulted in the near-instantaneous phase separation of the fluid condensate into two distinct phases — a continuous phase whose shape remained similar to that of the just-popped vesicle and numerous smaller, mobile droplets within the continuous phase that undergo fusion and ripening over a period of several seconds until the continuous phase hardens into a solid with micro- and nanoporous morphology reminiscent of the native plaque (**Figure 12**).[32] Biochemical analysis of the solid "microplaque" reveals that it comprises a different set of proteins than those that comprise the fluid condensate that remains in the pores.[32,80] It was possible to modulate phase separation and solidification process by changing the chemical properties of the buffer in which the vesicles were ruptured (**Figure 12**). In particular, it was observed that below pH 6, the vesicle contents initially undergo phase separation as in water, yet the proteins then quickly disperse and dissolve into the buffer after just a few seconds.[32] However, dissolution of the condensate phase can be inhibited under acidic conditions by the addition of high concentrations of sulfate anions, which is consistent with the fact that plaque vesicles and plaque pores are believed to contain concentrated inorganic sulfate[32,80] and the demonstration that other mussel (and non-mussel) proteins can form condensates in response to high levels of sulfate.[108] The fact that the critical pH for solidification was around pH 6 led researchers to propose an important role for a family of recently discovered putative histidine-rich proteins in the plaque, given histidine's pKa of ~6.5;[32,51] however, this hypothesis remains to be tested. While DOPA–metal-coordination cross-linking was not necessary for the initial solidification of plaque vesicle contents, spectroscopic analysis shows that Fe and V could be added to the already solidified protein to form reinforcing catechol-based cross-links (**Figure 12**).[32] These *in vitro* studies have not only enabled a whole new understanding of how the plaque assembly process works but also raised many more exciting questions that must be addressed in the future.

7. Larger context of byssus research

From the perspective of biomimetics, the complexity and refinement of byssal thread production appear daunting when viewed as a whole. However, mimicry of individual aspects of the process is well within the reach of modern manufacturing technology. Given their flexible nature, many simple synthetic polymers exhibit LLPS behavior under the right conditions[1,75] — more lyotropic LC systems are discovered every year, with plentiful collagen, chitin, and cellulose feedstocks promising a future of tough, biodegradable fibers;[85] even the membrane packaging of biological vesicles can be mimicked via microfluidics, thereby stabilizing materials for injectable applications.[109] While furthering our knowledge of the byssus has value in itself, the true objectives of bioinspired design are to address real-world engineering challenges. Over the past 15 years, mussel-inspired materials have become prominent in scientific literature, and with a few improvements, could be poised to graduate from specialized applications to large-scale green materials.[10,11,109] In particular, biocompatible mussel-inspired hydrogels have already been shown to exhibit excellent wet adhesion and have been used for a variety of applications, including hemostatic sealing of otherwise fatal arterial wounds.[10,49,110] Similarity between marine and physiological conditions (basic pH and high ionic content) makes byssus biomimetic materials attractive candidates for a slew of biomedical applications, but simply functionalizing a polymer with DOPA is insufficient given its vulnerability to oxidation in seawater and blood alike. Thus, the chemical interplay of the byssal proteins and their hierarchical organization must be fully understood to bring mussel-inspired materials to their full potential. For example, although there are some attempts to recreate the mechanical gradient of the fiber core, the lack of preCol-mimetic materials in the literature points out the difficulties associated with synthesizing semi-crystalline collagenous fibers.[111] However, recent elucidation and application of the binding mechanisms of preCol flanking and terminal domains,[66,96,112,113] as well as advances in recombinant collagen expression,[114] may signify the impending development of biomimetic byssus at industrial scale, to be used in cellular scaffolds, biomedical implants, or simply compete with synthetic spider silk for the title of toughest proteinaceous fiber.

Acknowledgments

The authors acknowledge support from a Canada Research Chair Award (CRC Tier 2 950-231953).

References

1. Abbas M, Lipiński WP, Wang J, Spruijt E. Peptide-based coacervates as biomimetic protocells. *Chem Soc Rev.* 2021;50(6): 3690–3705.
2. Harrington MJ, Mezzenga R, Miserez A. Fluid protein condensates for bio-inspired applications. *Nat Rev Bioeng.* 2023;2(3):260–278.
3. Shin Y, Brangwynne CP. Liquid phase condensation in cell physiology and disease. *Science.* 2017;357(6357):eaaf4382.
4. Adair GS, Davis HF, Partridge SM. A soluble protein derived from elastin. *Nature.* 1951;167(4250):605.
5. Malay AD, Suzuki T, Katashima T, Kono N, Arakawa K, Numata K. Spider silk self-assembly via modular liquid-liquid phase separation and nanofibrillation. *Sci Adv.* 2020;6(45):eabb6030.
6. Tan Y, Hoon S, Guerette PA, Wei W, Ghadban A, Hao C, et al. Infiltration of chitin by protein coacervates defines the squid beak mechanical gradient. *Nat Chem Biol.* 2015;11(7):488–495.
7. Baer A, Schmidt S, Haensch S, Eder M, Mayer G, Harrington MJ. Mechanoresponsive lipid-protein nanoglobules facilitate reversible fibre formation in velvet worm slime. *Nat Commun.* 2017;8(1):974.

8. Stewart RJ, Wang CS, Song IT, Jones JP. The role of coacervation and phase transitions in the sandcastle worm adhesive system. *Adv Colloid Interface Sci.* 2017;239:88–96.
9. Waite JH, Harrington MJ. Following the thread: Mytilus mussel byssus as an inspired multi-functional biomaterial. *Can J Chem.* 2021;100(3):197–211.
10. Harrington MJ, Mezzenga R, Miserez A. Fluid protein condensates for bio-inspired applications. *Nat Rev Bioeng.* 2024;2(3):260–278.
11. Rising A, Harrington MJ. Biological materials processing: Time-tested tricks for sustainable fiber fabrication. *Chem Rev.* 2023;123(5):2155–2199.
12. Baekeland LH. The synthesis, constitution, and uses of bakelite. *J Ind Eng Chem.* 1909;1(3):149–161.
13. Crespy D, Bozonnet M, Meier M. 100 years of bakelite, the material of a 1000 uses. *Angew Chem Int Ed.* 2008;47(18):3322–3328.
14. Geyer R, Jambeck JR, Law KL. Production, use, and fate of all plastics ever made. *Sci Adv.* 2017;3(7):e1700782.
15. Jambeck JR, Geyer R, Wilcox C, Siegler TR, Perryman M, Andrady A, et al. Plastic waste inputs from land into the ocean. *Science.* 2015;347(6223):768–771.
16. Ashby M. *Materials Selection in Mechanical Design.* 4th ed. Oxford: Butterworth-Heinemann; 2017.
17. Wirth C, Gleixner G, Heimann M. *Old-Growth Forests: Function, Fate and Value - An Overview.* Berlin: Springer-Verlag; 2009.
18. Singh TJ, Samanta S. Characterization of Kevlar fiber and its composites: A review. *Mater Today: Proc.* 2015;2(4):1381–1387.
19. Picken SJ, Sikkema DJ, Boerstoel H, Dingemans TJ, van der Zwaag S. Liquid crystal main-chain polymers for high-performance fibre applications. *Liq Cryst.* 2011;38(11-12):1591–1605.
20. Lucas F. Spider and their silk. *Discovery.* 1964;25:20–26.
21. Gosline J, Lillie M, Carrington E, Guerette P, Ortlepp C, Savage K. Elastic proteins: Biological roles and mechanical properties. *Philos Trans R Soc Lond B Biol Sci.* 2002;357:121–132.
22. Lefèvre T, Auger M. Spider silk as a blueprint for greener materials: A review. *Int Mater Rev.* 2016;61(2):127–153.

23. Swanson BO, Blackledge TA, Beltrán J, Hayashi CY. Variation in the material properties of spider dragline silk across species. *Appl Phys A*. 2006;82(2):213–218.
24. Arndt T, Greco G, Schmuck B, Bunz J, Shilkova O, Francis J, et al. Engineered spider silk proteins for biomimetic spinning of fibers with toughness equal to dragline silks. *Adv Funct Mater.* 2022;32(23):2200986.
25. Heim M, Keerl D, Scheibel T. Spider silk: From soluble protein to extraordinary fiber. *Angew Chem Intl Ed.* 2009;48(20):3584–3596.
26. Harrington MJ, Fratzl P. Natural load-bearing protein materials. *Prog Mater Sci.* 2021;120:100767.
27. Xu HT, Fan BL, Yu SY, Huang YH, Zhao ZH, Lian ZX, et al. Construct synthetic gene encoding artificial spider dragline silk protein and its expression in milk of transgenic mice. *Anim Biotechnol.* 2007;18(1):1–12.
28. Priemel T, Degtyar E, Dean MN, Harrington MJ. Rapid self-assembly of complex biomolecular architectures during mussel byssus biofabrication. *Nat Commun.* 2017;8:14539.
29. Killian ML, Cavinatto L, Galatz LM, Thomopoulos S. The role of mechanobiology in tendon healing. *J Should Elbow Surgy.* 2012;21(2):228–237.
30. Waite JH. The formation of mussel byssus: Anatomy of a natural manufacturing process. In: Case ST, editor. *Structure, Cellular Synthesis and Assembly of Biopolymers.* Berlin: Springer Berlin Heidelberg; 1992. p. 27–54.
31. Waite JH. Mussel adhesion - Essential footwork. *J Exp Biol.* 2017;220(4):517–530.
32. Renner-Rao M, Jehle F, Priemel T, Duthoo E, Fratzl P, Bertinetti L, et al. Mussels fabricate porous glues via multiphase liquid–liquid phase separation of multiprotein condensates. *ACS Nano.* 2022;16(12):20877–20890.
33. Bairati A, Zuccarello LV. The ultrastructure of the byssal apparatus of *Mytilus galloprovincialis* IV. Observation by transmission electron microscopy. *Cell Tissue Res.* 1976;166:219–234.
34. Filippidi E, DeMartini DG, de Molina PM, Danner EW, Kim J, Helgeson ME, et al. The microscopic network structure of mussel (*Mytilus*) adhesive plaques. *J RSoc Interface.* 2015;12(113): 20150827.

35. Aldred N, Ista LK, Callow ME, Callow JA, Lopez GP, Clare AS. Mussel (*Mytilus edulis*) byssus deposition in response to variations in surface wettability. *J R Soc Interface.* 2006;3(6):37–43.
36. Pandey N, Soto-Garcia LF, Liao J, Philippe Z, Nguyen KT, Hong Y. Mussel-inspired bioadhesives in healthcare: Design parameters, current trends, and future perspectives. *Biomater Sci.* 2020;8(5): 1240–1255.
37. Krauss S, Metzger TH, Fratzl P, Harrington MJ. Self-repair of a biological fiber guided by an ordered elastic framework. *Biomacromolecules.* 2013;14(5):1520–1528.
38. Waite J, Vaccaro E, Sun C, Lucas J. Elastomeric gradients: A hedge against stress concentration in marine holdfasts? *Philos Trans R Soc B.* 2002;357(1418):143–153.
39. Youssef L, Renner-Rao M, Eren ED, Jehle F, Harrington MJ. Fabrication of tunable mechanical gradients by mussels via bottom-up self-assembly of collagenous precursors. *ACS Nano.* 2023;17(3):2294–2305.
40. Qin XX, Waite JH. Exotic collagen gradients in the byssus of the mussel *Mytilus edulis. J Exp Biol.* 1995;198(3):633–644.
41. Reinecke A, Bertinetti L, Fratzl P, Harrington MJ. Cooperative behavior of a sacrificial bond network and elastic framework in providing self-healing capacity in mussel byssal threads. *J Struct Biol.* 2017;196:329–339.
42. Holten-Andersen N, Fantner GE, Hohlbauch S, Waite JH, Zok FW. Protective coatings on extensible biofibres. *Nat Mater.* 2007; 6(9):669–672.
43. Holten-Andersen N, Zhao H, Waite JH. Stiff coatings on compliant biofibers: The cuticle of *Mytilus californianus* byssal threads. *Biochemistry.* 2009;48(12):2752–2759.
44. Jehle F, Macías-Sánchez E, Sviben S, Fratzl P, Bertinetti L, Harrington MJ. Hierarchically-structured metalloprotein composite coatings biofabricated from co-existing condensed liquid phases. *Nat Commun.* 2020;11(1):862.
45. Holten-Andersen N, Mates TE, Toprak MS, Stucky GD, Zok FW, Waite JH. Metals and the integrity of a biological coating: The cuticle of mussel byssus. *Langmuir.* 2009;25(6):3323–3326.

46. Sivasundarampillai J, Youssef L, Priemel T, Mikulin S, Eren ED, Zaslansky P, et al. A strong quick-release biointerface in mussels mediated by serotonergic cilia-based adhesion. *Science.* 2023; 382(6672):829–834.
47. Tamarin A. An ultrastructural study of byssus stem formation in *Mytilus californianus. J Morphol.* 1975;145(2):151–177.
48. Krogsgaard M, Nue V, Birkedal H. Mussel-inspired materials: Self-healing through coordination chemistry. *Chem Eur J.* 2016;22(3): 844–857.
49. Lee BP, Messersmith PB, Israelachvili JN, Waite JH. Mussel-inspired adhesives and coatings. *Annu Rev Mater Res.* 2011;41:99–132.
50. McCartney MA. Structure, function and parallel evolution of the bivalve byssus, with insights from proteomes and the zebra mussel genome. *Philos Trans R Soc B Biol Sci.* 2021;376(1825):20200155.
51. DeMartini DG, Errico JM, Sjoestroem S, Fenster A, Waite JH. A cohort of new adhesive proteins identified from transcriptomic analysis of mussel foot glands. *J R Soc Interface.* 2017;14(131): 20170151.
52. Papov VV, Diamond TV, Biemann K, Waite JH. Hydroxyarginine-containing polyphenolic proteins in the adhesive plaques of the marine mussel *Mytilus edulis. J Biol Chem.* 1995;270(34): 20183–20192.
53. Waite JH, Qin XX. Polyphosphoprotein from the adhesive pads of *Mytilus edulis. Biochemistry.* 2001;40(9):2887–28893.
54. Zhao H, Robertson NB, Jewhurst SA, Waite JH. Probing the adhesive footprints of *Mytilus californianus* byssus. *J Biol Chem.* 2006;281(16):11090–11096.
55. Hwang DS, Zeng HB, Masic A, Harrington MJ, Israelachvili JN, Waite JH. Protein- and metal-dependent interactions of a prominent protein in mussel adhesive plaques. *J Biol Chem.* 2010; 285(33):25850–25858.
56. Zhao H, Waite JH. Proteins in load-bearing junctions: The histidine-rich metal-binding protein of mussel byssus. *Biochemistry.* 2006;45(47):14223–14231.
57. Yu J, Wei W, Danner E, Ashley RK, Israelachvili JN, Waite JH. Mussel protein adhesion depends on interprotein thiol-mediated redox modulation. *Nat Chem Biol.* 2011;7(9):588–590.

58. Priemel T, Palia R, Babych M, Thibodeaux CJ, Bourgault S, Harrington MJ. Compartmentalized processing of catechols during mussel byssus fabrication determines the destiny of DOPA. *Proc Natl Acad Sci U S A*. 2020;117(14):7613.
59. Nicklisch SCT, Waite JH. Mini-review: The role of redox in dopa-mediated marine adhesion. *Biofouling*. 2012;28(8):865–877.
60. Harrington MJ, Masic A, Holten-Andersen N, Waite JH, Fratzl P. Iron-clad fibers: A metal-based biological strategy for hard flexible coatings. *Science*. 2010;328:216–220.
61. Priemel T, Palia G, Förste F, Jehle F, Sviben S, Mantouvalou I, Zaslansky P, Bertinetti L, Harrington MJ. Microfluidic-like fabrication of a metal ion-cured bioadhesive by mussels. *Science*. 2021;374:206–211.
62. Valois E, Hoffman C, Demartini DG, Waite JH. The thiol-rich interlayer in the shell/core architecture of mussel byssal threads. *Langmuir*. 2019;35(48):15985–15991.
63. Waite JH, Qin X-X, Coyne KJ. The peculiar collagens of mussel byssus. *Matrix Biol*. 1998;17(2):93–106.
64. Hwang DS, Waite JH. Three intrinsically unstructured mussel adhesive proteins, mfp-1, mfp-2, and mfp-3: Analysis by circular dichroism. *Protein Sci*. 2012;21(11):1689–1695.
65. Schmitt CNZ, Politi Y, Reinecke A, Harrington MJ. Role of sacrificial protein-metal bond exchange in mussel byssal thread self-healing. *Biomacromolecules*. 2015;16(9):2852–2861.
66. Reinecke A, Brezesinski G, Harrington MJ. pH-responsive self-organization of metal-binding protein motifs from biomolecular junctions in mussel byssus. *Adv Mater Interfaces*. 2017;4(1):1600416. doi:10.1002/admi.201600416.
67. Zechel S, Hager MD, Priemel T, Harrington MJ. Healing through histidine: Bioinspired pathways to self-healing polymers via imidazole–metal coordination. *Biomimetics*. 2019;4(1):20.
68. Golser AV, Scheibel T. Biotechnological production of the mussel byssus derived collagen preColD. *RSC Adv*. 2017;7(61):38273–38278.
69. Choi J, Lee S, Ohkawa K, Hwang DS. Counterplotting the mechanosensing-based fouling mechanism of mussels against fouling. *ACS Nano*. 2021;15(11):18566–18579.

70. Rodriguez NRM, Das S, Kaufman Y, Israelachvili JN, Waite JH. Interfacial pH during mussel adhesive plaque formation. *Biofouling.* 2015;31(2):221–227.
71. Tamarin A, Lewis P, Askey J. Structure and formation of byssus attachment plaque in *Mytilus*. *J Morphol.* 1976;149(2):199–221.
72. Vitellaro-Zuccarello L. The collagen gland of *Mytilus galloprovincialis*: An ultrastructural and cytochemical study on secretory granules. *J Ultrastruct Res.* 1980;73:135–147.
73. Vitellaro-Zuccarello L. Ultrastructural and cytochemical study on the enzyme gland of the foot of a mollusk. *Tissue Cell.* 1981; 13(4):701–713.
74. Jehle F, Priemel T, Strauss M, Fratzl P, Bertinetti L, Harrington MJ. Collagen pentablock copolymers form smectic liquid crystals as precursors for mussel byssus fabrication. *ACS Nano.* 2021;15(4):6829–68238.
75. Yewdall NA, André AAM, Lu T, Spruijt E. Coacervates as models of membraneless organelles. *Curr Opin Colloid Interface Sci.* 2021;52:101416.
76. Shin Y, Chang Y-C, Lee DSW, Berry J, Sanders DW, Ronceray P, et al. Liquid nuclear condensates mechanically sense and restructure the genome. *Cell.* 2018;175(6):1481–1491.e13.
77. Bashirzadeh Y, Liu AP. Encapsulation of the cytoskeleton: Towards mimicking the mechanics of a cell. *Soft Matter.* 2019; 15(42):8425–8436.
78. Guo Q, Zou G, Qian X, Chen S, Gao H, Yu J. Hydrogen-bonds mediate liquid-liquid phase separation of mussel derived adhesive peptides. *Nat Commun.* 2022;13(1):5771.
79. Kim S, Yoo HY, Huang J, Lee Y, Park S, Park Y, et al. Salt triggers the simple coacervation of an underwater adhesive when cations meet aromatic pi electrons in seawater. *ACS Nano.* 2017;11(7):6764–6772.
80. Valois E, Mirshafian R, Waite JH. Phase-dependent redox insulation in mussel adhesion. *Sci Adv.* 2020;6(23):eaaz6486.
81. Tom JKA, Deniz AA. Complex dynamics of multicomponent biological coacervates. *Curr Opin Colloid Interface Sci.* 2021;56:101488.
82. Lu T, Spruijt E. Multiphase complex coacervate droplets. *J Am Chem Soc.* 2020;142(6):2905–2914.

83. Wei W, Tan Y, Martinez Rodriguez NR, Yu J, Israelachvili JN, Waite JH. A mussel-derived one component adhesive coacervate. *Acta Biomaterialia*. 2014;10(4):1663–1670.
84. Borcherds W, Bremer A, Borgia MB, Mittag T. How do intrinsically disordered protein regions encode a driving force for liquid–liquid phase separation? *Curr Opin Struct Biol*. 2021;67:41–50.
85. Rey AD. Liquid crystal models of biological materials and processes. *Soft Matter*. 2010;6(15):3402–3429.
86. Azzari P, Bagnani M, Mezzenga R. Liquid–liquid crystalline phase separation in biological filamentous colloids: Nucleation, growth and order–order transitions of cholesteric tactoids. *Soft Matter*. 2021;17(27):6627–6636.
87. Azzari P, Mezzenga R. LLPS vs. LLCPS: Analogies and differences. *Soft Matter*. 2023;19(10):1873–1881.
88. Onsager L. The effects of shape on the interaction of colloidal particles. *Ann N Y Acad Sci*. 1949;51(4):627–659.
89. Fraccia TP, Zanchetta G. Liquid–liquid crystalline phase separation in biomolecular solutions. *Curr Opin Colloid Interface Sci*. 2021;56:101500.
90. Besseau L, Giraud-Guille M-M. Stabilization of fluid cholesteric phases of collagen to ordered gelated matrices. *J Mol Biol*. 1995;251(2):197–202.
91. Nyström G, Arcari M, Mezzenga R. Confinement-induced liquid crystalline transitions in amyloid fibril cholesteric tactoids. *Nat Nanotechnol*. 2018;13(4):330–336.
92. Wen X, Meyer RB, Casper, DL. Observation of smectic-A ordering in a solution of rigid-rod-like particles. *Phys Rev Lett*. 1989;63: 2760–2763.
93. Petrone L, Kumar A, Sutanto CN, Patil NJ, Kannan S, Palaniappan A, et al. Mussel adhesion is dictated by time-regulated secretion and molecular conformation of mussel adhesive proteins. *Nat Commun*. 2015;6(1):8737.
94. Tamarin A, Keller PJ. Ultrastructural study of byssal thread forming system in *Mytilus*. *J Ultrastruct Res*. 1972;40(3–4):401–416.
95. Renner-Rao M, Clark M, Harrington MJ. Fiber formation from liquid crystalline collagen vesicles isolated from mussels. *Langmuir*. 2019;35:15992–16001.

96. Trapaidze A, D'Antuono M, Fratzl P, Harrington MJ. Exploring mussel byssus fabrication with peptide-polymer hybrids: Role of pH and metal coordination in self-assembly and mechanics of histidine-rich domains. *Eur Polym J.* 2018;109:229–236.
97. Holten-Andersen N, Jaishankar A, Harrington MJ, Fullenkamp DE, DiMarco G, He L, et al. Metal-coordination: Using one of nature's tricks to control soft material mechanics. *J Mater Chem B.* 2014;2(17):2467–2472.
98. Mesko M, Xiang L, Bohle S, Hwang DS, Zeng H, Harrington MJ. Catechol-vanadium binding enhances cross-linking and mechanics of a mussel byssus coating protein. *Chem Mater.* 2021;33(16): 6530–6540.
99. Holten-Andersen N, Harrington MJ, Birkedal H, Lee BP, Messersmith PB, Lee KYC, et al. pH-induced mussel metal-ligand cross-links yield self-healing polymer networks with near-covalent elastic moduli. *Proc Natl Acad Sci USA.* 2011;108(7):2651–2655.
100. Khare E, Holten-Andersen N, Buehler MJ. Transition-metal coordinate bonds for bioinspired macromolecules with tunable mechanical properties. *Nat Rev Mater.* 2021;6(5):421–436.
101. Yu J, Wei W, Danner E, Israelachvili JN, Waite JH. Effects of interfacial redox in mussel adhesive protein films on mica. *Adv Mater.* 2011;23(20):2362–2366.
102. Harrington MJ, Waite JH. pH-Dependent locking of giant mesogens in fibers drawn from mussel byssal collagens. *Biomacromolecules.* 2008;9(5):1480–1486.
103. Harrington MJ, Waite JH. How nature modulates a fiber's mechanical properties: Mechanically distinct fibers drawn from natural mesogenic block copolymer variants. *Adv Mater.* 2009; 21(4):440–444.
104. Bdolah A, Keller PJ. Isolation of collagen granules from the foot of the sea mussel, *Mytilus californianus. Comp Biochem Physiol B: Comp Biochem.* 1976;55(2):171–174.
105. Alfano C, Fichou Y, Huber K, Weiss M, Spruijt E, Ebbinghaus S, et al. Molecular crowding: The history and development of a scientific paradigm. *Chem Rev.* 2024;124(6):3186–3219.
106. Sun C, Lucas JM, Waite JH. Collagen-binding matrix proteins from elastomeric extraorganismic byssal fibers. *Biomacromolecules.* 2002;3(6):1240–1248.

107. Yoo HY, Song YH, Foo M, Seo E, Hwang DS, Seo JH. Recombinant mussel proximal thread matrix protein promotes osteoblast cell adhesion and proliferation. *BMC Biotechnol.* 2016;16(1):16.
108. Wu X, Sun Y, Yu J, Miserez A. Tuning the viscoelastic properties of peptide coacervates by single amino acid mutations and salt kosmotropicity. *Commun Chem.* 2024;7(1):5.
109. Amstad E, Harrington MJ. From vesicles to materials: Bioinspired strategies for fabricating hierarchically structured soft matter. *Philos Trans R Soc A.* 2021;379(2206):20200338.
110. Ma Y, Zhang B, Frenkel I, Zhang Z, Pei X, Zhou F, et al. Mussel-inspired underwater adhesives-from adhesion mechanisms to engineering applications: A critical review. *Prog Adhes.* 2021;6; 739–759.
111. Dewle A, Pathak N, Rakshasmare P, Srivastava A. Multifarious fabrication approaches of producing aligned collagen scaffolds for tissue engineering applications. *ACS Biomater Sci Eng.* 2020; 6(2):779–797.
112. Jehle F, Fratzl P, Harrington MJ. Metal-tunable self-assembly of hierarchical structure in mussel-inspired peptide films. *ACS Nano.* 2018;12(3):2160–2168.
113. Rammal M, Li C, Reeves J, Moraes C, Harrington MJ. pH-Responsive reversible granular hydrogels based on metal-binding mussel-inspired peptides. *ACS Appl Mater Interfaces.* 2023; 15(24):29004–29011.
114. Abdali Z, Renner-Rao M, Chow A, Cai A, Harrington MJ, Dorval Courchesne N-M. Extracellular secretion and simple purification of bacterial collagen from *Escherichia coli*. *Biomacromolecules.* 2022;23(4):1557–15568.

https://doi.org/10.1142/9789819806782_0006

Chapter 6

Bio-Instructive Supramolecular Assemblies

Victor A. Veenbrink, Fenna W.B. Craenmehr and Patricia Y.W. Dankers*

Biomedical Materials and Chemistry, Institute for Complex Molecular Systems, Department of Biomedical Engineering, Groene Loper 7, 5612 AP Eindhoven, The Netherlands
**Corresponding author: p.y.w.dankers@tue.nl*

Bio-instructive supramolecular materials are a class of materials that mimic the structural and functional properties of the natural extracellular matrix (ECM). They are made up of monomers that are held together by noncovalent interactions, such as hydrogen bonding, hydrophobic interactions, and π–π interactions. This results in dynamic, tunable, and modular materials with hierarchical structures. Over the past two decades, supramolecular materials have shown to be promising candidates for use in medical applications, such as tissue engineering and drug delivery. In tissue engineering, bio-instructive supramolecular materials can be used as scaffolds to provide mechanical and biochemical cues to direct cell behavior and differentiation during tissue regeneration. In drug delivery, these materials can be used to encapsulate and protect therapeutic agents for targeted and controlled release. Overall, the potential of bio-instructive supramolecular materials to act as mimics of the natural ECM and interact with cells and tissues in a specific and controlled manner makes them a promising area of research with many potential applications in medicine.

1. Introduction

Cells are supported by a class of materials called extracellular matrix (ECM).[1] The ECM forms through the hierarchical assembly of macromolecules into a transient, hydrogel-like fibrous network. It plays both structural and functional roles through mechanical support and biochemical cues, thereby regulating cellular behavior. As a result, it influences vital processes such as tissue growth and regeneration by surrounding, supporting, and giving structure to cells and tissues in the body.

The ECM differs between tissues with each tissue having a specific spatiotemporal distribution of macromolecular components and biochemical motifs.[2–4] These properties are strongly related to specific tissue functions. The difference in composition results in variations in mechanical properties, dynamic properties, and bioactivities. The natural ECM is composed of about 300 interconnected proteins and is well organized in its three-dimensional (3D) structure.[4] These macromolecules can be classified according to the categories of sequence-controlled and sequence-defined polymers, as defined by Lutz.[5]

> **Sequence-controlled polymer:** "A sequence-controlled polymer is a macromolecule in which monomer sequences are controlled to some degree. This control can be absolute but not necessarily. For example, an alternating copolymer prepared by radical chain-growth polymerization is a sequence-controlled polymer, although it is also a nonuniform polymer in which chains have different lengths and slightly different compositions."
>
> **Sequence-defined polymer:** "A sequence-defined polymer is a uniform macromolecule with an exact chain-length and a perfectly defined sequence of monomers. In other words, each monomer unit is at a defined position in the chain. Sequence-defined polymers constitute therefore a subclass of the field of sequence-controlled polymers."

Generally, the majority of macromolecules that constitute the ECM can be classified as sequence-defined polymers due to their

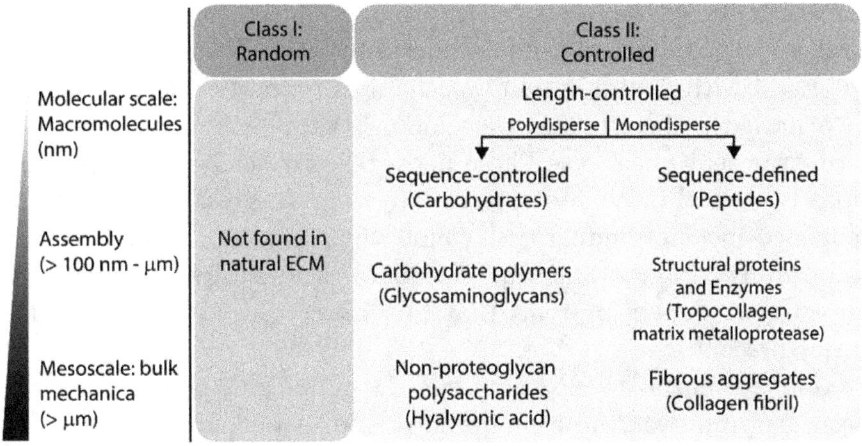

Figure 1. The constituents of the ECM are divided into sequence-controlled and sequence-defined categories.

well-defined monomers and controlled synthesis (**Figure 1**). However, not all constituents of the ECM have low polydispersity. For this reason, some constituents such as glycosaminoglycans do not belong to the class of sequence-defined polymers, but rather sequence-controlled polymers.[6] Macromolecules in the ECM that are not in the sequence-defined subclass usually fulfill a broader purpose. These purposes include attracting water for lubricating, cushioning, spacing, and binding and presenting signaling molecules.[7] While crucial to the functioning of specific tissue ECMs, these purposes lack the specificity of protein interactions. To obtain this level of specificity, nature relies on sequence-specific macromolecules. It is through a combination of both sequence-controlled and sequence-specific components, that is, tissue-specific material properties can be achieved.

The structure and organization of the ECM are the results of supramolecular interactions. Proteins and proteoglycans rely on sequence-specific motifs to interact with each other. Sequence specificity plays a major role in the organization of fibrous proteins in the ECM. These fibrous proteins include collagens,[8] fibronectins,[9] laminins,[10] and vitronectins.[11] Due to their sequence specificity, some fibrous proteins, such as collagens, can form hierarchical assemblies through supramolecular interactions, thereby imbuing the ECM with dynamical and mechanical properties that could not be achieved by macromolecules alone. Others, such as laminins, can form transient networks. These networks can facilitate the interactions between fibrous proteins and glycosaminoglycans through the sequence-specific binding sites found on the former.[12] Furthermore, cell–material interactions can take place through well-defined binding sites on matrix proteins that can be recognized by cell-specific surface receptors.

Interest in the development of ECM-mimicking biomaterials has been growing over the past decades. Model biomaterials facilitate the study of cell–cell and cell–material interactions and also contribute to a greater understanding of the molecular mechanisms of cellular behavior as regulated by ECM properties. Furthermore, they create possibilities to develop ECM mimics for biomedical applications. In nature, large variations can exist in the composition of the ECM, which makes it difficult to predict the effectiveness of a particular biomaterial for a specific tissue engineering purpose. Therefore, adaptability and adjustability of the material to the properties of a specific tissue type are required.

Over the past decade, a shift from traditional static bio-inert implants to covalent polymeric materials has been observed in the field of ECM-mimicking biomaterials.[13] Initially, ECM-mimicking biomaterials consisted of simple covalent polydisperse polymers with physical properties similar to those of a tissue of interest.[14] Bioactivity was introduced into these materials by functionalizing them with bioactive moieties, such as peptides, carbohydrates, and proteins. One drawback of the addition of bioactive cues in covalent polymeric materials is the increase in complexity, making the synthesis of the materials more challenging. Furthermore, including multiple bioactive motifs in a controlled fashion exacerbated this issue. As a result, bioactive covalent polymer materials became very polydisperse and poorly defined systems.

One of the ways nature has developed to avoid this issue is by relying on sequence-defined monomers that form larger-order structures through supramolecular interactions. This makes it possible to introduce a large amount of complexity through different supramolecular building blocks without losing the ability to fine-tune the functioning of the resulting assembly. Recently, synthetic supramolecular chemistry has been introduced to mimic this functionality.

Dynamic hydrogels based on supramolecular chemistry are increasingly preferred as synthetic ECM mimics due to their ability to adapt, self-regulate, and actively sense and respond to environmental stimuli, such as pH, temperature, light, and water uptake.[15,16] This adaptability allows matrix remodeling and cellular activities, such as cell spreading and migrating, to take place within the hydrogel matrix. The bioactive properties of supramolecular materials have been improved through the addition of bioactive motifs, such as sugar motifs and biomimetic peptide sequences.[17]

Within this chapter, we would like to address natural ECM as a well-defined material consisting of many different building blocks that can be mimicked using supramolecular chemistry.

2. Supramolecular materials

Supramolecular chemistry focuses on the study of noncovalent interactions between low-molecular-weight monomers.[18] These monomers can assemble in a variety of ways to form complex supramolecular structures through noncovalent interactions, such as hydrogen bonding, metal–ligand complexation, π–π interactions, and host–guest interactions (**Figure 2**). The interactions between supramolecular monomers are weak, reversible, and highly directional, giving them unique advantages over covalent chemistry. This is particularly useful in the field of biomaterials where the ability to create complex structures without the use of permanent bonds is highly desirable.

The use of supramolecular polymers as biomaterials has several advantages over traditional covalent chemistry. Notably, the dynamic nature of supramolecular interactions allows for stimuli-responsive materials that can rapidly respond to changes such as

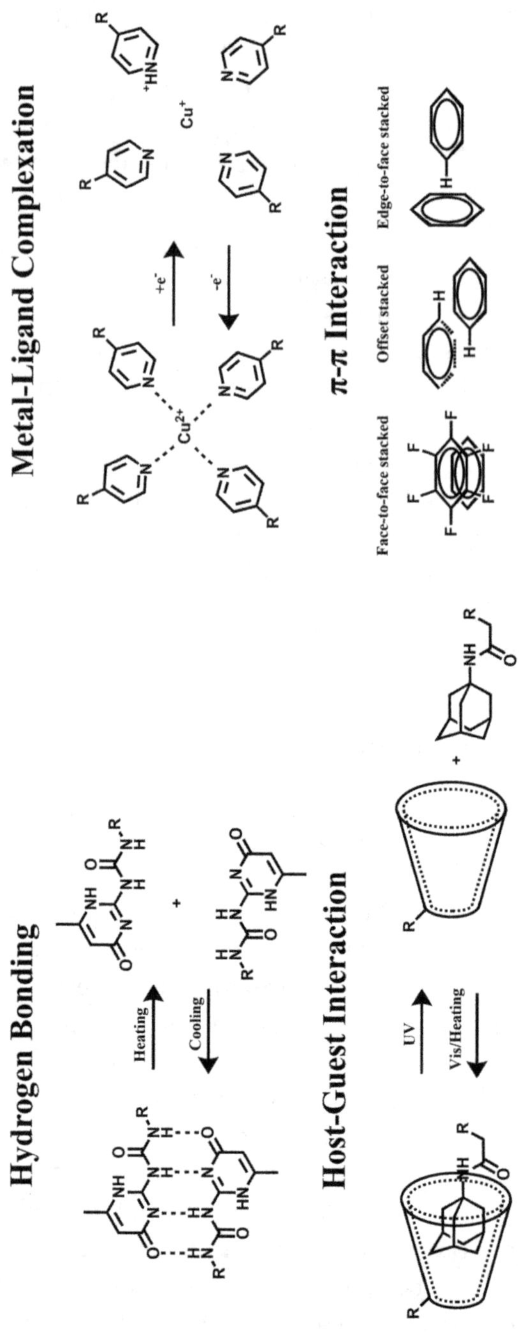

Figure 2. Examples of noncovalent interactions used in supramolecular materials.

heat, pH, and ultraviolet (UV) irradiation.[19] This allows for tunable and targeted assembly and disassembly of materials. The modular nature of supramolecular polymers facilitates the rapid creation of multicomponent systems. This means that the physical and biochemical properties of supramolecular materials can be tuned to the application without relying on additional complex synthetic steps. One of the properties most commonly associated with supramolecular materials is self-healing.[20,21] Due to the materials being held together by noncovalent interactions, often they can be damaged without breaking covalent bonds. As a result, many supramolecular materials can be repaired simply by connecting the broken parts together.

Much like natural macromolecules, supramolecular monomers can be assigned to the sequence-controlled and sequence-defined classes. Supramolecular monomers that consist of a polydisperse polymer functionalized with supramolecular cross-linkers belong to the sequence-controlled class. However, supramolecular monomers, which are well-defined small molecules that assemble into higher-order structures through supramolecular interactions, belong to the sequence-defined class. While there is no rule governing which supramolecular interactions belong to which class, certain interactions are more commonly used as cross-linkers in sequence-controlled supramolecular materials, whereas others are more commonly used to facilitate the creation of higher-order structures in sequence-defined supramolecular materials. Each of these will be discussed in greater detail in the following paragraphs.

Fischer postulated in 1873 that enzyme–substrate interactions take place in a "lock and key" mechanism.[22] This discovery paved the way for research into host–guest interactions. Host–guest interactions are one of the most important noncovalent interactions in biological systems. Prominent examples of host–guest interactions include the aforementioned enzyme–substrate interactions and ligand–receptor interactions found both inside and outside the cell. In these examples, nature relies on the complexity made possible by sequence-defined polymers.

There are several notable examples of host–guest interactions as supramolecular cross-linkers in biomaterials.[23] One such example was developed by Harada et al. in cyclodextrin (CD)-based supramolecular polymers.[24] The unique structure of CDs consists of a hydrophobic cavity and a hydrophilic exterior. This structure facilitates host–guest interactions within the hydrophobic cavity. By functionalizing a material with both cyclodextrins and a hydrophilic guest molecule such as adamantane, they can be used as cross-linkers to form supramolecular polymers (**Figure 3a**). CD-based supramolecular polymers have been developed for applications in biomedicine and industrial materials. In biomedicine, their potential as a drug delivery system has been investigated due to their ability to improve the stability and solubility of therapeutic agents and their ability to selectively target cancer cells.[25]

Coiled-coil forming domains are another example of supramolecular cross-linkers inspired by interactions found in nature. A prominent example comes from Jing et al.[26] They investigated self-assembling peptide–polymer hydrogels designed from the coiled-coil region of fibrin (**Figure 3b**). Here, they demonstrated that by functionalizing polyethylene glycol (PEG) with a peptide consisting of 37 amino acids that were derived from fibrin, they could form triblock gels with desirable physical properties. These gels were shown to gradually dissolve in phosphate-buffered saline solution over a period of several days. Furthermore, coiled-coil supramolecular biomaterials have been shown to trigger specific immune responses in mice.[27,28]

Besides host–guest and coiled-coil interactions, supramolecular cross-linking can also be performed through relatively simple hydrogen bonding interactions. Such interactions can be found in bisurea-(BU-) based materials developed by Koevoets et al.[29] BU molecules have been shown to function as hard blocks within thermoplastic elastomers with various polymeric soft blocks. BU is known to self-sort with matching motifs, based on bifurcated hydrogen bonding

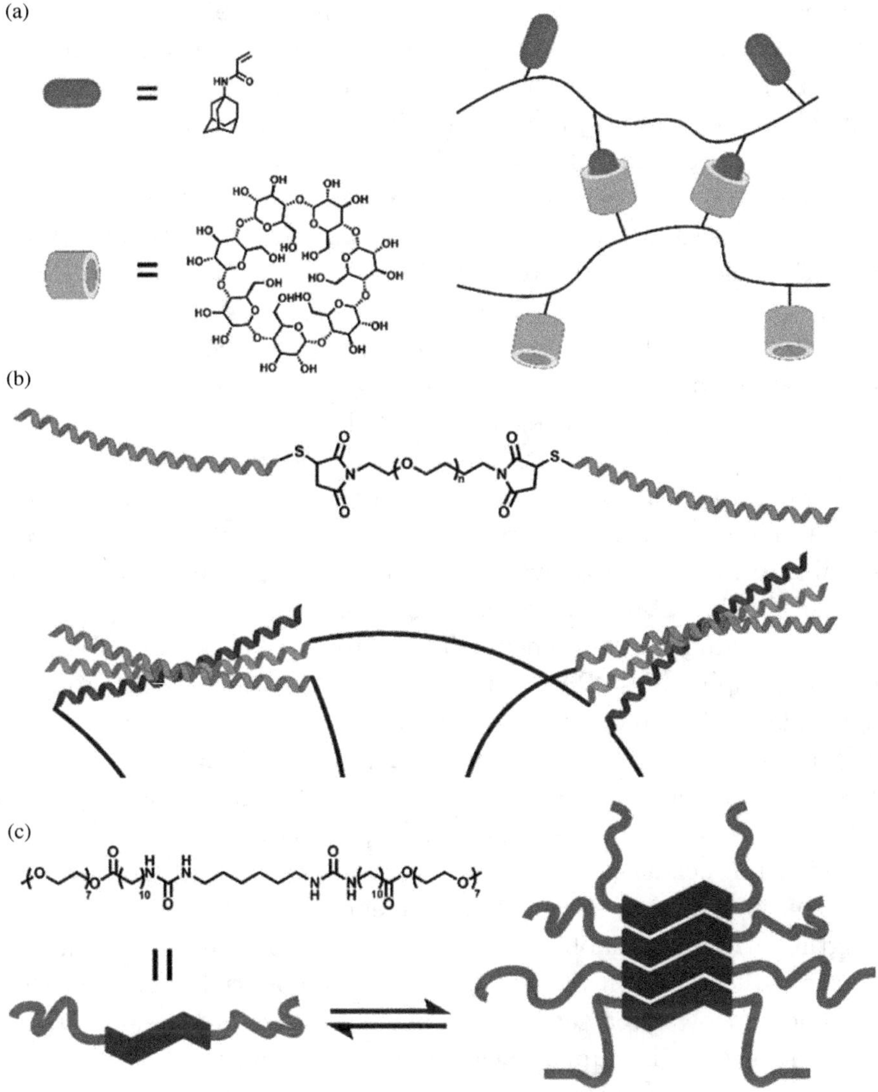

Figure 3. Supramolecular interactions used to cross-link sequence-controlled polymer systems. Polymer chains can be cross-linked using (**a**) host–guest interactions, (**b**) coiled-coil interactions, and (**c**) hydrogen bonding interactions.

between the urea groups (**Figure 3c**). The use of BU molecules shows promising results, as the strength of urea hydrogen bonding is known to exceed that of amide and urethane hydrogen bonds. They have been applied as an injectable, biocompatible, and elastic-segmented copolymer hydrogel for biomedical applications. The copolymer consists of hydrophobic BU hard segments and hydrophilic PEG segments.[30] One of the advantages is the tunability of the mechanical properties to match the properties of biological host tissues.

Another well-known example of a hydrogen bonding group is the ureido-pyrimidinone (UPy) moiety investigated by Dankers et al.[31] These will be discussed at length in the following subsection.

Host–guest, coiled-coil, and directed hydrogen bonding interactions can all be used to cross-link sequence-controlled supramolecular monomers. This gives rise to materials with dynamic and tunable properties. However, they lack the fine-tuned control of sequence-defined supramolecular systems.

One example of sequence-defined supramolecular assemblies is peptide amphiphiles (PAs).[32] PAs are small molecules containing a hydrophobic tail and a hydrophilic head consisting of a short peptide sequence (**Figure 4**). These molecules can self-assemble into nanostructures in aqueous solution through a balance in hydrophobic interactions between the tails and hydrogen bonding interactions between the heads.

This behavior is sequence dependent, as small changes to the length of the hydrophobic tail or selection of different amino acids in the β-sheet forming region can have pronounced effects on the morphology of the resulting supramolecular assemblies.[33] Tuning the sequence of the supramolecular monomer can result in the formation of fibers, vesicles, or monomers in solution.

Furthermore, PAs can be designed for specific applications, such as drug delivery, tissue engineering, and imaging. One approach to achieve this is through the inclusion of bioactive peptide epitopes that can grant biochemical signaling properties to the assembly. This makes it possible for the self-assembled peptide nanostructures to be used as scaffolds for tissue engineering, allowing for the

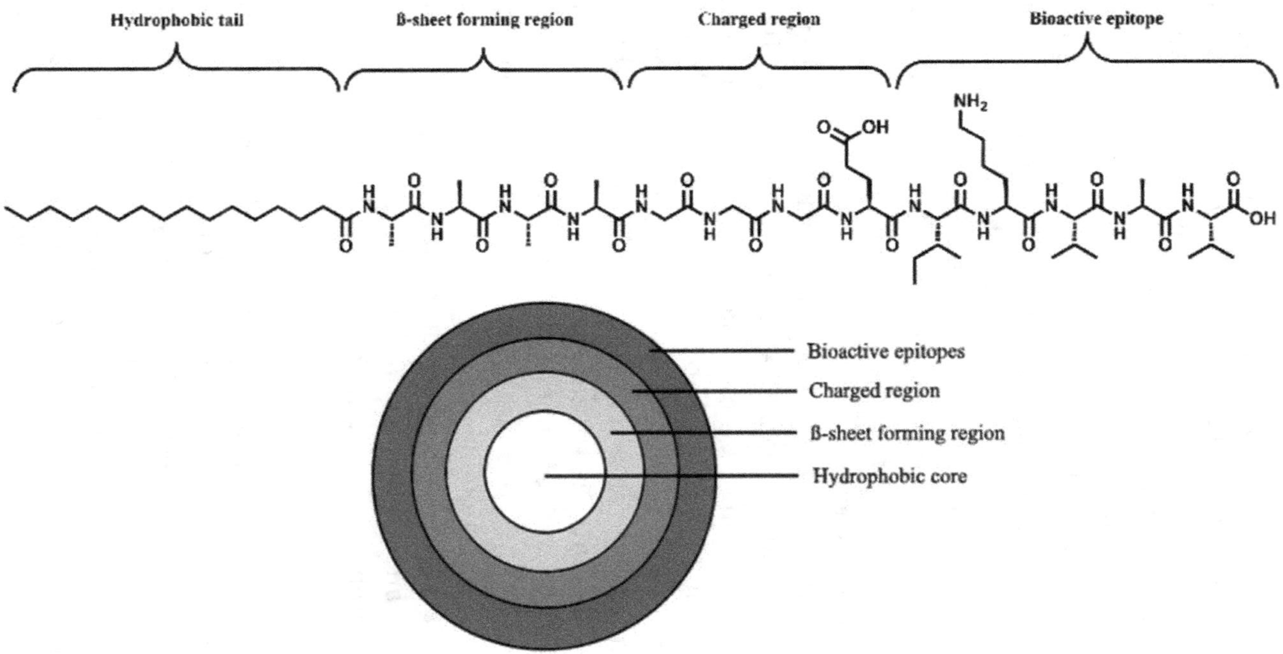

Figure 4. Schematic representation of a peptide amphiphile. Each PA consists of a hydrophobic core and a hydrophilic peptide region. The peptide region can contain β-sheet-forming sequences to increase stability, charged regions to tune dynamicity, and bioactive epitopes to promote cell adhesion.

controlled growth of cells and the formation of functional tissue. Furthermore, the assembly behavior of PAs can be controlled through triggers such as solvent hydrophobicity, temperature, and pH. These triggers make it possible to switch between monomers in solution, micelles, and fibers, thereby facilitating material assembly in a spatiotemporal fashion.

A notable example of PA scaffolds for tissue engineering comes from the work done by Stupp et al. They recently demonstrated that dynamic PAs functionalized with the laminin-111-derived Ile-Lys-Val-Ala-Val (IKVAV) and fibroblast growth factor 2–derived peptide sequences were able to stimulate neurite outgrowth in a rat model,[17] thereby demonstrating the potential of PA hydrogels in tissue regeneration.

A relative newcomer to the field of sequence-defined supramolecular biomaterials is benzene-1,3,5-tricarboxamide (BTA)–based materials developed by Roosma et al.[34] These monomers can form μm-long dynamic supramolecular fibers in solution through a combination of π–π interactions, hydrophobic interactions, and threefold hydrogen bonding (**Figure 5**). Since each BTA monomer has three potential sites for functionalization and fibers can be made by combining multiple monomers, it is possible to create highly customizable materials. BTA fibers have been successfully functionalized with peptides, charged groups, carbohydrates, and DNA, allowing for cellular recognition, intracellular delivery, and protein recruitment. These properties make BTA fibers a prominent candidate for ECM-mimicking supramolecular biomaterials.

It is important at this point to examine the advantages and disadvantages of both sequence-controlled and sequence-defined supramolecular systems. Sequence-controlled supramolecular materials are more reminiscent of their covalent counterparts. This makes them more approachable and easier to characterize with well-understood methods. Furthermore, materials consisting of sequence-controlled polymers with supramolecular crosslinkers are cheaper to produce than their sequence-defined

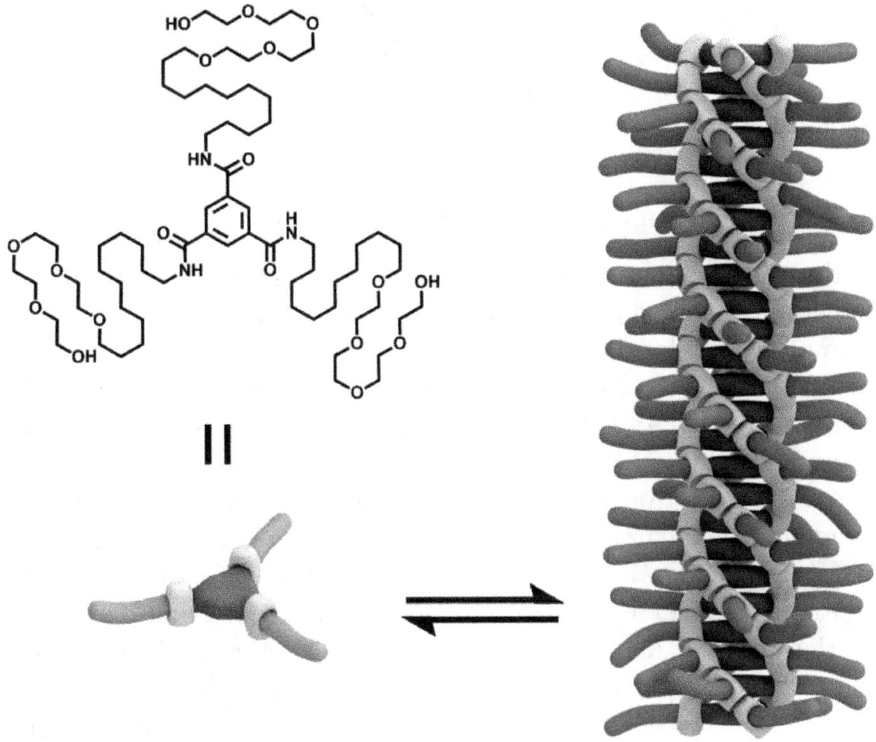

Figure 5. Schematic representation of supramolecular assemblies formed by the sequence-defined BTA supramolecular monomer.

counterparts. However, as the desired complexity of the material increases, so does its cost and difficulty to manufacture. This is especially problematic when the goal is to mimic the inherently complex ECM.

This is where sequence-defined supramolecular monomers have the advantage. While the monomers are often challenging to synthesize and expensive to produce, they can be used to make complex co-assemblies through simple mixing and solvent, temperature, or pH triggers. This makes it possible to create complex materials with both the desired physical and biochemical properties without sacrificing control over the material properties.

Furthermore, the modular nature of supramolecular assemblies facilitates the screening of enormous libraries of co-assemblies to find those with properties that most closely match the desired tissue ECM.

3. Analytical methods on different levels

Within the previous section, the definitions of sequence-defined and sequence-controlled polymers have been set, with a broad variety of examples provided. In order to work with supramolecular assemblies, in-depth molecular analysis at different length scales is required (**Figure 6**). Characterization and analysis are performed at different levels to obtain a complete overview of the biomaterial and its properties, from the microscopic to the macroscopic scale. Generally, the characterization of the synthesized monomers is the first step. Next, an analysis of the supramolecular monomers in solution to obtain a molecular understanding is performed, followed by the creation and characterization of a functional biomaterial, being an assembly in solution, a hydrogel, or a solid scaffold. The final step is the cellular readout, where the material is studied from the point of view of the cell.

Within this section, the different analytical methods will be discussed with a focus on assemblies in aqueous solutions and hydrogel materials.

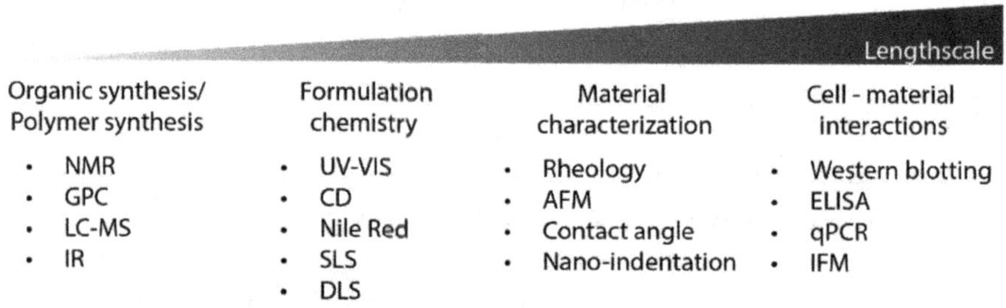

Figure 6. Analytical techniques used to characterize supramolecular biomaterials at different length scales.

3.1. *Analysis of monomers*

The design of macroscopic (bio)materials starts with the design and analysis of their sequence-defined monomers, synthesized via organic chemistry or polymer synthesis. Often, a combination of analysis techniques is used to obtain a complete, detailed overview of molecular properties. First, mass spectrometry (MS) is used to measure the mass-to-charge ratio (m/z) of the molecules present in a sample. These measurements can be used to calculate the exact molecular weight of the sample components. In liquid chromatography–mass spectrometry (LC-MS), the separation of molecules via LC is combined with the mass analysis capabilities of MS. Once the molecular weight of the monomer is confirmed, nuclear magnetic resonance (NMR) can in parallel be used to obtain physical, chemical, and structural information about the molecules. Using NMR, detailed information on the functional groups, dynamics, and 3D structure of molecules in solution and solid state can be obtained. Due to its usefulness in determining the structures of compounds and identifying them, infrared (IR) spectroscopy is one of the most common and widely used spectroscopic techniques. Chemical compounds have characteristic chemical properties due to the presence of different functional groups that can be detected by IR spectroscopy. Upon the transition from monomers to polymers, gel permeation chromatography (GPC) is often used for the measurement of polymer molecular weight, size, and structure. When characterizing sequence-controlled polymers, it is important to consider the dispersity (Đ) in addition to the molecular weight, which both can be determined via GPC.

After successful characterization of the monomers, supramolecular polymers can be formed that accordingly are used for water-based materials, such as assemblies in solution, hydrogels, and hydrogel coatings.

3.2. *Analysis of supramolecular assemblies in solution*

After analyzing and characterizing the monomers, a molecular understanding of the assembly behavior of these supramolecular

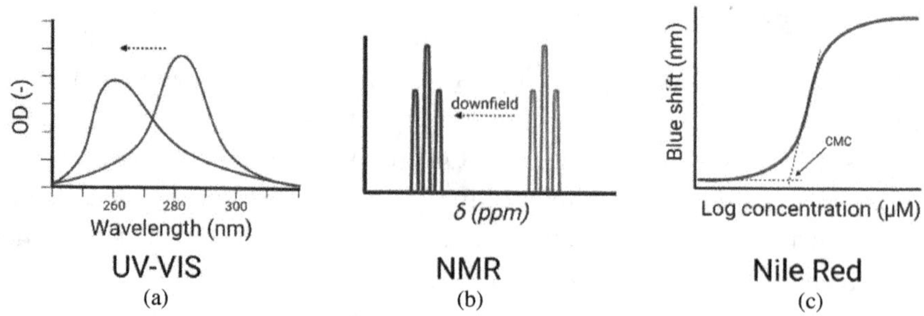

Figure 7. Overview of different analysis techniques for supramolecular assemblies in solution. The formation of supramolecular assemblies from molecularly dissolved building blocks can be observed in all techniques.

monomers (often in aqueous environments) is created by several analysis techniques, such as UV–vis spectroscopy, Nile red assay, NMR spectroscopy, hydrogen deuterium exchange (HDX)-MS, light scattering techniques, and cryogenic transmission electron microscopy (cryoTEM) (**Figure 7**). The results of different techniques to study assemblies need to be combined to obtain a complete overview of supramolecular structures in solution. Within this section, a detailed overview of the different techniques to study supramolecular assemblies in solution will be provided.

UV–vis spectrophotometry can be used to gain insight into the characteristics of a molecule through the way it interacts with light. The technique works by comparing the absorption of a range of wavelengths of light from the UV to the visible spectrum to that of a blank or reference sample. The absorption of light is influenced directly by the sample concentration and composition. In supramolecular molecules, UV–vis spectrophotometry is particularly of interest to study the assembly behavior in a specific solvent. Often, a reference sample is used, where it is known that the molecule is in its molecularly dissolved state. In general, the presence (or absence) of a local maximum (λ_{max}) at a specific wavelength, that is, characteristic for each molecule, can be indicative of the presence of defined supramolecular structures. For example, research by Fry et al. has proven UV–vis spectroscopy to be useful for the analysis of heme coordination by conformational changes in PA assemblies.[35]

The Nile red assay can be used to study (supramolecular) polymer formation through the creation of hydrophobic pockets upon assembly. This assay is often used to determine the critical aggregation concentration (CAC) of a specific molecule. Nile red is a solvatochromic fluorescent dye that intercalates in hydrophobic pockets within a molecule. Nile red has almost no fluorescent properties within a polar solvent. However, upon intercalation in a hydrophobic pocket, it becomes strongly fluorescent. Furthermore, a shift of excitation and emission peaks can be observed based on the solvent polarity. A solvent with a low polarity results in a shift toward lower wavelengths, also known as a blue shift. Generally, a dilution curve of the molecule of interest is measured to determine the CAC. The Nile red assay has been shown to be promising in the analysis of supramolecular dendrimer–surfactant assembly formation and the determination of total lipid content in microalgae.[36,37] In addition to Nile red, thioflavin T (ThT) is another fluorescent dye that can be used to gain insight into supramolecular assembly behavior. ThT displays enhanced fluorescence and a characteristic red shift upon binding to beta-sheet-rich structures, such as amyloid fibrils. This assay is useful in supramolecular chemistry as beta-sheets are a common supramolecular interaction. Literature shows several examples where the ThT assay has been used to monitor a-synuclein aggregation,[38] study kinetic aspects of fibrillation,[39] identify selective compounds for the stabilization of G-quadruplex structures,[40] and investigate structure–property–activity correlation for self-assembling peptides forming amyloids.[41]

HDX-MS can be used to investigate the dynamic behavior of supramolecular assemblies. By exposing the assemblies to a deuterated solvent, the exchange of hydrogen atoms for deuterium can be measured as a factor of time using MS. The rate at which this exchange occurs depends on the ability of the solvent to interact with hydrogen atoms on the supramolecular monomer. As such, monomers in solution will be rapidly deuterated, whereas hydrogen atoms buried in a supramolecular assembly will require more time to exchange, if they are able to exchange at all. HDX-MS has previously been used to investigate exchange processes in supramolecular (BTA) polymers.[42,43] More recently, it has been investigated that two types of hydrogen-deterium (H/D) exchange can occur for BU molecules: exchange due to monomer release and solvent penetration within less-ordered regions in the polymer.

Light scattering techniques can be used to gain insight into the morphology of supramolecular assemblies in solution. To this end, dynamic light scattering (DLS) and static light scattering (SLS) are the two most commonly employed techniques. DLS and SLS optical techniques provide information about the hydrodynamic size and the molecular weight of particles in solution, respectively. Light scattering has, among other things, been used to determine the size, shape, and aggregation number for supramolecular assemblies of chromophores,[44] kinetics of supramolecular copolymerization in water,[45] and single-chain polymeric nanoparticles.[46]

While NMR in the previous section has been discussed for the analysis of (supramolecular) monomers, it is also a very suitable technique for studying the formation of supramolecular assemblies. In 2000, Pons and Millet reviewed the use of NMR spectroscopy techniques as essential for understanding supramolecular structures.[47] NMR can be used to study events occurring on a wide time scale. The technique is able to detect changes as fast as in the picosecond range. For example, the host control of guest exchange within molecules such as cucurbituril and cavitand has been studied by NMR spectroscopy.

In line with NMR transmission, IR spectroscopy can also be used to characterize the internal structure of nanofibers.[48] Here, they present the spectroscopic analysis of PAs, suggesting the peptide amphiphile cylindrical nanofibers to be composed of betasheets oriented parallel to the long axis of the fibers.

Finally, cryo-TEM involves the examination of samples that have been cooled to cryogenic temperatures. It can be used to visualize the morphology of macromolecules at a nanometer scale. Furthermore, it allows for the characterization of assembled structures, such as fibers, at a micrometer scale. Cryo-TEM also allows in-depth analysis of the structures obtained, being able to discriminate single, double, and triple helices.[49] Cryo-TEM is an especially powerful method for uncovering the structure of soft nanostructured materials.[50]

In addition to the analysis of supramolecular assemblies in solution, these techniques can also be used to create additional insights into the assembly behavior upon changes in solvent, pH, or temperature. For BTA molecules, for example, UV–vis spectroscopy has been used as a tool to study the assembly behavior.[51] Often, the spectroscopic properties of a molecule are evaluated in a solvent in which the molecule is known to be molecularly dissolved (e.g., acetonitrile). Solutions in water can then be analyzed at different temperatures to investigate the self-assembly behavior and lower critical solution temperature (LCST). Research by Moyer et al. has shown that the CAC can increase upon a decrease in pH for specific PAs, while no effect is observed for other PAs.[52]

While the techniques described here are key to understanding and designing supramolecular monomers, forming different supramolecular assemblies in solution, they only describe the behavior in dilute, aqueous solutions at the micro- and mesoscale. In order to formulate ECM-mimicking biomaterials, the macroscopic properties of supramolecular materials have to be investigated, and the building blocks must allow for the formation of networks, forming either hydrogels or solid materials. The supramolecular assemblies based on sequence-defined structures, such as PAs and UPy monomers, often do not form hydrogels in solution. Previous research has shown that the addition of cross-linkers is necessary to form supramolecular materials. These cross-linkers can either be present within the molecule or added as an additional molecular building block. In the next section, the formulation of hydrogel materials will be discussed.

3.3. *Analysis of material properties*

After obtaining an in-depth overview of the molecular properties of the supramolecular monomers, it is important to analyze the properties of the materials they form. They can be used to formulate functional materials, ranging from hydrogels to coatings and scaffolds. The formulation protocol of different materials plays a crucial role, as it can have a profound effect on the morphology of the supramolecular polymers. Among others, temperature, solvent, pH, and concentration have shown to play an important role in the formulation of supramolecular assemblies and will therefore also affect the formulation of hydrogels or solid materials.

The mechanical properties of hydrogels are generally studied using rheology. In short, rheology uses mechanical force to characterize the deformation and flow behavior of a material. These properties are expressed as the elastic modulus (G', also known as the storage modulus) and the viscous modulus (G'', also known as the loss modulus). The elastic modulus reflects the elastic behavior of a material when deformed, describing the solid-state behavior of the sample. The viscous modulus reflects the flow of a material while it is deformed and describes the liquid-state behavior. Viscoelastic solids show $G' > G''$, so a higher storage modulus than loss modulus, indicating interactions inside the materials. These interactions include both chemical bonds and physical–chemical interactions. On the other hand, viscoelastic liquids with $G'' > G'$ have higher loss modules than storage modules because there are no strong bonds between the individual molecules in most of these materials.

The hydrophobic and hydrophilic properties of a surface can be determined using water contact angle measurements. By definition, a surface is hydrophobic when its static water contact angle Θ is $>90°$ and is hydrophilic when Θ is $<90°$, with the contact angle defined as the angle between a liquid surface and a solid surface at the point they meet. The water contact angle is an important measure of cell adhesion to materials, as cells generally are more likely to adhere to and grow on hydrophilic surfaces in comparison to hydrophobic surfaces.

To characterize the shape of assemblies and the surface topology of materials, atomic force microscopy (AFM) can be used. In AFM, the surface of a sample is scanned with a physical probe. This probe consists of a tip attached to a cantilever that reflects a laser, resulting in nanometer-resolution images of the material. Manipulation of the cantilever using defined force allows AFM to be used to determine the physical properties of a material, such as Young's modulus. AFM is particularly interesting for the characterization of solid materials and scaffolds. Kan et al. have confirmed using AFM imaging that the nanofiber structure of UPy molecules gradually disappears upon modifying the six positions of the pyrimidone, specifically upon increasing the steric hindrance of the substitution at this specific position.[53]

In addition to the surface topology of materials, the hardness, plasticity, and elastic modulus of materials are also important features for material characterization. Nanoindentation is one of the most common methods used to determine the mechanical characteristics of material. Similar to AFM, nanoindentation is performed on small areas of the material, bringing the advantage that it can determine local characteristics and provide spatial discrimination in heterogenous materials.

Often, a combination of the aforementioned techniques is required to obtain a solid understanding of the supramolecular (bio)material before the material can be used for cellular studies.

3.4. *Cellular readout*

Once a solid understanding of the characteristics of the supramolecular assemblies has been obtained, it is also important to gain insights into their effects at a cellular level. To this end, several techniques can be applied. These techniques are used to investigate material toxicity and cell–material interactions through cell morphology, protein expression, and gene expression.

Understanding the cytotoxic effects of supramolecular assemblies on cells is vital to their use as bio-instructive materials. These effects can be undesirable, such as when the material is intended for use as a regenerative scaffold. However, it is also possible for these toxic effects to be desirable. Examples of such scenarios include antimicrobial supramolecular assemblies and anticancer therapeutics that rely on targeted delivery of supramolecular monomers and on-site assembly behavior. In most cases, the toxic effects of a supramolecular material can be assessed through common techniques such as metabolic activity assays and immunofluorescence microscopy or flow cytometry analysis of apoptosis markers. It is also becoming more common to use live-cell microscopy to monitor cytotoxic effects. This makes it possible to observe apoptosis in real time and gain insights that would be lost in static techniques.

In cases where biomaterials are used as regenerative scaffolds, cell morphology is one of the most common readouts. As cells are prone to shape themselves according to their environment, information about their shape can give crucial insight into cell–material interactions. Cells exposed to stiffer matrices spread out more, thereby displaying a larger surface area, than cells on softer matrices. However, in supramolecular materials, this relationship is not always accurate. Due to the dynamic presentation of biochemical cues, the material behavior can be vastly different between the molecular level and the bulk material properties. Therefore, the balance between bulk material properties and monomer exchange rate can have a profound impact on cell fate. Furthermore, certain cell types such as neuronal cells and keratocytes will display very typical phenotypes as a result of being cultured on or in materials with the right physical and biochemical cues.

A common method to visualize cell morphology is to stain fixed cells for their cytoskeletal fibers, such as actin and vimentin. However, further insight into cell–material interactions can be obtained by staining for specific antigens, for example, those found in focal adhesions. This approach can also be applied to monitoring cell differentiation by staining for proteins that are known markers of specific cell types or pluripotency. Visualizing these proteins generally results in a qualitative readout.

Other techniques can be applied to obtain quantitative insights into protein expression as a result of cell–material interactions.

Common techniques used to this end include enzyme-linked immunosorbent assays (ELISA) and western blotting. Both of these techniques require cells to be removed from their material to be analyzed. Supramolecular materials are especially suited to this approach, as assembly behavior is often triggered and reversible. This means that the cells can be removed from the material without exposing them to stimuli that may otherwise affect the results. ELISA can be used to gain quantitative insights into the expression of specific proteins. Western blotting relies on a combination of protein separation through gel electrophoresis and targeted identification of proteins through antibody staining. This makes it possible to compare protein expression resulting from cell–material interactions on both a general and specific protein level. Insights obtained from these techniques can also be supported by readouts at a gene level. Techniques such as quantitative polymerase chain reaction (qPCR) can be used to this end to quantify the expression of mRNA related to well-known markers of cell processes.

Altogether, these techniques provide valuable insight into cell–material interactions at different levels. Some of these techniques are also especially suited to screening libraries of supramolecular materials. This topic will be discussed in greater detail in Section 5.

4. Ureido-pyrimidinone-based supramolecular biomaterials

In our group, we specialize in supramolecular materials based on the UPy moiety, developed by the groups of Meijer and Dankers. UPys were originally employed as a supramolecular crosslinker for sequence-controlled polymer systems due to their ability to form a quadruple hydrogen bond. This hydrogen bond has a binding constant of $6 \times 10^7\ M^{-1}$ and a lifetime of approximately 1 second in chloroform.[54] Under certain conditions, the UPy moiety exists as a mixture of two tautomers, the keto and enol forms (**Figure 8**).[55] The keto form demonstrates the highest dimerization constant due to its distribution of hydrogen bond donors and acceptors.

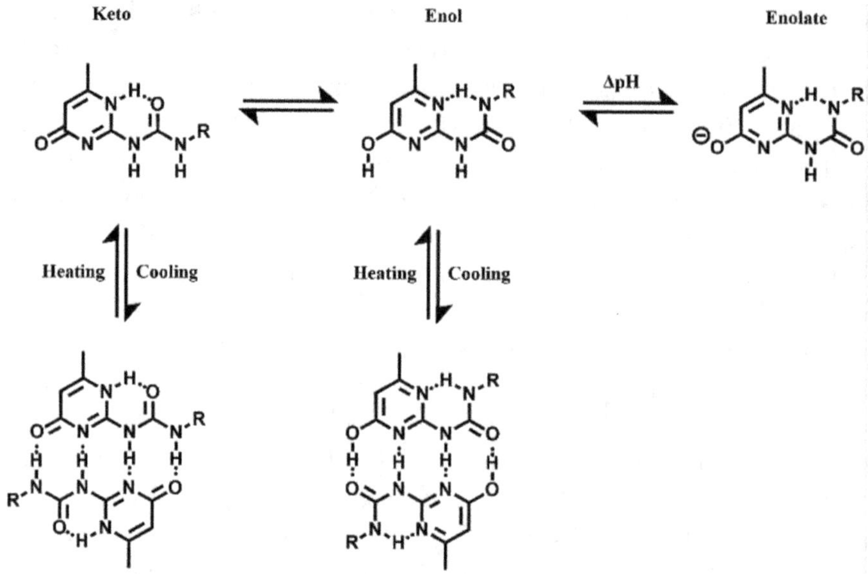

Figure 8. UPy tautomeric forms. The keto and enol forms can dimerize through quadruple hydrogen bonding, but the enolate form cannot.

One of the unique features of the UPy moiety is that its assembly behavior is tunable through both temperature and pH triggers.[56] At high temperatures, the hydrogen bonds are less stable, resulting in rapid reshuffling of the supramolecular assemblies. Furthermore, at basic pH, the enol tautomer can be deprotonated to form the enolate, thereby hindering the formation of hydrogen bonds.[57] This effect can be monitored using UV–vis spectroscopy. The ketone form capable of quadruple hydrogen bonding has a characteristic λ_{max} at 260 nm. The appearance of this maximum in the absorbance spectrum is therefore indicative of supramolecular assembly behavior.

By functionalizing both ends of a sequence-controlled polymer chain, such as polycaprolactone (PCL), a sequence-controlled supramolecular monomer is formed.[58] Materials made out of these monomers show tunable, dynamic, and self-healing properties. Furthermore, peptide-functionalized UPys can be incorporated into the material by simple mixing. Functionalizing a peptide with a UPy moiety results in it releasing from the material more gradually than a nonfunctionalized peptide.

Copolymers consisting of bifunctional UPy-modified polycaprolactone (PCLdiUPy) and peptide-functionalized UPys can be formed into various solid materials including, but not limited to, solvent-cast films, electrospun fibers, and meshes. Materials of this type have been used with promising results in both animal models and clinical trials, where they were employed as dialysis shunts,[59] heart valves,[60] and pelvic organ prolapse meshes.[61] However, more recently, the move has been made from solid materials made of sequence-controlled supramolecular monomers to hydrogels made of sequence-defined supramolecular monomers. It was demonstrated through cryo-TEM that polymer chains functionalized with UPy moieties formed one-dimensional (1D) fibers.[62] This led to the conclusion that UPy dimers were able to form stacks through π–π interactions.

Further optimization of the system resulted in a sequence-defined UPy monomer that consisted of a single UPy moiety covalently functionalized with a urea group flanked by two hydrophobic spacers.[63] These monomers are able to form robust, dynamic 1D fibers through a combination of hydrogen bonding, π–π interactions, and hydrophobic interactions. Furthermore, functionalizing the monomer with a PEG chain made the monomers water soluble. By combining the water solubility of PEG and the hydrophobic core of the UPy fiber, it is possible to form μm-long UPy fibers in water. These assemblies have been characterized using various techniques.

The hydrophobic alkyl spacers in the UPy molecules create a hydrophobic pocket upon the formation of linear supramolecular polymer stacks along the stacking axis. Therefore, the assembly of UPy monomers into fibrous structures can result in the intercalation of the solvatochromic Nile red dye into hydrophobic regions, causing a blue shift upon assembly. The relationship between the concentration and the blue shift yields the critical aggregation concentration. NMR has also been used to study the assembly of UPy monomers into supramolecular structures. A downfield shift of the

NH protons is observed upon increasing concentrations. This shift corresponds to the formation of UPy dimers as the environment near the urea bond changes. No difference in chemical shift is observed for protons that are not involved in hydrogen bonding between the two monomers. In addition, an up-field shift of alkylidene protons suggests the formation of larger aggregates, either micelles or fibers, in which both NH protons of the UPy molecule are involved in hydrogen bonding. Cryo-TEM has also contributed significantly to the understanding of UPy assemblies in both the analysis of fiber thickness and length upon modifications in either the sequence-defined supramolecular monomer or the introduction of UPy co-assemblies.

UPy fibers can be cross-linked by forming a co-assembly with monofunctional UPy monomers and bifunctional UPy monomers.[63] The resulting co-assembly forms a water-based polymeric network, also known as a hydrogel. The formulation of UPy-based supramolecular hydrogels requires several factors to be taken into account. First, the molecules need to be dissolved by heating the sample, dissolving the sample at basic pH, or a combination of both. Afterward, the sample is re-adjusted to neutral pH and room temperature. Finally, the individual components are mixed to form a co-assembly at room temperature. Previously described research by Diba et al.[63] has shown that the ratio of monovalent to bivalent UPy molecules affects the dynamic properties of the hydrogel (**Figure 9**).

UPy hydrogels formed in this fashion have several desirable properties. The mechanical properties of the gel can be modulated by tuning the ratio of monofunctional to bifunctional UPy compounds and by tuning the weight per volume percent of the UPy hydrogel. As a result, UPy hydrogels with mechanical properties similar to those of various natural ECMs can be formulated.

Furthermore, functionalized UPy compounds can be easily introduced into the hydrogels, resulting in, for example, bioactive cell-adhesive materials. By tuning the sequence of the UPy supramolecular monomer or the ratio of the components, the dynamics of the incorporation of functionalized UPys can be changed. As a

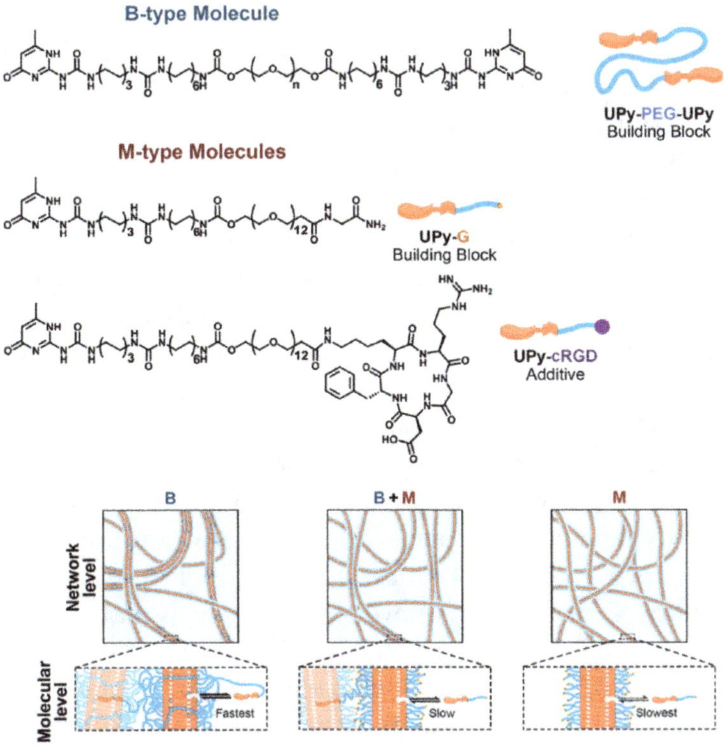

Figure 9. Fundamentals of the UPy–hydrogel system. UPy hydrogels consist of co-assemblies of monofunctional (M-type) molecules, including functional compounds such as cell-adhesive cRGD and nonfunctional compounds such as glycine. Assemblies are cross-linked by bifunctional (B-type) molecules. The ratio of M- to B-type molecules determines the dynamic behavior of the assembly at a molecular level.

result, functionalized UPys can be incorporated into supramolecular UPy fibers in a robust fashion or a dynamic fashion, depending on the purpose of the material. Several well-known cell-adhesive peptide epitopes have been used to functionalize UPy hydrogels, including fibronectin-derived GRGDS and PHSRN; laminin-111-derived IKVAV, YIGSR, AG32, and AG73; and collagen-derived GFOGER. Prominent applications of these hydrogels will be discussed in the following paragraphs.

Diba et al. investigated the effect of the dynamic properties of UPy hydrogels on cell adhesion.[63] They demonstrated that the exchange rate between UPys in an assembly and UPys in solution was dependent on the ratio between mono- and bifunctional UPys. Interestingly, increasing the amount of bifunctional cross-linkers caused the exchange rate to increase. As a result of this, hydrogels with an increasing amount of bifunctional cross-linker were softer and showed more complete stress relaxation. The exchange rate also influenced how cells interacted with cell-adhesive peptide-functionalized UPys incorporated into the hydrogel. To demonstrate this, the fibronectin-derived cyclic Arg–Gly–Asp (cRGD) peptide-functionalized UPy was used. More dynamic hydrogels, ergo hydrogels with a higher amount of bifunctional UPy cross-linker, showed a lower level of cell adhesion and spreading than their less dynamic counterparts. This indicated the importance of robustly incorporating peptide-functionalized UPys into a material. Furthermore, a method was developed for encapsulating cells and organoids in a 3D UPy hydrogel. This enabled the use of supramolecular UPy hydrogels as a modular platform for organoid culture.

Schotman et al.[64] investigated the use of UPy hydrogels as a cardiac drug delivery system. To increase the effective delivery of drugs to such a volatile organ, it is vital to increase their retention time. However, first, a robust method for monitoring the said retention time had to be created. To that end, a bioactive hydrogel for radioactive imaging in a large animal model was developed. It was demonstrated that incorporating a UPy moiety functionalized with a chelator complexed with radioactive indium-111 (UPy-DOTA-^{111}In) into a UPy–hydrogel made it possible to image the location of the hydrogel over several hours (**Figure 10**). Furthermore, incorporating the cell-adhesive RGD peptide into the labeled hydrogel further increased its retention time. The ability to track the location of the hydrogel also made it possible to visualize off-target localization. It was demonstrated that incorporating RGD decreased the distribution of the hydrogel to other organs. Importantly, this new method of monitoring hydrogel delivery allowed for dose

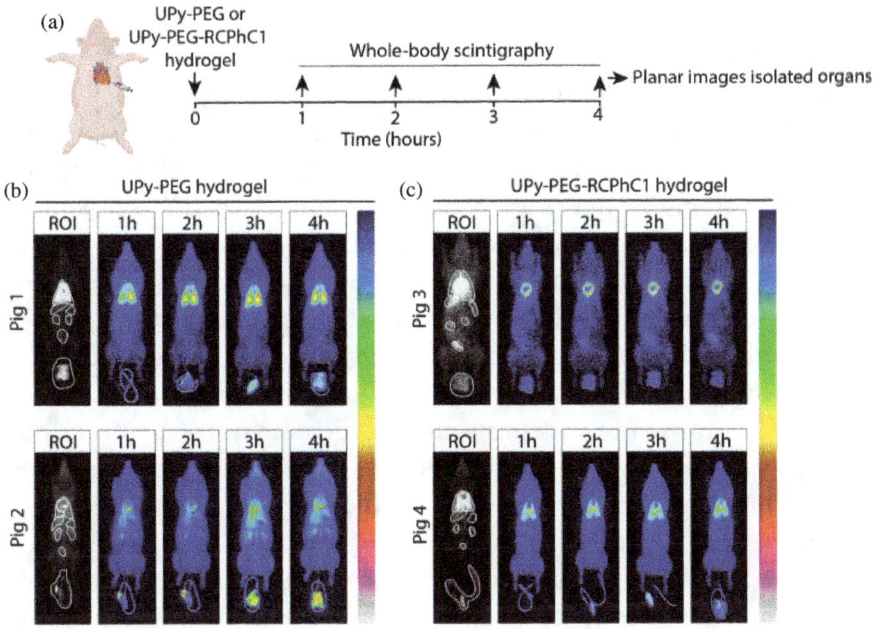

Figure 10. Radioactive imaging of UPy-DOTA-111 in functionalized UPy-hydrogels injected into porcine hearts. Mixing cell-adhesive RGD peptide-functionalized UPys into the hydrogel caused prolonged retention time and decreased off-target localization.

quantification. This can lead to an increased understanding of dose optimization and drug effectiveness.

As the UPy hydrogel system is state of the art, new applications are still in the works. In particular, experiments are ongoing in screening functionalized UPys and incorporating antimicrobial properties into UPy materials. These topics will be addressed at greater length in Section 5.

5. Epilogue

A major advantage of supramolecular chemistry in biomaterials research is its compatibility with high-throughput screening approaches. Producing materials with novel biochemical signals and tuning the mechanical properties of supramolecular materials are facilitated through their modular nature. As a result, immense libraries of materials can be rapidly fabricated.

Proof-of-concept studies of this kind have already been performed on solid materials based on BU chemistry.[65] Here, libraries of ECM–protein-derived short peptide-functionalized BU compounds were screened for their ability to induce functional monolayer formation in the renal epithelial cells. A library of 32 unique material compositions was formulated from a set of six functionalized BU compounds. This proof-of-concept study laid the foundation for later research.

A follow-up study was performed to demonstrate the breath of screening possible with supramolecular materials.[66] Here, a three dimensional design space was employed (**Figure 11**). Three different cell lines were exposed to two kinds of supramolecular materials (UPy and BU) functionalized with two cell-adhesive additives (cRGD and catechol). This resulted in 36 unique combinations, each of which provided insight into functional group presentation on supramolecular materials and the resulting cell–material interactions.

Materials based on supramolecular chemistry have shown to be eminently suitable for large studies and screening purposes, as the supramolecular monomers can be used in a mix-and-match manner yielding the desired material composition. Through this process, a feedback loop between chemistry, physical properties, and cellular

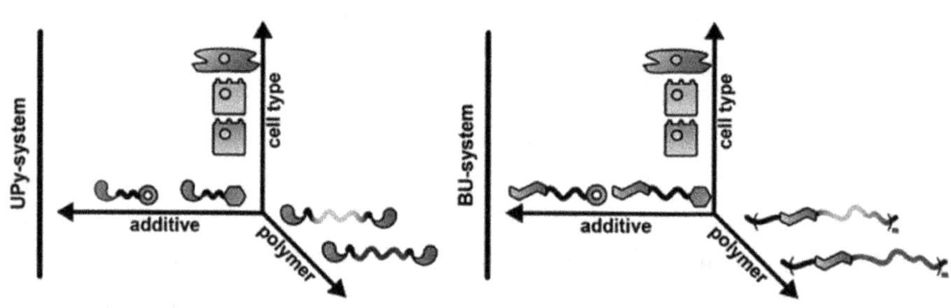

Figure 11. Three-dimensional design space for the screening of supramolecular biomaterials.

response is created to obtain a material with optimal and desired characteristics and output.

So far, the application of supramolecular materials has mainly been in tissue engineering and regenerative medicine using (human) eukaryotic cells. Supramolecular materials, however, also show great potential for combating bacterial infections.[67,68] Antimicrobial resistance (AMR) occurs when bacteria, fungi, and parasites change over time and no longer respond to medicines, making infections harder to treat and increasing the risk of disease spread, severe illness, and death. Resistance to antibiotics develops due to long-term or repeated use of the same class of antibiotic.

Over the past few years, natural antimicrobial peptides (AMPs) have attracted considerable interest as a new class of antimicrobial drugs. Among others, AMPs have a broad activity spectrum and are relatively selective toward their targets. An additional advantage of AMPs, due to their prevalent mechanism of action, which relies on their ability to permeabilize and/or to form pores within cytoplasmic membranes, is the high potential to act also on slow-growing or nongrowing bacteria.

However, the spatiotemporal delivery of AMPs is challenging. Therefore, the combination of AMPs with supramolecular materials could offer great potential. Previous research by Schmitz et al. has shown promising results. The UPy monomers were functionalized with AMPs to obtain UPy–AMP monomers. Due to the positive charge on the AMP, the self-assembling resulted in positively charged, stable, short fibers or rod-like UPy–AMP nanostructures. The antibacterial UPy–AMP assemblies demonstrated enhanced antimicrobial activity against both Gram-positive *Staphylococcus aureus* and Gram-negative *Escherichia coli* with modulable cytotoxicity.

Supramolecular AMP materials should be designed in such a way that cells survive while bacteria are killed. This might, for example, be achieved by a combination of cell-adhesive peptide sequences and AMPs. Ultimately, via in vivo and in vitro studies, this will lead to animal studies and clinical trials to make site-specifically designed supramolecular materials capable of promoting healthy regeneration and preventing systemic exposure to antimicrobial medicine.

References

1. Yue B. Biology of the extracellular matrix: An overview. *J Glaucoma.* 2014;23(8):S20–S23.
2. Xin T, Greco T, Myung P. Hardwiring stem cell communication through tissue structure. *Cell.* 2016;164(6):1212–1225.
3. Gattazzo F, Urciuolo A, Bonaldo P. Extracellular matrix: A dynamic microenvironment for stem cell niche. *Biochim Biophys Acta.* 2014;1840(8):2506–2519.
4. Nicolas J, Magli S, Rabbachin L, Sampaolesi S, Nicotra F, Russo L. 3D Extracellular matrix mimics: Fundamental concepts and role of materials chemistry to influence stem cell fate. *Biomacromolecules.* 2020;21(6):1968–1994.
5. Lutz JF. Defining the field of sequence-controlled polymers. *Macromol Rapid Commun.* 2017;38(24):1700582.
6. Pomin VH, Mulloy B. Glycosaminoglycans and proteoglycans. *Pharmaceuticals.* 2018;11(1):27.
7. Menezes R, Vincent R, Osorno L, Hu P, Arinzeh TL. Biomaterials and tissue engineering approaches using glycosaminoglycans for tissue repair: Lessons learned from the native extracellular matrix, *Acta Biomater.* 2023;163:210–227.
8. Ricard-Blum S. The collagen family. *Cold Spring Harb Perspect Biol.* 2011;3(1):2011.
9. Dalton CJ, Lemmon CA. Fibronectin: Molecular structure, fibrillar structure and mechanochemical signaling. *Cells.* 2021;10(9):2443.
10. Aumailley M. The laminin family. *Cell Adh Migr.* 2013;7(1):48–55.
11. Felding-Habermann B, Cheresh DA. Vitronectin and its receptors. *Curr Opin Cell Biol.* 1993;5(5):864–868.
12. Olsen BR. Matrix molecules and their ligands. In: Lanza R, Langer R, Vacanti J, editors. *Principles of Tissue Engineering.* 2nd ed. Oxford: Elsevier; 2014. p. 189–208.
13. Ciccone WJ, Motz C, Bentley C, Tasto JP. Bioabsorbable implants in orthopaedics: New developments and clinical applications. *J Am Acad Orthop Surg.* 2001;9(5):280–288.
14. Kalirajan C, Dukle A, Nathanael AJ, Oh TH, Manivasagam G. A critical review on polymeric biomaterials for biomedical applications. *Polymers.* 2021;13(17):3015.
15. Lim JYC, Lin Q, Xue K, Loh XJ. Recent advances in supramolecular hydrogels for biomedical applications. *Mater Today Adv.* 2019;3:100021.

16. Omar J, Ponsford D, Dreiss CA, Lee TC, Loh XJ. Supramolecular hydrogels: Design strategies and contemporary biomedical applications. *Chem Asian J.* 2022;17(9):e202200081.
17. Álvarez Z, Kolberg-Edelbrock AN, Sasselli IR, Ortega JA, Qiu R, Syrgiannis Z, et al. Bioactive scaffolds with enhanced supramolecular motion promote recovery from spinal cord injury. *Science.* 2021;374(6569):848–856.
18. Lehn J. *Supramolecular Chemistry.* Weinheim: VCH Publishers; 1995.
19. Yan X, Wang F, Zheng B, Huang F. Stimuli-responsive supramolecular polymeric materials. *Chem Soc Rev.* 2012;41(18):6042–6065.
20. Brunsveld L, Folmer BJB, Meijer EW, Sijbesma RP. Supramolecular polymers. *Chem Rev.* 2001;101(12):4071–4097.
21. Sinawang G, Osaki M, Takashima Y, Yamaguchi H, Harada A. Supramolecular self-healing materials from non-covalent crosslinking host–guest interactions. *Chem Commun.* 2020;56(32):4381–4395.
22. Fischer E. Einfluss der configuration auf die wirkung der enzyme. *Ber Dtsch Chem Ges.* 1894;27(3):2985–2993.
23. Mohamadhoseini M, Mohamadnia Z. Supramolecular self-healing materials via host-guest strategy between cyclodextrin and specific types of guest molecules. *Coord Chem Rev.* 2021;432:213711.
24. Harada A, Takashima Y, Yamaguchi H. Cyclodextrin-based supramolecular polymers. *Chem Soc Rev.* 2009;38(4):875–882.
25. Zou L, Braegelman AS, Webber MJ. Spatially defined drug targeting by in situ host-guest chemistry in a living animal. *ACS Cent Sci.* 2019;5(6):1035–1043.
26. Jing P, Rudra JS, Herr AB, Collier JH. Self-assembling peptide-polymer hydrogels designed from the coiled coil region of fibrin. *Biomacromolecules.* 2008;9(9):2438.
27. Conticello V, Hughes S, Modlin C. Biomaterials made from coiled-coil peptides. *Subcell Biochem.* 2017;82:575–600.
28. Rudra JS, Tripathi PK, Hildeman DA, Jung JP, Collier JH. Immune responses to coiled coil supramolecular biomaterials. *Biomaterials.* 2010;31(32):8475–8483.
29. Koevoets RA, Versteegen RM, Kooijman H, Spek AL, Sijbesma RP, Meijer EW. Molecular recognition in a thermoplastic elastomer. *J Am Chem Soc.* 2005;127(9):2999–3003.

30. Pawar GM, Koenigs M, Fahimi Z, Cox M, Voets IK, Wyss HM. Injectable hydrogels from segmented PEG-bisurea copolymers. *Biomacromolecules.* 2012;13(12):3966–3976.
31. Dankers PYW, Harmsen MC, Brouwer LA, Van Luyn MJA, Meijer, EW. A modular and supramolecular approach to bioactive scaffolds for tissue engineering. *Nat Mater.* 2005;4(7):568–574.
32. Hendricks MP, Sato K, Palmer LC, Stupp SI. Supramolecular assembly of peptide amphiphiles. *Acc Chem Res.* 2017;50(10): 2440–2448.
33. Pashuck ET, Cui H, Stupp SI. Tuning supramolecular rigidity of peptide fibers through molecular structure. *J Am Chem Soc.* 2010; 132(17):6041–6046.
34. Roosma J, Mes T, Leclère P, Palmans ARA, Meijer EW. Supramolecular materials from benzene-1,3,5-tricarboxamide-based nanorods. *J Am Chem Soc.* 2008;130(4):1120–1121.
35. Solomon LA, Kronenberg JB, Fry HC. Control of heme coordination and catalytic activity by conformational changes in peptide-amphiphile assemblies. *J Am Chem Soc.* 2017;139(25):8497–8507.
36. Orr V, Rehmann L. Improvement of the Nile Red fluorescence assay for determination of total lipid content in microalgae independent of chlorophyll content. *J Appl Phycol.* 2015;27(6):2181–2189.
37. WatkinsDM, Sayed-Sweet Y, Klimash JW, Turro NJ, Tomalia DA. Dendrimers with hydrophobic cores and the formation of supramolecular dendrimer-surfactant assemblies. *Langmuir.* 1997;13(12): 3136–3141.
38. Wördehoff MM, Hoyer W. α-Synuclein aggregation monitored by thioflavin t fluorescence assay. *Bio Protoc.* 2018;8(14):e2941.
39. Gade Malmos K, Blancas-Mejia LM, Weber B, Buchner J, Ramirez-Alvarado M, Naiki H, et al. ThT 101: A primer on the use of thioflavin T to investigate amyloid formation. *Amyloid.* 2017;24(1):1–16.
40. Jamroskovic J, Livendahl M, Eriksson J, Chorell E, Sabouri N. Identification of compounds that selectively stabilize specific G-quadruplex structures by using a thioflavin T-displacement assay as a tool. *Chemistry.* 2016;22(52):18932–18943.
41. Kaygisiz K, Rauch-Wirth L, Dutta A, Yu X, Nagata Y, Bereau T, et al. Data-mining unveils structure–property–activity correlation of viral infectivity enhancing self-assembling peptides. *Nat Commun.* 2023;14(1):1–17.

42. Lou X, Lafleur RP, Leenders CM, Schoenmakers SM, Matsumoto NM, Baker MB, et al. Dynamic diversity of synthetic supramolecular polymers in water as revealed by hydrogen/deuterium exchange. *Nat Commun.* 2017;8(1):1–8.
43. Lou X, Schoenmakers SM, van Dongen JL, Garcia-Iglesias M, Casellas NM, Fernández-Castaño Romera M, et al. Elucidating dynamic behavior of synthetic supramolecular polymers in water by hydrogen/deuterium exchange mass spectrometry. *J Polym Sci.* 2021;59(12):1151–1161.
44. Collings PJ, Gibbs EJ, Starr TE, Vafek O, Yee C, Pomerance LA, et al. Resonance light scattering and its application in determining the size, shape, and aggregation number for supramolecular assemblies of chromophores. *J Phys Chem B.* 1999;103(40):8474–8481.
45. Ahlers P, Fischer K, Spitzer D, Besenius P. Dynamic light scattering investigation of the kinetics and fidelity of supramolecular copolymerizations in water. *Macromolecules.* 2017;50(19):7712–7720.
46. Wijker S, Monnink R, Rijnders L, Deng L, Palmans ARA. Simultaneously controlling conformational and operational stability of single-chain polymeric nanoparticles in complex media. *Chem Commun.* 2023;59(36):5407–5410.
47. Pons M, Millet O. Dynamic NMR studies of supramolecular complexes. *Prog Nucl Magn Reason Spectrosc.* 2001;38(4):267–324.
48. Jiang H, Guler MO, Stupp SI. The internal structure of self-assembled peptide amphiphiles nanofibers. *Soft Matter.* 2007; 3(4):454–462.
49. Song F, Chen P, Sun D, Wang M, Dong L, Liang D. Cryo-EM study of the chromatin fiber reveals a double helix twisted by tetranucleosomal units. *Science.* 2014;344(6182):376–380.
50. Danino D. Cryo-TEM of soft molecular assemblies. *Curr Opin Colloid Interface Sci.* 2012;17(6):316–329.
51. Leenders CMA, Albertazzi L, Mes T, Koenigs MM, Palmans AR, Meijer EW. Supramolecular polymerization in water harnessing both hydrophobic effects and hydrogen bond formation. *Chem Commun.* 2013;49(19):1963–1965.
52. Moyer Finbloom JA, Chen F, Toft DJ, Cryns VL, Stupp SI. PH and amphiphilic structure direct supramolecular behavior in biofunctional assemblies. *J Am Chem Soc.* 2014;136(42):14746–14752.

53. Kan L, Zhang P, Jiang H, Zhang S, Liu Z, Zhang X. Microphase separation of a quadruple hydrogen bonding supramolecular polymer: Effect of the steric hindrance of the ureido-pyrimidone on their viscoelasticity. *RSC Adv.* 2019;9(16):8905–8911.
54. Söntjens SHM, Sijbesma RP, Van Genderen, Meijer EW. Stability and lifetime of quadruply hydrogen bonded 2-ureido-4[1H]-pyrimidinone dimers. *J Am Chem Soc.* 2000;122(31):7487–7493.
55. Beijer FH, Sijbesma RP, Kooijman H, Spek AL, Meijer EW. Strong dimerization of ureidopyrimidones via quadruple hydrogen bonding. *J Am Chem Soc.* 1998;120(27):6761–6769.
56. Nieuwenhuizen MML, De Greef TFA, Van Der Bruggen RLJ, Paulusse JM, Appel WP, Smulders MM. Self-assembly of ureido-pyrimidinone dimers into one-dimensional stacks by lateral hydrogen bonding. *Chem Eur J.* 2010;16(5):1601–1612.
57. Kieltyka RE, Pape AC, Albertazzi L, Nakano Y, Bastings MM, Voets IK, et al. Mesoscale modulation of supramolecular ureidopyrimidinone-based poly(ethylene glycol) transient networks in water. *J Am Chem Soc.* 2013;135(30):11159–11164.
58. Kim WH, Pan J, Charych D. Supramolecular polymer materials: Chain extension of telechelic polymers using a reactive hydrogen-bonding synthon. *J Mater Chem.* 2000;12(12):337.
59. Besseling PJ, Krebber MM, Fledderus JO, Teraa M, den Ouden K, van de Kaa M, et al. The effect of chronic kidney disease on tissue formation of in situ tissue-engineered vascular grafts. *APL Bioeng.* 2023;7(2):026107.
60. Brugmans M, Serrero A, Cox M, Svanidze O, Schoen FJ. Morphology and mechanisms of a novel absorbable polymeric conduit in the pulmonary circulation of sheep. *Cardiovasc Pathol.* 2019;38:31–38.
61. da Cunha MGMCM, Hympanova L, Rynkevic R, Mes T, Bosman AW, Deprest J. Biomechanical behaviour and biocompatibility of ureidopyrimidinone-polycarbonate electrospun and polypropylene meshes in a hernia repair in rabbits. *Materials (Basel).* 2019;12(7):1174.
62. Ramaekers M, de Feijter I, Bomans PH, Sommerdijk NA, Dankers PY, Meijer EW. Self-assembly of chiral supramolecular ureido-pyrimidinone-based poly(ethylene glycol) polymers via multiple pathways. *Macromolecules.* 2014;47(12):3823–3828.
63. Diba M, Spaans S, Hendrikse SI, Bastings MM, Schotman MJ, van Sprang JF, et al. Engineering the dynamics of cell adhesion cues in

supramolecular hydrogels for facile control over cell encapsulation and behavior. *Adv Mater.* 2021;33(37):2008111.

64. Schotman MJG, Peters MM, Krijger GC, van Adrichem I, de Roos R, Bemelmans JL, et al. In vivo retention quantification of supramolecular hydrogels engineered for cardiac delivery. *Adv Healthc Mater.* 2021;10(10):2001987.
65. Van Gaal, Vrehen AF, Van Sprang JF, Fransen PP, Van Turnhout MC, Dankers PY. Biomaterial screening of protein coatings and peptide additives: Towards a simple synthetic mimic of a complex natural coating for a bio-artificial kidney. *Biomater Sci.* 2021; 9(6):2209–2220.
66. van Gaal RC, Ippel BD, Spaans S, Komil MI, Dankers PYW. Effectiveness of cell adhesive additives in different supramolecular polymers. *J Polym Sci.* 2021;59(12):1253–1266.
67. Schmitz MGJ, Riool M, de Boer L, Vrehen AF, Bartels PA, Zaat SA, et al. Development of an antimicrobial peptide saap-148-functionalized supramolecular coating on titanium to prevent biomaterial-associated infections. *Adv Mater Technol.* 2023;8(13): 2201846.
68. Song J, Schmitz MG, Riool M, Guo S, Zaat SA, Dankers PY. Engineering antimicrobial supramolecular polymer assemblies. *J Polym Sci.* 2023;61:2866–2877.

https://doi.org/10.1142/9789819806782_0007

Chapter 7

Multiphasic Anisotropic Systems

Ninon Möhl and Laura De Laporte*

Institute for Technical and Macromolecular Chemistry,
RWTH Aachen University
52074 Aachen, Germany
DWI – Leibniz Institute for Interactive Materials
52074 Aachen, Germany
Institute of Applied Medical Engineering,
Department of Advanced Materials for Biomedicine,
RWTH Aachen University
52074 Aachen, Germany
**Corresponding author: delaporte@dwi.rwth-aachen.de*

The human body is composed of several highly organized anisotropic structures, such as tissues and organs. For the successful regeneration of such tissues or organs when injured, anisotropic scaffolds are needed. However, most of the established approaches use rigid scaffolds that need invasive surgeries. To overcome that, injectable 3D hydrogels, formed by cross-linking natural and/or synthetic macromolecules, have been developed, as they are minimally invasive. Depending on the injured native tissue, the hydrogel properties (such as stiffness, internal structure, and degradation rate) can be varied accordingly. However, conventional hydrogels do not comprise any directional cellular guidance cues, as they are isotropic. To provide control not only over the macroscopic properties but also over the microscopic properties, and to ensure anisotropy, multiphasic hydrogels have been developed. Here, different external triggers are employed to render the injectable hydrogels anisotropic through the addition of preprogrammed materials. The external triggers could be electrical stimuli, light, sound, or magnetic fields. Besides using preprogrammed materials, anisotropy in 3D hydrogels can also be achieved through mechanical deposition. This chapter aims to elucidate the most recent techniques and to discuss their advantages and disadvantages while considering the remaining challenges.

1. Introduction

The extracellular matrix (ECM) can be found in every tissue and organ, such as muscles, the heart, blood vessels, or nerves, and is sustained, rearranged, and shaped by the respective cells.[1] It does not only control different cellular processes, such as proliferation, differentiation, or adhesion, but also gives function to the native tissue.[2,3] The 3D network of the ECM is mainly composed of proteins, proteoglycans, and glycoproteins, resulting in a natural hydrogel. Its structural and spatial organization and composition is crucial, as it is characteristic to each tissue's function, as can be seen in **Figure 1**. Smooth muscle cells show a circular and longitudinal organization in order to regulate their contractile forces inside the gastrointestinal track.[4] Furthermore, smooth muscle cells are oriented circumferentially and form multiple layers in arterioles and arteries to provide vasoactivity, the ability to regulate blood pressure.[5] In order to generate force, the skeletal muscle is composed of oriented fibers that show viscoelastic and anisotropic behavior.[6–10] Some tissues exhibit a more complex ECM organization, such as the heart, where the structural architecture is one of the key factors to healthy pump function. Here, the myocardial cells are disposed into different directions to ensure anisotropic structure needed for contraction and pumping action.[11] In the cornea, the orthogonal disposition of collagen fibrils gives rise to optical transparency and its mechanical properties.[12] Following the rise in structural complexity, the ECM architecture in cartilage alters through different zones that have shown to be indispensable for proper tissue function.[13] Starting at the superficial zone, collagen

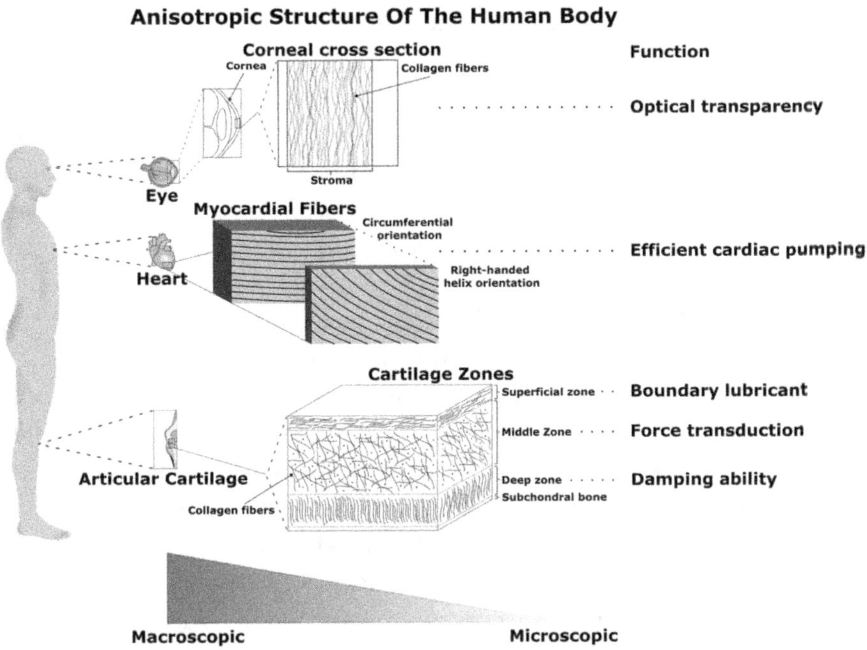

Figure 1. Anisotropic structure of the human body. In the cornea, the orthogonal disposition of collagen fibrils gives rise to optical transparency and its mechanical properties. In the heart, the myocardial fiber orientation ranges from a circumferential orientation to a right-handed helix orientation to ensure efficient cardiac pumping. In cartilage, the ECM exhibits different zonal orientations that provide enough damping and tensile strength. Adapted from Dwyer and Coulombe,[15] Bader and Lee,[17] Santra et al.,[18] and Paras Health.[19] The macroscopic function of the native tissue is given on the microscopic scale.

fibrils are positioned parallel to the articular surface, providing high tensile strength to endure tensile stresses linked to joint loading. Being flattened and at a high density, the cells form a boundary lubricant.[14] The middle zone transduces forces from the superficial to the deep zone through randomly oriented collagen fibers.[15] Finally, in the deep zone, collagen fibers are thicker and oriented orthogonally to the subchondral bone, providing the biggest damping ability to compressive forces.[15]

In tissue engineering and regenerative medicine, many studies use scaffolds made from macromolecules to mimic the hierarchical organized structure of the ECM.[20] Synthetic polymers, such as poly(glycolic acid) (PGA), poly(lactic acid) (PLA), and copolymers, poly(lactic-co-glycolic acid) (PLGA) are widely studied and established.[21,22] These polymers are U.S. Food and Drug Administration (FDA) approved; however, the scaffolds are usually rigid and produced through different fabrication methods, demanding the surgeon to make sufficiently large cuts to allow for the disposition of the scaffold at the needed site. Therefore, such constructs are not applicable for low-invasive delivery or high-throughput automated pipetting techniques as these techniques rely on injectable materials. Hydrogels have been extensively studied as potential injectable materials, as they are initially liquid, can cross-link after injection, and swell in water.[23] Additionally, they can resemble the structural and chemical components in the body and are biocompatible.[24] However, conventional hydrogels usually lack linear structural guidance for cell growth, as can be found in the ECM of the native tissue, due to their structural isotropy. Multiphasic hydrogels rendered anisotropic either through the exploitation of their inherent anisotropic properties or, more advanced, through the application of forces or the addition of (nano and micro) particles that can be triggered externally is a new research field. The latter offer control over many properties on the cellular scale and can be tuned according to the tissue's function. This enables the hydrogel precursors to be structured in vivo or be used in automated pipetting systems, as well as high-throughput applications, for clinical translation. The many possibilities of turning conventional isotropic hydrogels into anisotropic multiphasic hydrogels will be discussed intensively in the following chapter.

2. Multiphasic anisotropic hydrogels

The growth and regeneration of aligned tissues requires minimally invasive biomaterials to form 3D microenvironments resembling the complex structural architecture of the native tissue. Hydrogels fulfill these requirements, as they are injectable and can cross-link in situ.[25] Furthermore, hydrogels can be tuned with cell signaling and binding domains,[26] as well as growth factors,[27] needed for cellular attachment and proliferation. However, conventional synthetic injectable hydrogel systems lead to pore sizes less than 100 nm, although it is known that macropores in the range of tens to hundreds of microns, as well as pore interconnectivity support cell migration and proliferation.[28–30] Different approaches to ensure cell growth and migration have been studied, such as rendering the hydrogel degradable through hydrolysis or enzymatic degradation with peptide sequences sensitive to matrix metalloproteinases (MMPs).[31] To form 3D macroporous constructs for cell growth immediately after hydrogel formation, more advanced approaches have been tested based on leachable porogen,[32] phase separation, of interlinking of soft spherical or rod-shaped microgels.[33]

Besides the porosity, hydrogels are mostly isotropic and are thus lacking the ability to guide cells.[25,34] To combine microporosity and alignment within a hydrogel system, Rizzo et al.[35] developed a simple system where a precrosslinked bulk hydrogel is forced through a grid with tunable opening sizes, leading to high-aspect-ratio microstrands that are stabilized by a consecutive cross-linking step. This method has been shown to be cell instructive in silico and in vitro, inducing aligned muscle tissue maturation. Looking at methods that employ the inherent properties of the material, structural cellular guidance has been accomplished by the supramolecular assembly of oriented nanofibers.[36] Here, an aqueous solution of charged peptide amphiphiles (PAs) is used, which is heated and subsequently cooled down again to form aligned fiber bundles that have a diameter of approximately 40 nm. Human mesenchymal stem cells (hMSCs) and cardiomyocytes were cultured with the aligned fibers and were able to recognize the orientation of the fibers and align alongside them. The self-assembled PA can be used to generate oriented and injectable scaffolds; however, the alignment of the fibers is fixed by the direction of injection and cannot be altered by orthogonal means. In order to gain more control over physical and mechanical properties, external triggers (**Figure 2**),

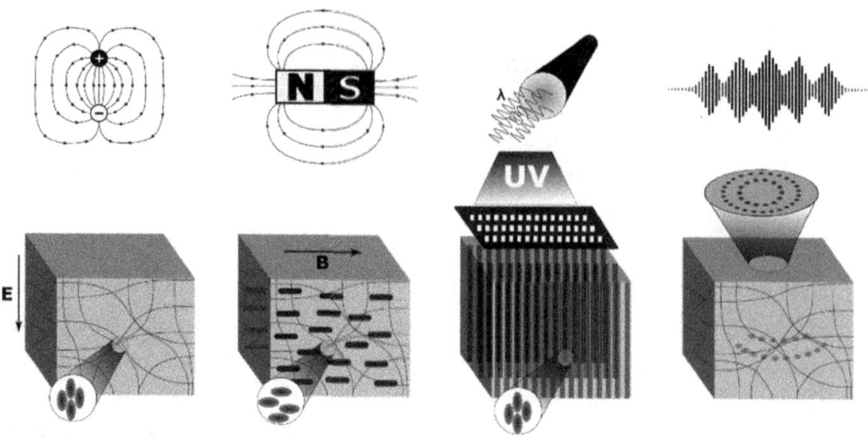

Figure 2. Triggers to externally induce anisotropy in 3D hydrogels. From left to right: electrically, magnetically, light- and sound-induced. Electrical stimuli,[37] magneto-responsive microparticles,[38] light,[39] and sound waves[40] induce the alignment of different cell types in 3D.

such as magnetic and electrical fields, light or sound waves have been introduced to obtain anisotropic features within the biological or synthetic hydrogels.

Using the diamagnetic properties of proteins or cellulose nanocrystals (CNCs), an external magnetic field has been employed to create material anisotropy using high magnetic fields (5–10 T).[41,42] Furthermore, the addition of magnetic iron oxide nanoparticles can generate polydisperse particle strings in the presence of low magnetic fields (< 500 mT) that guide cells in a hydrogel matrix.[43,44] Nevertheless, the latter demands a high concentration of iron oxide particles, and the dimensions of the strings cannot be adapted easily. Following the path of regulating physical and mechanical properties, the addition of magneto-responsive short fibers or rod-shaped microparticles or microgels into a hydrogel network gives the opportunity to decouple properties, such as stiffness, porosity, and dimensions, and control them independently, while enabling the combination of injectability and specific orientations.

The different production methods of such anisotropic microparticles are depicted in **Figure 3** and explained in the review paper by Babu et al.[45]

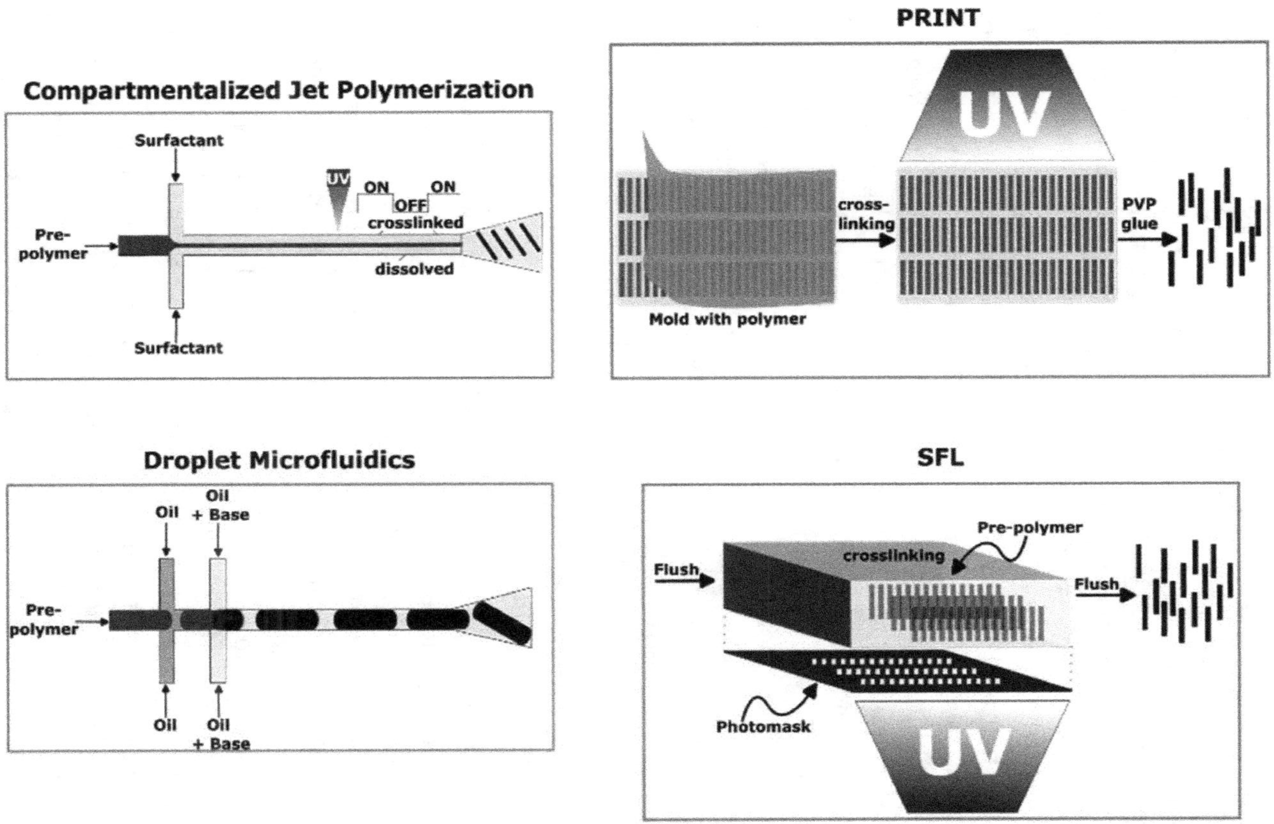

Figure 3. **Production methods of anisotropic microelements.** Adapted from Babu et al.[45]

This approach leads to multiphasic (hybrid) hydrogel systems, such as the Anisogel,[46,47] developed by the De Laporte group, with the ability to adjust structural and mechanical anisotropy in high resolution. The Anisogel is composed of rod-shaped microelements, for example, microgel rods or fibers, that are rendered magnetic through the addition of superparamagnetic iron oxide nanoparticles (SPIONs) in a low concentration. These microelements can be aligned in the direction of an external magnetic field (40–120 mT) inside a hydrogel precursor solution that can either be natural or synthetic, in a short amount of time. The microelements are fixed in their position as the hydrogel solution cross-links and forms a matrix, inhibiting any further movement of the aligned microelements once the magnetic field is removed. The magnetically induced alignment, as well as its possibilities and limitations, will be further discussed in Section 2.4. Besides the use of magnets to provide directionality inside a hydrogel, researchers have also explored light, sound, and electric fields to trigger anisotropy. Each trigger will be separately discussed in the following sections. Additionally, anisotropy through mechanical deposition will be addressed in Section 3.

2.1. *Electrically induced alignment*

It is well known that electrical stimulation (ES) controls basic cell behavior in the human body. At the single-cell level, ion fluxes through ion channels (voltage-activated Ca^{2+} channels) induce a transmembrane potential that is responsible for the regulation of excitable and nonexcitable cells.[48] Later, experimental data depicted an influence of externally applied electrical stimuli on transcriptional responses in cellular activities.[49] Fibroblasts are responsible for the formation of healthy connective tissues and the repair of injured tissues,[50,51] and they are well known to be sensitive to various signals, including mechanical, chemical, and electrical signals.[52–55] It has already been stated 150 years ago that disrupted epithelial layers at an injury site generate detectable electric fields, and it has been hypothesized that the transepithelial electrical potential of wound edges guides the migration of cells to the injury site.[56,57] In general, ES is a well-established technique to guide cells and is used in tissue engineering to regenerate harmed tissue.[58] Up to now, fibroblasts have been evaluated on 2D surfaces, for example, tissue-culture plastic, to observe how cells can be orchestrated to fill up a defect. However, this approach barely describes the 3D environment of a wound and has minimal physiological relevance. While biomaterials enable the formation of scaffolds in 3D, especially hydrogels have gained much interest because of their tunable mechanical, physical, and biochemical properties. Conductive hydrogel scaffolds have been prepared, where the cells are grown on top or inside the construct while applying external electrical stimuli. The electrical signals have demonstrated to guide, for example, neural progenitor cells (NPCs) and enhance their differentiation into astrocytes and myelinating oligodendrocytes.[59–62]

Soft (< 15 kPa) and viscoelastic hydrogel materials[63] have been developed to be employed as living electrodes to stimulate cellular growth for more than 12 weeks. The addition of electrical pulses supports neural cell differentiation and directs network formation.[62] Recently, conductive hydrogels made of alginate and carbon nanomaterials were combined to induce the growth of fibroblasts not only in 2D but also in 3D by applying external electrical stimuli.[37] Based on previous works,[62,63] porous conductive hydrogel scaffolds with different amounts of carbon nanomaterials (e.g., graphene flakes [GFs]; carbon nanotubes [CNTs]) inside an alginate matrix were fabricated. These highly porous scaffolds are beneficial because of their mechanical and electrical properties, keeping their viscoelastic properties due to the low amount of carbon nanomaterials needed to ensure conductivity. At the same time, metal-based conductive scaffolds have been developed, though resulting in stiffer networks compared to carbon nanomaterial composites, with the metal-based particles potentially impacting cell viability.[64] The hydrogel-based scaffolds containing different quantities of GF and CNT were prepared to obtain a total carbon nanomaterial concentration of 0.13%, 0.22%, or 0.31%, with the total amount being

< 0.3% of the weight owed to the high aspect ratio of the nanomaterial. The storage moduli (G') were determined using nanoindentation and decreased with a higher amount of carbon nanomaterial from ~10 kPa to ~6 kPa, thus imitating the mechanical modulus range of the dermis layer (~10 kPa) of the skin.[65] Simultaneously, the conductivity rises with the total amount of carbon nanomaterial. Fibroblasts were seeded inside the scaffolds, as well as on tissue-culture plastic wells as a control, and exhibited a high viability in both cases (~100%). This viability remained for each conductive hydrogel condition. Subsequently, the effect of carbon nanomaterial content on fibroblast growth after two days was analyzed using immunohistochemistry (IHC), with the same number of cells added to every scaffold. The scaffolds without a carbon nanomaterial resulted in the least amount of cells, whereas an increase in the carbon nanomaterial enhanced cell proliferation. Especially with 0.31% of carbon nanomaterial, the fibroblasts showed a spindle-like phenotype, resembling the fibroblasts' morphology in vivo.[66,67] The maximum applied voltage (volts, V) without harming cells was assessed to be 3 V, as at 4 V, most of the cells die. When the cells are distributed in 3D inside the scaffolds, electrical pulses throughout the scaffold tuned their spatial distribution. When the electrical field was applied from top to bottom, the cells were found mostly at the bottom, indicating a movement in the z-direction along the electric field lines.

Another way of rendering biomaterials conductive without the need of additional nanoparticles is the use of conductive polymers (CPs). CPs, including polyaniline (PANI), polypyrrole, and polythiophene, have been used to build electroactive culture systems due to their high stability, tunable electrical properties, and facile synthesis.[68] The unique feature about CPs is that, even without external triggers, CPs can emit (bio)electrical signals that have proven to enhance cellular activities.[69–71] The downside of these polymers is that they do not provide an adhesive surface for cell attachment, as they are hydrophobic and brittle, thus limiting their biomedical potential.[72] To circumvent that restriction, CPs have been added to scaffolds that were made of biomaterials, offering a suitable substrate for cellular interaction.[73,74] That way, both endogenous conductivity and biocompatibility could be ensured.

One such hybrid conductive hydrogel is made of water-soluble CP poly(aniline-co-*N*-(4-sulfophenyl)aniline (PASA), an altered version of PANI, and polyisocyanides (PICs). PICs form very soft hydrogels at physiological temperature and have been used as 3D matrices for in vitro experiments to analyze singe-cell[75,76] as well as organoid behavior.[77,78] In order to obtain the required mechanical properties and to mimic the native tissue, the molecular weight and polymer concentration could be varied.[79,80] These polymers are biofunctionalized with cell-adhesive Gly-Arg-Gly-Asp-Ser (GRGDS) and are shown to be nontoxic and biocompatible in vivo.[81,82] However, as PIC hydrogels are composed of neutral polymers, they barely provide any conductivity; thus, PASA was added to render these 3D constructs conductive. While PANI is biocompatible with L929 fibroblasts and myoblasts C2C12, enhancing their proliferation, and in the case of myoblasts, myogenesis,[83] PANI has a low solubility in water. Therefore, PANI has been modified with phenyl-4-sulfonate groups to be applicable in water, resulting in the CP PASA.

2.2. *Sound-induced alignment*

One of the most difficult features that really embody the success of in vitro generation of highly functional and manifold 3D biomimetics is the presence of a vascular network.[84] It is crucial to support cell–cell interaction and communication to enable the formation of tubule structures. The most established methods to develop in vitro vascularization can be categorized into bottom-up and top-down. In bottom-up systems, cells self-assemble through physical or chemical stimulation,[85,86] while in top-down systems, the vascular structure is determined through the biofabrication methods.[87–89] Furthermore, cell self-assembly has been induced using magnetic or acoustic fields, where the arrangement of the cells was set through the organization of the respective field lines.[90] Acoustic wave patterning is an unfolding technique to predefine cellular organization.[91] Here, an acoustically triggered pressure gradient in a liquid enables the spatial arrangement of microparticles and cells in a quick and contact-free manner. Compared to conventional bioprinting fabrications, the arrangement through acoustic patterning diminishes cell damage. There are many applications using acoustic vibrations, such as myotubes formation,[92] angiogenesis,[93] neural differentiation,[94] or cardiomyocyte differentiation.[91] Most of these fabrication methods rely on acoustophoretic devices that provide a fast process and a high resolution < 100 μm of the arrangements. However, despite the advantages, these devices are very complex and feature limitations, affecting the ECM properties and cell viability, such as an increase in temperature. Later, researchers have developed a new acoustic patterning technique employing Faraday waves at a liquid–air interface, generated through a low-frequency vibration. After only a few seconds of applying the acoustic standing surface waves, spheroids arrange accordingly in a fluid environment.[95] This technique offers new ways to achieve pattern formation, because of its tunability[96] and high degree of control,[97] compared to already established fabrication processes. It circumvents the challenge faced with acoustophoretic devices. Faraday waves have been used to organize spheroids into different complex geometries, as well as the formation of microtissues. This could be achieved through the vibration settings (frequency and amplitude), the properties of the fluid environment (viscosity, density, surface tension), and the shape of the reaction chamber.[95] Another acoustic based technology, called sound-induced morphogenesis (SIM), combined acoustic patterning with physiological self-assembly of cell to investigate the number of cells needed to form complex and perforable vascular networks.[40] To determine the functionality of the SIM device, it was tested with inorganic particles (tricalcium phosphate [TCP]) surrounded by gelatin methacryloyl (GelMA). TCP particles (diameter = 125–250 μm) were dispersed in a GelMA prepolymer (5% w v^{-1}, G′ = 240 Pa) while maintaining the amplitude and frequency constant at 0.5 g and

40 Hz, respectively. The patterning of the particles took about 10–15 seconds, and their position was fixed through cross-linking of GelMa by ultraviolet (UV) light. Different frequencies and amplitudes have been analyzed, resulting in different TCP geometrical arrangements, including concentric rings and a disposition of N x N dots. These preliminary experiments helped to assess the interaction between the frequency and the amplitude, and their effect on the generated pattern. For example, when augmenting the frequency and the amplitude, arrangements became more complex, moving from concentric rings to honeycomb-like patterns. Subsequently, self-assembly of cells was observed upon SIM patterning of endothelial cells (human umbilical vein endothelial cells [HUVEC]) together with hMSCs, surrounded by a fibrin gel. The self-assembly was evaluated with both single-cell dispersions and as spheroids. The SIM technology did not negatively impact cell viability, as at day 1 and day 4, a cell viability of 93% and 92% was determined, respectively. Especially interesting is the ability to obtain vessel-like structures. When combining green fluorescent protein-expressing human umbilical vein endothelial cells (GFP-HUVEC) with hMSC cells and inducing a formation of circular patterns through a frequency of 80 Hz and an amplitude of 2.5 g, the concentric rings allow for cell patterning and the connection of these rings to each other owned to the sprouting of the HUVEC. This is a fast, single-step procedure, inducing complex vascular networks within five days. Additionally, using SIM to pattern vascular networks offers the possibility to use low cell concentrations (0.4×10^{6} cells ml^{-1}), making it very interesting for cell therapy and tissue engineering.

Alternatively, standing waves have recently been used to build highly organized anisotropic arrays of chondrocytes that resemble the architecture of the deep zone in an articular cartilage.[98] The specific organization of chondrocytes within the articular cartilage requires high resolution, which can be met with acoustic cell patterning. High-frequency (6.7 MHz) ultra-sound waves are used to arrange bovine chondrocytes into aligned arrays within the entire volume of an agarose hydrogel. This organization is stable throughout 35 days, during which time the cells generate cartilaginous ECM comprising oriented collagen fibers. It is also shown that treating the chondrocytes with ultrasound does not impact their viability nor their ability to form hyaline cartilage. In the future, the assessed collagen alignment would need to be studied further regarding the mechanical loading ability, as this is required to have a fully functional tissue mimetic. Furthermore, the combination of all different cartilage zones, for example, middle zone and superficial zone, through acoustic patterning is a promising tool to build 3D articular cartilage tissue. Here, it is important to account for the different mechanical properties within each zone, as the elastic modulus increases from the kPa range in the superficial zone to the MPa range in the deep zone.[99] Using only one component, such as agarose, as a 3D matrix will not be able to fulfill these requirements. The combination of different polymers to reinforce the structure where necessary could be one approach.[100]

2.3. *Light-induced alignment*

The creation of complex cellular architectures employing light-based projection techniques has gained importance in the field of biomimetic tissues[101–103] in the past years. Especially in volumetric bioprinting, a fast (within seconds) biofabrication of geometrically complex cellular architectures has been developed through optical tomography, based on visible light projection.[104–106] However, one of the remaining challenges here is the printing resolution (> 100 μm),[92,107] which limits the engineering of anisotropic scaffolds with smaller feature, for example, for skeletal muscle.[108] Methods that are able to obtain structures in cell-scale (< 30 μm) resolution, for example, two-photon polymerization and ultrahigh resolution digital light processing, on the other hand, are limited by the incoherent light sources and can only produce large tissue-engineered constructs when a layer-by-layer approach is applied.[109,110] Within these techniques, there is a trade-off between speed and scalability, to produce anisotropic constructs for tissue engineering.

Looking at the cellular alignment and organization, topological cues comprising different dimensions have been proven to influence cell behavior in and on the substrate. For example, polydimethylsiloxane (PDMS)-based anisotropic topographies (wrinkles) have been used to investigate the osteogenic differentiation of human bone marrow-derived mesenchymal stem cells (hBM-MSCs).[111] The chosen dimensions are based on the nanoscale features (0.05 μm) and the microscale features (0.7 μm and 3.5 μm) found in the ECM in vivo. They show that wrinkles with amplitudes of 0.7 μm and 3.5 μm result in an orientation of the cells of 69% and 94% respectively, compared to 0.05 μm with 36%. This means that dimensions on the cellular scale are needed to guide cells. The addition of cell-derived matrices (CDMs) to these wrinkles enhances the capacity of osteogenesis and also provides more Yes-associated proteins (YAPs) inside the nucleus. These findings indicate the importance of topography and the ECM in cellular fate decisions. Additionally, the aspect ratio of the substrates plays an important role. Discrete guiding structures with aspect ratios (> 20) can be achieved through micropatterning techniques of hydrogels surfaces and show elongated and aligned cells along the direction of the patterned surface.[112] In a high-throughput micropatterning approach using two-photon lithography techniques, the minimal physical guidance cues needed for induced nerve growth along the axis of discrete anisotropic microelements with an aspect ratio of 20 was studied. The spatial distribution of these patterns (horizontal spacing vs. vertical spacing) regulates the alignment of the axons, showing a stronger neurite alignment along the long axis of the microelements, when the horizontal and vertical distance is reduced.[113] Interestingly, increasing the vertical distance has a more detrimental effect on neurite alignment, while for smaller vertical distance, an increase in the horizontal distance still induces oriented growth. Besides the alignment, it is well known that an elongated cell nucleus has an impact on cell behavior,[114,115] such as differentiation and gene expression, and can be found in anisotropic native tissues, for example, tendons.[116] To provide cell alignment and nucleus deformation in 3D constructs, filamented light (FLight) biofabrication was developed to generate macroscopic hydrogels comprising unidirectional microfilament networks that have diameters in the size range of single cells, within seconds.[39] The method is based on optical modulation instability (OMI),[110,117] which is the separation of a uniform optical beam into multiple light beams during the propagation inside optical nonlinear media. A material class that features optical nonlinearity is photoreactive polymers, as their polymerization kinetics and in return, their refractive index (RI) alter upon the change in the intensity of the cross-linking light.[118–120] The uniaxial microfilament network is formed through the irradiation of a photoresin-containing vial using a partially spatially coherent light beam. A partially coherent light beam features an intensity noise pattern, which creates local maxima in the intensity of light, and upon incidence with the photoresin, generates microfilaments, as a maximum intensity of light provides faster cross-linking kinetics. This effect causes a local maximum RI that functions as optical traps, proceeding in the propagation of micropatterned polymerization across the entire resin volume through self-focusing waveguides. The intensity noise pattern does not only lead to local maxima but also to local minima of the light intensity, impeding cross-linking of the resin, equivalent to the presence of void spaces. Once the uncrosslinked photoresin is washed away, channel-like void spaces (microchannels) comprising ultrahigh aspect ratios are formed. It needs to be stated that not the entire uncrosslinked photoresin is washed away, as some parts do actually slightly cross-link (hence, there is no sharp boundary between cross-linked and uncrosslinked resin due to noise in the light intensity). These slightly cross-linked regions ensure the stability of the entire hydrogel construct. These slightly cross-linked parts are not existent between each microfilament, resulting in a more porous construct. Using this technique, cells can be encapsulated, without damage, into the 3D anisotropic hydrogel microfilament matrix in a short amount of time (< 10 s). It represents a fast approach to mimic anisotropic tissues and provide guidance cues on the cellular scale while offering a cell-friendly method. The dimensions of the microchannels and aligned structures inside the hydrogel can be adapted on a cellular scale, ranging from 2 to 30 μm, by adjusting the spatial coherence lengths of the light beams.[121] This feature makes FLight suitable for multiple aligned tissues, comprising ECM fibers from 1 to 100 μm.[122,123] Different photosensitive resins have been analyzed, though the focus has been on the norbornene-thiol reaction so far due to the outstanding cross-linking kinetics that enable the generation of microfilaments within seconds while keeping the potential cell damage at a minimum. Regarding cell guidance and migration, a conventional hydrogel network does not offer cell guidance cues and also prevents efficient oxygen and nutrient transport when the tissue grows too dense. The hydrogel network with a nano mesh size has to degrade to enable cell growth. To provide efficient cell migration, proliferation, and vascularization in 3D, the void spaces ought to be in the micrometer scale.[124–126] As most of the FLight hydrogel matrix is composed of microchannels that show an ultrahigh aspect ratio (> 700:1), cells are able to migrate and align with elongated nuclei. To make sure that cell migration is not physically restricted due to a pore size smaller than 3 μm,[127] the channel diameter was adjusted to be > 3 μm by reducing the coherence length. Normal human dermal fibroblasts (NHDFs) were enclosed in FLight matrices during the fabrication process, and after 1 hour, the NHDFs showed spreading and an elongated morphology alongside a deformed nucleus, with an increase in the cells' aspect ratio during cultivation for seven days. It is shown that the cells can move through interconnected FLight microchannels, either perpendicular to the direction or parallel to the direction of the microchannels. Furthermore, FLight hydrogels allow for the migration of fibroblasts from the top surface inside the matrix after one day already, in contrast to conventional bulk hydrogels. The cell migration parallel, as well as perpendicular, to the microchannels is especially important for guaranteeing cell–cell contact in a 3D matrix, which is, for example, needed in gap junction-mediated intercellular communication.[128] Gap junctions are a special type of membrane contact that is build up by multiple microchannels, allowing small molecules to travel into the cytoplasm of adjacent cells, ensuring communication.[129] Further, alignment and migration of cells can be influenced by the presence of such microchannels, facilitating, for example, the merging of myoblasts for the regeneration of muscle fibers.[130] Moreover, growing myoblasts in ultra-high-aspect-ratio microchannels results in an increase in gene expression of the mechanosensitive ion channel Piezo1, which is associated with ECM production,[131–134] whereas in 2D microbeams and 3D bulk hydrogels, this was not the case. However, after three days, a decrease in gene expression was seen and was attributed to a loss of mechanical stimulation, possibly owed to the degradation of the microfilaments. The degradation of the material after three days limits the potential of this material approach to be applied for tissue mimetics, where a tissue maturation of several weeks is needed. Furthermore, the local mechanical properties cannot be assessed or tuned independently as it is one connected 3D construct. In the future, it would, therefore, be interesting to translate this method to synthetic materials that exhibit less loss of mechanical support over time.

2.4. *Magnetic alignment*

Applying a magnetic field to generate anisotropic tissue mimetics has been done early on through the magnetic alignment of collagen or fibrin fibrils using their diamagnetic properties, to be used for the regeneration of peripheral nerves. Here, a very strong magnetic field (~9.4 T) needs to be employed limiting the ability of this approach to be used as an injectable system in vivo. Therefore, nerve guidance conduits were rendered anisometric through the addition of the magnetically aligned gels before the execution of implantations.[41,135] The implementation of SPIONs into injectable systems enabled the possibility to be aligned in vivo. The SPIONs were added to a collagen matrix, which then, under the influence of an external magnetic field (25.5 mT), accumulate, resulting in long, aligned strings along the magnetic field lines. In the meantime, the collagen cross-links, holding the particle strings in place, and at the same time, the collagen fibers are aligned. The combination of both effects was used to align neurons.[43] Later, this approach was transferred to Matrigel and HA hydrogels employing protein-coated magnetic nanoparticles.[136] The downside of these methods is the use of high amounts of nanoparticles that could have cytotoxic effects, making it difficult to be applied clinically.[137,138] Another way of employing magnetic nanoparticles to align collagen is 4D bioprinting, where a magnetic field is applied during the bioprinting process. Here, streptavidin-coated iron nanoparticles are integrated in printable bioinks containing agarose and type I collagen.[139] A magnet is placed at one side of the solution to attract the particles and induce linear movement to orient the collagen fibers during the printing process. Cells were added to the bioink before printing, and after 21 days of culture, constructs comprising layers of aligned and random fibers indicated a higher expression of collagen II than constructs with only randomly oriented fibers.[139] Here, the majority of the nanoparticles would be found at the edge of the construct, making it easier to be potentially removed after printing, which makes this approach more suitable in vitro.

As the addition of magnetic nanoparticles may cause undesired side effects in vivo, anisotropy can also be generated through CNCs obtained from natural material that can be aligned perpendicular to an external magnetic field owing to their intrinsic diamagnetic properties. Their dimensions range from 100 to 200 nm in length and 5 to 20 nm in width, resulting in high aspect ratios. Through the physical entrapment of the particles dispersed in poly(oglioethylene glycol methacrylate) (POEGMA) and their strong interfacial interactions, CNC aggregation is reduced inside the hydrogel phase and allows for the release of individual nanoparticles.[140,141] CNCs were mixed with a POEGMA hydrogel prepolymer solution and subsequently injected in the presence of a magnetic field (1.2 T). The particles were oriented perpendicular to the magnetic field lines, and simultaneously hydrazide and aldehyde-functionalized POEGMA covalently cross-linked to form the surrounding hydrogel. Mouse myoblasts were cultured inside the 3D construct. While they did not align during the first two days, they followed the oriented disposition of CNCs upon differentiation into myotubes, resulting in aligned structures.[42] As the cultured cells are much larger than the nanometer sized CNCs, the alignment of myotubes is probably owed to the cells feeling the anisotropic features regarding stiffness and surface properties of the particles. Hydrogels comprising aligned CNCs have been shown to enhance fibroblast proliferation within the hydrogel in vivo, compared to hydrogels with unaligned CNCs. Additionally, CNC–POEGMA hydrogels indicate only mild acute inflammatory responses, as well as a tunable degradation rate depending on the polymer composition and the amount of dispersed CNCs.[140] In terms of alignment of fibroblasts or other cell types, oriented CNCs are not sufficient, as the required dimensions to mimic extracellular fibers, such as collagen, in the range of 1–20 μm[122,123] are not met. Importantly, CNCs do not degrade, and their fate in vivo is still under investigation.

In order to provide aligned directional cues with specific dimensions within an injectable 3D hydrogel, a new class of materials called Anisogel has been developed in the De Laporte lab, with the aim to regenerate the spinal cord after a spinal cord injury (SCI).[38]

Anisogels consist of rod-shaped microelements that are monodisperse, magneto-receptive, and are surrounded by a natural fibrin[38] or a synthetic[46,47] hydrogel matrix. Small concentrations of SPIONs are dispersed inside the microelements, which enable them to be aligned in a unidirectional way in the presence of an external magnetic field (40–120 mT). During the alignment of the microelements (< 60 s),[142] the matrix around them cross-links and fixes them in their position, even after removing the magnetic field. The low amount (1.8 ng μL^{-1} < 10 μM iron) of SPIONs is sufficient to align the microelements[143] due to their high aspect ratio, which minimizes the cytotoxic effects, as much higher concentrations in the order of 1.5 mM iron diminish the viability of sensitive cells and minimize the ability to form neurites.[144]

The preparation of the Anisogel is illustrated in **Figure 4**. The magneto-responsive microelements are first mixed with the matrix precursor. In the presence of a low external magnetic field, the

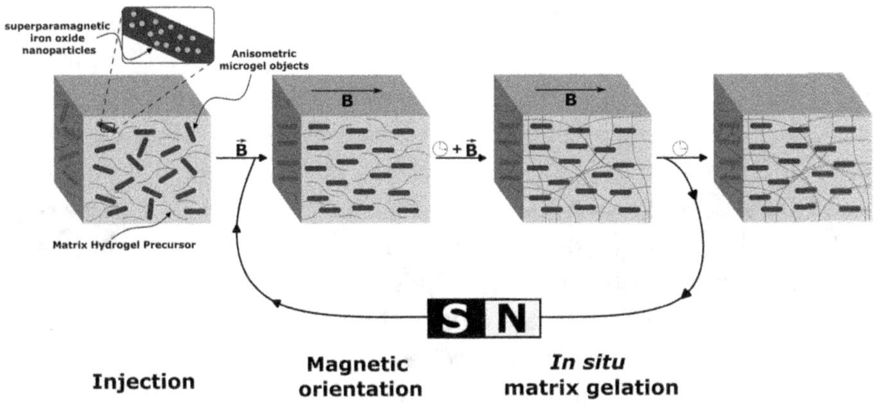

Figure 4. Anisogel preparation. Adapted from Licht et al.[47]

microelements orient along the direction of the field lines, while the matrix around them cross-links, forming an Anisogel.

It has been shown that fibroblasts and neurons cultured in Anisogels orient along the aligned rod-shaped microgels or short fibers.[46] The microgels are prepared using bioinert poly(ethylene glycol) (PEG) through the use of a soft lithography approach called PRINT (particle replication in nonwetting templates),[145] while the short fibers are made from poly(lactide-*co*-glycolide) using an electrospinning technique. PEG is one of the most established synthetic polymers when producing tissue-regenerative constructs, as it is biocompatible and approved for clinical use.[144,146,147] The ability to guide cells and cellular behavior strongly depends on the concentration, properties, and the type of microelements, as well as the cell type itself and the surrounding matrix used to fix the microelement orientation. When PEG microgels are being used as guiding elements, the aligned cells primally engage with the cell-adhesive surrounding hydrogel. Integrins are the main receptors on cells to bind ECM proteins — such as collagen, fibronectin, and laminin[148] — and it is believed that they are used in the case of Anisogels to pull on the surrounding matrix, sensing distinct mechanical responses from the hydrogel in the direction of or perpendicular to the aligned microgels. When the surrounding hydrogel is made of fibrin, fibroblasts align with 1.5 vol% of microgels comprising dimensions of 5 × 5 × 50 μm, leading to a mean intermicrogel distance of approximately 28 μm, whereas neurites only need 1 vol% microgels and a mean intermicrogel distance of approximately 34 μm to grow alongside the microgels' orientation. Mechano-biological differences could be observed when the concentration of microgels is increased and a higher amount of nuclear YAP was found, indicating that cells are influenced by the global anisotropic mechanical characteristics of the Anisogel. More information on YAP can be found in **Infobox 1**. Additionally, the fibronectin generated by the oriented fibroblasts matched the alignment of the microgels, giving rise to the possibility of a naturally aligned ECM matrix adopting the task of the microgels gradually, upon their degradation. The advantage of Anisogels as anisotropic hybrid hydrogels is their ability to vary the physical, mechanical, and biomechanical properties of each component independently from one another, giving a lot of control over the system at a high resolution. One could, for example, biochemically modify the PEG-based rod-shaped microgels with GRGDS, rendering the microgels cell adhesive. GRGDS-adapted microgels have shown a bigger interaction between cells and microgels but did not induce any stronger alignment of fibroblasts. This magnified interaction even resulted in a lower fibronectin production, meaning that a replacement of the synthetic Anisogel in the

> Cells do not only communicate through soluble signals but also through mechanical and physical signals coming from their surroundings, for example, the stiffness of the ECM. Cells can then translate these cues into biochemical signals, highly influencing their behavior. Yorkie-homologues Yes-associated protein (YAP) has been attributed to the mechanical stimuli transmitted due to ECM rigidity and cell shape. YAP has been found to function as the sensors of physical[149] and mechanical cues given by the cellular microenvironment.[150]

Infobox 1. Yes-associated protein (YAP) and its importance in cell signals.

course of tissue regeneration by the naturally produced oriented ECM may then be less likely.[46] Therefore, in the case of guided alignment, a stronger stimulation of the cells with biochemical signals could have counterproductive effects, meaning that the least amount needed to activate the regeneration could be the most efficient approach. This was further supported when the microgels were modified with ECM molecules inside the surrounding PEG-based hydrogels, also reducing the overall alignment of neurites.[45]

In addition, in the case of nerves, microgels with a reduced width of 2.5 μm induce more neurite extension than microgels with a width of 5 and 10 μm. A larger width of microgels, hence, a higher cross section, could hinder the axons' growth. Using only half of the microgel width and additionally only half of the length, and at the same time, decreasing the intermicrogel distance by 30% to approximately 24 μm, contributed to an enhanced neurite alignment in a fibrin-based Anisogel. Microgels with the dimensions 2.5 × 2.5 × 25 μm consumed only 0.45 vol% of the whole Anisogel.[143] However, these conditions do not lead to neurite alignment in a PEG-based Anisogel. The local physical, biochemical, and mechanical properties affecting the orientation of nerve growth have, therefore, been studied extensively to understand what promotes oriented neurite growth in a synthetic PEG-based Anisogel in vitro.[45] Compared to fibrin-based Anisogels, microgels inside the hydrogel need to comprise a higher concentration and a higher aspect ratio. Fibrin gels already provide guidance cues through fibrous domains that are not present inherently in homogenous PEG gels. Considering the microgels' stiffness, probing different stiffnesses (E-modulus: 35–950 kPa, measured with colloidal atomic force microscopy (AFM)) did not show significant differences in neurite orientation or growth. It needs to be stated that even the softest microgel used in this study is still 100 times stiffer than the surrounding hydrogel network. In order to analyze whether there would be a threshold in softness of the rod-shaped microgels to align the cells, one could use microgels obtained from compartmentalized jet polymerization in microfluidics.[151] As implementing certain ECM proteins to render the microgels cell adhesive reduced the orientation of neurites, though it induced a stronger outgrowth, it is indicated that directional guidance cues are strongly depended on the mechanical anisotropy in the system, whose signals can be disrupted by the addition of local cell-adhesive site on the microgels, attracting the cells in the opposite direction. To ensure neurite outgrowth while conserving their alignment, one should opt for the biomodification of the surrounding matrix instead. A covalent coupling of the microgels to the hydrogel matrix resulted in an impaired neurite orientation and growth. This could be due to a difference in force propagation.

Preliminary experiments on a mouse cadaver demonstrated that the magneto-receptive microgels can align with an external magnetic field being placed closed to the spinal cord and were then fixed inside the cross-linked hydrogel.[143] In the future, it might be interesting to replace linear integrin-mediated cell-adhesive peptides with their cyclic counterpart in vivo. Linear RGD peptides degrade faster than cyclic RGD peptides and do not provide integrin selectivity. It has been shown that mouse fibroblasts and primary nerve cells from embryonic chick dorsal root ganglions (DRGs) grow better with bicyclic integrin-selective RGD peptides, compared to monocyclic or linear RGD peptides.[152]

The same approach as the Anisogel was transferred to another system using rod-shaped CNCs consisting of magnetic nanoparticles inside a gelatin hydrogel. The rod-shaped nanocrystals were coated with polydopamine and PEG polymer brushes to provide colloidal stability and biocompatibility, resulting in CNCs with a diameter of 19 nm and aspect ratios reaching from 10 to 20.[153] When incorporating only 0.1–0.5 wt% of the CNCs inside a gelatin hydrogel, they were aligned in a low magnetic field of 108 mT along the magnetic field lines, making them applicable as directional microstructures with anisotropic mechanical properties.

3. Alignment through mechanical deposition

In the field of tissue engineering, it is known that bioprinted constructs provide significant advances compared to the nonprinted, conventional bulk samples, as they are customizable and tuneable in terms of material and cells placement in 3D.[154,155] The construct can be adjusted not only on a large scale in building 3D geometries but also on a small scale controlling the internal architecture, such as the pore size.[156,157] During the development of bioprinting techniques, scientists face different challenges as both the printability and cell compatibility need to be accounted for.[158] A good printability is characterized by the control of the flow properties, meaning the reproducibility and accuracy of the printed material. Furthermore, the materials need to be stable enough to retain the shape until it is fixed through a cross-linking mechanism.[159] A higher polymer concentration enhances printability; however, it also forms a denser network affecting cell viability and spreading.[160] The stability of the printed shapes of most bioinks is moderate because of their high water content weighing down the constructs.[161] Most efforts to improve printability impair the biological properties of the material.[162] Recent approaches to overcome these difficulties is the use of dynamic bioinks,[163] such as supramolecular materials comprising guest–host interactions. These materials exhibit shear thinning, shear recovery, and self-healing properties, as well as cytocompatibility.[164–166] However, the transient nature of the bonds eventually leads to a decomposition of the material over time and the lack of porosity (similar to bulk hydrogels), limiting their potential.[163] Another innovative approach is the implementation of microgels as a bioink, accomplishing both requirements, printability and biological compatibility.[167,168] Microgel bioinks show little intrinsic flow behavior and provide shear thinning and recovery owed to weak particle–particle interactions.[167] Another advantage is the microporosity of the printed structures due to void spaces during jamming,[168,169] supporting cell migration and proliferation. Most established microgel bioinks use spherical microgels that require secondary cross-linking to be stable in aqueous media and additionally do not provide any cellular directional guidance cues because of their isotropy. The latter is limiting the biological applications. Alternatively, a high-volume, cell-friendly bioink composed of high-aspect-ratio microelements was established, reporting the first implementation of elongated microgels in 3D bioprinting.[170] For this method, a bulk hydrogel is pushed through a grid with micron-sized (40–100 μm) holes, resulting in individual randomly entangled microstrands. The advantage of this technique is its simplicity, as it is fast and straightforward and can be applied to a wide range of materials. Especially interesting for clinical translation is the high scalability of the fabrication method (volumes of liters). Different natural polymers have been tested with this method, such as hyaluronan-methacrylate (HA-MA), gelatin, collagen, and so forth, as well as different types of cross-linking degrees. The stability of the printed constructs was assessed by incubating them at 37°C for seven days in phosphate-buffered saline (PBS) in an agitated environment. All entangled constructs remained stable and did not need any secondary cross-linking, compared to granular microgels. In tissue engineering, porosity plays a major role, as it is essential for transport of nutrients, gas exchange, and cell behavior.[171,172] The pore size is particularly important for vascularization and cell migration. Here, the pore size has been assessed through fluorescent high-molecular-weight dextran dye that infiltrates the void space, but not the microstrands. The amount of void space was acquired through image analysis and resulted, for example, in ~2% using the grid with bigger holes (100 μm) and ~7% of void space using the grid with smaller holes (40 μm). The void fraction can be adapted through the mesh size and the cross-linking density of the microstrands themselves as this will affect their entanglement. Another important feature of the printed microstrands is their alignment along the 3D printing direction, compared to conventional bulk hydrogels. An anisotropic alignment along one axis makes the material applicable for cellular guidance, mimicking the native tissue's function, as seen in cartilage or muscle fibers. The cytocompatibility of the printing process was analyzed in two different approaches. In the first approach, the cells were mixed inside the bulk hydrogel before pushing it through the mesh into microstrands, offering guidance cues during tissue maturation. Doing so, multinuclear myotubes oriented in the same direction as the microstrands were generated. In the second approach, the cells were suspended in the entangled microstrands and were then found to be in the void space. The latter was used to engineer artificial cartilage tissue using bovine chondrocytes mixed with HA-MA microstrands. The chondrocytes had a viability of > 90% and were round shaped after printing. After culturing them for 7 days up do 42 days, they showed an elongated shape. What was a translucent construct after printing became white after 42 days in culture. This indicated a dense formation of ECM and staining for cartilage-specific markers was conducted to prove this finding. Collagens I and II were both present after 21 days, though collagen I decreased after 42 days. This may be due to the dedifferentiation of chondrocytes.[173] Here, dedifferentiation refers to chondrocytes that do not provide differentiated functions, thus not producing cartilage proteoglycan and collagens. In addition to collagen expression, a significant increase in the compression modulus from 2.7 kPa to 212 kPa after 21 days and to 780 kPa after 42 days was observed. The reached value after 42 days is lower than the compression modulus of native articular bovine cartilage with 1829 kPa but is still a strong increase. It needs to be stated that the void fraction of the entangled microstrands is very low compared to microgel-based scaffolds.[33,159,174] For example, when 10 μm rods with aspect ratios of 5, 10, or 20 are chemically interlinked, void spaces of 65%, 75%, or 90% are achieved.[174] Therefore, the macroporosity in between the microstrands could be improved to ensure better cell migration and spreading. Furthermore, characterizing the individual microstrands has not been possible so far, as their separation would lead to breakage. Additionally, the fabrication method can, to date, only be applied to soft hydrogels, restricting the range of materials.

Another approach to ensure the needed mechanical and bioactive properties in 3D bioprinting is to introduce multiple materials.[175,176] Chitosan (CHT), a natural polycationic linear polysaccharide, is an established material that provides structural similarity to glycosaminoglycans (GAGs), which are an important part of the ECM.[177,178] As this polysaccharide is of cationic nature, it cross-links through ionic interactions[179] and demands a secondary polymer with anionic charges, such as the biopolymer gellan gum

(GG). The combination of both has been used to form scaffolds.[180,181] No fibrous structures have been generated using the above-described composition, as methods such as electrospinning are not possible because of the great charge repulsion. To overcome this challenge, microfluidic devices have been employed to produce aligned microstructures.[182,183] A review by Sun et al.[184] elucidates the advantages of combining microfluidics with 3D bioprinting, such as control over the microenvironment or the functionality, thus impacting cell viability. Regarding the microfluidic device itself, 3D-printed microfluidic devices have been studied to generate new materials, as they offer a high controllability and tunability of the flow, compared to conventional PDMS molding techniques using, for example, silicon or glass wafers.[185–188] In order to obtain good printability and the right biomechanical properties to, for example, replicate the complex organization of collagen fibers during bioprinting, two cross-linking mechanisms were combined. The biomaterial ink consisted of methacrylated GG (GG-MA, polyanion)[189] and CHT (polycation) and is first cross-linked ionically during mixing in the microfluidic device. The material is then photocrosslinked in a second step inside the microfluidic nozzle in situ before deposition to reinforce the printed structure.[190] This approach was later used to form polyelectrolyte complex (PEC) microstructures.[191,192] The PECs performed better in terms of their stability with temperatures ranging from 5 to 40°C, compared to both components used individually.[193] Thus, the printer set up did not need the implementation of a temperature control. The printed structures, obtained from extrusion through the printer nozzle using the microfluidic approach, had an elastic modulus of 1.04 MPa, similar to collagen tissue in the body. Employing scanning electron microscopy (SEM), it can be seen that the hydrogel structure bundles (~ 700 μm) are composed of multiple microstructures aligned in a parallel manner, with a diameter of the microfibrils of 1–2 μm.[190] This technique can in the future be used to encapsulate cells in situ and guide them with the aligned structures. However, the cytocompatibility would need to be assessed.

Cell-laden structures are on the rise for regenerative biomaterials,[194,195] especially when it comes to muscle tissue regeneration. Myoblasts have been mixed with natural hydrogels, such as alginate, GelMa, or nanocellulose, to function as bioinks employing different fabrication techniques.[196] Bioprinting is a promising tool, as it allows for the production of multiscalable structures in high resolution. Furthermore, when adding cells inside the bioink, it enables the formation of structures and cell deposition in 3D.[197,198] One 3D bioprinting method employing in situ cross-linking to encapsulate cells has been successfully used to produce cell-laden straddles, without any loss of cell viability.[199] However, this strategy does not provide any topographical cues and thus no cellular guidance, which is needed for the alignment of myofibers in a parallel manner to guarantee the tissue's function. Other bioprinting processes have managed cell alignment and uniaxially patterned myotube formation via applying wall shear stress within an extruding nozzle, but this lacks mechanical stability after printing.[200] Additionally, the fabrication parameters, such as the printing time (< 60 min) or the volume flow rate, are restricted to a narrow range to ensure high cell viability[201] and exhibit loss of cells through leaching.[202,203] To overcome these limitations, another technique has been developed based on the in situ cross-linking system from Ouyang et al.,[199] where UV cross-linking happens at the printer nozzle. In the more advanced method, cell-laden structures made of GelMa with C2C12 or GelMa with human adipose stem cells (hASCs) cells are produced to obtain oriented muscle constructs for tissue regeneration after volumetric muscle loss (VML), by adjusting the in situ cross-linking procedure.[204] Here, two UV irradiation systems and a glass nozzle are mounted to the printing barrel. Having multiple UV sources enables an equidimensional strand, and the printed material showed a low extrudate swelling, resulting in a preservation of the structure after the extrusion process. Furthermore, the printed structures were morphologically aligned, owed to shear stress. The mechanical properties of the material have an influence on the shear-induced alignment of the hydrogels. Therefore, two different temperatures were analyzed, where at 4°C, the hydrogel exhibited more elastic properties, and at 20°C, due to a sol–gel transition, more viscous properties. At 4°C, the printing wall shear stress was five times higher compared to 20°C, when aiming for comparable volume flow rates. Still, both temperatures resulted in similar cell viability (> 90%), but at 4°C, the cytoskeleton was more oriented after seven days of culturing compared to the cells printed at 20°C. This was most likely because of the much higher wall shear stress that induced an orientation of the GelMa fibers and cells inside the nozzle.[205] A better alignment of the nuclei was also observed likely due to a better cell–cell interaction. The cellular orientation and myogenic differentiation factors became constant at a wall shear stress of 625 Pa. To identify the regenerative capacity, the myogenic gene expression of the cell-laden construct was analyzed and compared to normally printed cell-laden GelMa structures. Cellular guidance and myotube heavy chain (MHC) were much higher in in situ cross-linked (ISC) constructs than in conventionally cross-linked (CC) constructs. Bioprinted constructs with hASCs were then also implanted in a mouse model and, based on the results, indicated a high degree of myogenesis. These findings depict the high potential of the method for regeneration after VML; however, bioprinting at 4°C impacts cell viability, setting the maximum printing duration to 1 hour. This limits the ability of printing larger constructs and thus being translated to clinical applications.

References

1. Humphrey JD, Dufresne ER, Schwartz MA. Mechanotransduction and extracellular matrix homeostasis. *Nat Rev Mol Cell Biol.* 2014;15(12):802–812.
2. Frantz C, Stewart KM, Weaver VM. The extracellular matrix at a glance. *J Cell Sci.* 2010;123(24):4195–4200.
3. Järveläinen H, Sainio A, Koulu M, Wight TN, Penttinen R. Extracellular matrix molecules: Potential targets in pharmacotherapy. *Pharmacol Rev.* 2009;61(2):198–223.
4. Sanders KM, Koh SD, Ro S, Ward SM. Regulation of gastrointestinal motility — Insights from smooth muscle biology. *Nat Rev Gastroenterol Hepatol.* 2012;9(11):633–645.

5. Chan-Park MB, Shen JY, Cao Y, Xiong Y, Liu Y, Rayatpisheh S, et al. Biomimetic control of vascular smooth muscle cell morphology and phenotype for functional tissue-engineered small-diameter blood vessels. *J Biomed Mater Res A*. 2009;88A(4):1104–1121.
6. Morrow DA, Haut Donahue TL, Odegard GM, Kaufman KR. Tensile material properties of skeletal muscle tissue in longitudinal and transverse directions. In: *Summer Bioengineering Conference*. New York: American Society of Mechanical Engineers; 2008.
7. Morrow DA, Haut Donahue TL, Odegard GM, Kaufman KR. Transversely isotropic tensile material properties of skeletal muscle tissue. *J Mech Behav Biomed Mater*. 2010;3(1):124–129.
8. Böl M, Kruse R, Ehret AE, Leichsenring K, Siebert T. Compressive properties of passive skeletal muscle — The impact of precise sample geometry on parameter identification in inverse finite element analysis. *J Biomech*. 2012;45(15):2673–2679.
9. Takaza M, Moerman KM, Simms CK. Passive skeletal muscle response to impact loading: Experimental testing and inverse modelling. *J Mech Behav Biomed Mater*. 2013;27:214–225.
10. Pietsch R, Wheatley BB, Haut Donahue TL, Gilbrech R, Prabhu R, Liao J, et al. Anisotropic compressive properties of passive porcine muscle tissue. *J Biomech Eng*. 2014;136(11):111003.
11. Wong J, Kuhl E. Generating fibre orientation maps in human heart models using poisson interpolation. *Comput Methods Biomech Biomed Engin*. 2014;17(11):1217–1226.
12. Meek KM, Knupp C. Corneal structure and transparency. *Prog Retin Eye Res*. 2015;49:1–16. (1873–1635 (Electronic)).
13. Klein TJ, Malda J, Sah RL, Hutmacher DW. Tissue engineering of articular cartilage with biomimetic zones. *Tissue Eng Part B Rev*. 200915(2):143–157.
14. Marcelino J, Carpten JD, Suwairi WM, Gutierrez OM, Schwartz S, Robbins C, et al. CACP, encoding a secreted proteoglycan, is mutated in camptodactyly-arthropathy-coxa vara-pericarditis syndrome. *Nat Genet*. 1999;23(3):319–322.
15. Fox AJS, Bedi A, Rodeo SA. The basic science of articular cartilage: Structure, composition, and function. *Sports Health*. 2009; 1(6):461–468.
16. Dwyer KD, Coulombe KLK. Cardiac mechanostructure: Using mechanics and anisotropy as inspiration for developing epicardial therapies in treating myocardial infarction. *Bioact Mater*. 2021; 6(7):2198–2220.

17. Bader D, Lee D. Chapter 4 - Structure – Properties of soft tissues articular cartilage. In: Elices M, editor. *Pergamon Materials Series.* Pergamon; 2000. p. 75–103.
18. Santra M, Liu Y-C, Jhanji V, Yam GHF. Human SMILE-derived stromal lenticule scaffold for regenerative therapy: Review and perspectives. *Int J Mol Sci.* 2022;23(14):7967.
19. Paras Health. *Osteoarthritis knee.* 2022. [cited 2023 April 5]. https://www.parashospitals.com/blogs/osteoarthritis-knee.
20. Kim BS, Mooney DJ. Development of biocompatible synthetic extracellular matrices for tissue engineering. *Trends Biotechnol.* 1998;16(5):224–230. doi:10.1016/s0167-7799(98)01191-3. (0167-7799 (Print)).
21. Harris LD, Kim BS, Mooney DJ. Open pore biodegradable matrices formed with gas foaming. *J Biomed Mater Res.* 1998 Dec 5; 42(3):396–402. (0021-9304 (Print)).
22. Thomson RC, Mikos AG, Beahm E, Lemon JC, Satterfield WC, Aufdemorte TB, et al. Guided tissue fabrication from periosteum using preformed biodegradable polymer scaffolds. *Biomaterials.* 1999;20(21):2007–2018. doi:10.1016/s0142-9612(99)00103-9. (0142-9612 (Print)).
23. Lee KY, Mooney DJ. Hydrogels for tissue engineering. *Chem Rev.* 2001;101(7):1869–1880.
24. Jhon MS, Andrade JD. Water and hydrogels. *J Biomed Mater Res.* 1973;7(6):509–522.
25. Kondiah PJ, Choonara YE, Kondiah PPD, Marimuthu T, Kumar P, Du Toit LC, et al. A review of injectable polymeric hydrogel systems for application in bone tissue engineering. *Molecules.* 2016 Nov 21;21(11):1580. (1420-3049 (Electronic)).
26. Vacanti JP, Vacanti CA. Chapter 1 — The history and scope of tissue engineering. In: Lanza RP, Langer R, Vacanti J, editors. *Principles of Tissue Engineering.* 2nd ed. San Diego: Academic Press; 2000. p. 3–7.
27. Whitaker MJ, Quirk RA, Howdle SM, Shakesheff KM, et al. Growth factor release from tissue engineering scaffolds. *J Pharm Pharmacol.* 2001;53(11):1427–1437.
28. Lien SM, Ko L-Y, HuangT-J. Effect of pore size on ECM secretion and cell growth in gelatin scaffold for articular cartilage tissue engineering. *Acta Biomater.* 2009;5(2):670–679. (1878-7568 (Electronic)).

29. Griffon DJ, Sedighi MR, Schaeffer DV, Eurell JA, Johnson AL. Chitosan scaffolds: Interconnective pore size and cartilage engineering. *Acta Biomater.* 2006;2(3):313–320. (1742-7061 (Print)).
30. Yang Y, De Laporte L, Zelivyanskaya ML, Whittlesey KJ, Anderson AJ, Cummings BJ, et al. Multiple channel bridges for spinal cord injury: Cellular characterization of host response. *Tissue Eng Part A.* 2009;15(11):3283–3295. (1937–335X (Electronic)).
31. He X, Jabbari E. Material properties and cytocompatibility of injectable MMP degradable poly(lactide ethylene oxide fumarate) hydrogel as a carrier for marrow stromal cells. *Biomacromolecules.* 2007;8(3):780–792.
32. Huebsch N, Lippens E, Lee K, Mehta M, Koshy ST, Darnel MC, et al. Matrix elasticity of void-forming hydrogels controls transplanted-stem-cell-mediated bone formation. *Nat Mater.* 2015; 14(12):1269–1277.
33. Rommel D, Mork M, Vedaraman S, Bastard C, Guerzoni LPB, Kittel Y, et al. Functionalized microgel rods interlinked into soft macroporous structures for 3D cell culture. *Adv Sci (Weinh).* 2022;9(10):e2103554.
34. Macaya D, Spector M. Injectable hydrogel materials for spinal cord regeneration: A review. *Biomed Mater.* 2012;7(1):012001. (1748–605X (Electronic)).
35. Rizzo R, Bonato A, Chansoria P, Zenobi-Wong M. Macroporous aligned hydrogel microstrands for 3D cell guidance. *ACS Biomat Sci Eng.* 2022;8(9):3871–3882.
36. Zhang S, Greenfield MA, Mata A, Palmer LC, Bitton R, Mantei JR, et al. A self-assembly pathway to aligned monodomain gels. *Nat Mater.* 2010;9(7):594–601.
37. Tringides CM, Mooney DJ. Conductive hydrogel scaffolds for the 3D localization and orientation of fibroblasts. *Macromol Biosci.* 2024;24(1):e2300044.
38. Rose JC, Cámara-Torres M, Rahimi K, Köhler J, Möller M, De Laporte L, et al. Nerve cells decide to orient inside an injectable hydrogel with minimal structural guidance. *Nano Lett.* 2017; 17(6):3782–3791.
39. Liu H, Chansoria P, Delrot P, Angelidakis E, Rizzo R, Rütsche D, et al. Filamented light (FLight) biofabrication of highly aligned tissue-engineered constructs. *Adv Mater.* 2022;34(45):e2204301.

40. Petta D, Basoli V, Pellicciotta D, Tognato R, Barcik J, Arrigoni C, et al. Sound-induced morphogenesis of multicellular systems for rapid orchestration of vascular networks. *Biofabrication.* 2021; 13(1):015004.
41. Dubey N, Letourneau PC, Tranquillo RT. Neuronal contact guidance in magnetically aligned fibrin gels: Effect of variation in gel mechano-structural properties. *Biomaterials.* 2001;22(10): 1065–1075.
42. De France KJ, Yager KG, Chan KJW, Corbett B, Cranston ED, Hoare T. Injectable anisotropic nanocomposite hydrogels direct in situ growth and alignment of myotubes. *Nano Lett.* 2017; 17(10):6487–6495.
43. Antman-Passig M, Shefi O. Remote magnetic orientation of 3D collagen hydrogels for directed neuronal regeneration. *Nano Lett.* 2016;16(4):2567–2573.
44. Alsberg E, Feinstein E, Joy MP, Prentiss M, Ingber DE. Magnetically-guided self-assembly of fibrin matrices with ordered nano-scale structure for tissue engineering. *Tissue Eng.* 2006; 12(11):3247–3256.
45. Babu S, Vedaraman S, Nava JLG, Licht C, Haraszti T, De Laporte L, et al. How do the local physical, biochemical, and mechanical properties of an injectable synthetic anisotropic hydrogel affect oriented nerve growth? *Adv Funct Mater.* 2022;32(50):2202468.
46. Rose JC, Gehlen DB, Haraszti T, Köhler J, Licht CJ, De Laporte L, et al. Biofunctionalized aligned microgels provide 3D cell guidance to mimic complex tissue matrices. *Biomaterials.* 2018;163: 128–141.
47. Licht C, Rose JC, Anarkoli AO, Blondel D, Roccio M, Haraszti T, et al. Synthetic 3D PEG-anisogel tailored with fibronectin fragments induce aligned nerve extension. *Biomacromolecules.* 2019; 20(11):4075–4087.
48. Penner R, Neher E. The role of calcium in stimulus-secretion coupling in excitable and non-excitable cells. *J Exp Biol.* 1988;139(1): 329–345.
49. Adams DS, Levin M. Endogenous voltage gradients as mediators of cell-cell communication: Strategies for investigating bioelectrical signals during pattern formation. *Cell Tissue Res.* 2013; 352(1):95–122.

50. Ravikanth M, Soujanya P, Manjunath K, Saraswathi TR, Ramachandran CR, et al. Heterogenecity of fibroblasts. *J Oral Maxillofac Pathol: JOMFP.* 2011;15(2):247–250.
51. Zou M-L, Teng Y-Y, Wu J-J, Liu S-Y, Tang X-Y, Jia Y, et al. Fibroblasts: Heterogeneous cells with potential in regenerative therapy for scarless wound healing. *Front Cell Dev Biol.* 2021;9: 713605.
52. Kendall R, Feghali-Bostwick C. Fibroblasts in fibrosis: Novel roles and mediators. *Front Pharmacol.* 2014;5:123.
53. Urabe H, Akimoto R, Kamiya S, Hosoki K, Ichikawa H, Nishiyama T. Effects of pulsed electrical stimulation on growth factor gene expression and proliferation in human dermal fibroblasts. *Mol Cell Biochem.* 2021;476(1):361–368.
54. Liu H, Wu M, Jia Y, Niu L, Huang L, Xu F. Control of fibroblast shape in sequentially formed 3D hybrid hydrogels regulates cellular responses to microenvironmental cues. *NPG Asia Mater.* 2020;12(1):45.
55. Borgens RB, Vanable JW Jr, Jaffe LF. Bioelectricity and regeneration: Large currents leave the stumps of regenerating newt limbs. *Proc Natl Acad Sci U S A.* 1977;74(10):4528–4532.
56. Zhao M. Electrical fields in wound healing — An overriding signal that directs cell migration. *Semin Cell Dev Biol.* 2009;20(6): 674–682. Elsevier.
57. Cheah YJ, Buyong MR, Mohd Yunus MH. Wound healing with electrical stimulation technologies: A review. *Polymers(Basel).* 2021;13(21):3790.
58. Bhardwaj ND, Devi D, Mandal BB. Tissue-engineered cartilage: The crossroads of biomaterials, cells and stimulating factors. *Macromol Biosci.* 2015;15(2):153–182.
59. Green RA, Hassarati RT, Goding JA, Baek S, Lovell NH, Martens PJ, et al. Conductive hydrogels: Mechanically robust hybrids for use as biomaterials. *Macromol Biosci.* 2012;12(4):494–501.
60. Liu K, Wei S, Song L, Liu H, Wang T. Conductive hydrogels — A novel material: Recent advances and future perspectives. *J Agric Food Chem.* 2020;68(28):7269–7280.
61. Barberio C, Saez J, WithersA, Nair M, Tamagnini F, Owens RM. Conducting polymer-ECM scaffolds for human neuronal cell differentiation. *Adv Healthc Mater.* 2022;11(20):2200941.

62. Tringides CM, Boulingre M, Khalil A, Lungjangwa T, Jaenisch R, Mooney DJ, et al. Tunable conductive hydrogel scaffolds for neural cell differentiation. *Adv Healthc Mater.* 2023;12(7):2202221.
63. Tringides CM, Vachicouras N, de Lázaro I, Wang H, Trouillet A, Seo BR, et al. Viscoelastic surface electrode arrays to interface with viscoelastic tissues. *Nat Nanotechnol.* 2021;16(9):1019–1029.
64. Tringides CM, Boulingre M, Mooney DJ. Metal-based porous hydrogels for highly conductive biomaterial scaffolds. *Oxf Open Mater Sci.* 2024;3(1):itad002.
65. Joodaki H, Panzer MB. Skin mechanical properties and modeling: A review. *Proc Inst Mech Eng H.* 2018;232(4):323–343.
66. Novotny GE, Gnoth C. Variability of fibroblast morphology in vivo: A silver impregnation study on human digital dermis and subcutis. *J Anat.* 1991;177:195–207.
67. Hakkinen KM, Harunaga JS, Doyle AD, Yamada KM. Direct comparisons of the morphology, migration, cell adhesions, and actin cytoskeleton of fibroblasts in four different three-dimensional extracellular matrices. *Tissue Eng Part A.* 2011;17(5–6):713–724.
68. Namsheer K. Rout CS. Conducting polymers: A comprehensive review on recent advances in synthesis, properties and applications. *RSC Adv.* 2021;11(10):5659–5697.
69. Harris AR, Wallace G. Organic electrodes and communications with excitable cells. *Adv Funct Mater.* 2018;28(12):1700587.
70. Kaur G, Adhikari R, Cass P, Bown M, Gunatillake T. Electrically conductive polymers and composites for biomedical applications. *RSC Adv.* 2015;5(47):37553–37567.
71. Mawad D, Stewart EM, Officer DL, Romeo T, Wagner P, Wagner K, et al. A single component conducting polymer hydrogel as a scaffold for tissue engineering. *Adv Funct Mater.* 2012;22(13): 2692–2699.
72. Rocha I, Cerqueira G, Penteado FV, Córdoba de Torresi SI. Electrical stimulation and conductive polymers as a powerful toolbox for tailoring cell behaviour in vitro. *Front Med Technol.* 2021;3:670274.
73. Guo B, Finne-Wistrand A, Albertsson AC. Versatile functionalization of polyester hydrogels with electroactive aniline oligomers. *J Polym Sci A Pol Chem.* 2011;49(9):2097–2105.
74. Sukmana I. Bioactive polymer scaffold for fabrication of vascularized engineering tissue. *J Artif Organs.* 2012;15:215–224.

75. Zhang Z, Chen W, Tiemessen DM, Oosterwijk E, Kouwer PHJ. A temperature-based easy-separable (tempEasy) 3D hydrogel coculture system. *Adv Healthc Mater.* 2022;11(10):2102389.
76. Das RK, Gocheva V, Hammink R, Zouani OF, Rowan AE. Stress-stiffening-mediated stem-cell commitment switch in soft responsive hydrogels. *Nat Mater.* 2016;15(3):318–325.
77. Ye S, Boeter JWB, Mihajlovic M, Van Steenbeek FG, Van Wolferen ME, Oosterhoff LA, et al. A chemically defined hydrogel for human liver organoid culture. *Adv Funct Mater.* 2020; 30(48):2000893.
78. Zhang Y, Tang C, Span PN, Rowan AE, Aalders TW, Schalken JA, et al. Polyisocyanide hydrogels as a tunable platform for mammary gland organoid formation. *Adv Sci(Weinh).* 2020;7(18): 2001797.
79. Kouwer PH, Koepf M, Le Sage VAA, Jaspers M, van Buul AM, Eksteen-Akeroyd ZH, et al. Responsive biomimetic networks from polyisocyanopeptide hydrogels. *Nature.* 2013;493(7434):651–655.
80. Liu K, Wiendels M, Yuan H, Ruan C, Kouwer PHJ. Cell-matrix reciprocity in 3D culture models with nonlinear elasticity. *Bioact Mater.* 2022;9:316–331.
81. op't Veld RC, van den Boomen OI, Lundvig DMS, Bronkhorst EM, Kouwer PHJ, Jansen JA, et al. Thermosensitive biomimetic polyisocyanopeptide hydrogels may facilitate wound repair. *Biomaterials.* 2018;181:392–401.
82. Wang B, Booij-Vrieling HE, Bronkhorst EM, Shao J, Kouwer PHJ, Jansen JA, et al. Antimicrobial and anti-inflammatory thermo-reversible hydrogel for periodontal delivery. *Acta Biomater.* 2020;116:259–267.
83. Srisuk P, Berti FV, Da Silva LP, Marques AP, Reis RL, Correlo VM, et al. Electroactive gellan gum/polyaniline spongy-like hydrogels. *ACS Biomater Sci Eng.* 2018;4(5):1779–1787.
84. Bae HA, Puranik AS, Gauvin R, Edalat F, Carrillo-Conde B, Peppas NA, et al. Building vascular networks. *Sci Transl Med.* 2012;4(160):160ps23.
85. Hasan A, Paul A, Vrana NE, Zhao X, Memic A, Hwang YS, et al. Microfluidic techniques for development of 3D vascularized tissue. *Biomaterials.* 2014;35(26):7308–7325.
86. Jeon JS, Bersini S, Gilardi M, Dubini G, Charest JL, Moretti M, et al. Human 3D vascularized organotypic microfluidic assays to

study breast cancer cell extravasation. *Proc Natl Acad Sci U S A.* 2015;112(1):214–219.

87. Chrobak KM, Potter DR, Tien J. Formation of perfused, functional microvascular tubes in vitro. *Microvasc Res.* 2006; 71(3):185–196.
88. Golden AP, Tien J. Fabrication of microfluidic hydrogels using molded gelatin as a sacrificial element. *Lab Chip.* 2007;7(6):720–725.
89. Miller JS, Stevens KR, Yang MT, Baker BM, Nguyen DHT, Cohen DM, et al. Rapid casting of patterned vascular networks for perfusable engineered three-dimensional tissues. *Nat Mater.* 2012;11(9): 768–774.
90. Guven S, Chen P, Inci F, Tasoglu S, Erkmen B, Demirci U, et al. Multiscale assembly for tissue engineering and regenerative medicine. *Trends Biotechnol.* 2015;33(5):269–279.
91. Naseer SM, Manbachi A, Samandari M, Walch P, Gao Y, Zhang YS, et al. Surface acoustic waves induced micropatterning of cells in gelatin methacryloyl (GelMA) hydrogels. *Biofabrication.* 2017;9(1):015020.
92. Xing J, Liu N, Xu N, Chen W, Xing D. Engineering complex anisotropic scaffolds beyond simply uniaxial alignment for tissue engineering. *Adv Funct Mater.* 2022;32(15):2110676.
93. Kang B, Shin J, Park HJ, Rhyou C, Kang D, Lee SJ, et al. High-resolution acoustophoretic 3D cell patterning to construct functional collateral cylindroids for ischemia therapy. *Nat Commun.* 2018;9(1):5402.
94. Bouyer C, Chen P, Güven S, Demirtaş TT, Nieland TJF, Padilla F, et al. A bio-acoustic levitational (BAL) assembly method for engineering of multilayered, 3D brain-like constructs, using human embryonic stem cell derived neuro-progenitors. *Adv Mater.* 2016;28(1):161–167.
95. Chen P, Güven S, Usta OB, Yarmush ML, Demirci U, et al. Biotunable acoustic node assembly of organoids. *Adv Healthc Mater.* 2015;4(13):1937–1943.
96. Sheldrake M, Sheldrake R. Determinants of faraday wave-patterns in water samples oscillated vertically at a range of frequencies from 50–200 Hz. *Water.* 2017;9:1–27.
97. Huepe C, Ding Y, Umbanhowar P, Silber M. Forcing function control of Faraday wave instabilities in viscous shallow fluids. *Phys Rev E Stat Nonlin Soft Matter Phys.* 2006;73(1 Pt 2):016310.

98. Armstrong JPK, Pchelintseva E, Treumuth S, Campanella C, Meinert C, Klein TJ, et al. Tissue engineering cartilage with deep zone cytoarchitecture by high-resolution acoustic cell patterning. *Adv Healthc Mater.* 2022;11(24):e2200481.
99. Antons J, Marascio MGM, Nohava J, Martin R, Applegate LA, Bourban PE, et al. Zone-dependent mechanical properties of human articular cartilage obtained by indentation measurements. *J Mater Sci Mater Med.* 2018 May 4;29(5):57. (1573-4838 (Electronic)).
100. Utech S, Boccaccini AR. A review of hydrogel-based composites for biomedical applications: Enhancement of hydrogel properties by addition of rigid inorganic fillers. *J Mater Sci.* 2016; 51(1):271–310.
101. Lee M, Rizzo R, Surman F, Zenobi-Wong M. Guiding lights: Tissue bioprinting using photoactivated materials. *Chem Rev.* 2020; 120(19):10950–11027.
102. Lim KS, Galarraga JH, Cui X, Lindberg GCJ, Burdick JA, Woodfield TBF. Fundamentals and applications of photo-cross-linking in bioprinting. *Chem Rev.* 2020 Oct 14;120(19):10662–10694. (1520-6890 (Electronic)).
103. Moroni L, Burdick JA, Highley C, Lee SJ, Morimoto Y, Takeuchi S, et al. Biofabrication strategies for 3D in vitro models and regenerative medicine. *Net Rev Mater.* 2018;3(5):21–37. (2058-8437 (Print)).
104. Bernal PN, Delrot P, Loterie D, Li Y, Malda J, Moser C, et al., Volumetric bioprinting of complex living-tissue constructs within seconds. *Adv Mater.* 2019;31(42):1904209.
105. Rizzo R, Rutsche D, Zenobi M, Liu H. Optimized photoclick (bio)resins for fast volumetric bioprinting. *Adv Mater.* 2021;33(49): 2102900.
106. Bernal PN, Bouwmeester M, Madrid-Wolff J, Falandt M, Florczak S, Ginés Rodriguez N, et al. Volumetric bioprinting of organoids and optically tuned hydrogels to build liver-like metabolic biofactories. *Adv Mater.* 2022;34(15):2110054.
107. Chen Z, Khuu N, Xu F, Kheiri S, Yakavets I, Rakhshani F, et al. Printing structurally anisotropic biocompatible fibrillar hydrogel for guided cell alignment. *Gels.* 2022;8(11):685.
108. Gholobova D, Terrie L, Gerard M, Declercq H, Thorrez L. Vascularization of tissue-engineered skeletal muscle constructs. *Biomaterials.* 2020;235:119708. (1878-5905 (Electronic)).

109. Zhang J, Hu Q, Wang S, Tao J, Gou M. Digital light processing based three-dimensional printing for medical applications. *Int J Bioprint.* 2019;6(1):242.(2424-8002 (Electronic)).
110. Streppe U, Michaelis D, Kowarschik R, Brauer A. Modulational instability in systems with integrating nonlinearity. *Phys Rev Lett.* 2005;95(7):073901.
111. Yang L, Ge L, van Rijn P. Synergistic effect of cell-derived extracellular matrices and topography on osteogenesis of mesenchymal stem cells. *ACS Appl Mater Interfaces.* 2020;12(23): 25591–25603.
112. Hu Y, You J-O, Aizenberg J. Micropatterned hydrogel surface with high-aspect-ratio features for cell guidance and tissue growth. *ACS Appl Mater Interfaces.* 2016;8(34):21939–21945.
113. Vedaraman S, Perez-Tirado A, Haraszti T, Gerardo-Nava J, Nishiguchi A, De Laporte L. Anisometric microstructures to determine minimal critical physical cues required for neurite alignment. *Adv Healthc Mater.* 2021;10(20):e2100874.
114. Jevtić P, Edens LJ, Vukovic LD, Levy DL. Sizing and shaping the nucleus: Mechanisms and significance. *Curr Opin Cell Biol.* 2014; 28:16–27.
115. Uhler C, Shivashankar GV. Regulation of genome organization and gene expression by nuclear mechanotransduction. *Nat Rev Mol Cell Biol.* 2017;18(12):717–727.
116. Islam A, Younesi M, Mbimba T, Akkus O. Collagen substrate stiffness anisotropy affects cellular elongation, nuclear shape, and stem cell fate toward anisotropic tissue lineage. *Adv Healthc Mater.* 2016;5(17):2237–2247.
117. Kip D, Soljacic M, Segev M, Eugenieva E, Christodoulides DN. Modulation instability and pattern formation in spatially incoherent light beams. *Science.* 2000;290(5491):495–498.
118. Hosein ID. Light-directed organization of polymer materials from photoreactive formulations. *Chem Mater.* 2020;32(7):2673–2687.
119. Biria S, Malley PPA, Kahan TF, Hosein ID. Tunable nonlinear optical pattern formation and microstructure in cross-linking acrylate systems during free-radical polymerization. *J Phys Chem C.* 2016;120(8):4517–4528.
120. Anyfantakis M, Loppinet B, Fytas G, Pispas S. Optical spatial solitons and modulation instabilities in transparent entangled polymer solutions. *Opt Lett.* 2008;33(23):2839–2841.

121. Madrid-Wolff J, Forero-Shelton M. 4f Koehler transmitted illumination condenser for teaching and low-cost microscopic imaging. *The Biophysicist.* 2020;1(2):3–14.
122. Prince E, Kumacheva E. Design and applications of man-made biomimetic fibrillar hydrogels. *Nat Rev Mater.* 2019;4(2):99–115.
123. Mouw JK, Ou G, Weaver VM. Extracellular matrix assembly: A multiscale deconstruction. *Nat Rev Mol Cell Biol.* 2014;15(12):771–785.
124. Gaharwar A, Singh I, Khademhosseini A. The author declares no competing interests. *Nat Rev. Mater.* 2020;5:686–705.
125. Nikolova MP, Chavali MS. Recent advances in biomaterials for 3D scaffolds: A review. *Bioact Mater.* 2019;4:271–292.
126. Zhu M, Li W, Dong X, Yuan X, Midgley AC, Chang H, et al. In vivo engineered extracellular matrix scaffolds with instructive niches for oriented tissue regeneration. *Nat Commun.* 2019;10(1):4620.
127. Wolf K, Te Lindert M, Krause M, Alexander S, Te Riet J, Willis AL, et al. Physical limits of cell migration: Control by ECM space and nuclear deformation and tuning by proteolysis and traction force. *J Cell Biol.* 2013;201(7):1069–1084.
128. Egerbacher M, Gabner S, Battisti S, Handschuh S. Tenocytes form a 3-D network and are connected via nanotubes. *J Anat.* 2020;236(1):165–170.
129. Liu W, Cui Y, Wei J, Sun J, Zheng L, Xie J. Gap junction-mediated cell-to-cell communication in oral development and oral diseases: A concise review of research progress. *Int J Oral Sci.* 2020;12(1):17.
130. McNally EM, Demonbreun AR. Resealing and rebuilding injured muscle. *Science.* 2021 Oct 15;374(6565):262–263. (1095-9203 (Electronic)).
131. Zhou T, Gao B, Fan Y, Liu Y, Feng S, Cong Q, et al. Piezo1/2 mediate mechanotransduction essential for bone formation through concerted activation of NFAT-YAP1-ß-catenin. *eLife.* 2020;9:e52779.
132. Fernandez-Yague MA, Trotier A, Demir S, Abbah SA, Larrañaga A, Thirumaran A, et al. A self-powered piezo-bioelectric device regulates tendon repair-associated signaling pathways through modulation of mechanosensitive ion channels. *Adv Mater.* 2021; 33(40):2008788.
133. Wang L, You X, Lotinun S, Zhang L, Wu N, Zou W. Mechanical sensing protein PIEZO1 regulates bone homeostasis via osteoblast-osteoclast crosstalk. *Nat Commun.* 2020;11(1):282.

134. Fu Y, Wan P, Zhang J, Li X, Xing J, Zou Y, et al. Targeting mechanosensitive piezo1 alleviated renal fibrosis through p38MAPK-YAP pathway. *Front Cell Dev Biol.* 2021;9:741060.
135. Ceballos D, Navarro X, Dubey N, Wendelschafer-Crabb G, Kennedy WR, Tranquillo RT. Magnetically aligned collagen gel filling a collagen nerve guide improves peripheral nerve regeneration. *Exp Neurol.* 1999;158(2):290–300.
136. Kim J, Staunton JR,Tanner K. Independent control of topography for 3D patterning of the ECM microenvironment. *Adv Mater.* 2016;28(1):132–137. (1521-4095 (Electronic)).
137. Feng Q, Liu Y, Huang J, Chen K, Huang J, Xiao K. Uptake, distribution, clearance, and toxicity of iron oxide nanoparticles with different sizes and coatings. *Sci Rep.* 2018;8(1):2082.
138. Singh N, Jenkins GJS, Asadi R, Doak SH. Potential toxicity of superparamagnetic iron oxide nanoparticles (SPION). *Nano Rev.* 2010;1(1):5358.
139. Betsch M, Cristian C, Lin YY, Blaeser A, Schöneberg J, Vogt M, et al. Incorporating 4D into bioprinting: Real-time magnetically directed collagen fiber alignment for generating complex multilayered tissues. *Adv Healthc Mater.* 2018;7(21):1800894.
140. De France KJ, Badv M, Dorogin J, Siebers E, Panchal V, Babi M, et al. Tissue response and biodistribution of injectable cellulose nanocrystal composite hydrogels. *ACS Biomater Sci Eng.* 2019;5(5):2235–2246.
141. De France KJ, Cranston ED, Hoare T. Mechanically reinforced injectable hydrogels. *ACS Appl Polym Mater.* 2019;2(3): 1016–1030.
142. Rose JC, Fölster M, Kivilip L, Gerardo-Nava JL, Jaekel EE, Gehlen DB, et al. Predicting the orientation of magnetic microgel rods for soft anisotropic biomimetic hydrogels. *Polym Chem.* 2020;11(2):496–507.
143. Rose JC, Gehlen DB, Omidinia-Anarkoli A, Fölster M, Haraszti T, Jaekel EE, et al. How much physical guidance is needed to orient growing axons in 3D hydrogels? *Adv Healthc Mater.* 2020; 9(21):e2000886.
144. Pisanic II TR, Blackwell JD, Shubayev VI, Fiñones RR, Jin S. Nanotoxicity of iron oxide nanoparticle internalization in growing neurons. *Biomaterials.* 2007;28(16):2572–2581.

145. Rolland JP, Maynor BW, Euliss LE, Exner AE, Dension GM, DeSimone JM. Direct fabrication and harvesting of monodisperse, shape-specific nanobiomaterials. *J Am Chem Soc.* 2005;127: 10096–10100.
146. Allen MS, Wood DE, Hawkinson RW, Harpole DH, McKenna RJ, Walsh GL, et al. Prospective randomized study evaluating a biodegradable polymeric sealant for sealing intraoperative air leaks that occur during pulmonary resection. *Ann Thorac Surg.* 2004;77(5):1792–1801.
147. Alfieri F, Hruby G, Alfieri F, Kneebone A, Eade T. SpaceOAR hydrogel in dose-escalated prostate cancer radiotherapy: Rectal dosimetry and late toxicity. *Clin Oncol (R Coll Radiol).* 2014; 28(10):e148–e154. European Congress of Radiology-2014 CSM.
148. Alberts B, Johnson A, Lewis J, Raff M, Roberts K, Walter P. Integrins. In: Alberts B, Bray D, Hopkin K, Johnson A, Lewis J, Raff M, et al., editors *Molecular Biology of the Cell.* 4th ed. New York: Garland Science; 2022. p. 1103–27.
149. Omidinia-Anarkoli A, Ephraim JW, Rimal R, De Laporte L. Hierarchical fibrous guiding cues at different scales influence linear neurite extension. *Acta Biomater.* 2020;113:350–359.
150. Dupont S, Morsut L, Aragona M, Enzo E, Giulitti S, Cordenonsi M, et al. Role of YAP/TAZ in mechanotransduction. *Nature.* 2011; 474(7350):179–183.
151. Krueger AJ, Bakirman O, Guerzoni LPB, Jans A, Gehlen DB, Rommel D, et al. Compartmentalized jet polymerization as a high-resolution process to continuously produce anisometric microgel rods with adjustable size and stiffness. *Adv Mater.* 2019; 31(49):e1903668.
152. Vedaraman S, Bernhagen D, Haraszti T, Licht C, Castro Nava A, Omidinia Anarkoli A, et al. Bicyclic RGD peptides enhance nerve growth in synthetic PEG-based Anisogels. *Biomater Sci.* 2021; 9(12):4329–4342.
153. Araújo-Custódio S, Gomez-Florit M, Tomás AR, Mendes BB, Babo PS, Mithieux SM, et al. Injectable and magnetic responsive hydrogels with bioinspired ordered structures. *ACS Biomater Sci Eng.* 2019;5(3):1392–1404.
154. Daly AC, Riley L, Segura T, Burdick JA. Hydrogel microparticles for biomedical applications. *Nat Rev Mater.* 2020;5(1):20–43.

155. Prasopthum A, Deng Z, Khan IM, Yin Z, Guo B, Yang J. Three dimensional printed degradable and conductive polymer scaffolds promote chondrogenic differentiation of chondroprogenitor cells. *Biomater Sci.* 2020;8(15):4287–4298.
156. Prasopthum A, Cooper M, Shakesheff KM, Yang J. Three-dimensional printed scaffolds with controlled micro-/nanoporous surface topography direct chondrogenic and osteogenic differentiation of mesenchymal stem cells. *ACS Appl Mater Interfaces.* 2019;11(21):18896–18906.
157. Mani MP, Sadia M, Jaganathan SK, Khudzari AZ, Supriyanto E, Saidin S, et al. A review on 3D printing in tissue engineering applications. *J Polym Eng.* 2022;42(3):243–265.
158. Chimene D, Lennox KK, Kaunas RR, Gaharwar AK. Advanced bioinks for 3D printing: A materials science perspective. *Ann Biomed Eng.* 2016;44(6):2090–2102.
159. Bhattacharjee T, Zehnder SM, Rowe KG, Jain S, Nixon RM, Sawyer WG, et al. Writing in the granular gel medium. *Sci Adv.* 2015;1(8):e1500655.
160. Habib A, Sathish V, Mallik S, Khoda B. 3D printability of alginate-carboxymethyl cellulose hydrogel. *Materials(Basel).* 2018;11(3):454.
161. Hiller A, Borchers K, Tov GEM, Southan A. Impact of intermediate UV curing and yield stress of 3D printed poly (ethylene glycol) diacrylate hydrogels on interlayer connectivity and maximum build height. *Addit Manuf.* 2017;18:136–144.
162. Malda J, Visser J, Melchels FP, Jüngst T, Hennink WEH, Dhert WJA, et al. 25th Anniversary article: Engineering hydrogels for biofabrication. *Adv Mater.* 2013;25(36):5011–5028.
163. Morgan FL, Moroni L, Baker MB. Dynamic bioinks to advance bioprinting. *Adv Healthc Mater.* 2020;9(15):e1901798.
164. Ouyang L, Highley CB, Rodell CB Sun W, Burdick JA. 3D printing of shear-thinning hyaluronic acid hydrogels with secondary cross-linking. *ACS Biomater Sci Eng.*2016;2(10):1743–1751.
165. Wang H, Zhu D, Paul A, Cai L, Enejder A, Yang F, et al. Covalently adaptable elastin-like protein–hyaluronic acid (ELP–HA) hybrid hydrogels with secondary thermoresponsive crosslinking for injectable stem cell delivery. *Adv Funct Mater.* 2017; 27(28):1605609.

166. Yesilyurt V, Webber MJ, Appel EA, Godwin C, Langer R, Anderson DG. Injectable self-healing glucose-responsive hydrogels with pH-regulated mechanical properties. *Adv Mater.* 2016;28(1):86–91.
167. Highley CB, Song KH, Daly AC, Burdick JA. Jammed microgel inks for 3D printing applications. *Adv Sci.* 2019;6(1):1801076.
168. Riley L, Schirmer L, Segura T. Granular hydrogels: Emergent properties of jammed hydrogel microparticles and their applications in tissue repair and regeneration. *Curr Opin Biotechnol.* 2019;60:1–8.
169. Truong NF, Kurt E, Tahmizyan N, Lesher-Perez SC, Chen M, Darling NJ, et al. Microporous annealed particle hydrogel stiffness, void space size, and adhesion properties impact cell proliferation, cell spreading, and gene transfer. *Acta Biomater.*2019;94:160–172.
170. Kessel B, Lee M, Bonato A, Tinguely Y, Tosoratti E, Zenobi-Wong M. 3D Bioprinting of macroporous materials based on entangled hydrogel microstrands. *Adv Sci.* 2020;7(18):2001419.
171. O'Brien FJ, Harley BA, Yannas IV, Gibson LJ. The effect of pore size on cell adhesion in collagen-GAG scaffolds. *Biomaterials.* 2005;26(4):433–441.
172. Zeltinger J, Sherwood JK, Graham DA, Mueller R, Griffith LG. Effect of pore size and void fraction on cellular adhesion, proliferation, and matrix deposition. *Tissue Eng.* 2001;7(5):557–572.
173. Benya PD, Shaffer JD. Dedifferentiated chondrocytes reexpress the differentiated collagen phenotype when cultured in agarose gels. *Cell.* 1982;30(1):215–224.
174. Suturin AC, Krüger AJD, Neidig K, Klos N, Dolfen N, Bund M, et al. Annealing high aspect ratio microgels into macroporous 3D scaffolds allows for higher porosities and effective cell migration. *Adv Healthc Mater.* 2022;11(24):e2200989.
175. Pedde RD, Mirani B, Navaei A, Styan T, Wong S, Mehrali M, et al. Emerging biofabrication strategies for engineering complex tissue constructs. *Adv Mater.* 2017;29(19):1606061.
176. Dinoro J, Maher M, Talebian S, Jafarkhani M, Mehrali M, Orive G, et al. Sulfated polysaccharide-based scaffolds for orthopaedic tissue engineering. *Biomaterials.* 2019;214:119214.
177. Talebian S, Mehrali M, Mohan S, Balaji Raghavendran H, Mehrali M, Mohammad Khanlou H, et al. Chitosan (PEO)/bioactive glass

hybrid nanofibers for bone tissue engineering. *Rsc Adv.* 2014; 4(90):49144–49152.

178. Thakur VK, Thakur MK. Recent advances in graft copolymerization and applications of chitosan: A review. *ACS Sustainable Chem Eng.* 2014;2(12):2637–2652.
179. Talebian S, Mehrali M, Taebnia N, Pennisi CP, Kadumudi FB, Foroughi J, et al. Self-healing hydrogels: The next paradigm shift in tissue engineering? *Adv Sci.* 2019;6(16):1801664.
180. de Oliveira AC, Sabino RM, Souza PR, Muniz EC, Popat KC, Kipper MJ, et al. Chitosan/gellan gum ratio content into blends modulates the scaffolding capacity of hydrogels on bone mesenchymal stem cells. *Mater Sci Eng C Mater Biol Appl.* 2020;106:110258.
181. Coutinho DF, Sant S, Shakiba M, Wang B, Gomes ME, Neves NM, et al. Microfabricated photocrosslinkable polyelectrolyte-complex of chitosan and methacrylated gellan gum. *J Mater Chem.* 2012;22(33):17262–17271.
182. Jun Y, Kang E, Chae S, Lee S-H. Microfluidic spinning of micro- and nano-scale fibers for tissue engineering. *Lab Chip.* 2014;14(13):2145–2160.
183. Waheed S, Cabot JM, Macdonald NP, Lewis T, Guijt RM, Paull B, et al. 3D printed microfluidic devices: Enablers and barriers. *Lab Chip.* 2016;16(11):1993–2013.
184. Sun H, Jia Y, Dong H, Dong D, Zheng J. Combining additive manufacturing with microfluidics: An emerging method for developing novel organs-on-chips. *Curr Opin Chem Eng.* 2020;28:1–9.
185. Au AK, Hutuh W, Horowi LF, Folch A. 3D-printed microfluidics. *Angew Chem Int Ed Engl.* 2016;55(12):3862–3881.
186. Li J, Baxani DK, Jamieson WD, Xu W, Garcia Rocha V, Barrow DA, et al. Formation of polarized, functional artificial cells from compartmentalized droplet networks and nanomaterials, using one-step, dual-material 3D-printed microfluidics. *Adv Sci.* 2020; 7(1):1901719.
187. Chan HN. *The Applications of 3D Printing in Microfluidics and Bioengineering.* Hong Kong: Hong Kong University of Science and Technology; 2016.
188. Bhattacharjee N, Urrios A, Kang S, Folch A. The upcoming 3D-printing revolution in microfluidics. *Lab Chip.* 2016;16(10): 1720–1742.

189. Bratlie KM, Xu Z, Li Z, Jiang S. Chemically modified gellan gum hydrogels with tunable properties for use as tissue engineering scaffolds. *ACS Onega.* 2018;3(6):6998–7007.
190. Robinson TM, Talebian S, Foroughi J, Yue Z, Fay CD, Wallace GG. Fabrication of aligned biomimetic gellan gum-chitosan microstructures through 3D printed microfluidic channels and multiple in situ cross-linking mechanisms. *ACS Biomater Sci Eng.* 2020; 6(6):3638–3648.
191. Guzzi EA, Tibbitt MW. Additive manufacturing of precision biomaterials. *Adv Mater.* 2020;32(13):e1901994.
192. Sant S, Coutinho DF, Gaharwar AK, Neves NM, Reis RL, Gomes ME, et al. Self-assembled hydrogel fiber bundles from oppositely charged polyelectrolytes mimic micro-/nanoscale hierarchy of collagen. *Adv Funct Mater.* 2017;27(36):1606273.
193. Buriuli M, Kumari WG,Verma D. Evaluation of hemostatic effect of polyelectrolyte complex-based dressings. *J Biomater Appl.* 2017;32(5):638–647.
194. Quarta M, Cromie M, Chacon R, Blonigan J, Garcia V, Akimenko I, et al. Bioengineered constructs combined with exercise enhance stem cell-mediated treatment of volumetric muscle loss. *Nat Commun.* 2017;8(1):15613.
195. Choi Y-J, JunYJ, Kim DY, Yi HG, Chae SH, Kang J, et al. A 3D cell printed muscle construct with tissue-derived bioink for the treatment of volumetric muscle loss. *Biomaterials.* 2019;206:160–169.
196. Shapira A, Asulin M, Noor N, Dvir T. Stabilization strategies in extrusion-based 3D bioprinting for tissue engineering. *Appl Phys Rev.* 2018;5(4):041112.
197. Rhee S, Puetzer JL, Mason BN, Reinhart-King CA, Bonassar LJ. 3D bioprinting of spatially heterogeneous collagen constructs for cartilage tissue engineering. *ACS Biomater Sci Eng.* 2016; 2(10):1800–1805.
198. Tatman PD, Gerull W, Sweeney-Easter S, Davis JI, Gee AO, Kim DH. Multiscale biofabrication of articular cartilage: Bioinspired and biomimetic approaches. *Tissue Eng Part B: Rev.* 2015; 21(6):543–559.
199. Ouyang L, Highley CB, Sun W, Burdick JA. A generalizable strategy for the 3D bioprinting of hydrogels from nonviscous photocrosslinkable inks. *Adv Mater.* 2017;29(8):1604983.

200. Kim W, Kim G. 3D bioprinting of functional cell-laden bioinks and its application for cell-alignment and maturation. *Appl Mater Today.* 2020;19:100588.
201. Kim W, Kim G. A functional bioink and its application in myoblast alignment and differentiation. *Chem Eng J.* 2019;366: 150–162.
202. Kim W, Lee H, Lee J, Atala A, Yoo JJ, Lee SJ, et al. Efficient myotube formation in 3D bioprinted tissue construct by biochemical and topographical cues. *Biomaterials.* 2020;230:119632.
203. Lee H, Kim WJ, Lee JU, Park KS, Yoo JJ, Atala A, et al. Self-aligned myofibers in 3D bioprinted extracellular matrix-based construct accelerate skeletal muscle function restoration. *Appl Phys Rev.* 2021;8(2):021405.
204. Hwangbo H, Lee H, Jin EJ, Lee J, Jo Y, Ryu D, Kim G. Bioprinting of aligned GelMa-based cell-laden structure for muscle tissue regeneration. *Bioact Mater.* 2021;8:57–70.
205. Distler T, Solisito AA, Schneidereit D, Friedrich O, Detsch R, Boccaccini AR. 3D printed oxidized alginate-gelatin bioink provides guidance for C2C12 muscle precursor cell orientation and differentiation via shear stress during bioprinting. *Biofabrication.* 2020;12(4):045005.

https://doi.org/10.1142/9789819806782_0008

Chapter 8

Dynamic Biomaterials from Reversible Covalent Interactions

Lucien Cousin and Mark W. Tibbitt*

Macromolecular Engineering Laboratory, Institute of Energy and Process Engineering, Department of Mechanical and Process Engineering, ETH Zurich, Sonneggstrasse 3, 8092 Zurich, Switzerland

**Corresponding author: mtibbitt@ethz.ch*

Biomaterials are a transformative technology in biomedicine, providing strategies for improved tissue repair, the design of microphysiological systems, and advanced drug delivery systems. Traditional polymeric biomaterials are engineered using biocompatible polymers that often contain or interact through permanent covalent bonds. Incorporating reversible covalent interactions into biomaterials design is increasingly used to introduce dynamic properties to engineered biomaterials. Reversible covalent biomaterials can enable minimally invasive application through injection, are viscoelastic, and exhibit stimuli responsive behavior. In this chapter, we survey recent advances in the fundamental understanding of reversible covalent networks from chemical design to physical understanding. We then link this to several prominent biomedical applications of reversible covalent biomaterials.

1. Introduction

Biomaterials that interface with living tissues are a valuable tool in biomedical sciences to improve tissue healing and repair, safely deliver molecular therapeutics, and to assemble organ models outside of the body, among other uses.[1–4] One common material class for the design and assembly of biomaterials is polymers and plastics, including thermoplastics or engineered hydrogels.[5,6] Polymeric biomaterials are often produced using permanent covalent linkages between monomers or polymer chains, resulting in static materials that present challenges in their processing and degradation.[7] To design resorbable or degradable biomaterials, permanent covalent linkages have been replaced with degradable linkages based on hydrolytically, enzymatically, or photolytically cleavable bonds. Building on the utility of degradable biomaterials, researchers have also explored the use of noncovalent interactions, including physical or supramolecular assembly of polymer chains.[8] Such networks can often be processed and degraded under ambient conditions, without the need for an external stimulus. Physical and supramolecular networks are cross-linked via noncovalent interactions, including hydrogen bonding, host–guest chemistries, Van der Waals forces, and hydrophobic interactions.[8] Dynamic biomaterials from reversible interactions between polymer chains exhibit many attractive properties, including improved processability, responsiveness, and degradability.[7] However, when these are formed from physical or supramolecular interactions, they are often weaker than covalently cross-linked materials.

One strategy to combine the benefits of covalent linkages with exchangeable cross-links is through the use of dynamic covalent chemistries at the junctions of reversible polymeric networks.[9] Dynamic covalent chemistries are a particularly useful class of reversible bonds that form covalent bonds that are both strong and exchangeable.[10] In general, dynamic covalent bonds can unbind and reform either under ambient conditions or via stimuli, including changes in temperature, pH, or the presence of a catalyst. The reversible nature of the bonds imparts similar attractive properties as seen in physical or supramolecular polymer networks, allowing them to dissipate energy, flow upon applied stress, and reform intact networks after rearrangement.[11] Notably, as the bonds are covalent in nature (i.e., reacting partners share electrons), they often form strong bonds that can render the materials robust and with similar properties to traditional, irreversible covalent networks.[10] Yet, when the reversibility is engaged, the networks can be processed, which is not possible with traditional thermosets. The utility of this approach has been demonstrated effectively by "vitrimers," which exhibit mechanical robustness of thermosets while allowing for recycling via bond exchange at elevated temperatures.[12,13] With the chemical development of vitrimers and other dynamic covalent networks (also referred to as covalent adaptable networks), there has been a growing interest in understanding the mechanical properties of this class of materials and linking these properties to the underlying features of the network.[14] As discussed below, the polymer physics and materials science communities are expanding our fundamental understanding of these materials, which is enabling further applications in advanced materials design, recyclable plastics, and biomedicine.[7]

As the community better understands how to engineer dynamic networks using reversible covalent interactions, dynamic biomaterials are being increasingly used for biomedical applications. The stimuli-responsive nature of dynamic covalent chemistries can enable changes in the network properties in response to external cues, allowing them to serve as biosensors, for example, in the case of glucose-sensing hydrogels. The network's ability to adapt its structure as bonds break and reform and to dissolve in a controlled manner over time also makes dynamic covalent biomaterials attractive as drug delivery systems. Controlled viscoelasticity is proving increasingly useful in the ability to mimic cellular microenvironments found in the human body, enabling directed cell response and tissue maturation, and is making dynamic hydrogels interesting as three-dimensional (3D) cell culture scaffolds. In addition, the elegant rheological properties of these materials — the ability to flow upon applied stress and rapidly restore full mechanical properties — make them suitable candidates as injectable or printable biomaterials. We elaborate on the fundamentals of dynamic biomaterials formed via reversible covalent interactions and then provide a high-level introduction to the use of these materials as biosensors, drug delivery vehicles, scaffolds for cell culture, and injectable or printable biomaterials.

2. Fundamentals of dynamic biomaterials from reversible covalent interactions

Reversible covalent networks can be thought of as an intermediate between traditional, permanent covalent networks and physical networks. In permanent covalent networks, the architecture of the network can be tuned precisely, by controlling the type or rate of polymerization, or by choosing a specific architecture for the polymeric precursors (e.g., star polymers, telechelic, and linear with binding sites along the backbone, and so on). However, the permanent nature of the bonds makes it challenging to process or reprocess permanent covalent networks without inducing irreversible damage. In contrast, physical networks can be formed using physical interactions, for example, Van der Waals interactions, hydrogen bonds, or hydrophobic interactions, which are weaker than permanent covalent interactions and can form and detach under ambient conditions. This makes physical networks very easy to process, either by the application of mechanical force, or by another stimulus such as heat. In principle, these networks can be processed and reprocessed indefinitely, as only the physical interactions break and then reform. However, it can be harder to control the topology of the formed networks as physical interactions are not always well defined in space. In addition, these interactions are often weak, and the formed networks may exhibit low mechanical strength.

Reversible covalent networks combine several advantages of both physical and permanent covalent networks. They preserve aspects of the well-defined topology of permanent covalent networks, as they can be engineered in a similar way due to the specificity and the small size of the binding partners. Due to the low activation energy of the reversible bonds, they can easily be reprocessed, with the same methods as for physical networks. Despite their reversible nature, the covalent nature of the bonds allows them to withstand higher forces than physical networks, and in some situations, they exhibit similar mechanical resistance as permanent covalent networks. It is important to mention that here, we only focus on reversible covalent networks when the bonds are breaking and reforming on experimental timescales. Many reversible networks require an external stimulus, such as light, heat, or the introduction of a catalyst, to make the bonds reversible and behave like permanent covalent networks without this stimulus. For such networks, the concepts we discuss here focus on their behavior when this stimulus is applied, that is, when the bonds are actively undergoing dynamic exchange.

2.1. *Common chemistries*

Reversible reactions comprise chemical transformation that can proceed in both directions: the equilibrium state will be composed of both unreacted species and the formed product, with the respective proportion determined by chemical equilibrium. This occurs when the activation energies of the forward and backward reactions are sufficiently low such that the reaction can proceed at room temperature. While all transformations are technically reversible with this definition, it is generally assumed that only reactions with a chemical equilibrium constant between 10^7 and 10^{-7} are noticeably reversible on relevant timescales. Reactions that fall outside of this range will lead to trace amounts of the unbound reactants or the bound product, and will therefore either not form a network or form such a stable network that rearrangement does not occur. Therefore, the main challenge with designing reversible reactions for dynamic network formation is to engineer reactions with a chemical equilibrium constant in this range. The most widely used reversible reactions fall into one of the following three categories:

- Reactions that are reversible at room temperature and only require the reactants or the products to proceed.
- Reactions that are reversible under the presence of an external stimulus: for instance, light, stress, or the presence of an enzyme or a catalyst.
- Metal–ligand coordination reactions (not strictly dynamic covalent bonds).

While the various chemistries used in the field of reversible covalent materials have already been the subject of several reviews, we will summarize here several of the key points and commonly used chemistries.[15,16]

2.1.1. *Spontaneous reversible reactions*

Cycloadditions are one class of reactions that have attracted a lot of attention as reversible covalent reactions. The most prominent class are Diels–Alder reactions, where a diene and a dienophile react to form a cyclohexene. The benefits of this class of reactions include the high number of dienes and dienophiles that have been identified to engage in reversible covalent reactions, some of them being already quite extensively characterized. However, high temperatures are often required for the reversible reaction to occur, with most reactions exhibiting a reversible behavior between 80°C and 150°C.[16,17] Nonetheless, these can be attractive to engineer reprocessable materials or biomaterials that can tolerate elevated temperatures. Furan and maleimide are the most common examples for a Diels–Alder reaction.[18,19] The bond between anthracene and maleimide has been shown to be reversible under the application of mechanical force.[20] Cyclopentadiene is both a diene and a dienophile and can thus react with itself to form dicyclopentadiene, allowing the use of only one reactive species instead of two in most other cases.[21] Fulvene and cyanoolefin show reversible behavior at room temperature, while fulvene and dichloromaleic acid show reversible behavior at 37°C, which can be of interest for biomedical applications.[22,23] Another type of reversible cycloaddition are acylhydrazones, which result from the combination of an aldehyde with a hydrazine. While the imine group is responsible for the reversible covalent bond, the neighboring amide group provides additional stability via hydrogen bonding.[16] Acylhydrazones have also been combined with disulfides to create reversible covalent polymer networks.[24]

Transesterification reactions are another commonly used class of reactions to design dynamic materials. Transesterification of carboxylic acids is the basis of vitrimers, a material class pioneered by Leibler and colleagues that comprise thermoset networks that become malleable at high temperatures due to bond exchange.[12,13] However, the transesterification reaction in canonical vitrimers is an associative exchange; this is in contrast to a dissociative exchange where a bond breaks prior to finding new or the same binding partner(s), which characterizes the reactions discussed up to this point. In associative exchanges, a functional group can only detach when another functional group binds. In this manner, the proportion of unreacted functional groups does not depend only on the chemical equilibrium and is difficult to control, contrary to dissociative exchanges. This substantial difference between associative and dissociative exchanges should be kept in mind when discussing the physics and applications of reversible covalent networks. The transesterification of boronic acids with diols is a dissociative exchange that can often proceed spontaneously under ambient conditions and is increasingly explored in advanced material and biomaterial design.[25] The reaction is reversible at room temperature with a strong dependence on temperature and pH. Many different types of boronic acids have been employed, with varying reactivity with pH and temperature. In addition, different diols have different binding strength with each boronic acid. This allows for facile tuning of the bond exchange and, thus, material properties when using boronate ester bonds.[26]

2.1.2. *Reactions with external triggers*

In addition to reactions that proceed spontaneously under ambient conditions, there are many bonds that can be made reversible with external triggers. Sulfur-based chemistries are commonly used in the field of dynamic covalent reactions. They can take the form of disulfide exchange, thiol–disulfide exchange, reduction–oxidation reactions, or radical-mediated disulfide fragmentation. The disulfide bond formed between two sulfur atoms is reversible, and the chemical equilibrium depends on the adjacent chemical constituents as well as external stimuli, including temperature, the presence of a catalyst, or radicals.[27] In dynamic network design, disulfides are often used as cross-links between multifunctional thiols, but they can also be incorporated within the backbone of the constituent polymer chains. A wide variety of strategies exist to trigger the reversible reaction of sulfur-based dynamic covalent bonds. In the case of disulfide bridges, heating is sufficient to observe the onset of exchange, with temperatures around 60°C being commonly reported.[16] Reduction–oxidation reactions require the presence of a catalyst to cleave the S-S bridge, and a second catalyst to reform it, which can be air.[28] Primary radicals can also cleave disulfide bonds to produce thiyl radicals, which then engage in a reversible exchange reaction with disulfide bridges. Similarly to vitrimers, this is an associative and not a dissociative process. However, the proportion of unbound reactive groups (here, thiyl radicals) can be controlled and varied, for instance, via the concentration of primary radicals. Radical-mediated reactions can be made photoresponsive via the use of a photoinitiator.[29] Finally, it is worth noting that replacing sulfur with selenium considerably lowers the activation energy of the reversible covalent bond, shifting to lower operating stimuli (e.g., decreased temperature or primary radical concentration).[30,31] While still in its infancy, this technique holds great promise for the future of reversible covalent chemistry.

[4+4] and [2+2] cycloadditions can also be made reversible in the presence of light.[32,33] Common compounds used for [2+2] cycloadditions are coumarins and cinnamates. In these cases, irradiation with a wavelength greater than 280 nm triggers dimerization, while wavelengths less than 280 nm induce cleavage. Anthracene and its derivatives are used for [4+4] cycloadditions, with similar wavelengths. [4+4] cycloadditions tend to be more robust than [2+2] cycloadditions due to their greater stability against hydrolysis and isomerization.

Like the disulfide bonds, other reversible covalent reactions can be activated in the presence of radicals, as is the case with the C-O-N bond in alkoxyamines. This is the basis of nitroxide-mediated radical polymerization: under specific conditions, including the presence of an initiator, the bond cleaves into a carbon-centered radical and a nitroxide radical. The reactions require rather high temperatures, typically above 100°C.[16] A limitation of this kind of reaction is that the carbon-centered radicals can recombine with themselves to form a permanent covalent bond, resulting in the progressive loss of reversibility in the material with the number of cycles.

2.1.3. *Metal–ligand coordination*

A further class of reversible yet strong interactions used in dynamic polymeric materials design are metal–ligand coordination reactions, which can be achieved with a wide variety of metal ions and ligands. Coordination reactions are a specific type of reaction where the two bond-forming electrons come from the same donor atom, usually a metal. Most applications in the field of polymer network design consist of a free metallic ion that can interact with two or more ligands functionalized on polymer chains.[34] While metal-ligand coordination reactions are not strictly reversible covalent reactions, they share several important similarities, and all aspects covered in the rest of this chapter are also valid for metal-ligand coordination reactions. The activation energies for metal-ligand coordination reactions range from low values similar to physical interactions up to high values, almost similar to permanent covalent reactions. In addition, binding occurs at very well-defined positions along the polymer backbone, where the ligands are present, which is one distinctive feature of reversible covalent reactions as compared with physical interactions. However, the metal ion can often bind to more than two ligands, enabling network fabrication in which the cross-links (with functionality of at least 3) are reversible. Many network designs have been comprised of so-called associative networks, consisting of telechelic chains where the functional groups can bind with multiple other functional groups, and not to be confused with associate exchange reactions. Associative networks have been particularly useful in numerical simulations of reversible polymer networks.

A wide variety of combinations of metal and ligand exists, with various properties. We refer the reader to other reviews for a detailed description and mention briefly the most common.[34–36] Most ligands are compatible with several metals as the critical parameter is the number of free electrons in the metal ion. Catechol-metal bonds can exhibit mono-, bis-, or tris-coordination bonds depending on the pH.[37] Histidine-metal bonds are inspired by living systems, in particular by mussels, where they serve a self-healing purpose in mussel adhesion. The coordination number of the bond depends on the metal used, with Zn^{2+} showing different behavior than Ni^{2+}.[38] Pyridinyl ligands are also widely used as they can form bonds with nearly all metal ions.[35]

2.2. *Tuning dynamic chemistries*

An important feature of the common chemistries used in the design of dynamic polymeric materials is that they can be tuned to alter their equilibrium binding constants or forward and backward reactions rates. Tailoring the bond behavior has a direct impact on the formed materials properties.[7,39]

Here, we briefly describe how to tune the effective properties of the junction, and therefore, the network without altering the chemical nature of the bond.

One strategy is to introduce a competitor.[40] A competitor is a molecule, usually of low molecular weight, that competes with either or both functional groups for binding to the other functional group. A simple choice in most cases is to use the same molecule as for one of the functional groups. Introducing a competitor in the network will reduce the number of functional groups bound to each other. For instance, in a network based on boronate esters, introducing a diol in the network will loosen it.[41]

As described above, another strategy is to modulate the environmental conditions, for example, pH, temperature, as the energy landscape of many reversible bonds is affected by these conditions. For example, in the case of boronate-ester-based materials, the relaxation time and plateau modulus can be tailored over an order of magnitude by adjusting the pH of the network.[39,42] Similarly, the equilibrium constant of dynamic covalent bonds varies with temperature. For most reactions, this equilibrium constant is described by Arrhenius scaling: if we write $\Delta G = \Delta G_{bound} - \Delta G_{unbound}$ (the free energy difference of the bond between the bound and unbound state), k_B (Boltzmann's constant). and T (temperature), the equilibrium constant scales as

$$K_{eq} = e^{-\frac{\Delta G}{k_B T}}$$

This allows a quantitative understanding of how temperature affects binding equilibria, and thus, material properties. Our increasing understanding of mechanochemistry highlights how the tension on individual network strands can also influence the binding equilibrium of reversible junctions, highlighting how molecular-scale, and macroscale, strain, for example, via swelling, can influence binding properties.[43]

A further strategy is to engineer the architecture of the network precursors.[44–46] Ideal dynamic covalent networks have been synthesized with polymer stars with the functional groups at the end of the arms, or, as is often the case with disulfides, with a small linker to bind the stars.[29] Others use linear polymers with the functional groups grafted along the polymer backbone, either at random or well-defined locations.[47] Block copolymers have also been used, where the monomers themselves are the functional groups.[48] Dynamic networks have also been made from telechelic chains, requiring that the functional groups can bind with more than one other group in order to form a network.[49] It is also possible to combine multiple chemistries and architectures to modulate the material properties beyond what one dynamic chemistry or topology offers.[50] The precise influence of the architecture of the polymer precursors on the behavior of the formed gel is still an active and interesting area of research.

Overall, the range of dynamic covalent and metal-ligand coordination chemistries, network topologies, and external control mechanisms provide a broad toolbox to engineer reversible covalent networks for a range of applications. In order to exploit these materials most effectively, it is important to understand the aspects of their underlying physics. In the following sections, we provide a high-level overview of how reversible covalent networks behave and how bond behavior relates to material properties.

2.3. *Polymer physics of reversible covalent networks*

Reversible covalent networks are similar to permanent covalent networks, with the central difference that a fraction of the bonds of the network are reversible or dynamic. These bonds can either be the cross-links or part of the network strands between the cross-links. In this discussion, we use the term reversible to refer to bonds that can break and reform spontaneously during use or processing. In most cases, the activation energy for the disruption of the formation of the bonds is on the order of a few $k_B T$, and the thermal energy is sufficient for bond rearrangement. In other systems, the reversible nature is achieved by a third component, for instance, light or the presence of a catalyst.

This reversible nature of the bonds means that, at any moment, not all bonds will be formed. If we note τ_f and τ_d, the characteristic timescales for a bond to be formed and to be disrupted, or detached, the proportion p of bonds formed at any moment will be

$$p = \frac{\tau_d}{\tau_d + \tau_f} \tag{1}$$

Importantly, this network is not fixed but dynamic: while the proportion p will remain fixed in the absence of perturbation, individual bonds constantly break and other bonds form at all times. Thus, the set of bonds that are formed varies with time, even in the absence of external perturbation. In the following, we will use the term "functional group" to designate the molecules, attached to the polymers, responsible for the reversible covalent interaction. This term excludes enzymes or any other molecule involved in the interaction but not in the bond. We use the term "open sticker" to designate functional groups that can bind to another functional group, and the term "bond" to refer to two or more functional groups bound to each other.

Another feature that governs the properties of reversible covalent networks is the network topology, which can be controlled to some extent in a similar manner as for permanent covalent networks. This can be achieved by designing the molecular architecture of the network precursors and the location of the reactive functional groups. For example, star polymers with end-functional groups can be used to realize homogeneous, ideal reversible covalent networks.[39,42] The formation of such networks is represented in **Figure 1a**. Four-arm stars are end-functionalized with two types of functional groups that can bind in a reversible covalent fashion. Mixing the two types of stars results in a network in which not all bonds are formed. Linear chains with regularly spaced functional groups can form high-modulus and heterogeneous networks.[51] The bonds can consist of two functional groups, in which case they are part of the network strand between the cross-links, or more than two functional groups, in which case they act as cross-links themselves. For a given topology of the network, the proportion p governs the macroscopic behavior of the network. Importantly, p, combined with the network topology, determines whether the network is above or below the percolation, or gelation, threshold.[42,52] Therefore, quantifying p under different network preparation methods will determine whether or not the network can exhibit solid-like behavior. It should be noted that the distinction between the liquid and solid states is more subtle in the case of reversible covalent

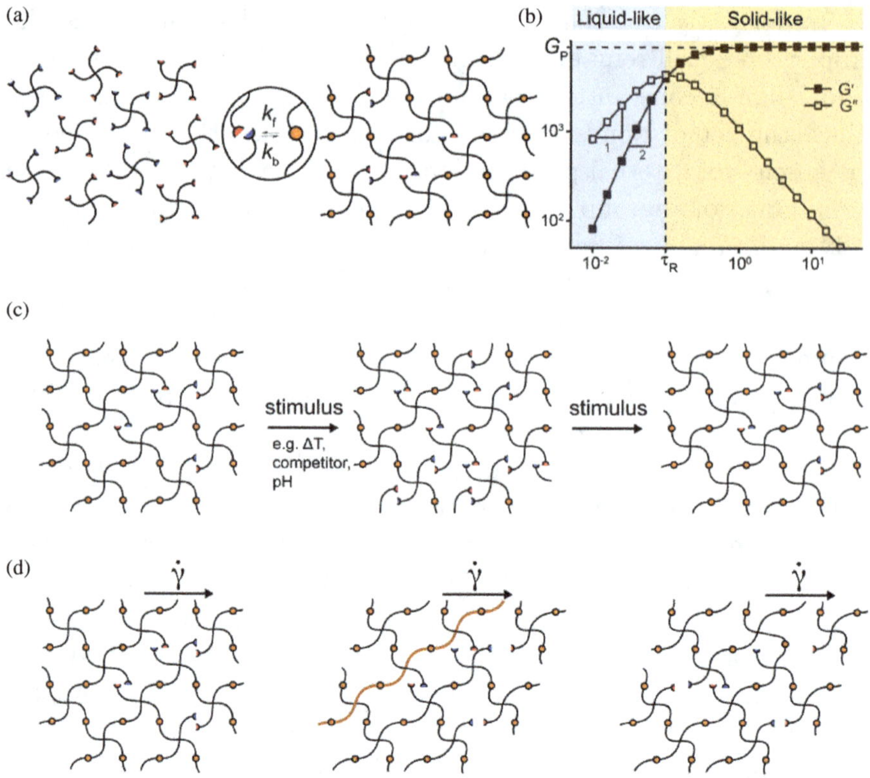

Figure 1. The physics of reversible covalent networks. (**a**) Reversible networks are formed by cross-linking polymeric precursors with reversible covalent reactions. Here, four-arm polymer stars end-functionalized with two types of functional groups are represented, and the properties of the materials are governed by the proportion of formed bonds, p, which is related to the chemical equilibrium. (**b**) Reversible covalent networks can be modeled as Maxwell materials, which exhibit liquid-like behavior at long timescales and solid-like behavior at short timescales. (**c**) Processing of reversible covalent networks is enabled by the exchange of bonds. (**d**) Shear flow of reversible covalent networks leads to an increase in tension in the chains in the direction of shear (orange). This tension leads to bond detachment, while new bonds are being formed resulting in a constant p.

networks than for permanent covalent networks, as will be highlighted below. Above the percolation threshold, p also determines the plateau modulus (or elasticity) of the network.[39,42] We provide more examples of the importance of p in the rest of this section.

The central role of p in governing the network properties in reversible covalent networks highlights the importance of understanding the underlying chemistry and how it varies with operating conditions. The fraction of bound bonds p can be a difficult quantity to measure directly, but relatively easy to calculate provided one knows some parameters of the interaction between the functional groups. For example, p can be calculated from the chemical equilibrium constant of the reaction that leads to the formation of the bond.[53] In the case where two functional groups, both with molar concentration, c, react together, one can write:

$$p = 1 + \frac{1}{2cK_{eq}} - \sqrt{\left(1 + \frac{1}{2cK_{eq}}\right)^2 - 1} \tag{2}$$

In more complex cases, one can solve, analytically or numerically, the system of equations obtained with the equations of mass balance and the definition of K_{eq} to derive an expression for p.

Due to the dynamic nature of the bonds, the same reversible covalent network can exhibit both liquid-like and solid-like behavior depending on the timescale of observation. The reason for this behavior lies in the dynamic nature of the bonds: the two functional groups forming a bond can detach and form a new bond with other functional groups. When the system is probed at very long timescales, the bonds have time to break, and the resulting open stickers can form new bonds that better accommodate the stress, leading to stress relaxation and network flow. In this case, the network shows liquid-like behavior: the amplitude of the deformation does not matter because the chains can rearrange according to the deformation. In contrast, when the system is probed at timescales too short for the bonds to break, the bonds behave like permanent bonds and do not appear as reversible. The network then shows solid-like, elastic behavior.

These features of reversible covalent networks were first predicted by Semenov and Rubinstein in a series of papers, and then shown experimentally by several groups.[39,42,54–57] More precisely, Semenov and Rubinstein predicted that ideal dynamic covalent networks should behave like ideal Maxwell materials, which consist of a spring and a dashpot in series.[58,59] The rheological behavior of a Maxwell material is represented in **Figure 1b**. Semenov and Rubinstein, and later, Sheridan et al. also showed how the characteristic timescale that separates the liquid-like and the solid-like behavior depends on the parameters of the system.[54,55,60] This timescale is called the mechanical relaxation time τ_R. If p is such that the system is below the percolation threshold, noted p_c, the network always exhibits liquid-like behavior. If $p > p_c$, the mechanical relaxation time is proportional to the distance to the percolation threshold and the characteristic time for a bond to break:

$$\tau_R \propto \tau_d(p - p_c) \tag{3}$$

The reasoning behind this formula is that, in order for the network to flow, a given number of strands has to break so that the percolation is lost. This number is proportional to $p - p_c$. In addition, the characteristic time for a bond to break is τ_d. The relation follows as a product of these two terms.

In the solid-like regime (high frequency; low timescales), the response of a Maxwell material is primarily dictated by the spring element. The elastic modulus is then independent of the frequency and is called the plateau modulus. In the case of reversible covalent networks, the spring element consists of all the chains that are elastically active. Remarkably, since the reversible nature of the bonds is effectively lost in the solid-like regime, the elasticity of the network is that of a permanent covalent network that would have the same internal topology and number density of cross-links. Hence, one can use the classic rubber elasticity theories developed for permanent covalent networks to understand the elasticity of reversible covalent networks in this regime. Parada et al., and later, Marco-Dufort et al., adapted the phantom network model, a commonly used model to predict the elasticity of a polymer network, to the case of reversible covalent networks.[39,42] The main modification is that reversible covalent networks have $p << 1$, that is, many of the chains that compose the network are not bound to any other chain and are thus elastically inactive. The expression for the number of elastically active chains was derived as a function of p, and then used in the phantom network model to predict the elasticity of reversible covalent networks.

As was highlighted above, the properties of the network depend mostly on the kinetic parameters of the bond, τ_d and τ_f, and on the proportion of bonds that are formed, p. The kinetic parameters can be tuned by changing the chemistry of the functional groups, the chemical environment (e.g., pH, buffer type, presence of a competitor), or the temperature. These changes will also often affect the chemical equilibrium constant K_{eq} involved in the calculation of p (see Eq.(2)). In addition, p can be varied without changing the properties of the bond. Increasing the density of functional groups will shift the chemical equilibrium, leading to an increase in p. Conversely, adding a competitor in the network will lead to a decrease in p. Temperature has a non-straightforward effect on the global behavior of the network: while an increase in temperature will generally lead to a decrease in p, it will also lead to an increase in the elasticity of the polymer chains via thermoelasticity. If $p \lesssim 1$, the biggest effect of an increase in temperature will be on the elasticity of the chains, and the network will become stiffer. If $p << 1$, the biggest effect will be on p, and the network will get softer. Finally, changing p independently of K_{eq} or τ_d changes not only the elasticity but also the mechanical relaxation time, according to Eq. (3).[60]

In total, understanding the chemical equilibrium and the kinetics of the reversible bond as well as the network topology in reversible covalent networks allows us to provide a general understanding of their macroscale properties. Nonetheless, developing a quantitative understanding of the physics of these networks, building on the foundational work of Semenov and Rubinstein and others in the community, remains a major focus in the community. Ideally, we would have tools to directly relate the bond behavior and polymer constituents to engineered macroscale properties and advances in characterizing these materials are bringing this closer to reality.

2.4. *Flow and processing properties of reversible covalent networks*

One unique feature of reversible covalent networks is that, despite their relatively high mechanical strength, they enable flow and processing owing to continual bond rearrangement. Processing of reversible covalent networks refers here to the application of a stimulus to the network in order to change some of its properties, and the possible subsequent release of this stimulus. The applied stimulus will change the energetic landscape of the chemical reaction that leads to bond formation, change the likelihood of the reaction, or introduce anisotropy in bond rearrangement. Examples of stimuli include heating, cooling, changing the pH or the chemical environment, and applying stress or strain on the network. Changing the temperature will leave the energetic landscape intact but shift the likelihood of the reaction (K_{eq}). Changing all other quantities mentioned will change the energetic landscape of the bonds. In the case of applied strain, the elastic energy stored in the chains lowers the energy required from thermal motion to disrupt the bond. Chains with an end-to-end vector parallel to the direction of stretch will be more likely to detach and rearrange, and thus bond rearrangement becomes anisotropic.

A crucial property of reversible covalent networks is that they are self-healing, meaning here that the broken bonds will reform once the stimulus is removed. This property stems from the low activation energy of the reversible bond: because the activation energy for the disruption of the bonds is much lower than the activation energy that leads to the disruption of all other covalent bonds in the system, the reversible bonds will always break first. When the material relaxes, the reversible covalent bonds will spontaneously reform, because the activation energy for bond formation is on the order of a few $k_B T$.[16,61] **Figure 1c** illustrates the response of reversible covalent networks to a stimulus and the subsequent self-healing behavior. In this manner, self-healing (or elastic recovery) is intrinsic in the case of reversible covalent networks — it occurs because of the chemical nature of the network, not because of the introduction of external self-healing agents or the application of a self-healing procedure.

While several studies have focused on how the chemical equilibrium depends on the external stimulus, much less is known about the transition states, their duration, and the underlying physical phenomena. This is due in part to the difficulty of modeling out-of-equilibrium, dynamic phenomena. Some advances have been made in this direction by elucidating the mechanisms behind the diffusion of the polymeric precursors in dynamic polymeric materials. Two dominant mechanisms of motion are proposed: "walking" and "hopping".[62,63] Walking refers to the motion of the precursor when at least one sticker remains attached, and this is a slow mode of diffusion constrained by the dynamics of the bonds. Hopping refers to the free, unhindered motion of the precursor that can only occur when all stickers are unbound. Here, the precursor diffuses as if it had no stickers until it binds again in a new location. The transition from walking to hopping has been linked to the superdiffusive behavior of the polymeric precursors, a type of diffusion faster than predicted by Fick's law.[62,64] While having all stickers unbound can seem unrealistic for a chain with a high number of stickers, the binding of two or more stickers within a single chain considerably reduces the conformational entropy of the chain.[63] Therefore, events where a precursor has one or no sticker bound may be more frequent than expected from a simple binomial distribution.[65]

The study of how reversible covalent networks flow is directly related to material processing. Flow can arise by applying a constant shear stress to the network. Similarly to other types of processing, this will induce changes in the proportion of bonds that are formed (p), and which bonds are formed. Tanaka and Edwards pioneered theoretical developments in this field, proposing a transient network framework applied to polymers with reversible interactions.[66] This model was then refined by many, including Vaccaro and Marrucci,[67] Wang,[68] and Sing et al.[69] All of the theoretical approaches to understand reversible network flow share some common hypotheses. First, they all assume affine deformations of the chains, that is, the local and global deformations are the same.

Second, they all consider that the dynamics of the network are governed by the dynamics of the bonds: the relaxation of the polymer chains is assumed to be much faster than the relaxation of the bonds. Finally, they all model strain-induced disruption of the bonds by a change in the energetic landscape of the bond under strain: when a chain is stretched, the activation energy for the disruption of the bond is lower, and this effect becomes more pronounced when the strain is increased. **Figure 1d** represents these common hypotheses: the chains all deform affinely in response to shear; this deformation leads to an increase in tension in some chains; this tension leads to the detachment of some bonds, which locally releases the tension, while new bonds are being formed due to the motion of dangling chains.

With this set of assumptions, Tanaka and Edwards reproduced the observed shear-thinning behavior in reversible networks at high shear rates and explained it by a decrease in the proportion of bonds formed. Wang later refined this framework by including free, unbound chains similar to the hopping chains mentioned above.[68] Vaccaro and Marrucci considered non-Gaussian stretching of highly stretched chains, which explained the onset of shear-thickening in associative polymer networks at high shear rates.[67] Sing et al. included local inhomogeneities in chain density, with dense clusters linked by bridges. By considering different length probability distributions for the chains belonging to clusters, bridges, or dangling chains, they came to different predictions for the decrease in viscosity in the shear-thinning regime, which were later experimentally verified.[69,70] While these models have all been successful in certain experimental cases, no one has been able to fully reproduce all experimental observations for the flow of reversible covalent networks.[70,71]

Some surprising features of reversible networks under flow have been shown experimentally. Shear banding has been observed at sufficiently high shear rates, both numerically and experimentally.[72,73] Shear banding refers to the formation of two, or more, bands forming in the direction perpendicular to flow, with differences in shear rate or density. This effect has been observed both in experiments and in simulations. While the underlying physical phenomena are unclear, it has been proposed that an instability due to the local coupling between shear rate and polymer concentration might trigger shear banding in some cases.[74–76] Gelation has also been observed in associative polymer networks sheared at high shear rates.[77,78] The molecular explanation for this behavior remains unclear.

2.5. *Open questions in the engineering of reversible covalent networks*

It has become clear that reversible covalent networks are governed primarily by the properties of the bonds. The kinetic properties of the bond control how dynamic the network is, and the thermodynamic properties of the bond control the degree of cross-linking of the network. However, this only allows in most cases a qualitative knowledge of the network's behavior. Indeed, these properties can only be directly correlated to the local behavior of the network, that is, the behavior at a microscopic scale. At a scale much greater than the characteristic distance between two bonds, the internal topology of the network, or its structure, also influences the network's mechanical properties. This remains a challenge for both experimental and theoretical studies.[79,80]

Several groups have tried to derive experimental methods to infer the topology of permanent covalent networks. Sakai and colleagues used small-angle neutron scattering (SANS) and static light scattering (SLS) to probe networks formed from four-arm star precursors and concluded that these networks have a nearly ideal structure, close to a diamond structure.[81–83] Olsen and coworkers created networks that can be cross-linked and then disassembled at different locations. Analyzing the products of the dissolved network was used to infer the proportion of dangling chains and loops present in the formed network.[84,85] However, these two techniques are difficult to apply in the case of dynamic networks. SANS requires sample freezing, thereby suppressing the reversible behavior. Network disassembly would have to be done on timescales much faster than the detachment time of a bond to avoid interference with the network, which is inherently challenging in reversible networks. More recently, Saalwächter and colleagues used multi-quantum nuclear magnetic resonance (NMR) to measure the proportion of dangling chains and small loops in polymer networks.[86–88] This technique and the analysis of the results are highly complex and have not been used for reversible networks to date, but hold promise to better understand the in situ nature of dynamic covalent networks.

Theoretical investigations have focused on understanding the role of defects in polymer networks. This is not specific to reversible covalent networks. However, contrary to permanent covalent networks, reversible covalent networks cannot exhibit a fully ideal structure, because they inherently contain a non-negligible proportion of dangling chains. The elasticity of reversible covalent networks can thus only be quantitatively described by taking into account the defects of the network. In addition, other defects such as loops and density inhomogeneities can be present, like in other nonideal structures. The effect of dangling chains and low-order loops, that is, loops involving two or three chains, has been investigated by Zhong et al. for permanent networks.[89] However, their analysis only focused on the limit case where $p \lesssim 1$ and the defects can be considered isolated. In reversible covalent networks, one often has $p << 1$ and defects can become as frequent if not more frequent than defect-free junctions. Other groups have tried to adapt the phantom network model or to use the effective medium approximation to derive the elasticity of polymer networks with low values of p, focusing only on dangling chains.[42,90] However, these models have only been qualitatively applied and verified on reversible covalent networks.

Density inhomogeneities can also be present in reversible covalent networks. Several groups have shown the emergence of microphase separation in such networks.[91–93] These structures are difficult to precisely quantify, because of their small scale and the low difference of refractive index between the concentrated and nonconcentrated phases. Commonly used methods include small-angle X-ray scattering (SAXS), broadband dielectric spectroscopy (BDS), transmission electron microscopy (TEM), and rheology.[91–93] The conditions leading to this morphology and how the microphases influence the network's behavior still remain to be fully understood.[93,94] In addition, density inhomogeneities make it more difficult to obtain ideal, homogeneous reversible covalent networks to test theoretical predictions. Recently, Nian et al. have shown that the apparent binding energy of reversible bonds seems to depend on the density of stickers in networks exhibiting no microphase separation.[95] This surprising result suggests that obtaining homogeneous networks may be crucial to obtain accurate measurements in reversible covalent networks.

Inhomogeneities in the connectivity of the network have been shown in reversible covalent networks. Omar and Wang have shown with numerical simulations that reversible covalent networks under flow exhibit inhomogeneities not only in the density of the precursors but also in the degree of connectivity of these precursors to the overall network.[74] Such predictions, while very challenging to verify experimentally, offer a possible explanation for the shear-thinning and shear-banding behavior observed in reversible covalent networks.

Very little has been investigated on the transitory aspects of reversible covalent networks. These networks can adapt to external stimuli and self-heal, as highlighted in the previous section. While the equilibrium states before and after the stimuli are understood to some extent, very little is known about the rate of the transition and the molecular rearrangements that occur during the transition.

3. Biomedical applications of reversible covalent biomaterials

Building on the developments in chemical design and physical understanding of reversible covalent biomaterials, this class of polymer networks is increasingly applied in biomedical applications. This is motivated primarily by their unique properties, including stimuli-responsiveness, processability, controlled degradation, and viscoelasticity, which are all common features of living systems. When applying dynamic covalent biomaterials in biomedical settings, additional constraints arise. The system (and therefore, the linking chemistry) should be compatible with aqueous conditions, meaning that water-swollen hydrogels or neat polymer networks are often used. The backbone polymers and the dynamic chemistry should not cause cell toxicity and ideally should not irreversibly affect biomolecules in the surrounding milieu. If used in the body, the material should not elicit an inflammatory or foreign body reaction and should lead to minimal or no activation of resident immune cells. These constraints are very similar to any polymeric material used in biomedical applications. The knowledge gained using permanent covalent biomaterials can then be leveraged to adapt reversible covalent networks to the particular constraints faced by biomaterials.

In this section, we describe four prominent biomedical uses of dynamic covalent biomaterials: biosensors, drug delivery systems, cell culture scaffolds, and injectable and printable materials. In each case, we highlight a few recent examples of how the physical properties of dynamic covalent biomaterials enabled each of the uses. This is not meant to be comprehensive, and there are many other specific demonstrations in each case and other applications of this class of materials envisioned or currently pursued. The aim is to motivate how physical insight and careful material design with dynamic covalent biomaterials can complement many of the existing biomaterials present in the literature and in the clinic.

3.1. *Biosensors*

Biosensors comprise the broad class of devices that can convert biological signal, such as glucose concentration or enzyme activity, into an output signal that can be easily measured and, in the best case, quantified.[96] Prominent examples of devices with biosensing capabilities include continuous glucose monitors and lateral flow antigen test strips (e.g., for Covid or pregnancy). Reversible covalent networks can interact with biological molecules via their bonds and do this in a reversible way, meaning that their lifetime is not affected by the presence of the molecule of interest. This makes them an attractive choice for designing biosensors or incorporating them into biosensing devices. A common implementation of reversible covalent chemistries in the design of biosensors has been to functionalize surfaces with one or more binding partners that can selectively interact with one or more analytes.[97] For example, surfaces have been functionalized with fluorophore-coupled boronic acid derivatives to engineer fluorescence-based glucose sensors. The operating principle is that the fluorophore optical spectrum shifts when the boronic acid binds to glucose or other diols.[98] Because the number of active fluorophores increases with glucose concentration, this measurement can be made quantitative with a calibration curve. Building on this concept, reversible covalent biomaterials have also been integrated into biosensing devices where the interaction with a biomolecule of interest leads to material changes, such as swelling or optical properties.[99]

Glucose sensing biomaterials have been designed using dynamic covalent hydrogels that incorporate boronic acid functionalities into the network backbone. In one example, holographic optical diffusing microstructures were fabricated into the surface of a glucose-responsive hydrogel during photopolymerization.[100] This construct served as a transducer of local glucose concentration, as binding between glucose and boronic acid derivatives in the hydrogel resulted in gel swelling and quantitative deformation of the holographic optical diffusing microstructures. A similar approach was exploited to design a hydrogel-interlayer radio-frequency resonator that enabled glucose sensing via swelling-induced changes in the resonant frequency of the device.[101] Standard radio-frequency sensors were able to read out the glucose concentration noninvasively based on calibrated changes in the frequency of the device. One specific challenge is to develop continuous glucose measurements in perioperative patients and patients in the intensive care unit. To address this, a material-based approach was developed that incorporates a glucose-responsive reversible hydrogel at the tip of an optical fiber (**Figure 2a**).[99] The gel contracted with rising

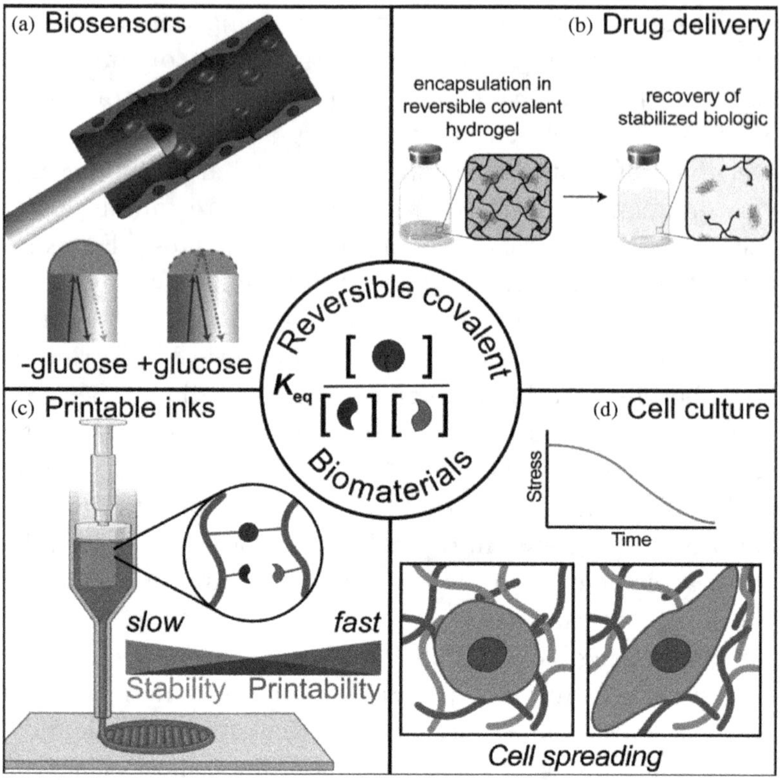

Figure 2. Biomedical applications of reversible covalent biomaterials. (**a**) Glucose sensors were designed using boronate-ester-based hydrogels that changed swelling in the presence of glucose, allowing for interferometric measurements of glucose concentration in blood. (**b**) Boronate-ester-based hydrogels were also used for the encapsulation, thermal stabilization, and on-demand release of biologics, taking advantage of the diol-responsive nature of the reversible networks. (**c**) Reversible biomaterials have also been designed as bioinks for extrusion-based 3D printing, exhibiting a trade-off between stability and printability depending on the rate of bond rearrangement. (**d**) Reversible covalent biomaterials are also being developed as viscoelastic 3D cell culture platform that relax stress, allowing cells to remodel the local environment to spread and migrate.

glucose concentrations, which can be measured using interferometry as the gel contraction alters the path length of light passing through the optical fiber-gel construct. Fluorophores coupled to boronic acids have also been immobilized within hydrogel networks to enable targeted glucose sensing within integrated devices.[102] Boronate-ester-based hydrogels have also been adapted for sensing of other molecules via different sensing mechanisms. Acrylamide hydrogels containing 3-(acrylamido)phenylboronic acid as a co-monomer were designed to bind glycated hemoglobin A1c, a glycemic biomarker, and binding resulted in a chromogenic change in the material for quantitative reading.[103] While sensors based on dynamic covalent biomaterials enable unique opportunities, care must be taken for potential or known cross-reactivity with other molecules that can be present in biological media. In the context of boronate-ester-based materials, it is known that most boronic acids are promiscuous to many diol-containing sugars and can even bind more strongly to fructose or mannitol as compared with glucose.[41] As with any sensor design, robustness and selectivity needs to be validated for each reversible biomaterial-based sensor in the use case of interest.

3.2. *Drug delivery*

In addition to their use as biosensors, the responsive nature of reversible covalent biomaterials renders them further useful as drug delivery systems. Polymeric biomaterials made from degradable or nondegradable biocompatible materials are commonly used as controlled release systems to deliver molecular therapeutics with the right dosage at the right time, the right rate, and to the right place.[2] Building on the general advances in drug delivery, dynamic covalent chemistries can be incorporated into these systems for reversible activation and release, thermal stabilization, and the design of on-demand or responsive drug delivery systems.[9,39,104] Because the reversible nature of the bonds allows the polymeric precursors to diffuse in the surrounding medium, most reversible covalent biomaterials will eventually dissolve in the body and are inherently biodegradable. This makes them additionally attractive for in vivo applications.

Within the field of drug delivery, reversible covalent biomaterials have been used to self-assemble nonviral vectors for gene delivery. In this context, pH-responsive reversible bonds, for example, hydrazones and oximes, have been used to assemble polymers that can complex with nucleic acid therapies to package them and facilitate cell entry.[105,106] Upon entry into the cell and passage to the late endosomes (acidic pH), the reversible bonds can disassemble and liberate the encapsulated nucleic acids. Reversible complexes assembled using dynamic covalent chemistry have also been applied for enzyme activation. For example, reversible imine reactions between multifunctional aldehydes and amines formed dynamic matrices around enzymes, including carbonic anhydrase and lipase, increasing enzyme activity in the matrix-bound state.[107,108] In a similar approach, reversible covalent hydrogels based on boronate esters have also been used as a platform to encapsulate, thermally stabilize, and release on-demand sensitive biologics (**Figure 2b**).[104] Here, the presence of a formed gel was required to prevent the biologics from aggregating during storage, and the dynamic covalent boronate esters allowed facile release in sugar-based solutions at the time of use.

One additional prominent use of reversible covalent biomaterials in the design of drug delivery systems is the engineering of glucose-responsive insulin release systems. This approach builds on work in the biosensing community that demonstrated how boronate esters can be employed in glucose sensing materials design. In this approach, the glucose-binding dependent changes in material properties are used to control the release of encapsulated insulin. Star-PEG molecules end-functionalized with boronic acids or diols were combined to form injectable boronate-ester-based gels that released insulin in a glucose-responsive manner.[109] In the presence of glucose, the equilibrium binding shifted to the unbound state, facilitating the diffusion of entrapped insulin. This general approach has since been explored by many research groups, aiming to refine the dynamic covalent chemistry and network structure to improve the release kinetics and glucose responsiveness.[110] Recent developments in the design of diboronate binding partners enable improved selectivity and specificity in dynamic covalent binding, as demonstrated via the design of nanocarriers for glucose-responsive insulin release.[111]

One inherent challenge to the use of a material where the backbone interactions are interchanging on experimental timescales is the propensity for the encapsulated cargo to leak out of the network prior to use. This can be further exacerbated if the reversible bonds can react with other molecules present in the biological media. This is the case, for example, with many glucose-responsive insulin delivery systems based on boronate esters, as boronic acids can bind to fructose or other diols that are present in the body. Therefore, additional work is needed to identify dynamic covalent drug delivery systems that can keep the encapsulated cargo in place until it is needed, exhibit reversible and responsive behavior once the conditions are met, and maintain the ability to biodegrade once the delivery system is no longer needed. In principle, biomaterials designed using dynamic covalent chemistries can meet all of these demands, but a more robust understanding of the reaction selectivity and material properties is needed.

3.3. *Injectable and printable materials*

The dynamic nature of the junctions in reversible covalent biomaterials also enables viscoelastic flow for injectable and printable applications.[7] In contrast with traditional covalent networks, the continued exchange of bonds after formation allows stress dissipation and ultimately flow upon applied shear. Further, the materials can self-heal or elastically recover once the shear application is stopped and maintain a fixed shape in space. This enables broad applications in the design of injectable biomaterials that can be prepared, loaded into a syringe, and applied in the body in a minimally invasive manner, or for the preparation of biomaterial inks for extrusion-based 3D printing.[112] In each case, the rheological requirements (shear-thinning during extrusion and elastic recovery after ejection) are similar and can be engineered by tailoring the network properties and bond chemistry of the networks.

In biomedical applications of injectable or printable biomaterials, the extrusion pressures and velocities are often limited by the inclusion of cells or interactions with biological tissues.[113] In particular, imposing excessive shear stresses, shear rates, or pressures can result in permanent damage of the biological components of the biomaterial. The knowledge of these quantities, often hard to measure, is crucial to ensure that the material retains its benefits once injected. While the precise mechanisms of viscoelastic flow for dynamic covalent hydrogels remain a topic of study,[114] there are many frameworks to relate macroscale rheological properties to flow properties in dynamic polymer networks.[113,115] Many of the dynamic covalent biomaterials discussed above exhibit suitable rheology and yield behavior for use in biomedical applications. Here, we highlight a few examples of how dynamic chemistries can be used to design injectable or printable biomaterials.

As described in the previous section, boronate-ester-based hydrogels have been developed as injectable drug delivery systems.[109] The shear-thinning and self-healing properties of the material allowed it to be loaded into a syringe and subsequently injected into subcutaneous tissue in mice. The glucose-responsive delivery system could then form a depot in the back of the mice. Over the course of 30 days, the dynamic hydrogel allowed for cell infiltration and was ultimately resorbed and replaced by vascularized tissue. The ability to be injected and then degraded over the course of weeks highlights the unique advantage of using dynamic bonds in the design of biomaterials for in vivo use. A similar strategy was used to engineer an injectable dynamic covalent dextran-based hydrogel, using double Michael addition between thiols and alkynones, to form reversible dithiane linkages, as self-adjuvating peptide vaccine depots.[116] Injectable hydrogels for the local delivery of miRNA therapies or microspheres to the myocardium were engineered via dynamic covalent hydrazone cross-links.[117,118] Similar injectable biomaterials were used to deliver therapeutic cells,[119] and the viscoelastic properties of dynamic networks have been shown to improve cell viability during injection.[113,120]

Complex rheological fluids that can be loaded into a syringe and extruded also enable the design of biomaterial inks for extrusion-based printing.[121] The same rheological features that protect encapsulated cells from shear forces during flow can be employed in the design of cell-laden bioinks to 3D print biological constructs. Building on work in the design of shear-thinning protein-based hydrogels, Heilshorn and coworkers have incorporated dynamic covalent hydrazone linkages into peptide-functionalized biomaterials using aldehyde-functionalized hyaluronic acid (HA) to engineer a suite of printable biomaterials.[122] Varying the degree of functionality, the HA molecular weight, and the stoichiometric ratio of functional groups modulated the rheological properties of the dynamic biomaterials for tailored viscoelasticity and printability. In addition, the use of small molecule modulators that can catalyze or compete with bond formation allowed to tune the reactivity of the dynamic covalent bond and therefore the rheological properties of the formed material (**Figure 2c**).[123] This provided a handle to tailor bioink properties for improved printability and cell culture during the fabrication of 3D bioprinted tissue models. The rheology of dynamic covalent bioinks was also engineered by mixing dynamic cross-linkers that operate on different timescales.[124] The combination of hydrazone and oxime linkages at varying ratios in the same alginate-based hydrogel provided a simple method to tune the flow properties of the hydrogel over a wide range without altering the mechanical properties at rest, by keeping the total polymer content and cross-linker concentration constant. This enabled tailored printability, including as bioinks for the formation of fibroblast-laden biomaterials. Additional handles to tune the rheological properties of reversible covalent bioinks include the use of different backbone polymers (e.g., gelatin), and network precursors, (e.g., dendrimers).[125,126]

Overall, the unique potential of tailored rheology (shear-thinning and self-healing/elastic recovery) enabled by the inclusion of dynamic covalent linkages renders this class of materials attractive as injectable and printable biomaterials. Ongoing work is focused on advancing our control over the flow properties by tailoring bond chemistry, mixing various bonds, using modulators, as well as understanding how network topology and bond strength relate to viscoelastic flow in these materials at a more fundamental level. There is an increasing demand for minimally invasive, injectable drug delivery systems and extrudable inks for bioprinting, and dynamic covalent biomaterials will be part of this future.

3.4. *Viscoelastic substrates for 3D cell culture*

In the human body, cells are embedded within a 3D matrix, termed the extracellular matrix (ECM), that provides structural support as well as mechanical and biochemical inputs.[127] The ECM itself is a polymer network, composed of proteins, glycoproteins, and glycosaminoglycans, and exhibits viscoelastic material properties.[128] Because of this similarity, hydrogel biomaterials have been commonly used as defined mimics of the ECM and provide a framework to study biological processes outside of the body in more relevant contexts.[129] The viscoelastic nature of the native ECM has been increasingly demonstrated to influence cell function, including stem cell differentiation, cell morphology, cell migration, and tissue morphogenesis (**Figure 2d**).[130] This has motivated bioengineers to explore strategies to incorporate defined viscoelasticity into 3D hydrogel-based ECM mimics. For instance, divalent cation cross-linked alginate networks inherently exhibit viscoelasticity and have been a prominent material for these investigations.[131] The inclusion of dynamic covalent chemistries provides an additional and tunable handle to modulate the viscoelasticity of hydrogels used for 3D cell culture.

Viscoelastic hydrogels based on dynamic covalent hydrazone bonds were developed as scaffolds to encapsulate adult stem cells, highlighting how viscoelasticity and stress relaxation influenced the morphology of encapsulated murine myoblasts.[132] This approach was adapted to the culture of neural spheroids, and the viscoelastic properties were then leveraged to use the biomaterial as a sensor. Materials models were extrapolated to study the forces exerted during neurite extension in 3D matrices.[133] In a similar work, dynamic boronate ester bonds were also integrated with traditional covalent linkages to engineer the viscoelastic properties of 3D cell culture scaffolds.[134] Using a proportion of permanent covalent bonds above the percolation threshold prevented the degradation of the material caused by the leakage of the network precursors. Pure boronate ester gels were also used as dynamic and self-healing 3D culture scaffolds for the assembly of higher-order tissue structures, bringing separately encapsulated cancer cells and fibroblasts together to investigate cell-cell interactions.[135]

Beyond single-cell encapsulation systems, viscoelastic hydrogels based on dynamic covalent chemistries have been used for the encapsulation and growth of multicellular structures, including organoids. Engineered hydrogel matrices based in part on dynamic covalent hydrazone linkages within an HA-protein backbone enabled the culture of human patient-derived intestinal organoids.[136] Organoid formation depended on the plateau modulus as well as the stress relaxation rate, emphasizing the importance of tunable viscoelasticity during their culture. In a similar approach, dynamic covalent oxime linkages were incorporated into alginate-based hydrogels for the culture of kidney organoids.[137] Soft and dynamic hydrogels fostered the development of all cellular subtypes and architecture of encapsulated kidney organoids, which were disrupted in stiff and static analogs.

As described in the section on drug delivery, an inherent advantage of reversible covalent hydrogels in biomedical use is their inherent ability to degrade as bonds break and chains disassemble. Yet, this same feature presents a challenge for long-term culture of cells and organoids. Therefore, additional work is needed to understand how to design stable viscoelastic culture systems that enable days to weeks culture without substantially altering the material's properties.

4. Outlook

The transformative potential of reversible covalent biomaterials in biomedicine is broad. Many foundational examples have demonstrated how the unique properties of reversible covalent biomaterials can be used in the design of reactive biosensors, responsive drug delivery systems, viscoelastic cell culture systems, and injectable/printable biomaterials. Translation of these materials has been enabled by robust understanding of how bond reactivity and network topology influence the macroscale properties in these networks. Future advances in this field will build on improved fundamental understanding and physical models that provide structure-dynamics-properties-function relationships for dynamic polymer networks. In addition, there are many groups working on this topic studying different chemistries and network architectures, and it would be useful for the community to collect much of this dispersed information to provide a more robust framework for the design and application of reversible covalent networks. Convergence of experts in polymer physics, macromolecular chemistry, materials science, rheology, computational modeling, and bioengineering is contributing to improved understanding in this field and will sharpen our understanding of how to design and apply reversible covalent biomaterials more broadly.

While the introduction of reversible binding interactions leads to unique properties in polymer networks, several open challenges remain. Biomedical use requires both mild reactants that do not cause toxicity in the body as well as selectivity and specificity of the binding pairs. This can be a challenge as the binding pairs are often based on chemistries commonly found in biology (e.g., diols, amines, aldehydes), which introduces inherent challenges to the selectivity and specificity of the binding partners. Identifying bioorthogonal reactants that can participate in reversible covalent bonding with minimal or no cross-reactivity with other species found in the body should remain a focus in the field. Further, biomedical applications often require material properties that balance stability and reversibility. The faster the kinetics (rapid reversibility and dissipation/flow in the material), the weaker the bond and the weaker the material. As the bond becomes more stable, the kinetics often slow. Identifying reversible chemistries and network structures that allow for tailorable reversibility and stability and effective compromises between these competing features should also be a priority in the community.

A similar challenge is to stabilize reversible covalent networks, that is, to transform them from reversible networks to permanent networks. The reversible nature of the bonds renders the materials as viscoelastic liquids, as they will eventually flow on sufficiently long timescales. The transition between reversible and permanent behavior can be achieved with networks that require a stimulus to exhibit reversible behavior, such as temperature, light irradiation, or pH change; however, these stimuli can cause damage to biological systems that the biomaterial interacts with. Techniques that can be used to permanently stabilize the chemical structure at arbitrary times while maintaining cytocompatibility would greatly expand the range of applications of reversible covalent biomaterials.

In addition to the open questions in the community, there are many future opportunities of reversible covalent biomaterials. The field of reversible covalent biomaterials is still relatively new, so it is likely that novel, creative uses remain to be found. We illustrate a few potential examples of such creative uses below. Several groups have engineered biomaterials that respond to different stimuli. Reversible networks can be selective in the type of stimulus they are sensitive to, and combining two (or more) different reversible networks in one biomaterial may allow property tuning in the biomaterial in response to more than one stimulus. For example, Chujo et al. developed hydrogels with both disulfide and anthracene bonds that respond to both light and redox stimuli.[138] Similarly, Deng et al. combined acylhydrazone and disulfide chemistries to create hydrogels that are reversible at pH 5 and 9 but behave as permanent covalent networks (not reversible) at pH 7.

In another direction, reversible covalent networks offer a unique opportunity to control the topology of the formed networks. Amamoto et al. engineered reversible networks based on disulfide bonds.[139] For this, they first prepared and cross-linked polymer networks, dried them, and dissolved them in either a good solvent or a nonsolvent. They then used ultraviolet (UV) light irradiation to induce reshuffling of the disulfide bonds. Subsequent swelling in toluene showed that the swelling degree of samples prepared in a good solvent was much greater than the swelling degree of samples prepared in a nonsolvent. This was attributed to a greater mesh size for the samples prepared in a good solvent. This swelling behavior was reversible: samples prepared in a good solvent could be reshuffled in a nonsolvent and showed similar degree of swelling compared to the samples first prepared in a nonsolvent. Adzima et al. used the ability of dynamic covalent bonds to be reversible or effectively permanent depending on the conditions to print 3D structures that could be made reversible.[140] The use of reversible covalent networks also enabled several improvements compared to permanent networks, including improved print resolution or simultaneous fabrication of structures in the case where more than one reversible covalent chemistry is used.

The field of reversible covalent biomaterials presents many unique opportunities. Continued advances in the fields of dynamic covalent chemistry, network design, polymer physics, and bioengineering will introduce new materials and application areas with both fundamental and clinical applications. Natural systems have evolved to incorporate reversible features into their design, and we are increasing our ability to introduce similar dynamics (in a controlled manner) into the biomaterials we engineer.

Acknowledgments

This work was supported by the Swiss National Science Foundation (200021_184697).

References

1. Tibbitt MW, Rodell CB, Burdick JA, Anseth KS. Progress in material design for biomedical applications. *Proc Natl Acad Sci USA*. 2015 Nov;112(47):14444–14451.
2. Tibbitt MW, Dahlman JE, Langer R. Emerging frontiers in drug delivery. *J Am Chem Soc*. 2016 Jan;138(3):704–717.
3. Li J, Mooney DJ. Designing hydrogels for controlled drug delivery. *Nat Rev Mater*. 2016 Dec;1(12):16071.
4. Zhang B, Korolj A, Lai BFL, Radisic M. Advances in organ-on-a-chip engineering. *Nat Rev Mater*. 2018 Aug;3(8):257–278.

5. Farah S, Anderson DG, Langer R. Physical and mechanical properties of PLA, and their functions in widespread applications — A comprehensive review. *Adv Drug Deliv Rev.* 2016 Dec;107: 367–392.
6. Lutolf MP, Hubbell JA. Synthetic biomaterials as instructive extracellular microenvironments for morphogenesis in tissue engineering. *Nat Biotechnol.* 2005 Jan;23(1):47–55.
7. Webber MJ, Tibbitt MW. Dynamic and reconfigurable materials from reversible network interactions. *Nat Rev Mater.* 2022 Jan; 7(7):541–556.
8. Webber MJ, Appel EA, Meijer EW, Langer R. Supramolecular biomaterials. *Nat Mater.* 2016 Jan;15(1):13–26.
9. Zhang Y, Qi Y, Ulrich S, Barboiu M, Ramström O. Dynamic covalent polymers for biomedical applications. *Mater Chem Front.* 2020 Feb;4(2):489–506.
10. Rowan SJ, Cantrill SJ, Cousins GRL, Sanders JKM, Stoddart JF. Dynamic covalent chemistry. *Angew Chem Int Ed Engl.* 2002 Mar; 41(6):898–952.
11. Kloxin CJ, Bowman CN. Covalent adaptable networks: Smart, reconfigurable and responsive network systems. *Chem Soc Rev.* 2013 Sep;42(17):7161–7173.
12. Montarnal D, Capelot M, Tournilhac F, Leibler L. Silica-like malleable materials from permanent organic networks. *Science.* 2011 Nov;334:965–968.
13. Zee NJV, Nicolaÿ R. Vitrimers: Permanently crosslinked polymers with dynamic network topology. *Prog Polym Sci.* 2020 May; 104:101233.
14. Guerre M, Taplan C, Winne JM, Du Prez FE. Vitrimers: Directing chemical reactivity to control material properties. *Chem Sci.* 2020 Apr;11(19):4855–4870.
15. Zou W, Dong J, Luo Y, Zhao Q, Xie T. Dynamic covalent polymer networks: From old chemistry to modern day innovations. *Adv Mater.* 2017;29(14):1606100.
16. Kuhl N, Bode S, Hager MD, Schubert US. Self-healing polymers based on reversible covalent bonds. In Hager MD, van der Zwaag S, Schubert US, editors. *Self-Healing Materials.* Cham: Springer International Publishing; 2016. p. 1–58.

17. Ratwani CR, Kamali AR, Abdelkader AM. Self-healing by Diels-Alder cycloaddition in advanced functional polymers: A review. *Prog Mater Sci.* Jan 2023;131:101001.
18. Kloxin CJ, Scott TF, Adzima BJ, Bowman CN. Covalent adaptable networks (cans): A unique paradigm in cross-linked polymers. *Macromolecules.* 2010;43(6):2643–2653.
19. Adzima BJ, Kloxin CJ, Bowman CN. Externally triggered healing of a thermoreversible covalent network via self-limited hysteresis heating. *Adv Mater.* 2010 Jul;22(25):2784–2787.
20. Wiggins KM, Syrett JA, Haddleton DM, Bielawski CW. Mechanically facilitated retro [4 + 2] cycloadditions. *J Am Chem Soc.* 2011;133(18):7180–7189.
21. Liu YL, Chuo TW. Self-healing polymers based on thermally reversible Diels–Alder chemistry. *Polym Chem.* 2013;4:2194–2205.
22. Reutenauer P, Buhler E, Boul P, Candau S, Lehn JM. Room temperature dynamic polymers based on Diels–Alder chemistry. *Chem Eur J.* 2009;15(8):1893–1900.
23. Wei Z, Yang JH, Du XJ, Xu F, Zrinyi M, Osada Y, et al. Dextran-based self-healing hydrogels formed by reversible Diels–Alder reaction under physiological conditions. *Macromol Rapid Commun.* 2013;34(18):1464–1470.
24. Zhang P, Deng F, Peng Y, Chen H, Gao Y, Li H. Redox- and pH-responsive polymer gels with reversible sol–gel transitions and self-healing properties. *RSC Adv.* 2014;4:47361–47367.
25. Guan Y, Zhang Y. Boronic acid-containing hydrogels: Synthesis and their applications. *Chem Soc Rev.* 2013 Sep;42:8106–8121.
26. Dufort BM, Tibbitt MW. Design of moldable hydrogels for biomedical applications using dynamic covalent boronic esters. *Mater Today Chem.* 2019 Jun;12:16–33.
27. Gyarmati B, Némethy A, Szilágyi A. Reversible disulphide formation in polymer networks: A versatile functional group from synthesis to applications. *Eur Polym J.* 2013 Jun;49:1268–1286.
28. Bapat AP, Ray JG, Savin DA, Sumerlin BS. Redox-responsive dynamic-covalent assemblies: Stars and miktoarm stars. *Macromolecules.* 2013 Mar;46:2188–2198.
29. Fairbanks BD, Singh SP, Bowman CN, Anseth KS. Photodegradable, photoadaptable hydrogels via radical-mediated disulfide fragmentation reaction. *Macromolecules.* 2011 Apr;44:2444–2450.

30. Ji S, Xia J, Xu H. Dynamic chemistry of selenium: Se-n and se-se dynamic covalent bonds in polymeric systems. *ACS Macro Lett.* 2016 Jan;5:78–82.
31. Ji S, Cao W, Yu Y, Xu H, Ji S, Cao W, et al. Visible-light-induced self-healing diselenide-containing polyurethane elastomer. *Adv Mater.* 2015 Dec;27:7740–7745.
32. Dillinq WL. Polymerization of unsaturated compounds by photocycloaddition reactions. *Chem Rev.* 1983;83:1–47.
33. Cardenas-Daw C, Kroeger A, Schaertl W, Froimowicz P, Landfester K. Reversible photocycloadditions, a powerful tool for tailoring (nano)materials. *Macromol Chem Phys.* 2012 Jan;213:144–156.
34. Khare E, Holten-Andersen N, Buehler MJ. Transition-metal coordinate bonds for bioinspired macromolecules with tunable mechanical properties. *Nat Rev Mater.* 2021 Feb;6:421–436.
35. Li CH, Zuo JL. Self-healing polymers based on coordination bonds. *Adv Mater.* 2020 Jul;32;1903762.
36. Brassinne J, Fustin CA, Gohy JF. Polymer gels constructed through metal–ligand coordination. *J Inorg Organomet Polym Mater.* 2012 Sep;23:24–40.
37. Lu L, Tian T, Wu S, Xiang T, Zhou, S. A pH-induced self-healable shape memory hydrogel with metal-coordination cross-links. *Polym Chem.* 2019 Apr;10:1920–1929.
38. Zechel S, Hager MD, Priemel T, Harrington MJ. Healing through histidine: Bioinspired pathways to self-healing polymers via imidazole-metal coordination. *Biomimetics.* 2019 Feb;4:20.
39. Marco-Dufort B, Iten R, Tibbitt MW. Linking molecular behavior to macroscopic properties in ideal dynamic covalent networks. *J Am Chem Soc.* 2020 Sep;142:15371–15385.
40. Cazzell SA, Holten-Andersen N. Expanding the stoichiometric window for metal cross-linked gel assembly using competition. *Proc Natl Acad Sci USA.* 2019 Oct;116:21369–21374.
41. Marco-Dufort B, Willi J, Vielba-Gomez F, Gatti F, Tibbitt MW. Environment controls biomolecule release from dynamic covalent hydrogels. *Biomacromolecules.* 2021 Jan;22:146–157.
42. Parada, GA, Zhao, X. Ideal reversible polymer networks. *Soft Matter.* 2018 Jun;14:5186–5196.
43. Metze FK, Sant S, Meng Z, Klok HA, Kaur K. Swelling-activated, soft mechanochemistry in polymer materials. *Langmuir.* 2023 Mar;39:3546–3557.

44. Miao W, Zou W, Jin B, Ni C, Zheng N, Zhao Q, et al. On demand shape memory polymer via light regulated topological defects in a dynamic covalent network, *Nat Commun.* 2020 Aug;11(1):1–8.
45. Zheng N, Xu Y, Zhao Q, Xie T. Dynamic covalent polymer networks: A molecular platform for designing functions beyond chemical recycling and self-healing. *Chem Rev.* 2021 Feb;121(3): 1716–1745.
46. Gu Y, Alt EA, Wang H, Li X, Willard AP, Joh son JA. Photoswitching topology in polymer networks with metal-organic cages as crosslinks. *Nature.* 2018 Aug;560(7716):65–69.
47. Zhang ZP, Rong MZ, Zhang MQ, Yuan C. Alkoxyamine with reduced homolysis temperature and its application in repeated autonomous self-healing of stiff polymers. *Polym Chem.* 2013;4:4648–4654.
48. Barthel MJ, Rudolph T, Teichler T, Paulus RM, Vitz J, Hoeppener S, et al. Self-healing materials via reversible crosslinking of poly (ethylene oxide)-block-poly(furfuryl glycidyl ether) (peo-b-pfge) block copolymer films. *Adv Funct Mater.* 2013;23(39):4921–4932.
49. Ghiassinejad S, Mortensen K, Rostamitabar M, Malineni J, Fustin CA, van Ruymbeke E. Dynamics and structure of metallo-supramolecular polymers based on short telechelic precursors. *Macromolecules.* 2021;54(13):6400–6416.
50. Grindy SC, Learsch R, Mozhdehi D, Cheng J, Barrett DG, Guan Z, et al. Control of hierarchical polymer mechanics with bioinspired metal-coordination dynamics. *Nat Mater.* 2015 Aug;14: 1210–1216.
51. Deng CC, Brooks WLA, Abboud KA, Sumerlin BS. Boronic acid-based hydrogels undergo self-healing at neutral and acidic pH. *ACS Macro Lett.* 2015 Feb;4(2):220–224.
52. Flory PJ. Molecular size distribution in three dimensional polymers. I. gelation1. *J Am Chem Soc.* 1941;63(11):3083–3090.
53. Zhang V, Kang B, Accardo JV, Kalow JA. Structure–reactivity–property relationships in covalent adaptable networks, *J Am Chem Soc.* 2022;49:22358–22377.
54. Semenov AN, Rubinstein, M. Thermoreversible gelation in solutions of associative polymers. 1. Statics. *Macromolecules.* 1998 Feb; 31(4):1373–1385.

55. Rubinstein, M, Semenov AN. Thermoreversible gelation in solutions of associating polymers. 2. Linear dynamics. *Macromolecules.* 1998 Feb;31(4):1386–1397.
56. Yount WC, Loveless DM, Craig SL. Small-molecule dynamics and mechanisms underlying the macroscopic mechanical properties of coordinatively cross-linked polymer networks. *J Am Chem Soc.* 2005 Oct;127:14488–14496.
57. Montano V, Picken SJ, Zwaag SVD, Garcia SJ. A deconvolution protocol of the mechanical relaxation spectrum to identify and quantify individual polymer feature contributions to self-healing. *Phys Chem Phys.* 2019 May;21:10171–10184.
58. Liu F, Li F, Deng G, Chen Y, Zhang B, Zhang J, et al. Rheological images of dynamic covalent polymer networks and mechanisms behind mechanical and self-healing properties. *Macromolecules.* 2012 Feb;45(3):1636–1645.
59. Rizwan M, Baker AE, Shoichet MS. Designing hydrogels for 3D cell culture using dynamic covalent crosslinking. *Adv Healthc Mater.* 2021 Jun;10(12):2100234.
60. Sheridan RJ, Bowman CN. A simple relationship relating linear viscoelastic properties and chemical structure in a model Diels-Alder polymer network. *Macromolecules.* 2012 Sep;45: 7634–7641.
61. Kilic R, Sanyal A. Self-healing hydrogels based on reversible covalent linkages: A survey of dynamic chemical bonds in network formation. *Adv Polym Sci.* 2020;285:243–294.
62. Rao A, Ramírez J, Olsen BD. Mechanisms of self-diffusion of linear associative polymers studied by Brownian dynamics simulation. *Macromolecules.* 2012 Dec;54:11212–11227.
63. Rapp PB, Omar AK, Silverman BR, Wang ZG, Tirrell DA. Mechanisms of diffusion in associative polymer networks: Evidence for chain hopping. *J Am Chem Soc.* 2018 Oct;140;14185.
64. Ramirez J, Dursch TJ, Olsen BD. A molecular explanation for anomalous diffusion in supramolecular polymer networks. *Macromolecules.* 2018 Apr;51:2517–2525.
65. Rasid IM, Holten-Andersen N, Olsen BD. Anomalous diffusion in associative networks of high-sticker-density polymers. *Macromolecules.* 2021 Feb;54:1354–1365.

66. Tanaka F, Edwards S. Viscoelastic properties of physically crosslinked networks. 1. Transient network theory. *Macromolecules.* 1992;25(5):1516–1523.
67. Vaccaro A, Marrucci G. A model for the nonlinear rheology of associating polymers. *J Non-Newton Fluid Mech.* 2000 Aug;92:261–273. ISSN 0377-0257.
68. Wang SQ. Transient network theory for shear-thickening fluids and physically crosslinked networks. *Macromolecules.* 1992;25(25): 7003–7010.
69. Sing MK, Wang ZG, McKinley GH, Olsen BD. Celebrating soft matter's 10th anniversary: Chain configuration and rate-dependent mechanical properties in transient networks. *Soft Matter.* 2015 Mar; 11:2085–2096.
70. Rasid IM, Ramirez J, Olsen BD, Holten-Andersen N. Understanding the molecular origin of shear thinning in associative polymers through quantification of bond dissociation under shear. *Phys Rev Mater.* 2020;4:055602.
71. Pellens L, Ahn KH, Lee SJ, Mewis J. Evaluation of a transient network model for telechelic associative polymers. *J Non-Newton Fluid Mech.* 2004 Aug;121:87–100.
72. Fielding SM, Olmsted, PD. Flow phase diagrams for concentration-coupled shear banding. *Eur Phys J E.* 2003 May;11:65–83.
73. Castillo-Tejas J, Manero O. Molecular dynamics of shear banding flow in associative polymers solutions: Effect of sticker groups location. *Rheol Acta.* 2021 Jul;60:293–307.
74. Omar AK, Wang ZG. Shear-induced heterogeneity in associating polymer gels: Role of network structure and dilatancy. *Phys Rev. Lett.* 2017 Sep;119:117801.
75. Rincoón E, Chávez AE, Herrera R, Manero O. Rheological modelling of complex fluids: A transient network model with microstates. *J Non-Newton Fluid Mech.* 2005 Nov;131:64–77.
76. Jin H, Kang K, Ahn KH, Dhont JK. Flow instability due to coupling of shear-gradients with concentration: Non-uniform flow of (hard-sphere) glasses. *Soft Matter.* 2014 Dec;10:9470–9485.
77. Wang J, Benyahia L, Chassenieux C, Tassin JF, Nicolai T. Shear-induced gelation of associative polyelectrolytes. *Polymer.* 2010 Apr;51:1964–1971.

78. Kjøniksen AL, Hiorth M, Roots J, Nyström B. Shear-induced association and gelation of aqueous solutions of pectin. *J Phys Chem B*. 2003 Jul;107:6324–6328.
79. Zhou H, Woo J, Cok AM, Wang M, Olsen BD, Johnson JA. Counting primary loops in polymer gels. *Proc Natl Acad Sci.* 2012;109(47):19119–19124.
80. Wang R, Alexander-Katz A, Johnson JA, Olsen BD. Universal cyclic topology in polymer networks. *Phys Rev Lett.* 2016 May; 116:188302.
81. Matsunaga T, Sakai T, Akagi Y, Chung U, Shibayama M. Structure characterization of tetra-peg gel by small-angle neutron scattering. *Macromolecules.* Feb., 2009;42:1344–1351.
82. Matsunaga T, Sakai T, Akagi Y, Chung U, Shibayama M. SANS and SLS studies on tetra-arm peg gels in as-prepared and swollen states. *Macromolecules.* 2009 Aug;42:6245–6252.
83. Akagi Y, Matsunaga T, Shibayama M, Chungs UI, Sakai T. Evaluation of topological defects in tetra-peg gels. *Macromolecules.* 2010 Jan;43:488–493.
84. Zhou H, Woo J, Cok AM, Wang M, Olsen BD, Johnson JA. Counting primary loops in polymer gels. *Proc Natnl Acad Sci USA*. 2012 Nov;109:19119–19124.
85. Zhou H, Schön EM, Wang M, Glassman MJ, Liu J, Zhong M, et al. Crossover experiments applied to network formation reactions: Improved strategies for counting elastically inactive molecular defects in peg gels and hyperbranched polymers, *J Am Chem Soc.* 2014 Jul;136:9464–9470.
86. Saalwächter K. Proton multiple-quantum NMR for the study of chain dynamics and structural constraints in polymeric soft materials. *Prog in Nucl Magn Reson Spectrosc.* 2007 Aug;51:1–35.
87. Saalwächter K, Gottlieb M, Liu R, Oppermann W. Gelation as studied by proton multiple-quantum NMR. *Macromolecules.* 2007 Mar;40:1555–1561.
88. Lange F, Schwenke K, Kurakazu M, Akagi Y, Chung U, Lang M, et al. Connectivity and structural defects in model hydrogels: A combined proton NMR and monte Carlo simulation study. *Macromolecules.* 2011 Dec;44:9666–9674.

89. Zhong M, Wang R, Kawamoto K, Olsen BD, Johnson JA. Quantifying the impact of molecular defects on polymer network elasticity. *Science.* 2016Sep;353(6305):1264–1268.
90. Nishi K, Noguchi H, Sakai T, Shibayama M. Rubber elasticity for percolation network consisting of gaussian chains. *J Chem Phys.* 2015 Nov;143(18):184905.
91. Ge S, Samanta S, Tress M, Li B, Xing K, Dieudonné-George P, et al. Critical role of the interfacial layer in associating polymers with microphase separation. *Macromolecules.* 2021 May;54: 4246–4256.
92. Ge S, Samanta S, Li B, Carden GP, Cao PF, Sokolov AP. Unravelling the mechanism of viscoelasticity in polymers with phase-separated dynamic bonds. *ACS Nano.* 2022 Mar;16:4746–4755.
93. Wang W, Zhang J, Jiang F, Wang X, Wang Z. Reprocessable supramolecular thermoplastic BAB-type triblock copolymer elastomers with enhanced tensile strength and toughness via metal-ligand coordination. *ACS Appl Polym Mater.* 2019 Mar;1:571–583.
94. Danielsen SP. Chemical compatibilization, macro-, and microphase separation of heteroassociative polymers. *Macromolecules.* 2023Aug;56(16):6527–6542.
95. Nian S, Patil S, Zhang S, Kim M, Chen Q, Zhernenkov M, et al. Dynamics of associative polymers with high density of reversible bonds. *Phys Rev Lett.* 2023 May;130:228101.
96. Turner APF. Biosensors: Sense and sensibility. *Chem Soc Rev.* 2013 Apr;42(8):3184–3196.
97. Williams GT, Kedge JL, Fossey JS. Molecular boronic acid-based saccharide sensors. *ACS Sens.* 2021 Apr;6(4):1508–1528.
98. Sun X, James TD. Glucose sensing in supramolecular chemistry. *Chem Rev.* 2015 Aug;115(15):8001–8037.
99. Skjaervold NK, Solligård E, Hjelme DR, Aadahl P. Continuous measurement of blood glucose: Validation of a new intravascular sensor. *Anesthesiology.* 2011 Jan;114(1):120–125.
100. Elsherif M, Hassan MU, Yetisen AK, Butt H. Glucose sensing with phenylboronic acid functionalized hydrogel-based optical diffusers. *ACS Nano.* 2018 Mar;12(3):2283–2291.
101. Dautta M, Alshetaiwi M, Escobar J, Tseng P. Passive and wireless, implantable glucose sensing with phenylboronic acid hydrogel-interlayer RF resonators. *Biosens Bioelectron.* 2020 Mar;151:112004.

102. Xu S, Sedgwick AC, Elfeky SA, Chen W, Jones AS, Williams GT, et al. A boronic acid-based fluorescent hydrogel for monosaccharide detection. *Front Chem Sci Eng.* 2020 Feb;14(1):112–116.
103. Han YD, Kim KR, Park YM, Song SY, Yang YJ, Lee K, et al. Boronate-functionalized hydrogel as a novel biosensing interface for the glycated hemoglobin a1c (HbA1c) based on the competitive binding with signaling glycoprotein. *Mater Sci Eng C Mater Biol Appl.* 2017 Aug;77:1160–1169.
104. Marco-Dufort B, Janczy JR, Hu T, Lütolf M, Gatti F, Wolf M, et al. Thermal stabilization of diverse biologics using reversible hydrogels. *Sci Adv.* 2022 Aug;8(31):eabo0502.
105. Kölmel DK, Kool ET. Oximes and hydrazones in bioconjugation: Mechanism and catalysis. *Chem Rev.* 2017 Aug;117(15): 10358–10376.
106. Bouillon C, Bessin Y, Poncet F, Gary-Bobo M, Dumy P, Barboiu M, et al. Biomolecular dynamic covalent polymers for DNA complexation and siRNA delivery. *J Mater Chem B Mater Biol Med.* 2018 Nov;6(44):7239–7246.
107. Zhang Y, Legrand YM, Petit E, Supuran CT, Barboiu M. Dynamic encapsulation and activation of carbonic anhydrase in multivalent dynameric host matrices. *Chem Commun.* 2016 Mar;52(21): 4053–4055.
108. Zhang Y, Feng WX, Legrand YM, Supuran CT, Su CY, Barboiu M. Dynameric host frameworks for the activation of lipase through h-bond and interfacial encapsulation. *Chem Commun.* 2016 Dec; 52(95):13768–13770.
109. Yesilyurt V, Webber MJ, Appel EA, Godwin C, Langer R, Anderson DG. Injectable self-healing glucose-responsive hydrogels with pH-regulated mechanical properties. *Adv Mater.* 2016 Jan; 28(1):86–91.
110. Xiang Y, Su B, Liu D, Webber MJ. Managing diabetes with hydrogel drug delivery. *Adv Ther.* 2023Jun;7(1):2300127.
111. Xian S, Xiang Y, Liu D, Fan B, Mitrová K, Ollier RC, et al. Insulin-dendrimer nanocomplex for multi-day glucose-responsive therapy in mice and swine. *Adv Mater.* 2023 Nov;36(5):e2308965.
112. Moroni L, Boland T, Burdick JA, De Maria C, Derby B, Forgacs G, et al. Biofabrication: A guide to technology and terminology. *Trends Biotechnol.* 2018 Apr;36(4):384–402.

113. Correa S, Grosskopf AK, Hernandez HL, Chan D, Yu AC, Stapleton LM, et al. Translational applications of hydrogels. *Chem Rev.* 2021;121(18):11385–11457.
114. Datta SS, Ardekani AM, Arratia PE, Beris AN, Bischofberger I, McKinley GH, et al. Perspectives on viscoelastic flow instabilities and elastic turbulence. *Phys Rev Fluids.* 2022 Aug;7(8):080701.
115. Bertsch P, Diba M, Mooney DJ, Leeuwenburgh SCG. Self-healing injectable hydrogels for tissue regeneration. *Chem Rev.* 2023 Jan;123(2):834–873.
116. Fan B, Torres Garćıa D, Salehi M, Webber MJ, van Kasteren SI, Eelkema R. Dynamic covalent dextran hydrogels as injectable, self-adjuvating peptide vaccine depots. *ACS Chem Biol.* 2023 Mar;18(3):652–659.
117. Yang H, Qin X, Wang H, Zhao X, Liu Y, Wo HT, et al. An in vivo miRNA delivery system for restoring infarcted myocardium. *ACS Nano.* 2019 Sep;13(9):9880–9894.
118. Suhar RA, Doulames VM, Liu Y, Hefferon ME, Figueroa O, Buabbas H, et al. Hyaluronan and elastin-like protein (HELP) gels significantly improve microsphere retention in the myocardium. *Biomater Sci.* 2022 May;10(10):2590–2608.
119. Wang H, Zhu D, Paul A, Cai L, Enejder A, Yang F, et al. Covalently adaptable elastin-like protein-hyaluronic acid (ELP-HA) hybrid hydrogels with secondary thermoresponsive crosslinking for injectable stem cell delivery. *Adv Funct Mater.* 2017Jul;27(28): 1605609.
120. Aguado BA, Mulyasasmita W, Su J, Lampe KJ, Heilshorn SC. Improving viability of stem cells during syringe needle flow through the design of hydrogel cell carriers. *Tissue Eng Part A.* 2012 Apr;18(7–8):806–815.
121. Guzzi EA, Tibbitt MW. Additive manufacturing of precision biomaterials. *Adv Mater.* 2020 Apr;32(13):e1901994.
122. de Paiva Narciso N, Navarro RS, Gilchrist AE, Trigo MLM, Aviles Rodriguez G, Heilshorn SC. Design parameters for injectable biopolymeric hydrogels with dynamic covalent chemistry crosslinks. *Adv Healthc Mater.* 2023 Oct;12(27):e2301265.
123. Hull SM, Lou J, Lindsay CD, Navarro RS, Cai B, Brunel LG, et al. 3d bioprinting of dynamic hydrogel bioinks enabled by small molecule modulators. *Sci Adv.* 2023 Mar;9(13):eade7880.

124. Morgan FLC, Fernández-Pérez J, Moroni L, Baker MB. Tuning hydrogels by mixing dynamic cross-linkers: Enabling cell-instructive hydrogels and advanced bioinks. *Adv Healthc Mater.* 2022 Jan;11(1):e2101576.
125. Lyu SAJ, Johnson M, Creagh-Flynn J, Zhou D, Lara-Sáez I, Xu Q, et al. Instant gelation system as Self-Healable and printable 3D cell culture bioink based on dynamic covalent chemistry. *ACS Appl Mater Interfaces.* 2020 Sep;12(35):38918–38924.
126. Wang Z, Gu J, Zhang D, Zhang J, Chen J. Structurally dynamic gelatin-based hydrogels with self-healing, shape memory, and cyto-compatible properties for 4D printing. *Biomacromolecules.* 2023 Jan;24(1):109–117.
127. Humphrey JD, Dufresne ER, Schwartz MA. Mechanotransduction and extracellular matrix homeostasis. *Nat Rev Mol Cell Biol.* 2014 Dec;15(12):802–812.
128. Hynes RO. The extracellular matrix: Not just pretty fibrils. *Science.* 2009 Nov;326(5957):1216–1219.
129. Tibbitt MW, Anseth, KS. Hydrogels as extracellular matrix mimics for 3D cell culture. *Biotechnol Bioeng.* 2009Jul;103(4):655–663.
130. Chaudhuri O, Cooper-White J, Janmey PA, Mooney DJ, Shenoy VB. Effects of extracellular matrix viscoelasticity on cellular behaviour. *Nature.* 2020 Aug;584(7822):535–546.
131. Nam S, Stowers R, Lou J, Xia Y, Chaudhuri O. Varying PEG density to control stress relaxation in alginate-PEG hydrogels for 3D cell culture studies. *Biomaterials.* 2019 Apr;200:15–24.
132. McKinnon DD, Domaille DW, Cha JN, Anseth KS. Hydrogels: Biophysically defined and cytocompatible covalently adaptable networks as viscoelastic 3D cell culture systems. *Adv Mater.* 2014; 6(26):821–821.
133. McKinnon DD, Domaille DW, Brown TW, Kyburz KA, Kiyotake E, Cha JN, et al. Measuring cellular forces using bis-aliphatic hydrazone crosslinked stress-relaxing hydrogels. *Soft Matter.* 2014 Dec;10(46):9230–9236.
134. Tang S, Ma H, Tu HC, Wang HR, Lin PC, Anseth KS. Adaptable fast relaxing boronate-based hydrogels for probing cell-matrix interactions. *Adv Sci.* 2018Sep;5(9);1800638.
135. Smithmyer ME, Deng CC, Cassel SS, LeValley PJ, Sumerlin BS, Kloxin AM. Self-healing boronic acid-based hydrogels for 3D co-cultures. *ACS Macro Lett.* 2018 Sep;7(9):1105–1110.

136. Hunt DR, Klett KC, Mascharak S, Wang H, Gong D, Lou J, et al. Engineered matrices enable the culture of human patient-derived intestinal organoids. *Adv Sci.* 2021 May;8(10):2004705.
137. Ruiter FAA, Morgan FLC, Roumans N, Schumacher A, Slaats GG, Moroni L, et al. Soft, dynamic hydrogel confinement improves kidney organoid lumen morphology and reduces epithelial-mesenchymal transition in culture. *Adv Sci.* 2022Jul;9(20):e2200543.
138. Chujo Y, Sada K, Nomura R, Naka A, Saegusa T. Photogelation and redox properties of anthracene-disulfide-modified polyoxazolines. *Macromolecules.* 1993;26:5611–5614.
139. Amamoto Y, Otsuka H, Takahara A, Matyjaszewski K. Changes in network structure of chemical gels controlled by solvent quality through photoinduced radical reshuffling reactions of trithiocarbonate units. *ACS Macro Lett.* 2012 Apr;1:478–481.
140. Adzima BJ, Kloxin CJ, Deforest CA, Anseth KS, Bowman CN. 3d photofixation lithography in Diels–Alder networks. *Macromol Rapid Commun.* 2012;33:2092–2096.

https://doi.org/10.1142/9789819806782_0009

Chapter 9

Design Considerations for Cell-Compatible Double-Network Hydrogels

Shoshana Weintraub and Dror Seliktar*

Faculty of Biomedical Engineering, Technion-Israel Institute of Technology, 32000 Haifa, Israel
**Corresponding author: bmdror@technion.ac.il*

Double-network hydrogels combine two independent polymer networks, one brittle and one ductile, to achieve significant mechanical strength and toughness. Multiple parameters must be considered when designing cell-laden double-network hydrogels, among them biocompatibility of materials and methods, desired mechanical and rheological behavior, environmental sensitivities, and degradation rate and by-products. This chapter will provide insight into the most important design parameters to account for when fabricating cell-compatible double-network hydrogels.

1. Introduction

The extracellular matrix (ECM) provides structural support to cells, playing a critical role in proliferation, differentiation, and migration processes. It contains various proteins that support different cellular processes, including collagen, elastin, fibronectin, and hyaluronic acid (HA).[1] ECM stiffness, determined by protein content, density, and cross-linking, affects the mechanical forces transmitted to the cells.[2]

Interpenetrating polymer networks (IPNs) combine independent, intertangled polymer networks that are not chemically cross-linked to one another. IPNs combine the properties of the polymer components, resulting in stronger mechanics and more controllable physical properties.[3,4] Double-network (DN) hydrogels are a class of strong and tough IPNs with a highly cross-linked, brittle primary network and loosely cross-linked, ductile secondary network.[5,6]

The challenge for many tissue engineering applications lies in achieving both the structural and mechanical properties of native tissue. The ECM has a heterogeneous structure that is not easily synthetically replicated. DN hydrogels synergistically combine the material properties of two polymers, endowing the resultant hydrogel with strength and toughness, better matching the properties of native tissue, particularly those of load-bearing soft tissues such as tendons and muscles.[7]

Numerous factors must be considered when designing a hydrogel system for biological applications, especially when incorporating cells. This chapter will provide insight into the most important design parameters for cell-compatible DN hydrogels.

2. Overview of DN hydrogels

DN hydrogels were first introduced by Jian Ping Gong's research group in 2003.[8] These gels are composed of a stiff, brittle primary network and a soft, ductile secondary network, which synergistically increase the toughness and stiffness.[8,9] The first DN system synthesized by the Gong laboratory, poly(2-acrlyamido-2-methylpropanesulfonic acid)–poly(acrylamide) (PAMPS–PAAm), could sustain stresses up to 43 times greater than and strains more than double a PAMPS single-network (SN) gel.[8]

As per Gong's original paper, there are two structural requirements for DN hydrogels. First, the soft polymer network should be the main component, with a polymer molar ratio of several tens (i.e., ratio of secondary to primary network constituents). The primary network serves as a sacrificial network, fracturing under stress, while the secondary network dissipates energy by deforming around the cracked bonds.[5,8,10,11] Under stress, the covalent bonds in the primary network break and form small islands within the secondary network, softening the material locally. The secondary network then deforms around the crack tip, preventing crack propagation. The combination of the two components imparts the material with both toughness and strength, properties lacking in conventional hydrogel materials.[5]

The second structural requirement pertains to the cross-linking density of the two networks: the primary network should be tightly cross-linked, while the secondary network should be loosely linked to allow for deformation. The crosslinker concentration of the secondary network determines the mechanical strength of the network. For example, Gong et al. found that the PAMPS/PAAm DN had the highest fracture stress and strain when the crosslinker concentration in the PAAm network was 0–0.1 mol%.[8] Subsequent work has suggested that in conventional DN hydrogels, unreacted crosslinker from the primary network forms covalent bonds between the polymer networks during the second cross-linking stage. These interconnected points allow for force transfer between the two networks, increasing the system's mechanical strength and toughness. When the two networks are completely independent, such as when the unreacted crosslinker is made inert before polymerization of the secondary network, there is a critical crosslinker concentration below which the gel lacks mechanical strength and above which the fracture energy decreases.[12]

3. Material and structural parameters

3.1. *Biocompatibility*

Biocompatibility was originally defined for medical devices as a material's ability to remain in contact with tissues without interaction, thus requiring biocompatible materials to be biologically and chemically inert. However, tissue engineering fundamentally requires materials to stimulate natural tissue to recruit cells to generate new tissue. As such, biocompatibility has been proposed to be "the ability of a material to perform with appropriate host response in a specific situation."[13] For tissue engineering, a biocompatible material "supports the appropriate cellular activity, including the facilitation of molecular and mechanical signaling, to optimize tissue regeneration, without eliciting any undesirable local or systemic responses in the eventual host."[13]

Biocompatible hydrogels may be formed from synthetic or natural polymers. Synthetic polymers have more controlled chemical compositions and molecular weight and superior mechanical properties to naturally derived polymers.[3] Poly(vinyl alcohol) (PVA), polyethyleneimine, poly(ethylene glycol) (PEG), and poly(lactic acid) (PLA) are some of the most commonly used synthetic polymers for biomedical hydrogels.[14,15] However, these materials do not sufficiently promote cell functionality or proliferation.[3,15]

Natural polymers offer advantages regarding cell interactions, degradability, and availability. Biopolymers derived from tissues, including polysaccharides (such as HA, chitosan, and alginate) and proteins (such as collagen, gelatin, and fibroin), contain hydroxyl, amine, and carboxylic groups that can be modified for chemical functionalization and cross-linking.[3,16–22] Modification of these biopolymers with methacrylate derivatives has been widely explored since methacrylate contains reactive groups that can be easily photocrosslinked.[15,23–26] Alginate, an anionic biopolymer, is also widely used due to its tunable cross-linking density and high swelling capacity.[3,20,27–29] Biocompatible and mechanically strong DN hydrogels can be obtained by combining natural and synthetic polymers (e.g., methacrylate-modified HA [MeHA]/DAAm) to harness the more controllable properties of synthetic polymers and inherent biocompatibility of natural polymers.[23]

3.1.1. *Fabrication*

When fabricating DN hydrogels for cell encapsulation and three-dimensional (3D) culture, it is important to ensure that all materials and methods used are cytocompatible. Many polymer syntheses employ the use of cytotoxic organic solvents or crosslinkers. Additionally, certain polymerization techniques, such as prolonged ultraviolet (UV) exposure, may negatively affect cell viability.[30,31]

Particularly for DN fabrication, which often includes two polymerization steps, it is critical to consider all materials and methods used during preparation.

Traditional DN fabrication is a two-step process: the primary network is tightly cross-linked and then immersed in a solution of the second monomer with a low concentration of crosslinker. Once the second monomer and crosslinker diffuse into the primary network, the second polymerization step is carried out.[5] One-pot methods employing bio-orthogonal click chemistries, thermoreversible sol–gel transitions, and orthogonal photoinitiated reactions have been developed to simplify and increase the cytocompatibility of DN fabrication.[32–36] Photopolymerization is a common technique for hydrogel cross-linking, offering advantages over other mechanisms, including solvent-free conditions and controllable kinetics and degree of cross-linking. The photoinitiator concentration, irradiation intensity, and length of irradiation can be tuned to ensure cytocompatibility.[30,31]

3.1.2. *Cell adhesion and proliferation*

Cell–matrix interactions are critical for cell growth and tissue regeneration. While synthetic polymers do not contain cell-binding sites, ECM-derived molecules or cell-adhesive peptides (such as the Arg-Gly-Asp [RGD] sequence) can be incorporated into the hydrogel network to improve cellular adhesion.[37,38] However, these integrations do not mimic the natural ECM topography.[3,39] DN hydrogels, on the other hand, can incorporate natural polymers that better mimic the ECM with synthetic polymers, allowing for better tunability of mechanical properties.

Collagen, the most prevalent material in the ECM, has been found to enhance cellular adhesion and spreading.[1] Brigham et al. found that hydrogels made from MeHA did not display significant cell spreading compared to DN collagen–MeHA gels.[19] Myung et al. modified PEG–PAA DN gels with type I collagen to improve cellular adhesion and spreading and achieved confluency within 48 hours.[40,41] Other natural polymers, including fibrin and gelatin, have been incorporated into DN gels to improve cell adhesion.[3,29,39,42] Vorwald et al. developed a fibrin–alginate IPN hydrogel in which the fibrinogen content enhanced the cell adhesion properties and the alginate cross-linking density determined the bulk mechanical properties and degradation rate such that both components influenced the rate and degree of cell spreading.[39] Jia et al. fabricated artificial blood vessels from PVA and carboxymethyl chitosan, incorporating chitosan into the hydrogel to enhance cell adhesion while retaining the PVA's superior mechanical properties.[43]

In addition to supporting cell adhesion, DN hydrogels must have sufficient porosity to support nutrient diffusion, waste removal, and cell spreading.[44–46] The inherent modularity of DNs enables the mesh size to be controlled by the gel composition and cross-linking density of each network.[47,48] For example, Liang et al. found that the pore size of a poly(N-isopropyl acrylamide) (PNIPAAm)–PEG and alginate DN hydrogel decreased with increasing PNIPAAm–PEG content.[48] Weng et al. similarly found that increasing the cross-linking density of the PDAAm network in a MeHA-PDAAm DN hydrogel resulted in a smaller mesh size.[23] Porous DN gels have also been fabricated using high-internal-phase emulsion (HIPE), salting-out, freeze-thaw, and porogen methods.[43,49–51] Porosity can also be introduced by selective degradation of one of the polymer networks.[3]

3.2. *Mechanical properties*

DN hydrogels have superior mechanical properties compared to their SN counterparts, which lack stress-dissipating mechanisms.[52] Conventional methods for strengthening SN gels, such as increasing the polymer concentration or cross-linking density, negatively affect cell growth and nutrient transport through the network.[3] DN gels can achieve mechanical strength, sometimes orders of magnitude greater than the SN gels, while supporting cell migration/infiltration.[52]

Matrix stiffness has been shown to play a role in cellular differentiation and tissue development.[53–55] Therefore, it is important to design hydrogels that can match the physiological stiffnesses of natural tissues. DN gels are particularly interesting for load-bearing tissue engineering, which have higher stiffnesses often not achievable with conventional SN hydrogels.[26] Cartilage, for example, has a Young's modulus on the order of megapascals and can withstand both tensile and compressive loads of up to 100 MPa.[56]

Whereas the mechanical properties of SN gels oftentimes cannot be tuned independently of one another, DN gels allow for the decoupling of certain properties. Berger et al. fabricated a gelatin methacryloyl (GelMA)/collagen IPN that decoupled matrix stiffness and fiber density, allowing them to study how each factor influenced breast cancer cell invasion and angiogenesis.[57] Rennerfeldt et al. studied the effects of varying agarose concentration, PEG-DA concentration, and PEG-DA molecular weight on the mechanical properties of an agarose-PEG-DA IPN and concluded that PEG-DA concentration determined the compressive and shear moduli, while PEG-DA molecular weight determined the failure properties.[52] Ledo et al. developed a collagen I/alginate IPN to independently tune the matrix stiffness without affecting the polymer concentration or adhesion ligand density.[58]

When designing a novel DN network, it is important to investigate the effects of each component on the hydrogel's mechanical properties. Understanding the role of each network's polymer concentration, cross-linking density, and molecular weight can aid in determining the optimal composition and properties of the final hydrogel. It is also important to consider the effects the composition may have on solution viscosity. Particularly for applications requiring low-viscosity solutions (such as 3D printing or injection-based applications), increasing the polymer concentration may increase the strength and toughness while also increasing the solution viscosity beyond a viable value.[32]

For cell-laden hydrogels, it is critical to investigate the effects of cell encapsulation on the system's mechanical properties. Rennerfeldt et al. found that acellular agarose-PEG-DA IPN gels could achieve fracture stresses of up to 3–4 MPa. However, the fracture strength of cell-laden gels of identical composition did not exceed 1.5 MPa, limiting the system's applicability for cartilage tissue engineering.[52]

3.2.1. *Swelling behavior*

Hydrogels have high equilibrium swelling fractions and can contain up to 99 wt% water. Both intrinsic factors, such as polymer concentration, cross-link density, charge, and hydrophobicity, and extrinsic factors, such as temperature, pH, and solvent type, influence the degree of swelling.[14,47] Increased cross-link density and polymer concentration decrease the degree of swelling, while increased hydrophobicity and ionic strength increase it.

Increased swelling results in lower mechanical strength and more brittle behavior. As such, the primary network of DN gels should have a high degree of swelling. While the initial DN fabrication methods required the first network to be strongly polyelectrolytic, approaches enabling DNs to be formed with neutral and weak polyelectrolytic polymers have been developed. The molecular stent method, developed by Gong's research group, increases the osmotic pressure in the first network by submersing it in a polyelectrolytic solution, thereby increasing its swelling ratio.[59]

DN gels can incorporate components to control the degree of swelling, thereby improving the mechanical properties. Akkineni et al. combined gellan gum with alginate to decrease the degree of swelling compared to pure alginate gels and improve the mechanical properties of the fabricated scaffolds.[27] Hwang et al. fabricated PEG/PAA DN hydrogels with pH sensitivity imparted by the PAA network such that the hydrogel's swelling behavior could be tuned by the cross-linking density of the PAA network, resulting in an improved storage modulus.[47]

3.3. *Rheological behavior*

As described earlier, DN gels can achieve incredible mechanical strength through the simultaneous fracture of the primary, brittle network and deformation of the secondary, ductile network. However, classic DN gels are susceptible to the Mullins effect, in which the gels soften during the first loading–unloading cycle.[60,61] During the first loading cycle, the first network undergoes permanent deformation, resulting in a significant hysteresis. Webber et al. suggested that this hysteresis is due to the heterogeneity of the cross-links in the primary network such that any deformation beyond 20% results in unrecoverable deformation.[61]

For DN applications requiring repeated loading and unloading and injectability, the primary network should possess self-healing properties that enable the broken cross-links to recover upon stress removal. Hydrogel injectability allows for minimally invasive delivery, desirable in many applications. Injectable hydrogels should be shear-thinning and self-healing, flowing under applied shear stress and rapidly recovering upon stress removal.[3,62] The exact mechanism of the reversible cross-linking network depends on the desired self-healing kinetics and system temperature, as well as the material requirements of the system.[62]

Various cross-linking mechanisms, including physical cross-linking (i.e., hydrogen bonding or ionic interactions) and dynamic covalent bonding, have been proposed for a self-healing primary network. Rodell et al. developed a DN hydrogel with guest–host cross-linking in the primary network to allow for rapid self-healing and injectable delivery. Repeated application of 80% strain to the methacrylated guest–host DN gel did not cause Mullins-type softening.[62] Mu et al. fabricated a gelatin nanoparticle and injectable platelet-rich fibrin DN hydrogel that was shear-thinning and self-healing before complete solidification of the hydrogel, granting the network moldability during synthesis.[17] Guo et al. developed a catechol-modified HA-alginate DN inspired by mussel-adhesive chemistry, in which catechol-mediated hydrogen bonding and π–π stacking interactions enabled self-healing behavior while the alginate network allowed for injectability.[63]

3.4. *Degradation rate and products*

Hydrogel degradation is a critical aspect of designing biomaterial systems for tissue engineering applications. The degradation mechanism, rate, and by-products are all important factors to consider.

When designing a DN hydrogel as a scaffold for cellular infiltration and tissue regeneration, the hydrogel scaffold should degrade at a rate that matches the tissue regeneration rate, allowing for cell spreading, vascularization, and ECM deposition.[64] Hydrogel degradation rate also plays a critical role in determining mesenchymal cell spreading and invasion, which can influence cell differentiation.[38,65] Lee et al. used a disulfide bond-containing crosslinker to make PEG/collagen DN hydrogel susceptible to glutathione secreted by adipose-derived stem cells, which gradually softened the hydrogel network and improved adipogenesis.[34]

Synthetic hydrogels can be designed to be hydrolytically or enzymatically degradable. Matrix metalloproteinase (MMP)-sensitive peptides can be incorporated as crosslinkers into hydrogel networks to enable cell-mediated proteolytic degradation.[65–68] Bai et al. used tannic acid as a crosslinker in PVA/collagen DN to enhance the biodegradability of the PVA network and simultaneously slow the degradation of the collagen network.[22] In hybrid DNs made from both synthetic and natural polymers, the natural polymer network can be modulated to control the gel's degradation profile, while the synthetic polymer provides structural support to the proliferating cells.[42]

In some applications, such as long-term implants, biodegradation is undesirable. For these applications, synthetic polymers may outperform natural polymers, though care should be taken to ensure biocompatibility and lack of an immunogenic response. Azuma et al. investigated the biodegradation of four DN hydrogels to evaluate their applicability as artificial cartilage by measuring their mechanical properties before and after implantation in subcutaneous tissue for six weeks.[69]

4. Conclusion

In comparison to SN hydrogels, the dual-polymer network in DN hydrogels allows for increased control over the gel properties. Their inherent modularity allows for the decoupling of mechanical and structural features such that each polymer network can be independently tuned to optimize the hydrogel performance. When designing hydrogels for cell encapsulation or tissue scaffolding, DNs provide enhanced control over cell growth and differentiation by allowing the hydrogel stiffness and mechanical properties to be decoupled from cellular adhesion and degradation properties.

While this chapter has provided insight into some of the most important parameters to consider when designing cell-compatible DN hydrogels, there are many others it has not touched upon, such as the design of smart hydrogels responsive to their environment and advanced fabrication methods.[3,14,21,68]

References

1. Kular JK, Basu S, Sharma, RI. The extracellular matrix: Structure, composition, age-related differences, tools for analysis and applications for tissue engineering. *J Tissue Eng.* 2014;5:2041731414557112.
2. Mai Z, Lin Y, Lin P, Zhao X, Cui L. Modulating extracellular matrix stiffness: A strategic approach to boost cancer immunotherapy. *Cell Death Dis.* 2024;15:307.
3. Dhand AP, Galarraga JH, Burdick JA. Enhancing biopolymer hydrogel functionality through interpenetrating networks. *Trends Biotechnol.* 2021;39:519–538.
4. Dragan ES. Design and applications of interpenetrating polymer network hydrogels. A review. *Chem Eng J.* 2014;243:572–590.
5. Gong JP. Why are double network hydrogels so tough? *Soft Matter.* 2021;6:2583.
6. Nakajima T, Gong JP. Double-network hydrogels: Soft and tough IPN. In: Kobayashi S, Müllen, K, editors. *Encyclopedia of Polymeric Nanomaterials.* Berlin, Heidelberg: Springer; 2021. p. 1–6.
7. Haque MdA, Kurokawa T, Gong JP. Super tough double network hydrogels and their application as biomaterials. *Polymer.* 2012;53:1805–1822.
8. Gong JP, Katsuyama Y, Kurokawa T, Osada, Y. Double-network hydrogels with extremely high mechanical strength. *Adv Mater.* 2003;15:1155–1158.
9. Nakayama A, Kakugo A, Gong JP, Osada Y, Takai M, Erata T, et al. High mechanical strength double-network hydrogel with bacterial cellulose. *Adv Funct Mater.* 2004;14:1124–1128.
10. Brown HR. A model of the fracture of double network gels. *Macromolecules.* 2007;40:3815–3818.

11. Tanaka Y. A local damage model for anomalous high toughness of double-network gels. *Europhys Lett.* 2007;78:56005.
12. Nakajima T, Furukawa H, Tanaka Y, Kurokawa T, Osada Y, Gong JP. True chemical structure of double network hydrogels. *Macromolecules.* 2009;42:2184–2189.
13. Williams DF. On the mechanisms of biocompatibility. *Biomaterials.* 2008;29:2941–2953.
14. Bashir S, Hina M, Iqbal J, Rajpar AH, Mujtaba MA, Alghamdi NA, et al. Fundamental concepts of hydrogels: Synthesis, properties, and their applications. *Polymers.* 2020;12:2702.
15. Simaan-Yameen H, Bar-Am O, Saar G, Seliktar D. Methacrylated fibrinogen hydrogels for 3D cell culture and delivery. *Acta Biomater.* 2023;164:94–110.
16. Shin H, Olsen BD, Khademhosseini A. The mechanical properties and cytotoxicity of cell-laden double-network hydrogels based on photocrosslinkable gelatin and gellan gum biomacromolecules. *Biomaterials.* 2012;33:3143–3152.
17. Mu Z, Chen K, Yuan S, Li Y, Huang Y, Wang C, et al. Gelatin nanoparticle-injectable platelet-rich fibrin double network hydrogels with local adaptability and bioactivity for enhanced osteogenesis. *Adv Healthc Mater.* 2020;9:e1901469.
18. Zhang Y, Heher P, Hilborn J, Redl H, Ossipov DA. Hyaluronic acid-fibrin interpenetrating double network hydrogel prepared in situ by orthogonal disulfide cross-linking reaction for biomedical applications. *Acta Biomater.* 2016;38:23–32.
19. Brigham MD, Bick A, Lo E, Bendali A, Burdick JA, Khademhosseini A. Mechanically robust and bioadhesive collagen and photocrosslinkable hyaluronic acid semi-interpenetrating networks. *Tissue Eng Part A.* 2009;15: 1645–1653.
20. Ma C, Choi JB, Jang YS, Kim SY, Bae TS, Kim YK, et al. Mammalian and fish gelatin methacryloyl-alginate interpenetrating polymer network hydrogels for tissue engineering. *ACS Omega.* 2021;6:17433–17441.
21. Sinad KVG, Ebubechukwu RC, Chu CK. Recent advances in double network hydrogels based on naturally-derived polymers: Synthesis, properties, and biological applications. *J Mater Chem B.* 2023;11: 11460–11482.

22. Bai Z, Wang T, Zheng X, Huang Y, Chen Y, Dan W. High strength and bioactivity polyvinyl alcohol/collagen composite hydrogel with tannic acid as cross-linker. *Polym Eng Sci.* 2021;61:278–287.
23. Weng L, Gouldstone A, Wu, Y, Chen W. Mechanically strong double network photocrosslinked hydrogels from N, N-dimethylacrylamide and glycidyl methacrylated hyaluronan. *Biomaterials.* 2018;29: 2153–2163.
24. Liang, J, Wang Z, Poot AA, Grijpma DW, Dijkstra PJ, Wang R. Enzymatic post-crosslinking of printed hydrogels of methacrylated gelatin and tyramine-conjugated 8-arm poly(ethylene glycol) to prepare interpenetrating 3D network structures. *Int J Bioprinting.* 2023;9:750.
25. Burdick JA, Prestwich GD. Hyaluronic acid hydrogels for biomedical applications. *Adv Mater.* 2011;23:H41–H56.
26. Yang J, Zhang YS, Yue K, Khademhosseini A. Cell-laden hydrogels for osteochondral and cartilage tissue engineering. *Acta Biomater.* 2017;57:1–25.
27. Akkineni AR, Ahlfeld T, Funk A, Waske A, Lode A, Gelinsky M. Highly concentrated alginate-gellan gum composites for 3D plotting of complex tissue engineering scaffolds. *Polymers.* 2016;8:170.
28. Kadri, R, Elkhoury K, Ben Messaoud G, Kahn C, Tamayol A, Mano JF, et al. Physicochemical interactions in nanofunctionalized alginate/GelMA IPN hydrogels. *Nanomater Basel Switz.* 2021;11:2256.
29. Marfoglia A, Tibourtine F, Pilloux L, Cazalbou S. Tunable double-network GelMA/alginate hydrogels for platelet lysate-derived protein delivery. *Bioeng Basel Switz.* 2023;10:1044.
30. Chiulan I, Heggset EB, Voicu ŞI, Chinga-Carrasco G. Photopolymerization of bio-based polymers in a biomedical engineering perspective. *Biomacromolecules.* 2021;22:1795–1814.
31. Mironi-Harpaz I, Wang DY, Venkatraman S, Seliktar D. Photopolymerization of cell-encapsulating hydrogels: Crosslinking efficiency versus cytotoxicity. *Acta Biomater.* 2012;8:1838–1848.
32. Dhand, AP, Davidson MD, Galarraga JH, Qazi TH, Locke RC, Mauck RL, et al. Simultaneous one-pot interpenetrating network formation to expand 3D processing capabilities. *Adv Mater Deerfield Beach Fla.* 2022;34:e2202261.
33. Chen Q, Zhu L, Zhao C, Wang Q, Zheng J. A Robust, One-pot synthesis of highly mechanical and recoverable double network

hydrogels using thermoreversible sol-gel polysaccharide. *Adv Mater.* 2013;25:4171–4176.

34. Lee, H, Hong HJ, Ahn S, Kim D, Kang SH, Cho K, et al. One-pot synthesis of double-network PEG/collagen hydrogel for enhanced adipogenic differentiation and retrieval of adipose-derived stem cells. *Polymers.* 2023;15:1777.
35. Truong VX, Ablett MP, Richardson SM, Hoyland JA, Dove AP. Simultaneous orthogonal dual-click approach to tough, in-situ-forming hydrogels for cell encapsulation. *J Am Chem Soc.* 2015; 137:1618–1622.
36. Chen F, Le P, Lai K, Fernandes-Cunha GM, Myung D. Simultaneous interpenetrating polymer network of collagen and hyaluronic acid as an in situ-forming corneal defect filler. *Chem Mater Publ Am Chem Soc.* 2020;32:5208.
37. Lau HK, Kiick KL. Opportunities for multicomponent hybrid hydrogels in biomedical applications. *Biomacromolecules.* 2015;16:28–42.
38. Anderson SB, Lin CC, Kuntzler DV, Anseth KS. The performance of human mesenchymal stem cells encapsulated in cell-degradable polymer-peptide hydrogels. *Biomaterials.* 2011;32:3564–3574.
39. Vorwald CE, Gonzalez-Fernandez T, Joshee S, Sikorski P, Leach JK. Tunable fibrin-alginate interpenetrating network hydrogels to support cell spreading and network formation. *Acta Biomater.* 2020;108:142–152.
40. Myung D, Koh W, Ko J, Noolandi J, Carrasco M, Smith A, et al. Characterization of poly(ethylene glycol)–poly(acrylic acid) (PEG–PAA) double networks designed for corneal implant applications. *Invest Ophthalmol Vis Sci.* 2005;46:5003.
41. Myung, D, Koh W, Bakri A, Zhang F, Marshall A, Ko J, et al. Design and fabrication of an artificial cornea based on a photolithographically patterned hydrogel construct. *Biomed Microdevices.* 2007;9:911–922.
42. Daniele MA, Adams AA, Naciri J, North SH, Ligler FS. Interpenetrating networks based on gelatin methacrylamide and PEG formed using concurrent thiol click chemistries for hydrogel tissue engineering scaffolds. *Biomaterials.* 2014;35:1845–1856.
43. Jia XY, Huang CF, Meng X, Zhu DY, Chen ZP, Jiang T, et al. Dynamically cross-linked double-network hydrogels with matched

mechanical properties and ideal biocompatibility for artificial blood vessels. *ACS Appl Mater Interfaces.* 2024;16:28134–28146.

44. Hwang CM, Sant S, Masaeli M, Kachouie NN, Zamanian B, Lee SH, et al. Fabrication of three-dimensional porous cell-laden hydrogel for tissue engineering. *Biofabrication.* 2010;2:035003.
45. Chan BP, Leong KW. Scaffolding in tissue engineering: General approaches and tissue-specific considerations. *Eur Spine J.* 2008;17: 467–479.
46. Goh, M, Kim Y, Gwon K, Min K, Hwang Y, Tae G. *In situ* formation of injectable and porous heparin-based hydrogel. *Carbohydr Polym.* 2017;174:990–998.
47. Hwang, U, Moon H, Park J, Jung HW. Crosslinking and swelling properties of pH-responsive poly(ethylene glycol)/poly(acrylic acid) interpenetrating polymer network hydrogels. *Polymers.* 2024;16:2149.
48. Liang Z, Liu C, Li L, Xu P, Luo G, Ding M, et al. Double-network hydrogel with tunable mechanical performance and biocompatibility for the fabrication of stem cells-encapsulated fibers and 3D assemble. *Sci Rep.* 2016;6:33462.
49. Wang, Z, Sun L, Wang W, Wang Z, Shi G, Dai H, et al. A double-network porous hydrogel based on high internal phase emulsions as a vehicle for potassium sucrose octasulfate delivery accelerates diabetic wound healing. *Regen Biomater.* 2024;11:rbae024.
50. Fan Z, Chen Z, Zhang H, Nie Y, Xu S. Gradient mineralized and porous double-network hydrogel effectively induce the differentiation of BMSCs into osteochondral tissue in vitro for potential application in cartilage repair. *Macromol Biosci.* 2021;21:2000323.
51. Gao M, Gawel K, Stokke BT. Swelling dynamics of a DNA-polymer hybrid hydrogel prepared using polyethylene glycol as a porogen. *Gels.* 2015;1:219.
52. Rennerfeldt DA, Renth AN, Talata Z, Gehrke SH, Detamore MS. Tuning mechanical performance of poly(ethylene glycol) and agarose interpenetrating network hydrogels for cartilage tissue engineering. *Biomaterials.* 2013;34:8241–8257.
53. Vazin T, Schaffer DV. Engineering strategies to emulate the stem cell niche. *Trends Biotechnol.* 2010;28:117–124.
54. Engler AJ, Sen S, Sweeney HL, Discher DE. Matrix elasticity directs stem cell lineage specification. *Cell.* 2006;126:677–689.

55. Atcha H, Choi YS, Chaudhuri O, Engler AJ. Getting physical: Material mechanics is an intrinsic cell cue. *Cell Stem Cell.* 2023; 30:750.
56. Fu, L, Li L, Bian Q, Xue B, Jin J, Li J, et al. Cartilage-like protein hydrogels engineered via entanglement. *Nature.* 2023;618:740–747.
57. Berger AJ, Linsmeier KM, Kreeger PK, Masters KS. Decoupling the effects of stiffness and fiber density on cellular behaviors via an interpenetrating network of gelatin-methacrylate and collagen. *Biomaterials.* 2017;141:125–135.
58. Ledo AM, Vining KH, Alonso MJ, Garcia-Fuentes M, Mooney DJ. Extracellular matrix mechanics regulate transfection and SOX9-directed differentiation of mesenchymal stem cells. *Acta Biomater.* 2020;110:153.
59. Nakajima, T, Sato H, Zhao Y, Kawahara S, Kurokawa T, Sugahara K, et al. A universal molecular stent method to toughen any hydrogels based on double network concept. *Adv Funct Mater.* 2012;22: 4426–4432.
60. Diani J, Fayolle B, Gilormini P. A review on the Mullins effect. *Eur Polym J.* 2009;45:601–612.
61. Webber RE, Creton C, Brown HR, Gong JP. Large Strain hysteresis and mullins effect of tough double-network hydrogels. *Macromolecules.* 2007;40:2919–2927.
62. Rodell CB, Dusaj NN, Highley CB, Burdick JA. Injectable and cytocompatible tough double-network hydrogels through tandem supramolecular and covalent crosslinking. *Adv Mater Deerfield Beach Fla.* 2016;28:8419–8424.
63. Guo Z, Xia J, Mi S, Sun W. Mussel-inspired naturally derived double-network hydrogels and their application in 3D printing: From soft, injectable bioadhesives to mechanically strong hydrogels. *ACS Biomater Sci Eng.* 2020;6:1798–1808.
64. Ashton RS, Banerjee A, Punyani S, Schaffer DV, Kane RS. Scaffolds based on degradable alginate hydrogels and poly(lactide-*co*-glycolide) microspheres for stem cell culture. *Biomaterials.* 2007; 28:5518–5525.
65. Patterson J, Hubbell JA. Enhanced proteolytic degradation of molecularly engineered PEG hydrogels in response to MMP-1 and MMP-2. *Biomaterials.* 2010;31:7836–7845.
66. Ali S, Saik JE, Gould DJ, Dickinson ME, West JL. Immobilization of cell-adhesive laminin peptides in degradable PEGDA hydrogels

influences endothelial cell tubulogenesis. *BioResearch Open Access.* 2013;2:241.

67. Li Y, Rodrigues J, Tomás H. Injectable and biodegradable hydrogels: Gelation, biodegradation and biomedical applications. *Chem Soc Rev.* 2012;41:2193–2221.
68. Foyt DA, Norman MDA, Yu TTL, Gentleman E. Exploiting advanced hydrogel technologies to address key challenges in regenerative medicine. *Adv Healthc Mater.* 2018;7:e1700939.
69. Azuma, C, Yasuda K, Tanabe Y, Taniguro H, Kanaya F, Nakayama A, et al. Biodegradation of high-toughness double network hydrogels as potential materials for artificial cartilage. *J Biomed Mater Res A.* 2007;81A:373–380.

https://doi.org/10.1142/9789819806782_0010

Chapter 10

Programmable DNA-based Nanomaterials for Biomedical Applications

Elisha Krieg

Division of Polymer Biomaterials Science,
Leibniz Institute of Polymer Research Dresden,
Hohe Straße 6, 01069 Dresden, Germany
Faculty of Chemistry and Food Chemistry, TU Dresden,
01069 Dresden, Germany
krieg@ipfdd.de

Deoxyribonucleic acid (DNA) plays a central role as the carrier of genetic information in all living organisms. However, its unique physicochemical properties also make the DNA molecule interesting for the field of materials science. This chapter describes a class of materials that exhibit "programmable" properties through the use of nanoscopic components made of DNA. The distinctive characteristics of DNA-based nanosystems and DNA-functionalized polymer materials are described, outlining the design strategies that allow the construction of sophisticated supramolecular architectures with predictable, tunable, and reconfigurable characteristics. These materials may enable more efficient diagnostic assays and improved procedures for personalized medicine. The most recent progress in three specific application areas is highlighted as follows: (1) printable and stimuli-responsive DNA-based hydrogels can serve as precision matrices for the in vitro study of human cells and organoids; (2) materials for the isolation and purification of nucleic acids from biological and clinical samples enable more efficient next-generation sequencing diagnostics; and (3) new DNA-based reaction cascades allow the sensitive identification of specific nucleic acid sequences for cost-effective point-of-care detection assays.

1. Introduction

Synthetic polymer materials have profoundly transformed human life over the past century. Polymers can be produced in large quantities with high purity, and they offer a wide variety of performance characteristics. Correspondingly, they are suitable for various applications ranging from packaging materials to medical devices. While their strong covalent bonds confer exceptional stability, these same bonds also limit their ability to self-heal, respond to stimuli, or be recycled efficiently.

Biological matter stands in stark contrast to our synthetic polymer world (**Table 1**). Cells orchestrate the assembly of materials

Table 1. Some key differences between typical synthetic polymers and biological matter.

	Synthetic	Biological
Composition	Only a few (well-defined) components	High compositional complexity
Production	Top-down fabrication	Bottom-up self-assembly
Control over properties	"Hard-coded" in the molecular structure: molecular weight, charge, functional groups; further controlled by fabrication method and/or assembly pathway	Genetically encoded, dynamically controlled, e.g., through homeostasis, regulation of transcription, and translation
Stability and regeneration	At thermodynamic equilibrium, highly stable but not adaptive or self-healing	Far from equilibrium, labile, self-healing, reconfigurable, and recyclable

that are compositionally complex. For example, several thousand proteins in the human proteome serve various types of structural purposes.[1] Highly specific and reversible noncovalent bonds drive the assembly of *supramolecular polymers*,[2,3] which constitute both the cytoskeleton (e.g., actin filaments and microtubules) and the extracellular matrix (e.g., fibronectin and collagen fibers). The assembled materials possess hierarchical architectures that are continuously remodeled. Their properties are controlled by *molecular programs* encoded in the organism's genome. Moreover, they are continuously monitored and adjusted through *regulatory circuits*. The complex responsiveness of these materials is crucial for many processes, such as cell differentiation, migration, transport, and other diverse forms of adaptation to external stimuli.[4,5]

For many years, chemists and materials scientists have aimed to engineer new forms of intelligent materials that mimic some of the intriguing characteristics of biological matter[6,7]: materials that self-organize, self-heal, and actively respond to their environment in a predictable manner.[8] These biologically inspired materials are of great interest for many biotechnological and medical applications.[9] Stimuli-responsive or "smart" polymers that exhibit intriguing switchable properties have been developed.[10] Moreover, a wide range of synthetic supramolecular polymers have been developed, whose properties resemble various phenomena found in nature, such as cooperative assembly,[11] multistimuli responsiveness,[12] dynamic instability,[13] and self-healing.[14,15]

However, creating more complex biological-like materials presents a significant challenge for chemists. Although synthetic chemistry has established a powerful set of reactions that allow the synthesis of nearly any desired target molecule, methods for making and breaking noncovalent bonds in a specific and predictable manner are highly limited.[2,8,16,17] Large molecules such as polymers are subject to a series of intra- and intermolecular interactions that are practically impossible to simulate computationally with high confidence. As a result, the properties of synthetic materials are often difficult to predict, let alone rationally design.

Can we engineer more sophisticated synthetic materials that leverage the advantages of molecular self-organization in a predictable or even programmable manner? A *programmable material* must exhibit a few key characteristics that go beyond mere stimuli responsiveness (see the text boxes that follow).[18,19] Like biological materials, the properties of synthetic programmable materials must be controlled by a set of instructions — *molecular programs* — that are added to the material to modulate its behavior in a predictable manner following simple design rules.

Although the concept of programmable materials appears futuristic, research over the past decades has shown that it can be realized — at least at the nanoscale — with deoxyribonucleic acid (DNA) molecules.[20–24] Owing to the exceptionally high binding selectivity of Watson–Crick base pairing, the complicated molecular structure of the DNA molecule can be abstracted down to a simple string of letters (i.e., the nucleobases A, T, C, and G), without losing critical information on the binding behavior: A binds to T and C binds to G. On the basis of this remarkable property, the field of DNA nanotechnology has flourished, allowing us to synthesize diverse types of programmable nanostructures, such as DNA origami, DNA-based tiles, bricks, and wireframe structures — all held together by the same types of DNA base pairs.[20,25–27]

Stimulus-responsive material

Definition: Material that undergoes a *reversible* and *significant* change in one of its properties (e.g., color, shape, stiffness) in *response* to a small change in an external *stimulus* (e.g., light, temperature, pH) in a *nontrivial* manner.

Example: *Poly(N-isopropylacrylamide) (PNIPAM)* exhibits stimuli responsiveness through a reversible phase transition at its lower critical solution temperature (LCST).[28,29] Below the LCST (~32°C), PNIPAM is hydrophilic and well soluble in water, but at temperatures above the LCST, it becomes hydrophobic, causing it to precipitate. PNIPAM's phase behavior can be considered nontrivial as most polymers do not exhibit an LCST.

Programmable material

> ***Definition:*** Material comprising exchangeable *modular* components, whose effects on the material's properties are *predictable* (i.e., they do not require trial-and-error testing) and are based on a simple set of *abstract* rules.
>
> ***Example:*** The *DNA origami* technique uses modular components (one long "scaffold" strand and many shorter "staple" strands). When mixed together, the components self-assemble into a predefined nanostructure. The origami design follows a simple set of abstract rules (e.g., base pairing rules) and does not require prior computational simulations or experimental trial-and-error testing.

DNA nanotechnology allows us not only to create custom nanoscale shapes but also to build programmable reaction cascades and molecular circuits.[30] Already in 1994, Leonard Adleman demonstrated that a mixture of synthetic DNA strands that interact with each other in solution could be used to solve the Hamiltonian path problem.[31] Since then, increasingly sophisticated schemes for DNA-based logic circuits,[32] data storage,[33] and even neural networks[34] have been devised. DNA strand displacement reactions can be used to translate and amplify molecular signals, indicating promising potential for diagnostic applications.[35] DNA-based nanodevices can be operated through diverse types of stimuli, including nucleic acids,[36] proteins,[37] electric current,[38] and mechanical force.[39]

To date, DNA nanostructures have been studied primarily in solution or on surfaces. However, as argued in subsequent chapters, the unique properties of DNA offer a large and mostly untapped potential for engineering modularly assembled, self-organizing, programmable, and reconfigurable *macroscopic* materials. These materials have intriguing emergent properties and a unique potential for various applications in biology and medicine.[18,40–43]

2. Concept of macroscopic DNA-based nanomaterials

The idea of using DNA to engineer macroscopic nanomaterials is as old as the field of DNA nanotechnology itself. In his seminal article from 1982, Ned Seeman first conceived DNA-based junctions to be used to build customizable porous crystal lattices.[44] Decades later, his group reported the successful growth of macroscopic 3D crystals with tunable pore sizes, which were self-assembled from nanoscopic DNA tensegrity triangles.[45] In the meanwhile, several other groups had used DNA to engineer *amorphous* materials — hydrogels. These materials were initially based on chemically cross-linked salmon sperm DNA,[46] later DNA-cross-linked polymers,[47,48] and all-DNA hydrogels.[49] Importantly, when DNA is used as a building block for bulk materials, its molecular-level programmability can be leveraged to achieve programmable physical properties at the macroscopic level. For instance, DNA cross-linkers can be designed to reversibly open or close via toehold-mediated strand displacement (TMSD), a process where an input DNA molecule binds to a single-stranded overhang (toehold) of a DNA construct, initiating a strand displacement reaction.[50,51] This mechanism can be exploited to control gel elasticity[47,52] or trigger the release of entrapped particles.[53]

Materials in which a synthetic polymer constitutes an inexpensive and independently tunable structural framework are particularly appealing[18,54]: Synthetic polymers offer a vast design space, as their properties can be tailored precisely to application-specific requirements (e.g., charge, chain flexibility, solubility, and biofunctionality). DNA "anchor" strands are covalently grafted onto the polymer backbone to provide binding sites for DNA modules. The latter can be "plugged in" to the polymer by self-assembly (and replaced at any time) to modulate the material's properties in diverse ways (**Figure 1**). The binding energy (ΔG), melting temperature (T_m),[55] association rate constant (k_a),[50] and nanomechanical stability (F_r)[56] of the DNA modules can be encoded in the nucleotide sequences. Examples of programmable functional DNA-based modules are shown in **Figure 2**. *Binding sites*[57–59] give the material the ability to capture specific target molecules with adjustable affinity and release them when needed. *Stimuli-responsive motifs*[58,60–65] allow for the modification of cross-linking and functionalization through external signals. *Nanomechanical sensors*[66,67] make deformations of the material visible at the microscopic level.

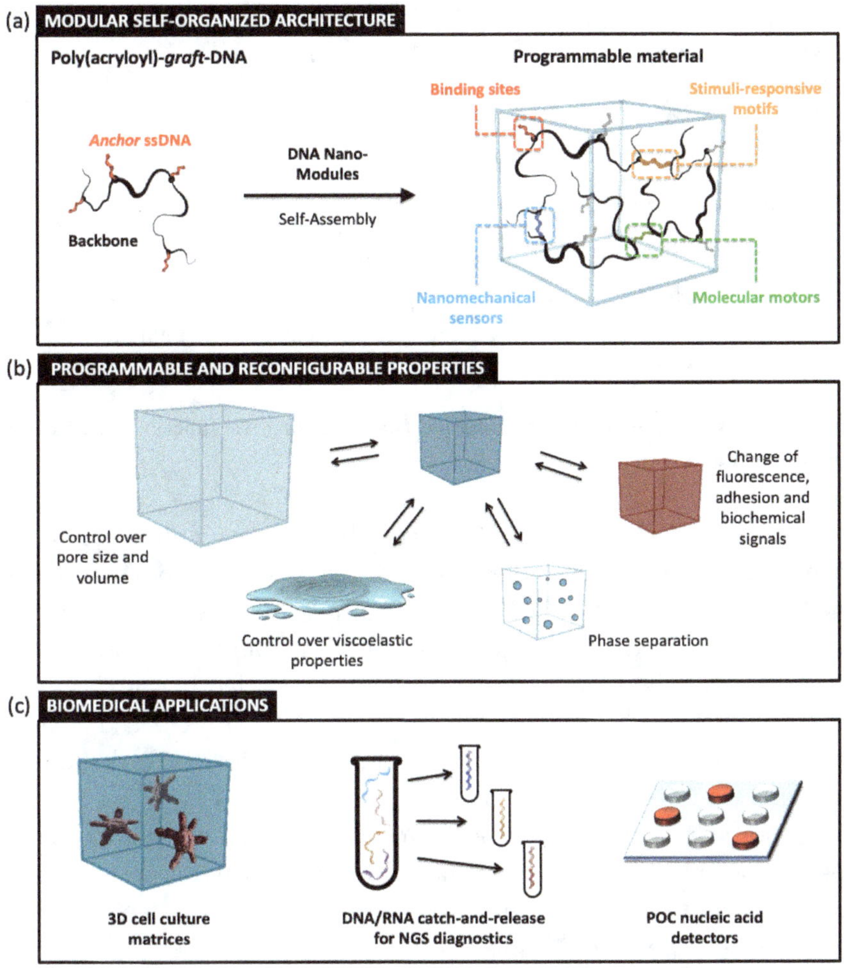

Figure 1. Self-organized structure, dynamic properties, and biomedical applications of DNA-functionalized polymer materials.

Enzyme-driven motors[68] enable active mechanical state changes in the material, which can be used for the release of enclosed components and as a signal for the detection of target molecules.

This material platform has great potential for various biomedical applications. The following sections describe three specifically intriguing fields that are exemplified by recent progress in our

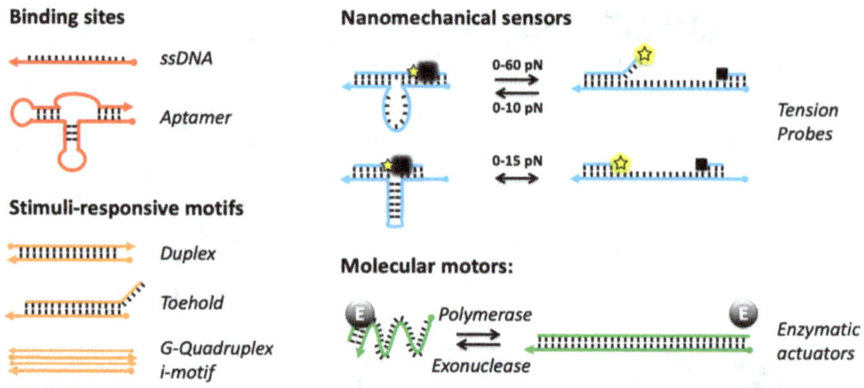

Figure 2. Examples of DNA modules that add diverse functionalities to the polymer scaffold.

research group (**Figure 1c**): (1) printable 3D cell culture matrices with sensory and switchable properties that can support, monitor, and guide the growth of cells and organoids; (2) assays for the purification and modification of DNA and ribonucleic acid (RNA) libraries that allow more sensitive and cost-effective next-generation sequencing (NGS) diagnostics; and (3) DNA-based sensors that may be used in cost-effective point-of-care (POC) devices for the detection of pathogens.

Despite the similarities between synthetic DNA nanomaterials and biological matter, there remains a noteworthy difference in the way programmability is currently implemented (**Figure 3**): apart from exotic edge cases,[69] cells use DNA exclusively as the carrier of genetic information rather than directly carrying out biological tasks. In eukaryotic cells, genomic DNA is constrained to the nucleus. Its molecular programs (i.e., genes) are first transcribed into RNA and then translated into proteins. The latter carry out specific structural and enzymatic functions. In contrast, in current DNA nanomaterials, DNA usually both encodes *and* performs its function. Notably, recent research toward the development of "synthetic cells" has harnessed in vitro transcription to produce RNA nanostructures from DNA templates within vesicles — a rare example of a synthetic material in which genetic information and function are indeed separated.[70]

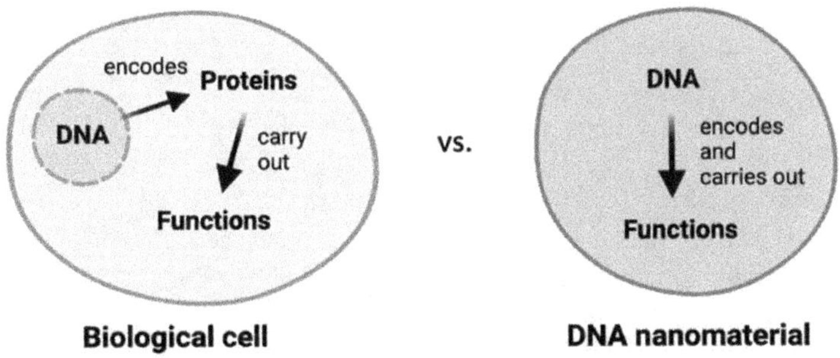

Figure 3. Programmability in biological cells versus most DNA-based materials. Created with BioRender.com.

3. 3D matrices for cell and organoid culture

The in vitro culture of living cells enables the study of tissue development and a wide range of diseases within a controlled environment. Specifically, 3D cell culture involves embedding cells or organoids — microscopic organ-like structures — into a hydrogel matrix. This matrix mimics the natural tissue environment and promotes cell development through biochemical and mechanical cues. Modeling human tissues and organs is indispensable not only in basic research but also increasingly in personalized medicine, disease research, and drug screening.

However, commonly used cell culture matrices are often of *animal origin*, which presents significant disadvantages: high batch-to-batch variability leads to inconsistent experimental results. Owing to their limited modifiability, adapting them to the specific needs of different cells is difficult. Moreover, they can have unpredictable effects on cell behavior, complicating the interpretation of experimental results. Finally, safety concerns and regulatory hurdles arise due to the risks associated with biological contaminants and allergens. *Fully synthetic* cell culture matrices solve many of these problems. However, they are often unsuitable for more complex multicellular structures, such as organoids. Moreover, they offer only limited options for adjusting their mechanical properties.

The mechanical characteristics of a cell culture matrix are indeed crucial factors for accurately mimicking native tissue environments. Early studies revealed that the stiffness of a cell

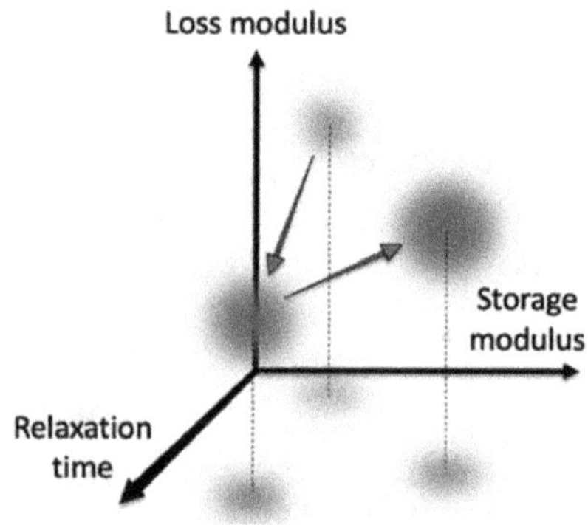

Figure 4. Parameter space of three important factors determining matrix mechanics. The storage modulus (G') describes the stiffness of the material, the loss modulus (G'') describes its viscosity, and the relaxation time (τ) measures the timescale on which straining the material results in permanent deformation. An ideal matrix material would allow the *independent* adjustment of all relevant mechanical parameters.

substrate affects cell spreading, migration, differentiation, tissue regeneration, and disease progression.[71–73] More recent research has revealed that several other mechanical features — such as matrix plasticity and nonlinear response — can also influence cell development and morphogenesis.[74,75] The ideal material allows its user to *independently* adjust multiple parameters of the mechanical properties of the matrix, such as the storage modulus, loss modulus, and stress-relaxation time (**Figure 4**). For studies into mechanobiology, these adjustments should not require major changes to the chemical composition of the material to avoid unintended effects on cells and distortions of experimental results.

Several research groups have explored the potential of DNA hydrogels as highly customizable and reconfigurable cell culture matrices. Examples include the encapsulation of Chinese hamster ovarian cells[49] and the culture of rat spinal cord cells[76] and HeLa cells in various DNA hydrogels.[77–80] Despite these promising developments, most cultures were performed with relatively undemanding cell types and over short periods of time. Most reported hydrogels also require large quantities of DNA, which makes these materials prohibitively expensive for high-throughput and large-scale cell culture studies. Moreover, high DNA concentrations may also lead to unintended interference with cellular development, for instance, by scavenging cytokines[81] or triggering inflammatory responses.[82]

To address these limitations, we recently developed a new cell culture matrix called *DyNAtrix*.[83] A synthetic poly(acrylamide-co-acrylic acid) backbone with an ultrahigh molecular weight (UHMW) serves as the structural backbone of the material (**Figure 5a**). It is equipped with DNA anchor strands as well as functional peptides. The latter contains the arginyl-glycyl-aspartic acid (RGD)-sequence motif that promotes the adhesion of embedded cells. The addition of DNA modules gives the user a high degree of control over various material properties:

1. *Combinatorial Cross-Linker Libraries (CCL)* comprise a mixture of DNA cross-linker strands. The complexity of the CCL (i.e., the number of distinct cross-linker pairs) affects the ratio of intra- versus intermolecular cross-links within the hydrogel network. As a result, the storage modulus of the matrix can be fine-tuned by adjusting the *diversity* of DNA sequences in the library. The combination of CCL with the UHMW backbone allows the assembly of stable hydrogels at very low weight fractions of DNA.
2. *Heat-Activated Cross-Linkers (HAC)* employ DNA-blocking strands that are annealed to the cross-linkers prior to hydrogel formation. The blocking strands allow the user to *engineer the kinetic barrier* for cross-link activation. This concept was used to optimize the temperature and rate of hydrogel formation to achieve a user-friendly cell culture workflow that matches the commonly used protocols for *Matrigel*, the most widely used commercial cell culture matrix.
3. *Stress-Relaxation Cross-Linkers (SRC)* are a set of cross-linkers in *DyNAtrix* that open reversibly under mechanical strain. Their nanomechanical stability can be adjusted through the DNA nucleotide sequence, which allows control over the macroscopic

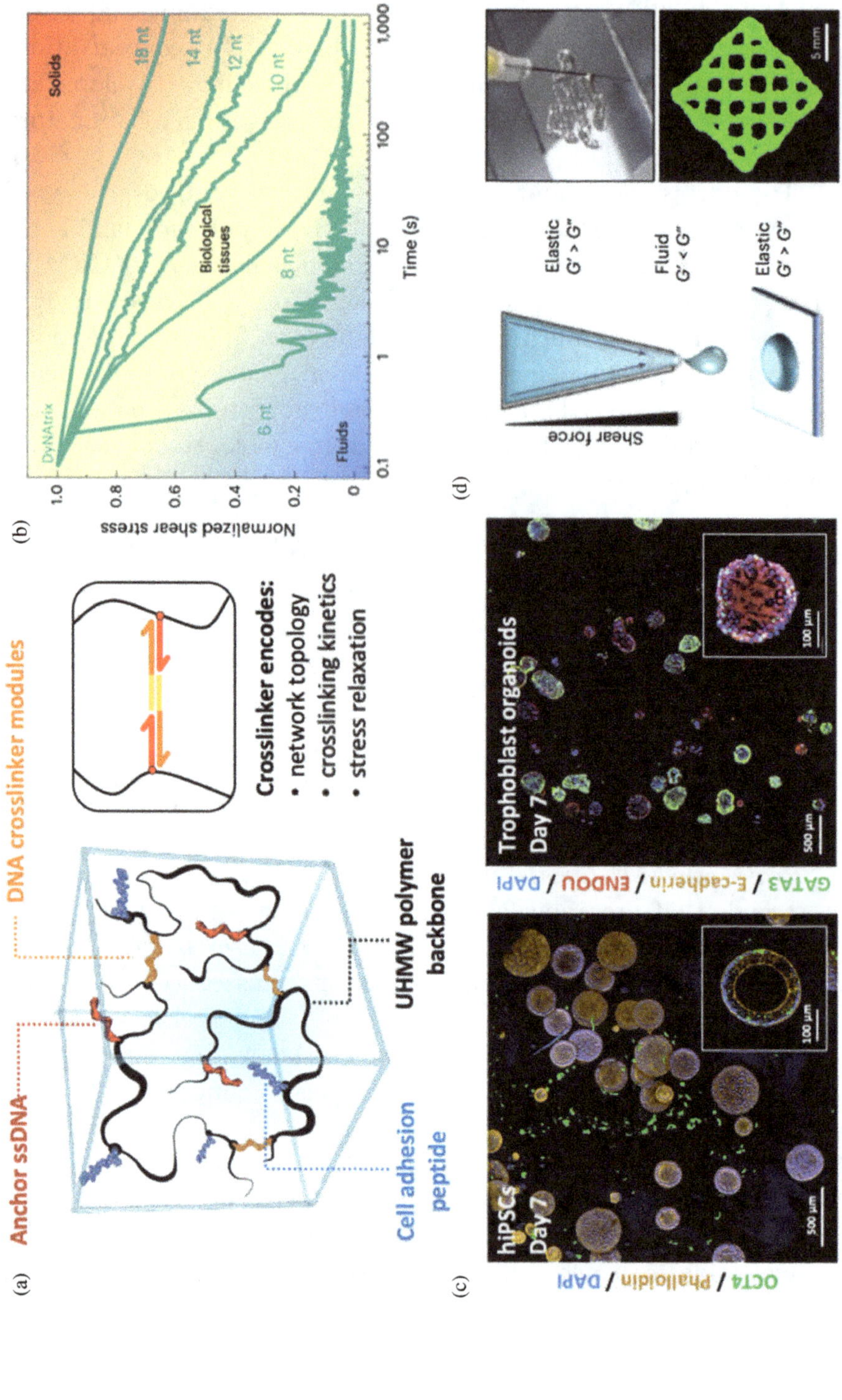
(a)
Anchor ssDNA
DNA crosslinker modules
Cell adhesion peptide
UHMW polymer backbone
Crosslinker encodes:
• network topology
• crosslinking kinetics
• stress relaxation
(b)
Normalized shear stress
Time (s)
Solids
Fluids
Biological tissues
DyNAtrix
6 nt
8 nt
10 nt
12 nt
14 nt
18 nt
(c)
OCT4 / Phalloidin / DAPI
hiPSCs Day 7
GATA3 / E-cadherin / ENDOU / DAPI
Trophoblast organoids Day 7
(d)
Shear force
Elastic G' > G"
Fluid G' < G"
Elastic G' > G"

Figure 5. Design and key properties of *DyNAtrix*. (**a**) *DyNAtrix* consists of an ultrahigh-molecular-weight (UHMW) polymer backbone that is decorated with DNA anchor strands and cell-adhesion peptides. The anchor strands are cross-linked by self-assembled DNA cross-linker modules. (**b**) The stress relaxation time of the material can be adjusted over several orders of magnitude. (**c**) *DyNAtrix* is compatible with various cell types and organoids, including human-induced pluripotent stem cells (hiPSCs) and human trophoblasts. (**d**) Custom 3D tissue scaffolds can be fabricated from *DyNAtrix* by extrusion printing. Adapted from Speed and Gupta at al.[18] and Peng et al.[83]

stress relaxation behavior of the gel (**Figure 5b**). They allow the stress relaxation time (τ) to be adjusted over a vast range (0.1 seconds $\leq \tau \leq$ 2 days), covering the entire spectrum of physiologically relevant relaxation times.

In our benchmark studies, *DyNAtrix* demonstrates its effectiveness compared to *Matrigel*, which is considered the gold standard for organoid cultures. Cell cultures based on *DyNAtrix* exhibited high cell viability (95–97% survival rate), comparable to or even surpassing that of *Matrigel*-based cultures. Culture times of up to 21 days, including multiple passages, were demonstrated. Notably, *DyNAtrix* is the first synthetic cell culture matrix that enables the culture of human trophoblast organoids (**Figure 5c**). Our studies also revealed that the gel can be used to control the development of kidney cysts through its dynamically adjustable mechanical properties. *DyNAtrix* also exhibited low activation of white blood cells and blood coagulation and demonstrated superior compatibility with the human immune system compared with medically approved Teflon. These properties make it interesting for applications in regenerative medicine.

Due to its reversible DNA cross-linking modules, *DyNAtrix* is self-healing and can be easily extruded through a fine nozzle and thus printed (**Figure 5d**). Embedded cells maintain high viability and proliferation after printing, indicating that *DyNAtrix* is suitable for additive manufacturing of complex 3D tissue scaffolds.

Programmable DNA-functionalized polymer matrices like *DyNAtrix* are still in their early development but show significant potential for use in advanced cell and organoid studies. Unintended nuclease degradation by enzymes present in serum-based culture media or in conditioned media still poses a challenge, which can be addressed by using chemically defined media, adding nuclease inhibitors,[83,84] or using nuclease-resistant DNA analogs.[85,86] The most recent advancements in the field suggest that this technology is nearing a breakthrough, with the potential to systematically control and interrogate cell and organoid development.

4. Pulldown polymers

Many human diseases are caused by defective genes or gene expression.[87] This applies not only to rare monogenetic diseases but also to many types of cancer, cardiovascular diseases, and diabetes. Early diagnosis and personalized treatment can greatly improve the quality of life for many patients. Expression analysis of tumor tissue, for example, can be used to identify the tumor type and predict the most effective therapy method. Rapid progress in the field of NGS suggests that sequencing methods will revolutionize medical diagnostics in this century.

One important step in the preparation of NGS samples from biological and clinical samples is to selectively bind, purify, label, and release the DNA and RNA molecules of interest. However, there is currently no *universally applicable* method for the modification and purification of nucleic acid sequencing libraries. Instead, various commercial kits that are specialized for the respective target molecules are used. These kits can be divided into two categories: (1) *enzymatic* methods and (2) microparticle-based *pulldown assays*. The most widely used commercial kits are the *NEBNext Depletion Kits* (enzymatic) from *New England Biolabs*, *Ribo-Zero* (previously microparticle-based, now enzymatic) from *Illumina*, and *Dynabeads* (microparticle-based) from *Thermo Fisher Scientific*. However, all currently available kits have major deficiencies: (1) limited application range, usually specialized for one type of target molecule; (2) very high cost; (3) relatively short shelf life; (4) low binding capacity of microparticles due to molecular crowding at the solid-liquid interface; and (5) limited tolerance for variations in buffer conditions.

We have proposed that stimuli-responsive DNA-grafted *polymers*[88–90] may be used as an alternative to enzymatic and microparticle-based techniques.[59] Using polymers to capture and release nucleic acids is appealing for several reasons. First, polymers can be optimized to dissolve homogeneously in water; thus, they do not offer macroscopic surfaces that are prone to nonspecific adsorption of nontarget molecules. Second, their binding is less limited by interfacial crowding and therefore provides a much higher potential binding capacity. Third, dissolved polymers do not spontaneously sediment and therefore support prolonged annealing times under stationary conditions (e.g., when incubated in a thermocycler). Fourth, they can have very high stability, supporting long shelf life and climate-friendly shipping without a cold chain. Fifth, their synthesis is usually scalable and inexpensive.

Early examples of targeted nucleic acid capture on polymers were based on DNA-functionalized polyethylene glycol (PEG)[91] and PNIPAM.[92–94] DNA and RNA targets were captured on the polymers either by hybridization or via triplex formation. The PEG-based system supported *affinity partitioning*, which required chromatography to isolate the captured target. In contrast, the temperature-induced precipitation of PNIPAM provided a convenient trigger to pull down the target. However, using temperature as a stimulus for this *affinity precipitation* is problematic since varying the temperature also affects nucleic acid hybridization. Thus, annealing the polymer to facilitate efficient binding of DNA probes with targets can lead to premature precipitation. Conversely, precipitation through heating can trigger dehybridization. Moreover, nonspecific coprecipitation of DNA can occur.

In 2018, we discovered that a DNA-grafted copolymer of acrylamide and acrylic acid can be precipitated from aqueous solution by the addition of methanol.[95] Importantly, its precipitation does not cause coprecipitation of unbound DNA even when the latter is several thousand kilobases long. The temperature-independent responsiveness and high binding specificity of this *methanol-responsive polymer (MeRPy)*, as it was later called,[59,96] make it suitable for a wide range of applications. Initially, *MeRPy* was used for the synthesis and purification of kilobase-long single-stranded DNA for DNA origami, fluorescent in situ hybridization, and CRISPR-Cas9 gene editing.[95,96]

More generally, *MeRPy* can be used to purify, sort, or deplete a wide range of DNA targets in solution for either preparative or analytical applications.[59] The concept is shown schematically in **Figure 6a**. The polymer's affinity is first programmed via the addition of *catcher strands* that hybridize to the *anchor strands* on the polymer. Each catcher strand contains a target binding site (blue) and — optionally — a target release site (green). DNA targets bind specifically to the target binding site, and the polymer is subsequently precipitated to pull down the targets. The polymer is then redispersed, and targets can be released via the addition of specific *release strands* that bind to the release site and trigger TMSD. We validated its function with various types of DNA, ranging from short single-stranded oligonucleotides (**Figure 6b**) to complex double-stranded mixtures from clinical NGS libraries. *MeRPy* depletion of highly abundant sequences in complementary DNA samples generated from human pancreatic tissues effectively increased the quality of transcription profiles, thereby permitting the identification of more than 1,000 additional transcripts that were not detected in unmodified control samples (**Figure 6c**).

Overall, *MeRPy* is a promising alternative to existing nucleic acid purification and selection methods in the NGS sample preparation pipeline. It exhibits advantageous performance characteristics at a much reduced cost compared to conventional kits. One limitation is that *MeRPy* pulldown is currently only suitable for DNA but not RNA targets since RNA exhibits an unexpected nonspecific affinity to the *MeRPy* precipitate. This limitation may be solved in the future through modification of the polymer backbone or via optimization of the pulldown mechanism.

5. DNA reaction cascades for POC detection

Healthcare is becoming an increasingly large burden on the economies of both developing and developed countries. The proportion of healthcare expenses with respect to gross domestic product (GDP) has risen globally from 8.6% in 2000 to 10.8% in 2020.[97] Moreover, global health risks often arise in regions where infrastructure and financial resources are insufficient. New medical technologies must

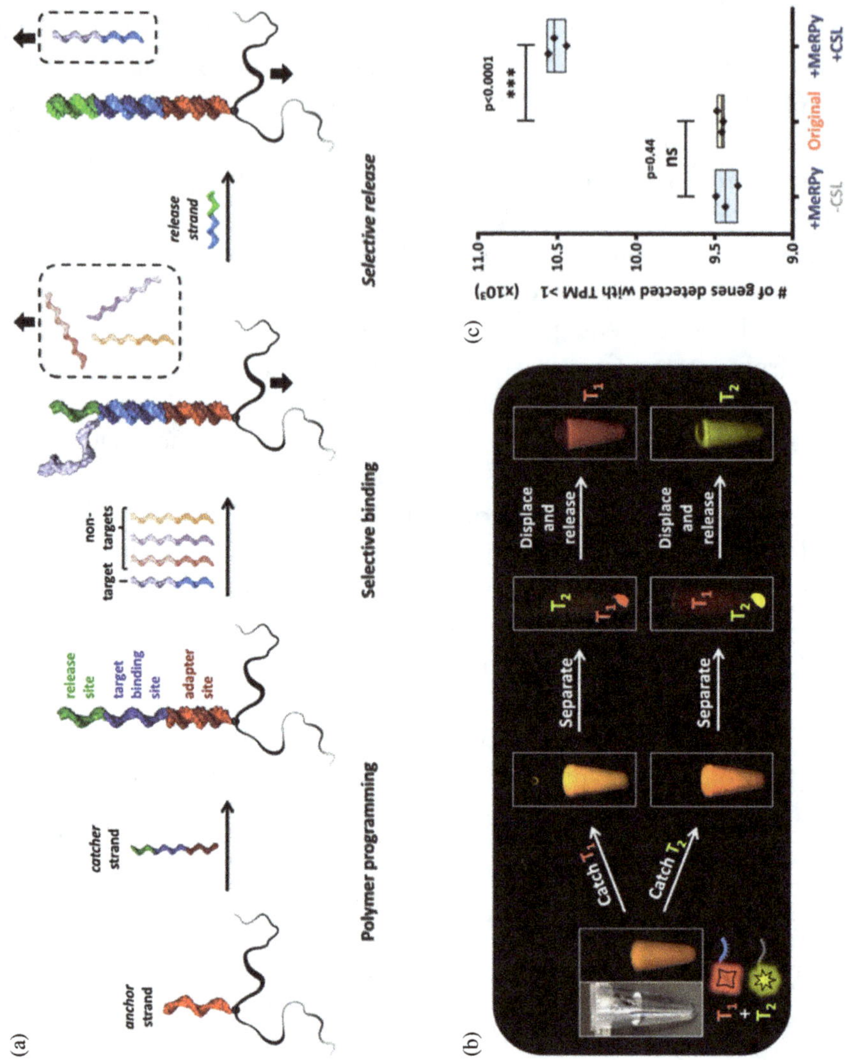
(a)
anchor strand
catcher strand
release site
target binding site
adapter site
target
non-targets
release strand
Polymer programming
Selective binding
Selective release
(b)
Catch T1
Catch T2
Separate
Displace and release
T1
T2
(c)
# of genes detected with TPM >1 (x10³)
11.0
10.5
10.0
9.5
9.0
p=0.44
ns
p<0.0001
***
+MeRPy -CSL
Original
+MeRPy +CSL

Figure 6. Principle and application of the *MeRPy* pulldown assay. (**a**) Schematic of the mechanisms for *MeRPy* programming, selective binding, and target release. Thick black arrows (up/down) indicate steps in which *MeRPy* is precipitated and separated from the supernatant. (**b**) Demonstration of the specific catch and release with fluorescently labeled target strands. (**c**) *MeRPy* pulldown can be used to remove high-abundance components from NGS libraries by programming the polymer with a *catcher strand library (CSL)*. This treatment significantly increased the number of detected transcripts. Adapted from Krieg et al.[59]

therefore not only outperform conventional methods but also remain affordable.

A critical element of the global healthcare system is disease surveillance. As recent epidemics have shown, rapid pathogen detection is important for containing disease outbreaks. Most infectious diseases can be diagnosed by detecting specific nucleic acid sequences. This is predominantly performed by polymerase chain reaction (PCR) or reverse transcription PCR (RT-PCR). Compared with antibody detection, PCR-based methods are superior, as they can also be used when a patient has not yet developed antibodies or has been previously infected with a similar virus. However, because expensive equipment is needed, PCR is not well suited for POC diagnostics.[98] For the development of inexpensive POC tests, reducing the cost of signal amplification and making the amplified sequence visible via simple means under isothermal conditions are critical.

Several isothermal amplification assays have been developed over the past decades, most notably loop-mediated isothermal amplification (LAMP),[99] recombinase polymerase amplification (RPA),[100] and assays based on CRISPR/Cas enzymes.[101–106] However, all these methods rely on complex mixtures of reagents and enzymes — especially for RNA detection — that are expensive, prone to supply shortages, require cold storage, and are not suitable for application in resource-limited settings.

New DNA-based reaction cascades — usually based on the widely used TMSD mechanism — have great potential to enable new, cost-effective diagnostic applications.[30,35,107] These systems are appealing because they consist of a well-defined mixture of synthetic oligonucleotides that can perform signal translation and amplification without enzymes.[36,108,109] However, TMSD systems commonly suffer from leakage, where unintended activation leads to false positives and high background signals.[109–111] Therefore, the sensitivity and specificity of pure TMSD reactions usually do not match real-life diagnostic requirements.

In an attempt to address these limitations, we recently developed the *Y-Switch*, a DNA-based biosensor designed to detect pathogen-specific nucleic acids with high sensitivity and specificity.[112] This system was inspired by previously reported transcriptional and translational switches, such as *genelets*[113] and *toehold switches*.[114] The *Y-Switch* uses a TMSD cascade reaction that is coupled to the in vitro transcription of a fluorescent signal (**Figure 7**). The system relies on two individually assembled DNA constructs (**Figure 7a**): a pathogen-recognizing construct, [P], and an aptamer-encoding construct, [S]. The [P] module contains a pathogen-specific sequence that serves as a toehold for a target DNA or RNA molecule. The [S] module contains a sequence encoding the fluorescent Spinach[115] aptamer and a single-stranded T7 RNA polymerase promoter domain that is designed to be inactive in the absence of the target. A premature interaction between [P] and [S] is prevented by biotinylated *blocking strands*, which are preannealed to both constructs separately. However, when a specific nucleic acid target is present, it binds to the pathogen-recognizing domain on [P], triggering a TMSD cascade that ultimately leads to the activation of the promoter and transcription of the Spinach aptamer.

The key features of the *Y-Switch* are its high performance for continuous transcription (creating a strong signal) and its tolerance to leakage (suppressing background signals). Leakage products can be selectively removed via pulldown with streptavidin-coated magnetic beads, effectively eliminating background fluorescence and increasing sensitivity. The system works under isothermal conditions and requires only a single enzyme (T7 RNA polymerase). Importantly, the *Y-Switch* can detect both DNA and RNA targets directly in a simple workflow without the need for reverse transcription or preamplification.

The *Y-Switch* was validated with nucleic acid sequences derived from the Zika virus (ZKV) and Severe Acute Respiratory Syndrome Coronavirus 2 (SARS-CoV-2). It shows high specificity with negligible interference from nontarget nucleic acids, such as human cytoplasmic RNA. This allows robust detection in the picomolar range with zero background signal in the absence of the target (**Figure 7b, c**). By exchanging the [P] module, the material can be quickly adapted to other pathogens without the need to

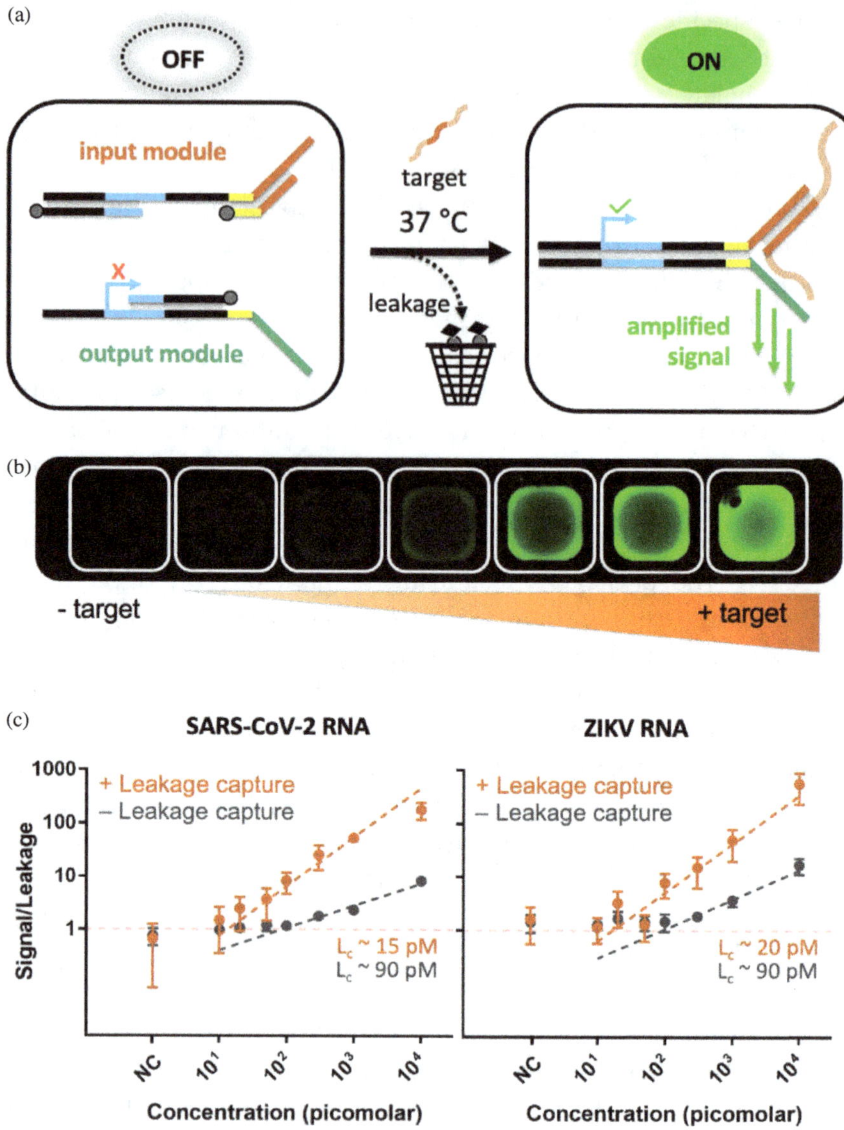

Figure 7. The *Y-Switch* allows direct detection of pathogen-specific DNA or RNA. (**a**) Schematic design of the *Y-Switch* working principle. The input module [P] and the output module [S] are mixed together in a metastable state. The binding of a specific target nucleic acid to the input module triggers a TMSD reaction that leads to the activation of a promoter, resulting in the amplification

Figure 7. (*Continued on facing page*) of a fluorescent aptamer that is encoded on the output module. (**b**) The assay generates a fluorescent signal when target molecules are present but produces practically no background signal in the absence of the target. (**c**) Sensitivity studies with SARS-CoV-2 and ZIKV-derived RNA sequences. The capture of leakage products via biotin pulldown makes the *Y-Switch* leakage-tolerant and thereby improves its sensitivity. Adapted from Gupta and Krieg.[112]

change its basic structure. Additionally, the system can be integrated into more elaborate DNA-based networks for advanced synthetic biology applications.

To date, the *Y-Switch* has been validated only in solution; however, there are several ways in which its performance may be improved in the future through the use of DNA-functionalized polymers. First, *MeRPy* could be used to pull down and strongly enrich a specific trigger molecule prior to *Y-Switch* activation, thus increasing its sensitivity by orders of magnitude. The *Y-Switch* could also find applications in multiplexed POC detectors that are capable of detecting and discriminating several pathogens simultaneously. Multiplexing is commonly achieved in microfluidic devices, where several millimeter-sized wells contain different reaction mixtures. However, the fabrication of microfluidic devices is relatively expensive. As a low-cost alternative, the *Y-Switch* could be immobilized within *DyNAtrix*, which can be subsequently extrusion-printed onto a cheap substrate such as paper,[116,117] dried and stored for long periods of time, and finally used as a multiplexed POC diagnostic device whenever needed.

6. Conclusions and future perspective

This chapter explored how DNA-based nanosystems, particularly DNA-functionalized polymers, can be used to create sophisticated supramolecular architectures and macroscopic materials with programmable properties. By leveraging the highly specific base-pairing interactions of DNA, scientists can create materials that self-organize and respond to stimuli in a manner akin to biological systems, potentially revolutionizing diagnostics and personalized medicine.

One significant application discussed is the development of *DyNAtrix*, a printable and stimuli-responsive DNA-functionalized polymer hydrogel designed for 3D cell and organoid culture. *DyNAtrix* will help elucidate unsolved questions in the fields of mechanobiology and the physics of life. Current studies with this new matrix focus on understanding the timescale of mechanical signaling in epithelial cells. The composition of *DyNAtrix* is highly complex (consisting of up to several hundred distinct DNA cross-linker strands), but unlike biologically derived materials, it is fully synthetic and compositionally defined. Therefore, this material platform helps increase reproducibility and reduce regulatory obstacles for translation into personalized medicine.

The concept of combinatorial cross-linker libraries deserves special attention since it provides a new level of control over the topology of polymer networks. In the case of *DyNAtrix*, a greater diversity of DNA cross-linker sequences leads to an increased degree of *inter*molecular cross-linking and thus greater stiffness. Interestingly, this concept can also be used "in reverse": by integrating a large number of combinatorial DNA recognition sequences into plasmid molecules, we recently found that *intra*molecular cyclization can be favored over *inter*molecular cross-linking. The creation of mixtures of DNA plasmids with more than 16,000 distinct recognition sequences made it possible to assemble an exotic type of soft material, called an *Olympic gel*, which is held together by mechanically interlocked cyclic molecules (**Figure 8**).[118] This elusive hydrogel structure was hypothesized by Pierre-Gilldes de Gennes nearly half a century ago,[119] yet its synthesis had proven challenging. Our plasmid-based *Olympic gel* exhibits intriguing material characteristics in agreement with computational predictions, including an unusual nonlinear viscoelastic response and nonclassical swelling behavior.[118] This material demonstrates that high compositional complexities can (1) be easily achieved in DNA-based materials by using combinatorial methods and (2) give rise to

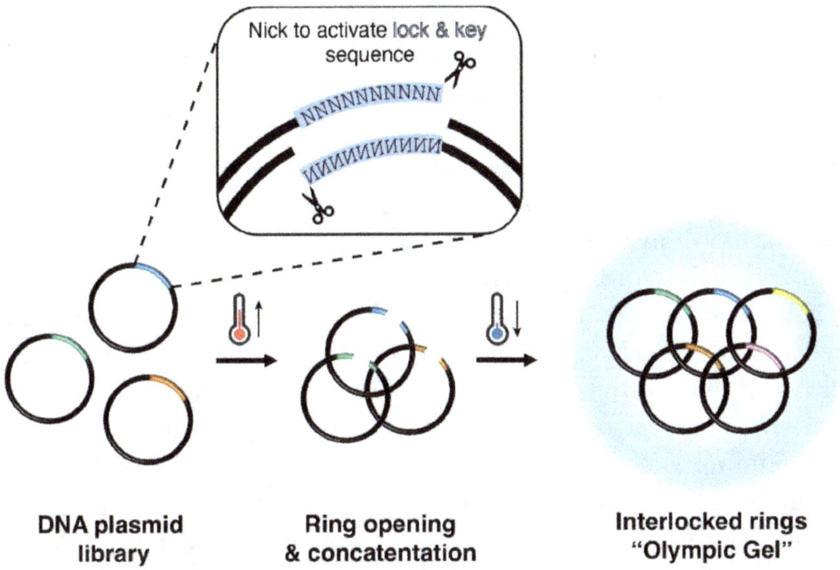

Figure 8. Synthetic route to *Olympic gels*. Large libraries of cyclic DNA molecules are generated by cloning combinatorial oligonucleotides into a plasmid backbone. The resulting DNA library can be activated by enzymatic nicking. When heated, the molecules open and entangle. Upon cooling, their reactive termini selectively bind the original binding partner and thus cyclize, thereby creating a mechanically interlocked molecular network. Reproduced from Speed et al.[118]

unique molecular architectures with emergent properties that are very different from those of conventional polymer materials.

Importantly, the same DNA-based material platform can be used for multiple applications because of the modularity of DNA assembly. For example, the synthetic polymer backbone of *MeRPy* is closely related to that of *DyNAtrix*. The crucial difference between these systems is the function of the supplemented DNA modules (cf. **Figures 1a, 5a, 6a**): *DyNAtrix* uses a variety of crosslinker modules to achieve programmable gelation, whereas *MeRPy* is supplemented with binding modules to achieve programmable capture of specific biomolecules. *MeRPy*-assisted depletion of undesired components in NGS libraries might decrease the cost of NGS library preparations, lower the use of plasticware, and reduce the risk of sample loss and cross-contamination. In the future, *MeRPy* may be upgraded with cross-linking modules to agglomerate and reversibly pull down the polymer under mild conditions, that is, without the use of methanol. Notably, RNA-based liquid condensates have recently been suggested as alternative biomolecule separation agents.[120] Like *MeRPy*, these systems provide the ability to partition specific target molecules, sort, and potentially separate them for analytical or preparative purposes.

The chapter also highlighted DNA-based reaction cascades for nucleic acid detection, exemplified by the *Y-Switch* sensor. This synthetic gene switch utilizes a TMSD reaction coupled with in vitro transcription to produce a fluorescent signal upon detection of specific nucleic acid sequences. The *Y-Switch* performs its function under isothermal conditions and requires only a single enzyme, making it a promising tool for cost-effective POC diagnostic assays. As described in Section 5, the *Y-Switch* could be used in conjunction with *MeRPy* target enrichment and grafted onto an extrusion-printable gel such as *DyNAtrix* to further increase its sensitivity and multiplexed detection performance.

References

1. The UniProt Consortium. UniProt: The universal protein knowledgebase in 2025. *Nucleic Acids Res.* 2025;53:609–617.
2. Krieg E, Bastings MMC, Besenius P, Rybtchinski B. Supramolecular polymers in aqueous media. *Chem Rev.* 2016;116:2414–2477.
3. Brunsveld L, Folmer BJB, Meijer EW, Sijbesma RP. Supramolecular polymers. *Chem Rev.* 2001;101:4071–4098.
4. Discher DE, Mooney DJ, Zandstra PW. Growth factors, matrices, and forces combine and control stem cells. *Science.* 2009;324: 1673–1677.
5. Gardel ML, Schneider IC, Aratyn-Schaus Y, Waterman CM. Mechanical integration of actin and adhesion dynamics in cell migration. *Annu Rev Cell Dev Biol.* 2010;26:315–333.
6. Sanchez C, Arribart H, Giraud Guille MM. Biomimetism and bioinspiration as tools for the design of innovative materials and systems. *Nat Mater.* 2005;4:277–288.

7. Bhushan B. Biomimetics: Lessons from nature — an overview. *Philos Trans R Soc Math Phys Eng Sci.* 2009;367:1445–1486.
8. Lehn JM. Toward complex matter: Supramolecular chemistry and self-organization. *Proc Natl Acad Sci U S A.* 2002;99:4763–4768.
9. Goor OJGM, Hendrikse SIS, Dankers PYW, Meijer EW. From supramolecular polymers to multi-component biomaterials. *Chem Soc Rev.* 2017;46:6621–6637.
10. Stuart MAC, Huck WT, Genzer J, Müller M, Ober C, Stamm M, et al. Emerging applications of stimuli-responsive polymer materials. *Nat Mater.* 2010;9:101–113.
11. Zhao D, Moore JS. Nucleation–elongation: A mechanism for cooperative supramolecular polymerization. *Org Biomol Chem.* 2003;1:3471–3491.
12. Krieg E, Shirman E, Weissman H, Shimoni E, Wolf SG, Pinkas I, et al. Supramolecular gel based on a perylene diimide dye: Multiple stimuli responsiveness, robustness, and photofunction. *J Am Chem Soc.* 2009;131:14365–14373.
13. Debnath S, Roy S, Ulijn RV. Peptide nanofibers with dynamic instability through nonequilibrium biocatalytic assembly. *J Am Chem Soc.* 2013;135:16789–16792.
14. Aida T, Meijer EW, Stupp, SI. Functional supramolecular polymers. *Science.* 2012;335:813–817.
15. Wang Q, Mynar JL, Yoshida M, Lee E, Lee M, Okuro K, et al. High-water-content mouldable hydrogels by mixing clay and a dendritic molecular binder. *Nature.* 2010;463:339–343.
16. Krieg E, Rybtchinski B. Noncovalent water-based materials: Robust yet adaptive. *Chem Eur J.* 2011;17:9016–9026.
17. Dastidar P. Supramolecular gelling agents: Can they be designed? *Chem Soc Rev.* 2008;37:2699–2715.
18. Speed SK, Gupta K, Peng YH, Hsiao SK, Krieg E. Programmable polymer materials empowered by DNA nanotechnology. *J Polym Sci.* 2023;61:1713–1729.
19. Cademartiri, L, Bishop, KJ. Programmable self-assembly. *Nat Mater.* 2015;14:2–9.
20. Rothemund PWK. Folding DNA to create nanoscale shapes and patterns. *Nature.* 2006;440:297–302.
21. Zhang DY, Turberfield AJ, Yurke B, Winfree E. Engineering entropy-driven reactions and networks catalyzed by DNA. *Science.* 2007;318:1121–1125.

22. Yin P, Choi HMT, Calvert CR, Pierce NA. Programming biomolecular self-assembly pathways. *Nature.* 2008;451:318–322.
23. Li J, Green AA, Yan H, Fan C. Engineering nucleic acid structures for programmable molecular circuitry and intracellular biocomputation. *Nat Chem.* 2017;9:1056–1067.
24. Jones MR, Seeman NC, Mirkin CA. Programmable materials and the nature of the DNA bond. *Science.* 2015;347:1260901.
25. Benson E, Mohammed A, Gardell J, Masich S, Czeizler E, Orponen P, et al. DNA rendering of polyhedral meshes at the nanoscale. *Nature.* 2015;523:441–444.
26. Douglas SM, Dietz H, Liedl T, Högberg B, Graf F, Shih WM. Self-assembly of DNA into nanoscale three-dimensional shapes. *Nature.* 2009;459:414–418.
27. Wei B, Dai M, Yin P. Complex shapes self-assembled from single-stranded DNA tiles. *Nature.* 2012;485:623–626.
28. Halperin A, Kröger M, Winnik FM. Poly(N-isopropylacrylamide) phase diagrams: Fifty years of research. *Angew Chem Int Ed.* 2015;54:15342–15367.
29. Heskins M, Guillet J. Solution properties of poly(N-isopropylacrylamide). *J Macromol Sci Part Chem.* 1968;2(8):1441–1455.
30. Simmel FC, Yurke, B Singh HR. Principles and applications of nucleic acid strand displacement reactions. *Chem Rev.* 2019;119: 6326–6369.
31. Adleman LM. Molecular computation of solutions to combinatorial problems. *Science.* 1994;266:1021–1024.
32. Woods D, Doty D, Myhrvold C, Hui J, Zhou F, Yin P, et al. Diverse and robust molecular algorithms using reprogrammable DNA self-assembly. *Nature.* 2019;567:366–372.
33. Goldman N, Bertone P, Chen S, Dessimoz C, LeProust EM, Sipos B, et al. Towards practical, high-capacity, low-maintenance information storage in synthesized DNA. *Nature.* 2013;494:77–80.
34. Qian L, Winfree E, Bruck, J. Neural network computation with DNA strand displacement cascades. *Nature.* 2011;475:368–372.
35. Jung C, Ellington AD. Diagnostic applications of nucleic acid circuits. *Acc Chem Res.* 2014;47:1825–1835.
36. Seelig G, Soloveichik D, Zhang DY, Winfree E. Enzyme-free nucleic acid logic circuits. *Science.* 2006;314:1585–1588.

37. Jung JK, Archuleta CM, Alam KK, Lucks JB. Programming cell-free biosensors with DNA strand displacement circuits. *Nat Chem Biol.* 2022;18:385–393.
38. Kopperger E, List J, Madhira S, Rothfischer F, Lamb DC, Simmel FC. A self-assembled nanoscale robotic arm controlled by electric fields. *Science.* 2018;359:296–301.
39. Shrestha P, Yang D, Tomov TE, MacDonald JI, Ward A, Bergal HT, et al. Single-molecule mechanical fingerprinting with DNA nanoswitch calipers. *Nat Nanotechnol.* 2021;16:1362–1370.
40. Li J, Mo L, Lu CH, Fu T, Yang HH, Tan W. Functional nucleic acid-based hydrogels for bioanalytical and biomedical applications. *Chem Soc Rev.* 2016;45:1410–1431.
41. Zhang Y, Tu J, Wang D, Zhu H, Maity SK, Qu X, et al. Programmable and multifunctional DNA-based materials for biomedical applications. *Adv Mater.* 2018;30:1703658.
42. Simmel FC, Schulman R. Self-organizing materials built with DNA. *MRS Bull.* 2017;42:913–919.
43. Ma Y, Duan X, Huang J. DNA hydrogels as functional materials and their biomedical applications. *Adv Funct Mater.* 2024;34:2309070.
44. Seeman NC. Nucleic acid junctions and lattices. *J Theor Biol.* 1982;99:237–247.
45. Zheng J, Birktoft JJ, Chen Y, Wang T, Sha R, Constantinou PE, et al. From molecular to macroscopic via the rational design of a self-assembled 3D DNA crystal. *Nature.* 2009;461:74–77.
46. Amiya T, Tanaka T. Phase transitions in crosslinked gels of natural polymers. *Macromolecules.* 1987;20, 1162–1164.
47. Lin DC, Yurke B, Langrana NA. Mechanical properties of a reversible, DNA-crosslinked polyacrylamide hydrogel. *J Biomech Eng.* 2004;126:104–110.
48. Nagahara S, Matsuda T. Hydrogel formation via hybridization of oligonucleotides derivatized in water-soluble vinyl polymers. *Polym Gels Netw.* 1996;4:111–127.
49. Um SH, Lee JB, Park N, Kwon SY, Umbach CC, Luo D. Enzyme-catalysed assembly of DNA hydrogel. *Nat Mater.* 2006;5:797–801.
50. Yurke B, Mills AP. Using DNA to power nanostructures. *Genet Program Evolvable Mach.* 2003;4:111–122.

51. Lee CS, Davis RW, Davidson N. A physical study by electron microscopy of the terminally repetitious, circularly permuted DNA from the coliphage particles of *Escherichia coli* 15. *J Mol Biol.* 1970;48:1–22.
52. Lin DC, Yurke B, Langrana NA. Inducing reversible stiffness changes in DNA-crosslinked gels. *J Mater Res.* 2005;20:1456–1464.
53. Liedl T, Dietz H, Yurke B, Simmel F. Controlled trapping and release of quantum dots in a DNA-switchable hydrogel. *Small.* 2007;3:1688–1693.
54. Whitfield CJ, Zhang M, Winterwerber P, Wu Y, Ng DY, Weil T. Functional DNA–polymer conjugates. *Chem Rev.* 2021;121: 11030–11084.
55. Zadeh JN, Steenberg CD, Bois JS, Wolfe BR, Pierce MB, Khan AR, et al. NUPACK: Analysis and design of nucleic acid systems. *J Comput Chem.* 2011;32:170–173.
56. Woodside MT, Behnke-Parks WM, Larizadeh K, Travers K, Herschlag D, Block SM. Nanomechanical measurements of the sequence-dependent folding landscapes of single nucleic acid hairpins. *Proc Natl Acad Sci U S A.* 2006;103:6190–6195.
57. Wei B, Cheng I, Luo KQ, Mi Y. Capture and release of protein by a reversible DNA-induced sol–gel transition system. *Angew Chem Int Ed.* 2008;47:331–333.
58. Zhu Z, Wu C, Liu H, Zou Y, Zhang X, Kang H, et al. An aptamer cross-linked hydrogel as a colorimetric platform for visual detection. *Angew Chem Int Ed.* 2010;49:1052–1056.
59. Krieg E, Gupta K, Dahl A, Lesche M, Boye S, Lederer A, et al. A smart polymer for sequence-selective binding, pulldown, and release of DNA targets. *Commun Biol.* 2020;3:369.
60. Kahn JS, Hu Y, Willner I. Stimuli-responsive DNA-based hydrogels: From basic principles to applications. *Acc Chem Res.* 2017;50:680–690.
61. Guo W, Lu CH, Orbach R, Wang F, Qi XJ, Cecconello A, et al. pH-Stimulated DNA hydrogels exhibiting shape-memory properties. *Adv Mater.* 2015;27:73–78.
62. Cangialosi A, Yoon C, Liu J, Huang Q, Guo J, Nguyen TD, et al. DNA sequence–directed shape change of photopatterned hydrogels via high-degree swelling. *Science.* 2017;357:1126–1130.
63. Afting C, Walther T, Drozdowski OM, Schlagheck C, Schwarz US, Wittbrodt J, et al. DNA microbeads for spatio-temporally

controlled morphogen release within organoids. *Nat Nanotechnol.* 2024;19(2):1–9.

64. Wang D, Hu Y, Liu P, Luo D. Bioresponsive DNA hydrogels: Beyond the conventional stimuli responsiveness. *Acc Chem Res.* 2017;50:733–739.
65. Shao Y, Jia H, Cao T, Liu D. Supramolecular hydrogels based on DNA self-assembly. *Acc Chem Res.* 2017;50:659–668.
66. Zhang Y, Ge C, Zhu C, Salaita K. DNA-based digital tension probes reveal integrin forces during early cell adhesion. *Nat Commun.* 2014;5:5167.
67. Merindol R, Delechiave G, Heinen L, Catalani LH, Walther A. Modular design of programmable mechanofluorescent DNA hydrogels. *Nat Commun.* 2019;10:528.
68. Bertrand OJN, Fygenson DK, Saleh OA. Active, motor-driven mechanics in a DNA gel. *Proc Natl Acad Sci U S A.* 2012;109: 17342–17347.
69. Papayannopoulos V. Neutrophil extracellular traps in immunity and disease. *Nat Rev Immunol.* 2018;18:134–147.
70. Tran MP, Chakraborty T, Poppleton E, Monari L, Giessler F, Göpfrich K. Genetic encoding and expression of RNA origami cytoskeletons in synthetic cells. *bioRxiv.* 2024:2024–2006.
71. Pelham RJ, Wang Y. Cell locomotion and focal adhesions are regulated by substrate flexibility. *Proc Natl Acad Sci U S A.* 1997;94:13661–13665.
72. Discher DE, Janmey P, Wang Y. Tissue cells feel and respond to the stiffness of their substrate. *Science.* 2005;310:1139–1143.
73. Engler AJ, Sen S, Sweeney HL, Discher DE. Matrix elasticity directs stem cell lineage specification. *Cell.* 2006;126:677–689.
74. Blache U, Ford EM, Ha B, Rijns L, Chaudhuri O, Dankers PY, et al. Engineered hydrogels for mechanobiology. *Nat Rev Methods Primer.* 2022;2:1–22.
75. Chaudhuri O, Cooper-White J, Janmey PA, Mooney DJ, Shenoy VB. Effects of extracellular matrix viscoelasticity on cellular behaviour. *Nature.* 2020;584:535–546.
76. Jiang FX, Yurke B, Firestein BL, Langrana NA. Neurite outgrowth on a DNA crosslinked hydrogel with tunable stiffnesses. *Ann Biomed Eng.* 2008;36:1565.
77. Yang B, Zhao Z, Pan Y, Xie J, Zhou B, Li Y, et al. Shear-thinning and designable responsive supramolecular DNA hydrogels based

on chemically branched DNA. *ACS Appl Mater Interfaces.* 2021;13:48414–48422.
78. Wang Y, Shao Y, Ma X, Zhou B, Faulkner-Jones A, Shu W, et al. Constructing tissuelike complex structures using cell-laden DNA hydrogel bricks. *ACS Appl Mater Interfaces.* 2017;9:12311–12315.
79. Shi R, Fern J, Xu W, Jia S, Huang Q, Pahapale G, et al. Multicomponent DNA polymerization motor gels. *Small.* 2020; 16(17):2002946.
80. Huang F, Chen M, Zhou Z, Duan R, Xia F, Willner I. Spatiotemporal patterning of photoresponsive DNA-based hydrogels to tune local cell responses. *Nat Commun.* 2021;12:2364.
81. Andrade H, Lin W, Zhang Y. Specificity from nonspecific interaction: Regulation of tumor necrosis factor-α activity by DNA. *J Biol Chem.* 2019;294:6397–6404.
82. Surana S, Shenoy AR, Krishnan Y. Designing DNA nanodevices for compatibility with the immune system of higher organisms. *Nat Nanotechnol.* 2015;10:741–747.
83. Peng Y-H, Hsiao SK, Gupta K, Ruland A, Auernhammer GK, Maitz MF, et al. Dynamic matrices with DNA-encoded viscoelasticity for cell and organoid culture. *Nat Nanotechnol.* 2023;18: 1463–1473.
84. Lazarides E, Lindberg U. Actin is the naturally occurring inhibitor of deoxyribonuclease I. *Proc Natl Acad Sci U S A.* 1974;71:4742–4746.
85. Beaucage SL, Caruthers MH. Deoxynucleoside phosphoramidites — A new class of key intermediates for deoxypolynucleotide synthesis. *Tetrahedron Lett.* 1981;22:1859–1862.
86. Singh SK, Koshkin AA, Wengel J, Nielsen P. LNA (locked nucleic acids): Synthesis and high-affinity nucleic acid recognition. *Chem Commun.* 1998;4:455–456.
87. Evans RG, Barer ML, Marmor TR, editors. *Why Are Some People Healthy and Others Not? The Determinants of Health Populations.* New York: Aldine Transaction; 1994.
88. Peng L, Zhang L, Xu T, Hou J, Li Z, Wang Q, et al. Engineering and applications of DNA-grafting polymer materials. *Chem Sci R Soc Chem.* 2013;4:1928–1938.
89. Kwak M, Herrmann A. Nucleic acid/organic polymer hybrid materials: Synthesis, superstructures, and applications. *Angew Chem Int Ed.* 2010;49:8574–8587.

90. Cobo I, Li M, Sumerlin BS, Perrier S. Smart hybrid materials by conjugation of responsive polymers to biomacromolecules. *Nat Mater.* 2015;14:143–159.
91. Jäschke A, Fürste JP, Erdmann VA, Cech D. Hybridization-based affinity partitioning of nucleic acids using PEG-coupled oligonucleotides. *Nucleic Acids Res.* 1994;22:1880–1884.
92. Costioli MD, Fisch I, Garret-Flaudy F, Hilbrig F, Freitag R. DNA purification by triple-helix affinity precipitation. *Biotechnol Bioeng.* 2003;81:535–545.
93. Lai JJ, Hoffman JM, Ebara M, Hoffman AS, Estournès C, Wattiaux A, et al. Dual magnetic-/temperature-responsive nanoparticles for microfluidic separations and assays. *Langmuir.* 2007;23:7385–7391.
94. Mori T, Umeno D, Maeda M. Sequence-specific affinity precipitation of oligonucleotide using poly(N-isopropylacrylamide)–oligonucleotide conjugate. *Biotechnol Bioeng.* 2001;72:261–268.
95. Krieg E, Shih WM. Selective nascent polymer catch-and-release enables scalable isolation of multi-kilobase single-stranded DNA. *Angew Chem Int Ed.* 2018;57:714–718.
96. Minev D, Guerra R, Kishi JY, Smith C, Krieg E, Said K, et al. Rapid in vitro production of single-stranded DNA. *Nucleic Acids Res.* 2019;47:11956–11962.
97. World Bank. World Bank Open Data. Available from: https://data.worldbank.org.
98. Niemz A, Ferguson TM, Boyle DS. Point-of-care nucleic acid testing for infectious diseases. *Trends Biotechnol.* 2011;29:240–250.
99. Notomi T, Okayama H, Masubuchi H, Yonekawa T, Watanabe K, Amino N, et al. Loop-mediated isothermal amplification of DNA. *Nucleic Acids Res.* 2000;28:E63.
100. Piepenburg O, Williams CH, Stemple DL, Armes NA. DNA detection using recombination proteins. *PLoS Biol.* 2006;4:e204.
101. Tong X, Zhang K, Han Y, Li T, Duan M, Ji R, et al. Fast and sensitive CRISPR detection by minimized interference of target amplification. *Nat Chem Biol.* 2024:1–9.
102. Gootenberg JS, Abudayyeh OO, Lee JW, Essletzbichler P, Dy AJ, Joung J, et al. Nucleic acid detection with CRISPR-Cas13a/C2c2. *Science.* 2013;356:438–442.

103. Kellner MJ, Koob JG, Gootenberg JS, Abudayyeh OO, Zhang F. SHERLOCK: Nucleic acid detection with CRISPR nucleases. *Nat Protoc.* 2019;14, 2986–3012.
104. Chen JS, Ma E, Harrington LB, Da Costa M, Tian X, Palefsky JM, et al. CRISPR-Cas12a target binding unleashes indiscriminate single-stranded DNase activity. *Science.* 2018;360, 436–439.
105. Azhar M, Phutela R, Kumar M, Ansari AH, Rauthan R, Gulati S, et al. Rapid and accurate nucleobase detection using FnCas9 and its application in COVID-19 diagnosis. *Biosens Bioelectron.* 2021;183:113207.
106. Li SY, Cheng QX, Wang JM, Li XY, Zhang ZL, Gao S, et al. CRISPR-Cas12a-assisted nucleic acid detection. *Cell Discov.* 2018; 4:20.
107. Zhang C, Chen J, Sun R, Huang Z, Luo Z, Zhou C, et al. The recent development of hybridization chain reaction strategies in biosensors. *ACS Sens.* 2020;5:2977–3000.
108. Zhang DY, Seelig G. Dynamic DNA nanotechnology using strand-displacement reactions. *Nat Chem.* 2011;3:103–113.
109. Srinivas N, Parkin J, Seelig G, Winfree E, Soloveichik D. Enzyme-free nucleic acid dynamical systems. *Science.* 2017;358:eaal2052.
110. Li S. Li P, Ge M, Wang H, Cheng Y, Li G, et al. Elucidation of leak-resistance DNA hybridization chain reaction with universality and extensibility. *Nucleic Acids Res.* 2020;48:2220–2231.
111. Olson X, Kotani S, Padilla JE, Hallstrom N, Goltry S, Lee J, et al. Availability: A metric for nucleic acid strand displacement systems. *ACS Synth Biol.* 2017;6:84–93.
112. Gupta K, Krieg E. Y-switch: A spring-loaded synthetic gene switch for robust DNA/RNA signal amplification and detection. *Nucleic Acids Res.* 2024;52:e80.
113. Kim J, White KS, Winfree E. Construction of an *in vitro* bistable circuit from synthetic transcriptional switches. *Mol Syst Biol.* 2006;2:68.
114. Green AA, Silver PA, Collins JJ, Yin P. Toehold switches: De-novo-designed regulators of gene expression. *Cell.* 2014;159: 925–939.
115. Paige JS, Wu KY, Jaffrey SR. RNA mimics of green fluorescent protein. *Science.* 2011;333:642–646.

116. Pardee K, Green AA, Takahashi MK, Braff D, Lambert G, Lee JW, et al. Rapid, low-cost detection of Zika virus using programmable biomolecular components. *Cell.* 2016;165:1255–1266.
117. Kalligosfyri PM, Cinti S. Paper-based materials for diagnostics. *ACS Mater Lett.* 2024;6:1184–1198.
118. Speed S, Atabay A, Peng YH, Gupta K, Müller T, Fischer C, et al. Assembling a true 'Olympic gel' from >16,000 combinatorial DNA rings. *bioRXiv.* 2024. Preprint at doi:10.1101/2024.07.12.603212.
119. De Gennes PG. *Scaling Concepts in Polymer Physics.* Ithaca (NY): Cornell University Press; 1979. p. 324.
120. Stewart JM, Li S, Tang AA, Klocke MA, Gobry MV, Fabrini G, et al. Modular RNA motifs for orthogonal phase separated compartments. *Nat Commun.* 2024;15:6244.

https://doi.org/10.1142/9789819806782_0011

Chapter 11

Smart Microgel and Nanogel Platforms for Biomedical Applications

Devlina Ghosh,[#] Ruichen Zhang,[#] Yanjing Ji[#] and Patrick van Rijn[*]

University of Groningen, University Medical Center Groningen, Department of Biomaterials and Biomedical Technology-FB40, Antonius Deusinglaan 1, 9713 AV Groningen, The Netherlands

**Corresponding author:p.van.rijn@umcg.nl*

#All authors have contributed equally.

Microgels/nanogels (μ/nGels) are soft, cross-linked polymeric particles with tunable properties, high water retention capacity, and inherently biocompatible by nature. In this chapter, we provide knowledge of understanding the realm of μ/nGels in detail by exploring the various traditional and advanced particle synthesis approaches along with the physicochemical techniques involved in particle characterization. By modifying the particle surfaces or incorporating specific reactive moieties, "smart" μ/nGels platforms are developed and represent promising tools to introduce unique chemical functionality and multi-stimuli responsiveness. Scientists can skillfully integrate the smart polymeric particles with the inherent physicochemical properties of the bulk material to form a biologically compatible interface. Therefore, utilizing various coating deposition techniques, the synthesized smart polymeric particles are applied as surface coatings and characterized to determine optimal characteristics for specific applications. In addition, the implementation of the synthesized μ/nGel particles as both suspensions and/or surface coatings is discussed. This has opened many horizons and provided an impulse to explore different biomedical applications in the domain of antifouling/antimicrobial, drug delivery, wound dressing, tissue engineering, biotribology, immunomodulation, biosensing, and imaging modalities.

1. Introduction

Smart polymeric particles such as microgels (μGels) and nanogels (nGels) have been extensively explored in the realm of biomedical applications.[1–5] μ/nGels are micro- and nanosized hydrogel particles that are distinguished by their three-dimensional (3D) cross-linked structures, colloidal properties, and hydrophilic nature. These polymeric particles are synthesized through a diverse range of methods, which allow for precise control over their physical and chemical properties, including size, shape, and cross-linking density.[6,7] The chemical composition is tailored by incorporating functional groups or bioactive molecules into the polymer matrix,[8,9] while modification techniques, such as grafting polymer chains or conjugating targeting ligands, enable the precise control of their interactions with biological systems.[10,11] Hence, these particles undergo physicochemical changes, such as swelling, shrinking, and dissociation, under suitable stimuli conditions. The applied stimuli can vary from temperature,[12] light,[13] pH,[14] redox,[15] magnetic strength[2] to ultrasound.[16] The particles find their biomedical applications in antifouling/antibacterial platforms, drug delivery, tissue engineering, diagnostic imaging, and as biosensors.

To substantiate these aforementioned novel biomedical applications, these particles can be implemented as suspensions or surface coatings. In a suspension, the particles are dispersed in a liquid, whereas when applied as a coating, the particles form the interface between the bulk material and its surroundings (here, the biological/physiological environment).

This chapter centers around the synthesis (traditional and advanced methods) of μ/nGel particles and their physical and chemical characterization in suspension. By conjugating versatile reactive moieties into the polymeric networks, these particles have the potential to be developed as smart or intelligent μ/nGels that are responsive to external physicochemical and biological cues. Furthermore, we delve into the translation of these smart particles from suspension to surface coatings using various coating deposition techniques and additionally explore various characterization techniques associated with coating. We dedicate the concluding section to understanding the wide-ranging biomedical applications based on μ/nGel suspensions and surface coatings.

2. Synthesis and physicochemical characterization of nanogels

nGels are synthesized through various methods, including (i) direct polymerization in homogeneous or heterogeneous mixtures of diverse monomers, (ii) self-assembled combinations, (iii) emulsion systems, template-based nanofabrication, or (iv) polymerization involving larger polymer structures as their building blocks. These techniques are extensively reviewed in the literature.[17–21] Typically, the synthesis entails processes such as grafting, cross-linking polymerization, and the formation of covalent bonds in the polymer backbone.[22,23] Cross-linking formation plays a vital role in the fabrication of μ/nGels, and thus, various cross-linking approaches are employed to stabilize the particles and control their swelling behavior.[24] One method is physical cross-linking, which relies on noncovalent interactions such as hydrogen bonding, ionic interactions, hydrophobic interactions, and crystallization.[25,26] These methods are reversible and are often used when stimuli-responsive μ/nGels are synthesized. Another approach involves chemical cross-linking, which entails the formation of covalent bonds between polymer chains using cross-linking agents or initiators. These methods generally provide stable and robust polymeric particles.[27,28]

During this procedure, either the monomer or the precursor of the polymer must possess at least one polymerizable functional group like (meth)acrylate, (meth)acrylamide, or styrene (St), which is susceptible to polymerization by free radicals generated within the reaction system.[29,30] (Meth)acrylate group is the functional group most often used in the polymerization to produce nGels.[31–34] In this section, we have focused on heterogeneous polymerization, which is one of the most widely used methods for synthesizing stable nGels.

2.1. *Traditional synthesis approaches*

Heterogeneous polymerization techniques, including inverse emulsion polymerization, inverse miniemulsion polymerization, inverse microemulsion polymerization, dispersion polymerization, and precipitation polymerization, are traditional synthesis methods, as shown in **Figure 1**.[35] In brief, inverse emulsion is a water-in-oil heterogeneous polymerization process that consists of oil-soluble surfactants forming "(nano/micro) reactors" (the aqueous droplets) in a continuous organic medium. It is carried out by polymerization of hydrophilic (ionic) monomers and cross-linkers in the aqueous droplets, which occurs after adding a polymerization initiator. Takeuchi et al. prepared uniform, ultrasmall cyclodextrin (γ-CyD) nGels with $\approx$10 nm particle size via inverse emulsion polymerization using [dilauryl(dimethyl)]ammonium bromide, a cationic surfactant.[19] The nanoparticles (NPs) produced by microemulsion polymerization are typically smaller in size (< 80 nm) compared to those produced by precipitation polymerization.[36] Precipitation polymerization is a method to prepare hydrophilic nGels. In water droplets without surfactants, the precursors produce nGel particles of determinate size during in situ cross-linking. In the dispersion polymerization process, all starting materials (i.e., monomers, initiators, stabilizers, and additives) are uniformly dispersed in a continuous phase at the same time and polymerized by initiators or other chemical means to form nanosized gel particles. The size distribution of nGels prepared by this method is narrow. Dispersion polymerization differs

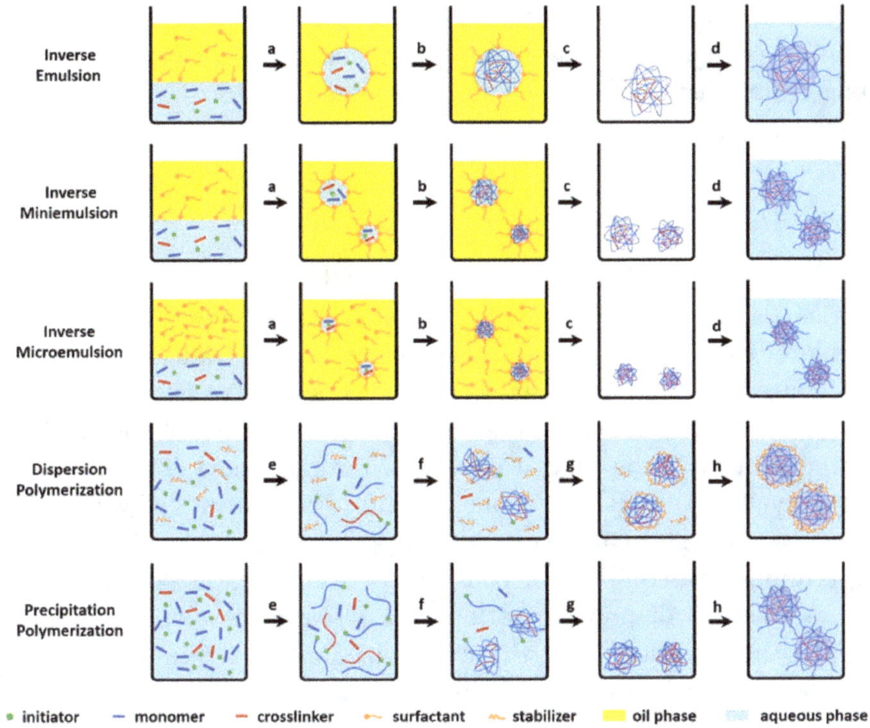

Figure 1. Schematic description of nGels formation by heterogeneous polymerization. Reproduced with permission from Keskin et al.,[35] copyright 2021. The Authors. Publishing services by Elsevier B.V. on behalf of KeAi Communications Co. Ltd.

from precipitation polymerization as it is in the presence of a second soluble polymer as a spatial stabilizer.[37]

2.2. *Advanced chemical approaches*

Traditional approaches often rely on conventional cross-linking methods and basic polymerization techniques to fabricate nGels. With advanced synthesis methodologies, such as controlled/living polymerization techniques and template-assisted synthesis, researchers are now endowed with unprecedented control over nGel structure, composition, and functionality. These advanced methods not only offer precise control but also enable the design of nGels with tailored properties to meet diverse application requirements, spanning from drug delivery and tissue engineering to diagnostics and environmental remediation.[38–40]

2.2.1. *Atom transfer radical polymerization*

Atom transfer radical polymerization (ATRP) is a reactive polymerization reaction wherein transition metal complexes are used to regulate the initiation and termination of free radicals, thereby achieving a controllable polymerization process. Precise size control is achieved by adjusting reaction conditions, such as polymer molecular weight and molecular weight distribution. There are five major components in ATRP: monomer, initiator, catalyst, solvent, and temperature. The catalyst is the key component of ATRP as it determines the equilibrium constant of the reaction, which determines the rate of polymerization.[41] In research developed by Jung et al., they fabricated OH-functionalized nGels cross-linked with disulfide linkages by introducing 2-hydroxyethyl acrylate (HEA) during the ATRP process of poly(oligo(ethylene oxide) monomethyl ether methacrylate) (OEOMA) (**Figure 2a**).[42] These nGels consisted of poly(oligo(ethylene oxide) monomethyl ether methacrylate) (POEOMA) and oligoethylene glycol, an analog of linear polyethylene oxide (PEO), which prevents proteins from binding to the nGels.

2.2.2. *Reversible addition-fragmentation chain transfer polymerization*

Among the major reactive/controllable radical polymerizations, reversible addition-fragmentation chain transfer (RAFT) polymerization is considered, by many, to be the most versatile polymerization technique.[45] The RAFT method differs from the free radical method by the additional use of a chain transfer agent (CTA) to mediate the polymerization. Zhang et al. reported a RAFT polymerization process initiated by enzymes; here, horseradish peroxidase

(HRP) served as a catalyst, with CTA1 in a pH 3.5–7 buffer solution, triggering the oxidation of acetylacetone (ACAC) with hydrogen peroxide to produce the ACAC radical (**Figure 2b**).[21] This radical effectively controlled the initiation of the polymerization in a buffered aqueous solution at room temperature when a suitable CTA was available.

2.2.3. *Electrostatic assembly-directed polymerization*

Electrostatic assembly-directed polymerization (EADP) is an emerging approach to synthesize materials whose basic principle involves electrostatic interactions to direct polymer formation. This approach typically involves two main steps: (i) charge electrostatic assembly and (ii) polymerization reaction. Based on EADP, Wang et al. recently developed cationic polyethylmethacrylate trimethylammonium chloride (PMETAC) nGels with controllable particle size and adjustable cross-linking degree.[43] The designed nGels were further used to load tetrachloroaurate ions ($AuCl_4^-$) through electrostatic interaction, and subsequent in situ reduction produced gold NPs (AuNPs) embedded in the nGels (**Figure 2c**).

2.3. *Microfluidic approach*

The clinical translation of nGels encounters challenges due to limitations of the batch synthesis process, particularly concerning its impact on payload encapsulation efficiency and release profile.[46] Therefore, to address the challenge of optimal therapeutic performance, the microfluidic approach is often employed. Although the application of microfluidics in the manufacturing of drug delivery systems is at an early stage, it has shown enormous potential.[47,48] Microfluidics involves the manipulation of fluid flow at a microscopic scale, providing precise control over characteristics of the flow and facilitating rapid and homogeneous mixing.[49,50] Within microfluidic devices, the nucleation and growth phases of nGels can be distinctly separated from the site of solution mixing.

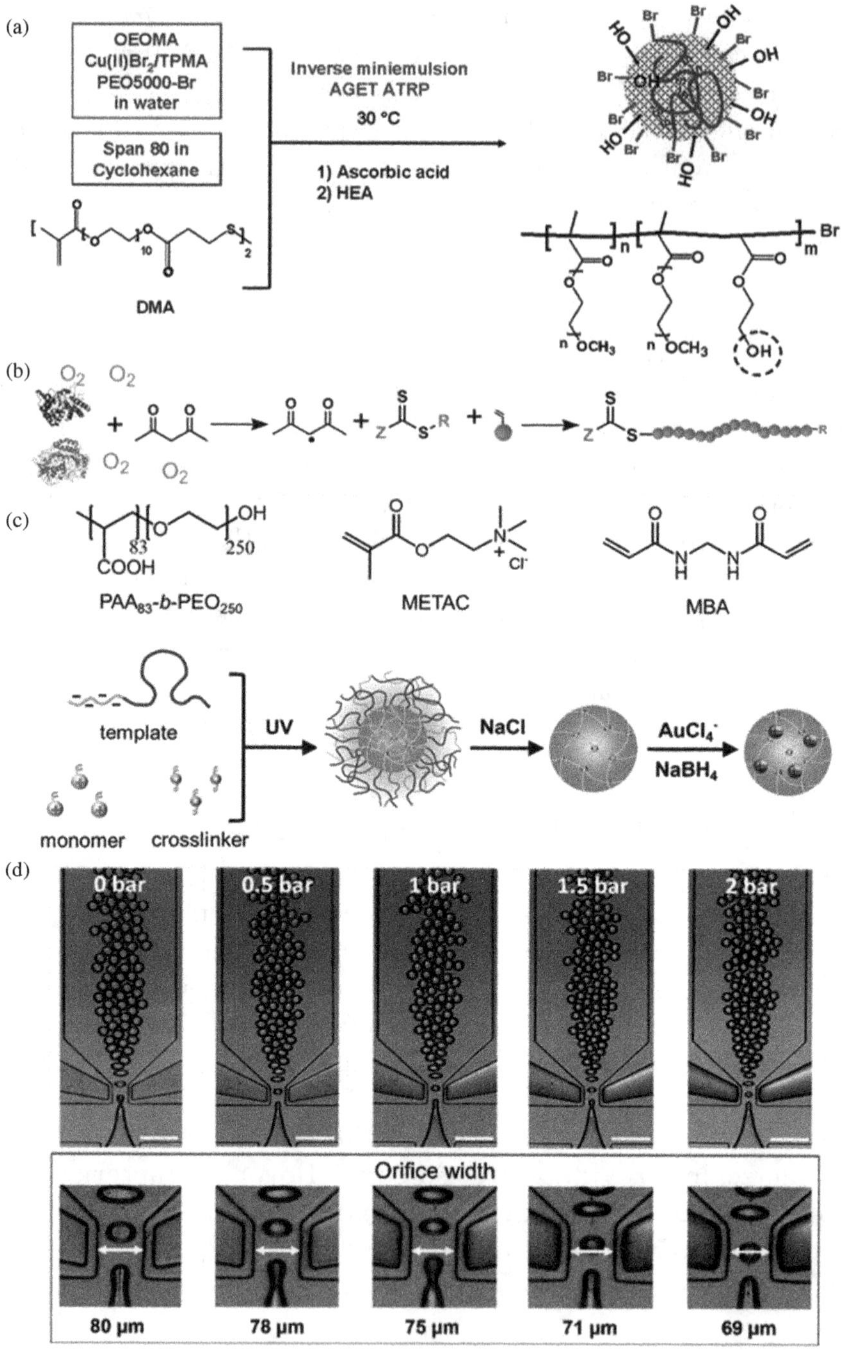
(a)
OEOMA
Cu(II)Br2/TPMA
PEO5000-Br
in water
Span 80 in
Cyclohexane
DMA
Inverse miniemulsion
AGET ATRP
30 °C
1) Ascorbic acid
2) HEA
(b)
(c)
PAA83-b-PEO250
METAC
MBA
template
monomer
crosslinker
UV
NaCl
AuCl4-
NaBH4
(d)
0 bar
0.5 bar
1 bar
1.5 bar
2 bar
Orifice width
80 μm
78 μm
75 μm
71 μm
69 μm

This separation allows for meticulous regulation of particle size and morphology and thus reduces the polydispersity index (PDI). Giannitelli et al. synthesized nGels composed of hyaluronic acid and polyethyleneimine (PEI) using a microfluidic flow-focusing device equipped with pressure-driven micromanipulators.[44] nGels ranging from 56.7 ± 0.4 μm to 44.2 ± 0.3 μm were synthesized by a droplet-based microfluidic method using a chip with microdrive characteristics. At 2 bar pressure, the volume is reduced by about 53% compared to 0 bar (**Figure 2d**). The key role of the actuator is to dynamically control the orifice geometry so that the droplet size can be changed on demand without changing the input flow rate of the two-phase immiscible.[3,20,51]

2.4. *Functionalization strategies*

One of the key attributes that contribute to the versatility of μ/nGels is their inherent chemical functionality, which allows for precise control over their properties and functions.[52] By varying the type and density of functional groups, the chemical composition and their structure can be tailored to suit different applications.[7,53] The chemical functionality of nGels and their potential for post-modification, which is to achieve further complexity and functionality, will be discussed.

Figure 2. (**a**) Biodegradable, cross-linked POEOMA NPs were synthesized from OEOMA by ATRP in a reversed-phase microemulsion in the presence of a disulfide-functionalized dimethacrylate (DMA) cross-linking agent at ambient temperature (30°C). Reproduced with permission from Jung et al.,[42] 2020 American Chemical Society. (**b**) HRP-initiated RAFT polymerization dimethylacrylamide (DMA) was performed. Reproduced with permission from Zhang et al.,[21] 2020 American Chemical Society. (**c**) Scheme of the nGels preparation and gold NPs (AuNPs) immobilization. Reproduced with permission from Wang et al.,[43] copyright 2023 by the authors. Licensee MDPI, Basel, Switzerland. (**d**) A representative microdroplet size distribution micrograph. Scale bar: 200 μm. Reproduced with permission from Giannitelli et al.,[44] copyright The Royal Society of Chemistry 2022.

2.4.1. *Chemical functionality*

The common chemical functionalities present in μ/nGels include hydroxyl, carboxyl, amino, thiol, and epoxy groups, among others, which provide sites for conjugation with various molecules and biomolecules.[35,54] Fu et al. reported a novel approach to prepare hydroxyl groups included nGels using poly(ethylene glycol) (PEG) as monomer via reverse (mini)emulsion copper(I)-catalyzed azide-alkyne cycloaddition (REM-CuAAC).[55] Singh et al. synthesized both carboxyl and amine functional groups containing nGels; acrylic acid (AA) was used as a comonomer along with fluorenylmethoxy-carbonyl (Fmoc)-protected PEG methacrylamide macromonomer via free radical precipitation polymerization.[56]

Saraswathy et al. synthesized thiol-functionalized nanostructured particles containing the monomer of the parent network within a highly cyclized, cross-linked structure that can then be dispersed and subsequently cross-linked to generate an optically clear, flexible network utilizing a one-pot, step-growth thiol-Michael addition-solution polymerization reaction under ambient conditions.[57] The epoxy-functional spherical NPs have been achieved by Hatton et al. using polymerization-induced self-assembly in aqueous solution.[58]

2.4.2. *Postmodification strategies*

Postmodification can be achieved through a variety of chemical reactions, including click chemistry, Michael addition, and Schiff base formation, among others. These reactions enable the attachment of a wide range of molecules, including drugs, targeting ligands, imaging agents, and stimuli-responsive moieties, thereby imparting new functionalities to the μ/nGels.

Chen et al. applied click chemistry to decorated biodegradable polymeric nGels with a folic acid–PEG conjugate via copper-free click chemistry, which was employed to bypass drug efflux pumps and enhance targeting efficiency of nGels, resulting in highly efficient killing of resistant cancer cells.[59] Wei et al. conjugated hemoglobin by Schiff base reaction under mild conditions with in situ cross-linked and aldehyde group-functionalized nGels from dextran-*g*-succinic anhydride-*g*-dopamine conjugate, suggesting the potential for oxygen carriers.[60]

2.5. *Characterization of nanogels in suspension*

Postsynthesis, the particles need to undergo a comprehensive characterization over several aspects such as physical, chemical, and biological properties using integrated techniques: transmission electron microscopy (TEM), dynamic light scattering (DLS), electrophoretic light scattering (ELS), and ultraviolet–visible (UV–Vis) spectroscopy, as shown in **Figure 3**. These techniques give information regarding morphology, particle size distribution, PDI, and zeta potential/surface charge.

TEM has been considered the gold standard for visualizing NPs since it provides structural details with nanometer resolution.[61] A contrast between the background and the nGel particles is created by negatively staining the particles with uranium acetate. As shown in **Figure 3a**, primary amine-functionalized core–shell nGels (Amine-NG) and methacryloyl-functionalized nGels (MA-NG) developed by Zu et al. appear as light-colored regions because they have a low electron scattering capacity, while the dense staining around them scatters more electrons and is darker in color.[62] As the stain does not penetrate the particle, thus its internal structural details cannot be inferred.

DLS is an established and precise measurement technique for characterizing particle size in liquid. The hydrodynamic diameters of both Amine-NG and MA-NG in water displayed a narrow size distribution at 20°C; the average hydrodynamic diameter is ~519 nm for the Amine-NG and 561 ± 7 nm for the MA-NG (**Figure 3b**). In addition, the light scattered by thermosensitive nGels particles is measured as a function of the scattering angle at temperatures below and above the volume phase transition temperature (VPTT). **Figure 3c** shows the variation of Amine-NG and

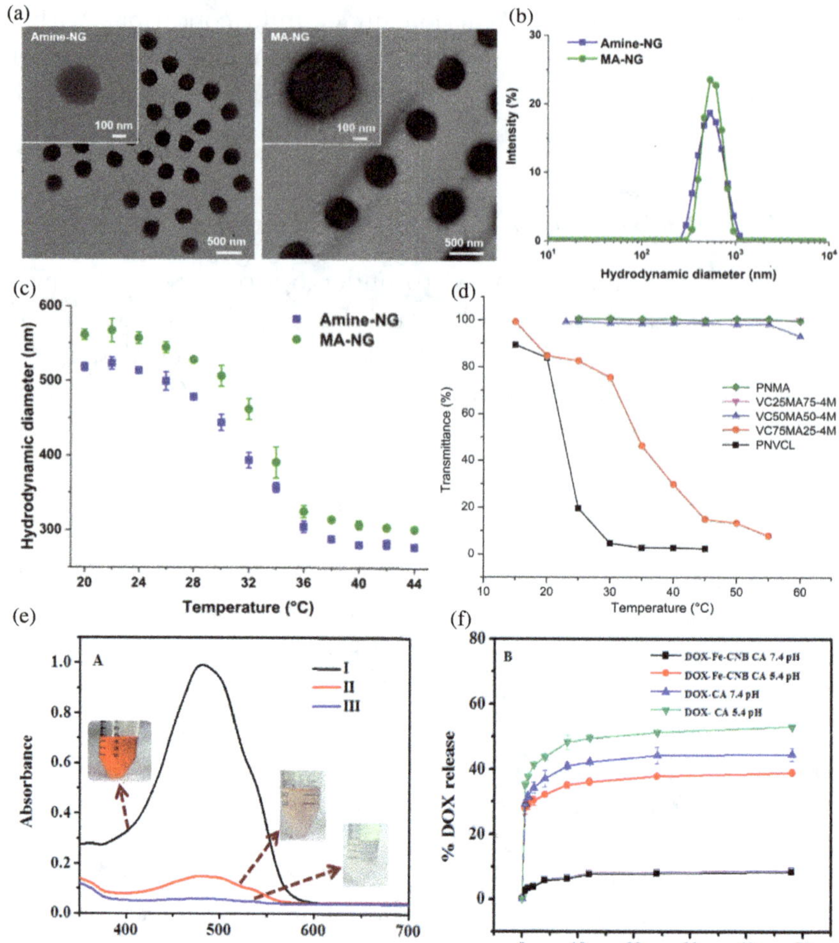

Figure 3. (**a**) TEM images of Amine-NG and MA-NG. (**b**) Size distribution by the intensity of Amine-NG and MA-NG in aqueous media, determined by DLS analysis in water at 20°C. (**c**) Hydrodynamic diameter as a function of the temperatures of Amine-NG and MA-NG, obtained from DLS measurements. Reproduced with permission from Zu et al.,[62] copyright 2021 by the authors, Licensee MDPI, Basel, Switzerland. (**d**) Thermoresponsive behavior of P(NVCL-*co*-NMA) nGels with different monomer feed ratios in 3 mg/mL phosphate buffer pH 7.4. Reproduced with permission from Nizardo et al.,[63] copyright 2022 The Author(s). Published by Informa UK Limited, trading as Taylor & Francis Group. (**e**) Drug loading by UV absorption peak at ($\lambda = 480$ nm), (I) standard

MA-NG in hydrodynamic diameter according to temperature since the poly(N-isopropylacrylamide) (p(NIPAM)) segment is temperature responsive.[62] This technique is also used to assess the PDI of the particles, which is a metric of the width of the nGels size distribution.[65] Lower PDI values indicate a more homogeneous particle size distribution, while higher PDI values imply a wider size distribution.[66] It has to be noted that TEM will provide a smaller diameter than DLS. In TEM, the particles are dehydrated, and therefore, in a shrunken state rather than a fully hydrated and expanded state as measured with DLS. However, TEM will be able to verify whether there are indeed particles present, and therefore, using several techniques in parallel is always advised.

ELS is used to measure the electrophoretic mobility of particles in dispersion or solution. When an electric field is applied to the charged particles, they migrate in opposite directions. By measuring the speed at which the nGels move in the electric field, their zeta potential can be deduced.[67]

UV–Vis is a common analytical technique widely used in the study of nGel cloud point transition temperature and drug delivery. Nizardo et al. investigated the thermosensitivity of poly(N-vinylcaprolactam-*co*-N-methylolacrylamide) [P(NVCL-*co*-NMA)] nGels by detecting its transmittance at 600 nm using UV–Vis spectrophotometer (**Figure 3d**).[63] nGels usually undergo structural changes during the volume transition, resulting in altered optical properties of the material. Such changes may affect the transmittance of nGels to light. Information about the volume shift of the nGels can be obtained by recording the transmittance versus temperature curve. UV–Vis is also utilized in encapsulating drugs within nGels to assess the capacity of nGels for encapsulation as well as drug release kinetics. Varghese et al. confirmed doxorubicin (DOX) loading on

Figure 3. (*Continued on facing page*) DOX solution, (II) decanted solution of the drug after DOX uptake by CA, (III) decanted solution of drug after DOX uptake by Fe–CNB CA. (**f**) The releasing of DOX from CA and nano Fe–CNB CA at 5.4 and 7.4 pH. (**e**) and (**f**) Reproduced with permission from Varghese et al.,[64] copyright The Royal Society of Chemistry 2020.

calcium alginate (CA) and iron-based magnetic nanobiocomposite CA by UV–Vis spectrophotometry and measured the release rate of DOX at 480 nm, as shown in **Figure 3e and 3f.**[64]

3. Smart microgels and nanogels

"Smart" μ or nGels, which respond to various stimuli, have shown tremendous potential in various biomedical and pharmaceutical fields.[1,68] Depending on the different responsiveness of active moieties contained within the μ and nGel, they can respond to specific or multiple combined stimuli such as chemical stimuli (pH and redox), physical stimuli (temperature and light), and biological stimuli (enzymes).[69–72] When exposed to these suitable stimuli, the obtained smart μ/nGels undergo a series of physicochemical changes, such as swelling, shrinking, and dissociation. The mechanisms and chemical structures of various stimuli-responsive μ/nGel are shown in **Table 1**.

3.1. *Chemical stimuli-responsive microgels and nanogels*

3.1.1. *pH-responsive microgels and nanogels*

The pH sensitivity of μ and nGels is usually associated with conjugating anionic or cationic groups, such as -COOH and -NH_2, within the polymeric networks. These ionizable groups experience protonation or deprotonation under various pH values, causing the swelling or collapsing behaviors of μ and nGel.[73] In an acidic environment, μ and nGels with amino groups swell due to the protonation of -NH_2 to -NH^{3+}. The ionized positively charged groups cause electrostatic repulsion and swelling of polymeric networks. In contrast, anionic μ and nGels (such as with a -COOH group) swell at higher pH owing to the deprotonation of carboxyl groups (from -COOH to -COO^-) in an alkaline medium.[74] Since the tumor tissue microenvironment is more acidic (pH = 6.5–7.0) compared to the normal tissue (pH = 7.4), μ/nGels with specific pH-stimuli responsibility allow the targeted drug delivery and minimize the side effects of the drug.[89]

Table 1. Types and mechanisms of stimuli-responsive μ and nGels.

Stimuli type	Polymer type/reactive moieties	Mechanism	Ref
pH-responsive	Polymers with ionizable anionic or cationic groups	Swelling or collapsing behaviors of polymers upon protonation or deprotonation of anionic/cationic moieties under various pH values	73–75
Redox-responsive	(a) Polymers with reduction-sensitive linkages such as disulfide (-S-S-) bonds, thioether, and diselenide bonds (b) Organometallic polymers bearing transition metals such as ferrocene units (Fc)	(a) Glutathione (GSH)-triggered reduction and hence cleavage of disulfide (-S-S-) linkages (b) Electrochemically or chemically triggered reversible oxidization of ferrocene (Fc) to positively charged ferrocenium ion (Fc+)	76–80
Temperature-responsive	PNiPAAm, PVCL, and PDEAAm-based polymers	Interactions between hydrophilic and hydrophobic groups change along with temperature variations	81–83
Light-responsive	Photosensitizer-incorporated polymers	(a) Photochemical changes (b) Photoisomerization (c) Photothermal conversion	84–86
Enzyme-responsive	Polymers containing enzyme-responsive moieties, such as protease-cleavable peptides	Enzyme-trigged physicochemical changes (swell/dehydration, bond dissociation, and degradation) of polymers	87,88

Pikabea et al. prepared cationic pH-sensitive and thermosensitive nGels based on poly(2-(diethylamino ethyl)methacrylate cross-linked with ethylene glycol dimethacrylate.[75] The systems have shown great pH sensitivity due to protonation (-NH^{3+}) of amine groups. As shown in **Figure 4a**, in an acidic environment below the dissociation constant (pKa) of amine, the prepared nGels showed increased particle size.

3.1.2. *Redox-responsive microgels and nanogels*

One of the characteristics of the tumor microenvironment is a redox gradient between its reducing intracellular (i.e., cytoplasm, endosomes, and lysosomes) and oxidizing extracellular environment.[90] The high concentration of glutathione (GSH) in the intracellular microenvironments of cancer cells acts as a reducing agent for redox-sensitive μ and nGels.[76] The sensitivity of μ and nGels to a redox environment is associated with incorporating reduction-sensitive linkages, such as disulfide (-S-S-), thioether, and diselenide bonds. Among them, the cleavable disulfide (-S-S-) bonds are the most common and effective linkages used in the redox-responsive systems: They are usually stable in the extracellular environments but can be reduced and dissociated to thiols (-SH) in the GSH-rich reductive tumor microenvironment.[77] Through response to the redox gradient between the tumor and normal cell, the μ and nGels can realize controlled delivery; thus, there is a high selectivity of drug accumulation in the tumor site. Qian et al. have prepared methacrylated carboxymethyl cellulose (MACMC)-based nGels with redox sensitivity by incorporating cystamine bisacrylamide (CBA) with reductive disulfide (-S-S-) bonds (**Figure 4b**).[78] The obtained nGels had a spherical morphology with a diameter of about 192 nm and can be utilized as the delivery vehicle for DOX to cancer cells. It was found that these redox-sensitive nGels disintegrated in the reducing high GSH concentration cancer tissue microenvironment and acquired the rapid release of the drug in the targeted sites.

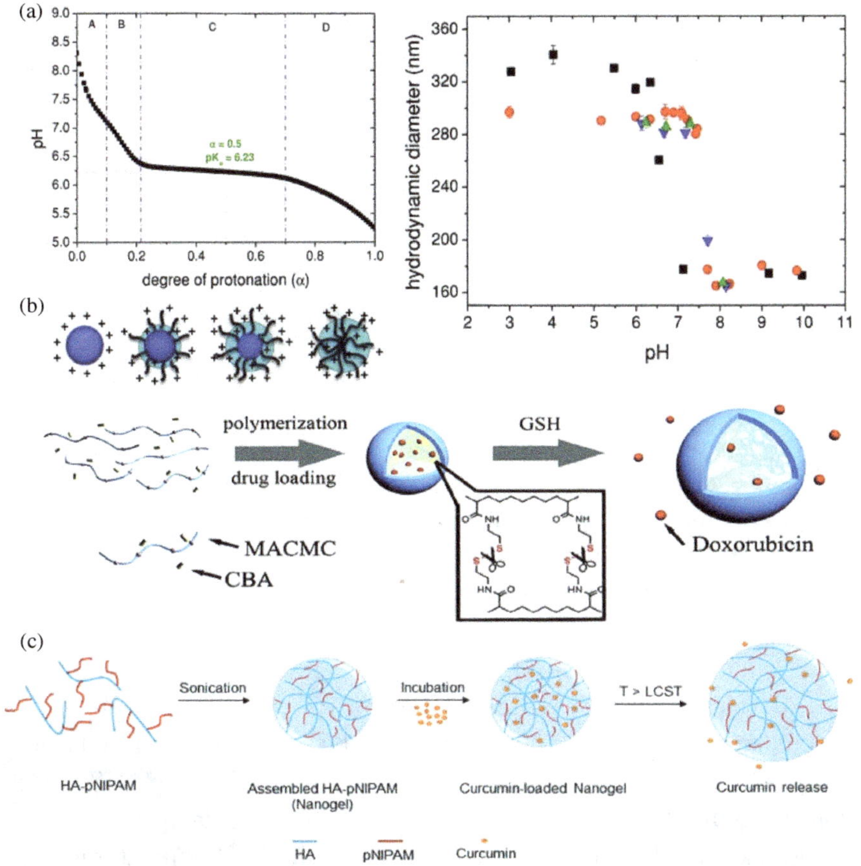

Figure 4. (**a**) pH as a function of the degree of protonation (α) of the nGel particles and average hydrodynamic particle diameter as a function of pH. Reproduced with permission from Pikabea et al.,[75] copyright 2013 WILEY-VCH Verlag GmbH & Co. KGaA, Weinheim. (**b**) Schematic diagram of the synthesis of MACMC–CBA nGels and the release of DOX triggered by GSH. Reproduced with permission from Qian et al.,[78] copyright 2014, The Royal Society of Chemistry. (**c**) The assembly of HA-g-pN nGel and curcumin release. Reproduced with permission from Luckanagul et al.,[83] copyright 2021 by the authors. Licensee MDPI, Basel, Switzerland. (**d**) Schematic description of the enzyme-responsive PAMAM nGels and the collapsed sketch triggered by enzyme line. Reproduced with permission from Wang et al.,[98] copyright 2016, American Chemical Society.

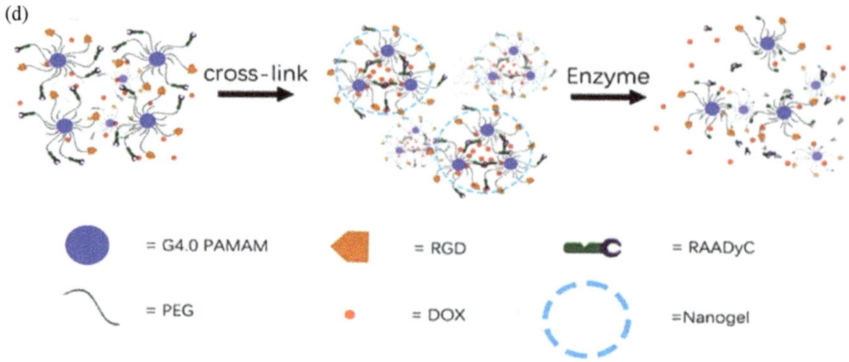

Figure 4. (*Continued*)

Besides introducing linkages such as disulfide bonds to design reduction-sensitive systems, the novel reversible redox-active polymers have gained significant research interest due to their sensitivity to both electrochemical and chemical stimuli and redox-reversibility.[79,80,91] The commonly used moieties are transition metals such as ferrocene (Fc) units, which can be reversibly oxidized to the positively charged Fc^+ units (ferrocenium ion) via chemical and electrochemical stimuli. The Fc-containing polymers swell in the oxidized state or shrink in the reduced state.[79,91] Though further combining the Fc units with host molecules such as β-cyclodextrins (β-CDs), Jung et al. synthesized supramolecular poly(N-vinylcaprolactam) (PVCL) μGels containing redox-triggered reversible dynamic cross-links.[80] The obtained μGels exhibit hydrodynamic radii in the range of 100 to 200 nm. The degradation of the μGels was observed under both chemical and electrochemical oxidation via the exclusion switching of oxidized-state Fc^+ and β-CD.

3.2 *Physical stimuli-responsive microgels and nanogels*

3.2.1. *Temperature-responsive microgels and nanogels*

The temperature-sensitive μ and nGels undergo a volume-phase transition (swelling/shrinkage) in response to temperature variation.[92]

The mechanism of thermoresponsibility is based on the hydrophilic and hydrophobic groups, such as -CONH and alkyl groups, incorporated in the polymeric networks.[81] Poly N-isopropyl acrylamide (PNiPAAm), PVCL, and p(N,N-diethylacrylamide) (PDEAAm) are the three most widely used polymers with various lower critical solution temperatures (LCSTs).[93] Below the LCST, these polymers are well hydrated and swell by forming hydrogen bonds between water and the hydrophilic groups, such as -CONH in PNiPAAm. While increasing the temperature above the LCST, the hydrophobic interactions between the hydrophobic alkyl chains dominate, causing the collapse of polymeric networks.[82] By adjusting the structures of polymers, the LCST of thermoresponsive polymers can be tailored correspondingly. This unique thermo-related characteristic leads to their applications in the temperature-triggered drug release owing to the higher temperature at the diseased sites. Luckanagul et al. synthesized a novel self-assembled thermoresponsive nGels based on the hyaluronic acid-grafted poly(N-isopropylacrylamide) (HA-pNIPAM) for delivering hydrophobic anticancer drugs (**Figure 4c**).[83] The resulting nGels exhibited enhanced water solubility and improved bioavailability of low-solubility drugs. Moreover, the systems with thermoresponsiveness can also be applied in the biomedical coating fields. Uhlig et al. synthesized a N-isopropylacrylamide (NIPAM) -based thermosensitive μGel. The prepared μGel suspensions were spin-coated on the PEI-modified glass substrate.[94] It was demonstrated that the obtained thermoresponsive μGel coatings can manipulate the adhesion and detachment of cells from cultivation substrates.

3.2.2. *Light-responsive microgels and nanogels*

Applying light as a stimulus to control the behavior of μ and nGels has been explored significantly in targeted drug delivery, diagnosis, and phototherapy due to its noninvasiveness, spatial and temporal control, and ease of use.[95] Triggered by irradiation of UV, visible (UV–Vis), or near-infrared (NIR) light, the photosensitive μ and nGels undergo three main physicochemical/conformational changes:

(**1**) photochemical changes, such as bond cleavage, oxidation, and photoinitiated polymerization; (**2**) photoisomerization involving the conformational changes of photosensitized moieties; and (**3**) photothermal conversion.[68,96,97]

Azo-benzenes (-N=N- group between two phenyl groups) are commonly used photosensitive moieties in light-responsive systems. Upon UV irradiation, it transforms from trans to cis conformation reversibly and alters its physicochemical properties accordingly. Patnaik et al. prepared azobenzene-grafted dextran nGels through self-aggregation of azo-dextran polymers.[84] Regulated by photoisomerization of the azobenzene moiety (trans–cis) under UV radiation (365 nm), the noncovalently cross-linkers dissociated, realizing the controlled release of the entrapped drug (aspirin).

The photothermal-based smart μ and nGels convert the photon energy into thermal energy when irradiated by light. They usually consist of two parts: The first component is a light-reactive moiety, such as Au particles or graphene oxide, which generates heat and increases the temperature upon receiving the photon energy. The second part involves the temperature-responsive polymers with a VPTT, which experience swell/shrinkage under induced temperature changes.[85] Chen et al. designed a novel nanosized lubricant coating by using polydopamine (PDA) and MoS_2@chitosan@pNIPAM composite. By utilizing the photothermal conversion ability of MoS_2 to NIR light, the surface temperature can be adjusted.[13] Therefore, owing to the swelling/deswelling properties of thermal-sensitive pNIPAM, this PDA–pNIPAM coating can realize the friction adjustment by the presence or absence of the hydration layer. Such coating has tremendous potential in the fields of artificial joints, biosensing, soft robots, and microfluidic devices.

3.3. *Biological stimuli-responsive microgels and nanogels*

Biological stimuli, such as specific enzymes, provide new opportunities for designing smart μ and nGels in the biomedical fields.[72] The enzymes are vital in many metabolic and biological processes. The overexpression of specific enzymes in tumors or other inflammatory sites also allows the targeted delivery by the enzyme-sensitive μ or nGels. In the presence of a specific enzyme, the enzyme-responsive μ and nGels undergo physicochemical changes (swelling/dehydration, bond dissociation, and degradation) to target drug delivery and control the release of loaded drugs.[88] Wang et al. designed an enzyme-responsive nGel based on p(amidoamine) (PAMAM) dendrimers for drug delivery (**Figure 4d**).[98] The obtained nGels had a sphere-like shape with a relatively uniform size of 50 nm. When exposed to the elastase enzyme, the size of nGels with embedded DOX hydrochloride decreased, and the structures decomposed, leading to a controlled release of the drug. Aslzad et al. reported an efficient enzyme-responsive nGels-based carrier for DOX delivery prepared by chitosan/gelatin and AuNP (CS/AuNPs@Gel).[87] The mean size of the obtained gels was around 119.3 nm, with a zeta potential of +31.9 mV. It was found that controlled DOX release from CS/AuNPs@Gel-DOX nGel can be enhanced by gelatinase via an enzymatic degradation process.

3.4. *Multi-stimuli-responsive microgels and nanogels*

Lately, researchers have focused on designing more advanced and sophisticated μ and nGels responsive to more than one type of stimuli. The "smarter" multi-stimuli-responsive systems hold improved sensitivities to the surroundings, more versatile control options, and increased therapeutic efficacy, which present broader potential in biomedical fields than the single stimuli-responsive ones.[99–101]

Cao et al. synthesized a multi-stimuli-responsive nGel composed of hydrophilic pH and thermoresponsive poly(2-(dimethylamino) ethyl methacrylate) and hydrophobic photocleavable o-nitrobenzyl linkage.[69] The resulting nGels can deliver hydrophobic and hydrophilic drugs and realize the selective release upon different stimuli (temperature, pH, UV light, and redox reagent). Chen et al. prepared a triple-responsive nGel based on the spiropyran, AA, and

N,N'-bis(acryloyl)cystamine compounds for controlled release of drug.[70] It was reported that the obtained nGels acquired controlled release over 24 hours under NIR irradiation, an acidic environment, or a reducing agent. Under mild triple stimulation via NIR light, pH 6, and 4-mM reducing agent, the loaded drug could be released more efficiently. Lekjinda et al. investigated a facile and green approach to synthesize the trimethyl chitosan (TMC)–shell nGels with multi-stimuli responsibility as a potential nanocarrier for photothermal chemotherapy.[71] The TMC nGels exhibited a pH-dependent reversible upper critical solution temperature (UCST)-type volume phase transition around 30°C in an acidic aqueous environment (pH = 4.5–5). Triggered by light, the encapsulated drugs (honokiol) were released due to the incorporation of sodium copper chlorophyllin as a bio-based photosensitive agent.

4. Applications of nanogels as surface coatings

As discussed in the previous section, postmodification of smart polymeric particles is attained by introducing multiple functionalities/bioactive agents to form a highly applicable, facile surface coating without altering the overall (bulk) material. Hence, translation of nGel particles in suspension to surface coatings can open a plethora of biomedical applications and demonstrate high potential for clinical applications. The coatings serve as an intermediate layer that acts as a protective layer or adds a functional attribute to the intrinsic properties of the biomaterial in question.

4.1. *Coating deposition techniques*

The commercially available coating deposition techniques can be primarily grouped as spray-based coating and liquid-based deposition methods. The different spray-based and liquid-based coating deposition techniques, along with the basic parameters involved, their advantages, and their limitations, are summarized in **Table 2**.

Table 2. Summary of the different coating deposition techniques.

(a) Spray-based coating approach	
(1) Spray coating	**(2) Approach**
An atomizer with a spray nozzle converts the liquid into mist, which is sprayed on the material surface.[102,103]	Basic concept
- Nozzle/spray-related factors:[104,105] - Dimensions of the nozzle, applied pressure, speed of spraying, working distance, and number of coating cycles. - Rheological properties: - Surface tension, viscosity, and density. - Substrate related factors: - Physical properties (roughness, wettability), pretreatment.	Parameters influencing the coating thickness
- Contactless approach. - Cost and time efficient. - Useful in coating complex structures. - Able to achieve variability in coating thickness within the same substrate.[106,107]	Advantages
- Clogging of the nozzle.[108] - Exposure to harmful mist during the spray. - Over-excessive manual spraying can alter the overall thickness and roughness.	Disadvantages
(b) Liquid-based coating approach	
(1) Dip coating	**(2) ESD**
Material is fully immersed into the liquid, allowing the particles to attach on the surface, and is guided by gravitational force, surface tension, and viscous drag in the upward direction as the material immerses into it.[112,113]	Combination of manual spray-coating approach under the influence of an electric field.[105,109]

(*Continued*)

Table 2. (*Continued*)

(b) Liquid-based coating approach	
- Rheological properties:[114–116] Surface tension, viscosity, and density. - Deposition factors: Rate of advancement and withdrawal of the material. Number of immersion cycles, immersion time in every cycle.	- Substrate related factors:[110] Temperature and charge of the underlying substrate, working distance between the charged droplet and the substrate. - Other factors: Rheological properties such as particle size, density, and distribution can be controlled by the electric field intensity, conductivity, and flow rate of the mist.
- Useful in coating complex structures as it allows the particles to rearrange themselves on the material surface for the formation of uniform coating. - Coating in bulk quantities.	- Precision over coating deposition at nanoscale. - High deposition efficiency under the electric fields. This attracts the droplets toward the charged substrate and prevents overspraying to ensure minimal wastage.[111]
- Prolonged coating deposition time ~20 minutes.[117] - Large volume of liquid is necessary, which might not be cost-effective. - Risk of contamination from successive dipping or residual waste.	- Contaminants in the liquid get charged under the electric field, so purity of the coating can be compromised. - Generation of heat can be unsuitable for heat-sensitive materials, and this limits their usability. - Expensive instrumentation.
(3) Spin coating	**(4) Drop coating**
In addition to the drop-coating technique, the spin-coater is spun at a high speed and the dispensed liquid spreads on the material surface via acting centrifugal forces.[105,123]	A small droplet of liquid is dispensed onto a specific region on the material surface. The surface coating is formed by the spreading of the liquid by physical forces (interfacial tension and gravity).[118,119]

Table 2. (*Continued*)

(b) Liquid-based coating approach	
- Variable factors: Angular velocity of the spin-coater. Rheological properties of the liquid such as the concentration, viscosity and surface tension of the liquid. - External factors such as temperature, pH, and rate of evaporation of the solvent.[119,124]	Variable factors: Composition and viscosity of the liquid, external temperature for drying, drying time, and temperature of the solvent.[120,121]
- Facile and cost-efficient approach. - More time efficient than the dip-coating technique, as deposition of each layer takes ~1 minute. - Applicable for producing thin, uniform films on planar surfaces.[117]	- Facile implementation. - Cost-effective. - Favorable for small-sized materials as there is better control over thickness and the homogeneity of the coating.[122]
- Wastage of resources as dispensed liquid gets disposed of during the high spin. - High viscous liquid prevents its uniform spreading. - Not ideal for large and asymmetrical geometrical structures.[125]	- Coffee ring effect formation, which can alter the overall distribution and reproducibility of the coating. Use of anisotropic particles, superhydrophobic surfaces, and Marangoni flow of the liquid are strategies to mitigate the problem.[108,122]

4.1.1. *Spray-based coating*

The spray-coating technique has been integrated into biomedical and commercial industries for preparing functional coatings, protective coatings, and decorative coatings.[126,127] Here, an atomizer with a spray nozzle converts the liquid into mist, which is sprayed on the surface of the pretreated material, allowing precise control over the volume of dispensed liquid.[128] In the research of Ghosh et al., they spray-coated on 11 unique clinically relevant pretreated, plasma-oxidized materials to deposit poly(N-isopropylacrylamide–co-N-(3-aminopropyl)methacrylamide dihydrochloride) (p(NIPAM-co-APMA)) nGel suspension. Atomic force microscopy (AFM) characterization highlighted the presence of a homogeneous and uniform coating on the material surfaces.[102] In addition to being a time-efficient and inexpensive approach, this method is ideal for sensitive materials owing to its contactless coating approach. It is particularly useful for coating complex architectural structures and controlling the variability in thickness or the coated areas within the same material.[106,107] Several variables such as nozzle dimensions, applied pressure, speed of spraying, working distance between the nozzle and the substrate, number of coating cycles, spraying pattern; surface tension, viscosity, and density of the liquid; and pretreatment on the material and its physical attributes (roughness, wettability) are responsible for determining the thickness of the coating on the substrate.[104,105,129] The variability in functionalized surface coating presents as one of the main challenges of sensor performance.[104,108] Struchkov et al. addressed this by tailoring the abovementioned parameters within the spray-coating technique to deposit graphene oxide films with well-controlled thickness for sensing applications.[104] Some concerns associated with this technique are clogging of the nozzle, exposure to harmful mist during the spray, and its over-excessive application leading to wastage of resources, therefore potentially altering the overall thickness of the coating and forming nonuniform texture and topography.[108]

A variation to the conventional spray deposition technique is the electrostatic spray deposition (ESD) or electrospray method, which integrates the utilization of the traditional spray-coating approach under the influence of an electric field. The electric field is generated at the nozzle under high voltage power, which imparts charge to the particles, and consequently, an electrostatic field is generated between the nozzle and the grounded material. The electrospray is directed toward the heated substrate, so on deposition of the particles, the liquid solidifies by solvent evaporation, and chemical reactions/modifications occur simultaneously for the formation of the coating.[105,109,130] The influential parameters are the applied electric field, concentration of the solution, flow rate, the geometry of the tip, working distance between the nozzle tip and the substrate, temperature of the underlying substrate, and its conductivity.[131,132] Previously, Yule et al. showed the direct correlation between the spray angle and the charging voltage,[133] while Maski et al. investigated the impact of charging voltage on the spray and established that the moderate speed of application and high charged voltage were the determining factors for the deposition of the droplet.[134]

Unlike the conventional spray-coating strategy, which is handled manually, here, the particle size, density, and their distribution are finely tuned by manipulating the abovementioned parameters.[110,131] Likewise, almost 100% deposition efficiency is achieved when the "to be" coated surface is larger than the spray cloud. This is attributed to the presence of the electric fields, which attract the droplets toward the charged surface. This prevents overspraying and ensures minimal waste of the coating components.[111,135] Therefore, the ESD approach facilitates the formation of highly controlled nanoscale particle morphology and uniform coatings with high coating efficiency. Moreover, the quality of the deposited coating is high and without any defects.[136,137] The downsides of this technique include initial investment in expensive instrumentation and require training of personnel. In addition, the contaminants in the liquid also get charged under the electric field, which can interfere with the purity of the coating on the substrate. Furthermore, the heat generated might be unsuitable for some materials, therefore limiting its usability.[105]

4.1.2. *Liquid-based coating*

4.1.2.1. Dip coating

For the dip-coating approach, the "to-be coated" material is completely submerged (dipped) into the liquid, and the particles attach to its surface, followed by evaporation of the liquid and heat treatment of the coated substrate.[112,113] Theoretically, the setup is simple; however, there are multiple physical forces performing simultaneously. The viscous drag of the liquid acting in the upward direction as the material immerses into it, the downward gravitational force, along with the surface tension of the liquid, and the interplay among these different elements contribute to the formation of the surface coating.[138,139] The coating thickness is influenced by the density, surface tension, and viscosity of the liquid; rate of advancement and withdrawal of the material from the liquid, number of immersion cycles, and immersion time in every cycle; and environmental conditions such as pH, temperature, and humidity.[114,116] The deposition step is prolonged, with each layer requiring ~20 minutes to allow the adhesion of particles to the material surface. Therefore, this method can be very time-consuming, especially to fabricate a layer-by-layer assembly of coatings.[117] Highly viscous liquids require fewer dipping cycles to achieve the desired coating thickness. Thicker coatings produce agglomerations, leading to breaks or clumps. This effect is prevented by utilizing less viscous liquid and allowing multiple dipping cycles.[112,140] Zarubica demonstrated a linear relationship between the withdrawal rate of the substrate and the coating thickness,[141] with faster withdrawal rate forming thicker coatings due to higher viscous and surface forces acting on it. In the work of Xia et al., the variability in the packing density of poly(N-isopropylacrylamide-co-styrene) (PNIPAAmSt) μGel particles was showcased, noting that the highest coating density was observed at the slowest withdrawal rate and highest suspension density. Coatings were applied as a dip coating with a varied range of 0.25–1% w/w suspension density and 1–10 μm/s withdrawal rates.[178] This process is highly advantageous for coating similar materials in large quantities and with complex architecture (holes or intricate design)[112] and gives enough time for the particles in the liquid to organize themselves and form a stable, homogenous coating. A large volume of liquid is required to coat the entire material, which might pose a limitation in the case of expensive biological particles/molecules, in addition to the risk of contamination due to successive dipping of different materials and retention of residual wastes from preceding materials.[142]

4.1.2.2. Drop coating

Contrary to the spin-coating technique (discussed in Section 4.1.2.3), which is guided by interfacial stabilization, the drop-coating method allows the physical absorption of the dispensed liquid onto the surface, governed by weak interaction forces.[118,119] Also referred to as the drop-casting method, here, a small volume of liquid in the form of a droplet is dispensed onto a localized/specific region on the substrate, while it spreads over the surface by acting physical forces such as surface tension and van der Waals forces, and the excess liquid evaporates, leaving behind the coated layer.[119,143] In this method, the properties of the particles such as chemical composition, particle charge, and degree of cross-linking (softness); inherent characteristics of the dispensed liquid such as concentration and viscosity; its interaction with the underlying substrate; and external factors such as deposition temperature and drying time are crucial determining factors on the adsorption kinetics, thickness, and uniformity of the coating.[120,121] Quint et al. integrated the drop-casting method with interfacial tension by adding ethanol to the aqueous-based PNIPAM μGel suspension. This approach facilitated the migration of the μGel particles in the suspension to the air/liquid interface and resulted in the fabrication of highly ordered, hexagonally close-packed coating.[144] Based on this observation, in another study, Minato et al. studied the effect of charge on μGel assembly.[145] To this end, they incorporated 1 mol% AA to synthesize p(NIPAM-*co*-AA) μGel suspension. The anionic-charged particles showed more mobility as they repelled each other due to electrostatic interactions, which disrupted their transfer to an adsorbent material and formed packing defects. This issue was addressed by increasing the ionic strength of the surrounding liquid in proportion to its monomeric constituents (NIPAM/AA). Consequently, this reduced the interparticle repulsive forces and formed a closely packed coating on the substrate.

This technique presents several advantages such as surface modification and implementation, which are simple, cost-effective, and easily accessible for small-scale experiments in lab settings, as they do not require any complex, expensive equipment or specialized skills. Moreover, it is favorable for smaller substrates, as it enables better control on the coating thickness and its homogeneity.[122] A major drawback is the formation of the coffee ring effect, which occurs after the solvent evaporation from the surface and can alter the overall distribution and reproducibility of the coating; however, several strategies including the use of anisotropic particles, superhydrophobic surfaces, Marangoni flow of the liquid, or the addition of acoustics are adopted to mitigate the underlined problem.[108,122] The packing density of the coating can vary from the center to the edge of the surface, as evaporation of the excess solvent is the sole driving force for drying and coating formation. This variation can be mitigated by employing multiple washing steps to remove any residual particles.[146]

4.1.2.3. Spin coating

The spin-coating technique is a facile, time-efficient, and cost-efficient approach. There are three key steps in this technique that are followed in a sequential order — dropping, spinning, and drying of the liquid. The dispensed liquid on the material is spun at a regulated high speed, resulting in its spreading radially outward and thinning across the surface by the acting centrifugal forces. As the thickness decreases, the liquid becomes more concentrated and viscous and counteracts the centrifugal force.[147] Once the liquid stops spreading radially, the volatile excess solvent is evaporated, leaving behind the dried uniform layer of desired thickness.[105,123] The spinning rate of the spin-coater (angular velocity); concentration of the solute (here, the nGel particles); viscosity and surface tension of the liquid; surface properties of the underlying substrate; and external factors such as temperature, pH, and rate of evaporation play a pivotal role in determining the thickness and roughness of the coating. This indicates that the parameters governing the coating thickness are similar to the dip-coating method, with an additional variable of the spinning rate here.[119,124] Furthermore, this technique addresses the long deposition time required in the dip-coating method, as deposition of each layer takes ~1 minute.[117] The measured thickness is directly proportional to the second power of the solution concentration and inversely proportional to the square root of the spinning speed. Therefore, following this mathematical relationship, thicker coatings are achieved at higher solute concentrations and lower spinning rates. Previously studied literature has highlighted that at lower spinning rates, the coating has a nonuniform distribution, whereas low coating efficiency occurs at higher spinning rates; therefore, a balance has to be achieved to obtain the optimal thickness and efficiency.[125,148] Approximately 95–98% of the liquid is flung/disposed of during the high spin, and only ~2–5% is effectively applied on the substrate, leading to wastage of resources (both substrate and the coating liquid).[112,149] Non-Newtonian fluid or highly viscous liquid is not ideal for this technique as it prevents them from uniformly spreading on the substrate.[124,150]

4.2. *Physicochemical characterization of micro/nanogel-based coatings*

μ/nGels exhibit marked differences in their physical characteristics in suspension and as surface coatings.[119,151,152] Keskin et al. observed a shift in the hydrodynamic radius of N-isopropymethacrylamide (P(NIPMAM)) μGels particles in suspension at 30°C (below the VPTT), decreasing from ~101 nm to ~31 nm in the surface-coated dry state. The cross-linking density/stiffness of the particles influences the particle size as highly cross-linked polymeric particles are strongly interconnected and rigid, thereby limiting their swelling/deswelling capacity (swelling ratio).[128] **Figure 5** outlines the key factors essential for applying a μ/nGel coating on the material surface along with different characterization techniques. Before biomedical applications, the applied coating is subjected to comprehensive physicochemical characterization techniques,[153] such as chemical analysis, mechanical properties, thickness of the coating, and its

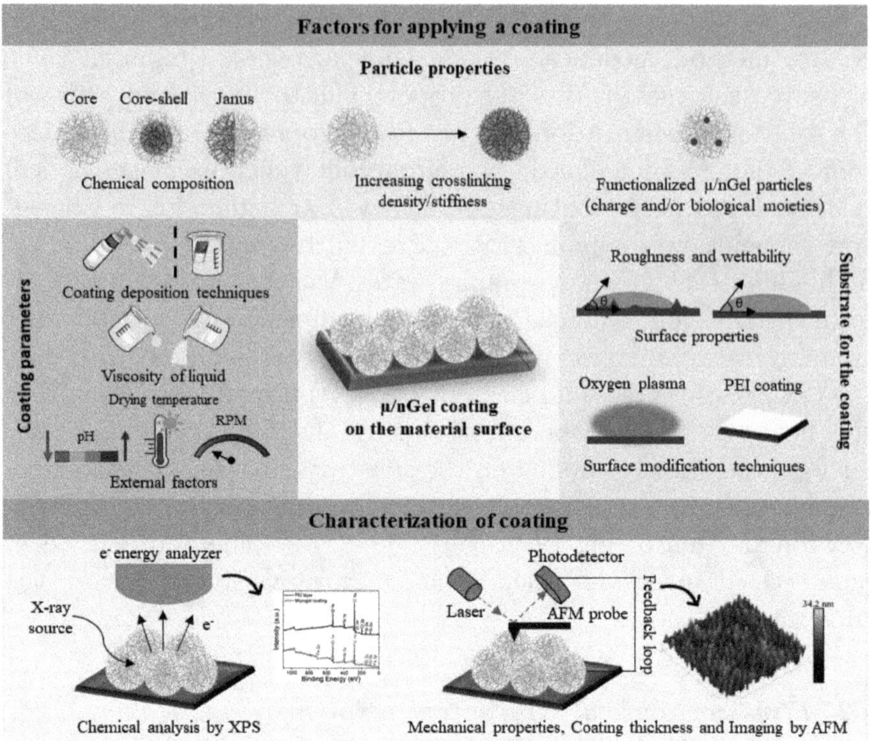

Figure 5. Schematic illustration demonstrating the key factors for applying a µ/nGel coating on the material surface along with the different techniques involved in their physicochemical characterization. Partly adapted from Chen et al.,[154] copyright 2021 American Chemical Society, and partly created by Biorender.com.

imaging to validate the presence of the optimal properties in the coating, as presented in **Figure 6**.

4.2.1. *Chemical analysis*

X-ray photoelectron spectroscopy (XPS), Fourier transform infrared (FTIR) spectroscopy, and scanning electron microscopy with energy-dispersive X-ray spectroscopy (SEM–EDX) have great significance in the domain of material science, tribology, coatings, and nanoscience.[155] XPS is based on photoionization and energy dispersive analysis of the ejected photoelectrons from the atom's inner shell.

It is also known as the fingerprint technique, which gives elemental information associated with their distinguished core-level peaks, thus allowing identification of the elemental distribution, concentration, oxidation state, and chemical bonding.[156,157] In addition, knowledge of variations in the elemental composition of the coating across its thickness, coverage of the coating, orientation of the absorbed molecules, and details of the interface is also available. FTIR works on the principle of detecting the infrared rays absorbed by the coating, providing a wealth of information regarding the coating's functional groups, molecular structure, and chemical composition.

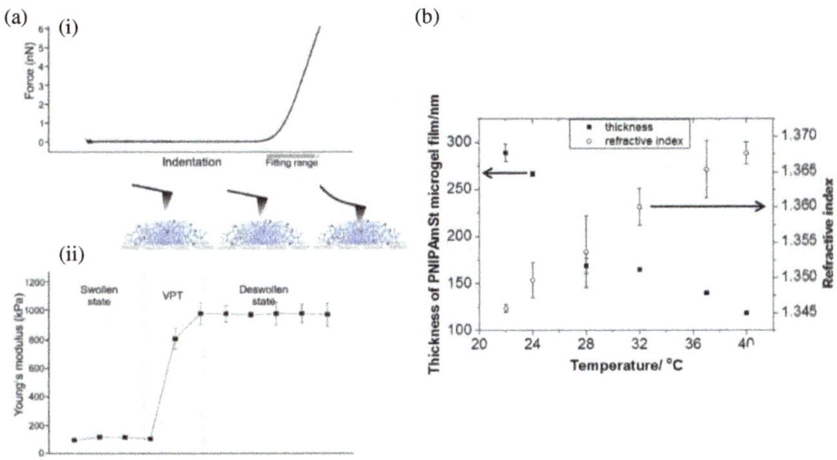

Figure 6. Coating characterization to determine the (a). (i) Schematic illustration of the force–indentation curve measured at a constant temperature. It corresponds to the three distinct indentation regimes ranging from the AFM tip before indentation and the calculated force is 0 nN to the tip indenting into the particle (as shown in the force–indentation curve). (ii) Utilizing this data and applying the Hertzian Model, the Young's modulus curve is computed at different temperatures. Adapted with permission from Wellert et al.,[152] copyright 2014 Walter de Gruyter Berlin/Boston. (**b**) Ellipsometric thickness of PNIPAAmSt µGel layer and refractive index plotted against temperature ranging from 20°C to 40°C. Adapted from Xia et al.,[146] copyright 2013 American Chemical Society. (**c**) Coating characterization to determine the imaging by IVIS and AFM to analyze the presence and hence the in vitro stability of the FITC-labeled nGel coating on exposure to PBS and FBS. Reproduced with permission from Lončarević and Čupić,[114] copyright 2023 The Authors. Advanced NanoBiomed Research published by Wiley-VCH GmbH.

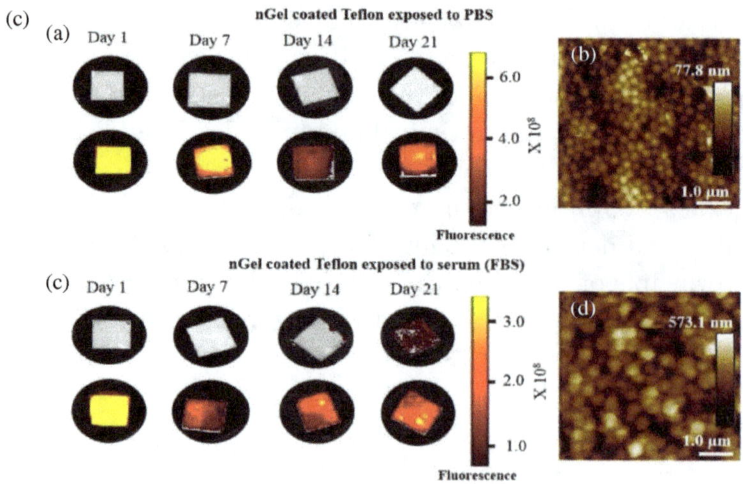

Figure 6. (*Continued*)

Depending on the penetrability of the infrared radiation, FTIR only gives an insight into the bulk material to a few micrometers.[158] Rahimi et al. compared the noncoated and coated magnetic NPs by detecting the presence of silane and NIPAM from the FTIR spectra analysis.[159] SEM–EDX is a combination of two powerful techniques; SEM scans a focused electron beam on the surface and utilizes the secondary electron beam emitted from the surface to generate topographical images of the coating. Simultaneously, EDX bombards the sample with high-energy electrons and detects the emitted X-rays to provide information on the chemical composition and distribution of the coating by elemental mapping, recognition of the functional groups, and the chemical bonds.[160]

4.2.2. *Mechanical properties*

The mechanical properties of the coating determine the threshold (point of failure) at which the coating starts to exhibit physical defects or loss of functionality. The chemical constituents of polymeric particles such as the co-monomer and cross-linker used in the synthetic reaction, along with external (environmental) factors such as temperature, pH, and ionic strength, influence the mechanical properties of the stimuli-responsive coatings.[152,161] AFM is a versatile, high-precision imaging tool used for surface (coating) characterization in both dry and wet conditions at the micro and nanoscale levels.[162] With no prior sample preparation, the probe scans over the region of interest to analyze surface topography, including the morphology of the particles, roughness, and thickness of the coated surface; coverage of the coating; and characterizes mechanical properties such as the size, stiffness, softness, and elasticity of the coating.[102,128,163] Young's modulus (or the elastic modulus) provides valuable details regarding the stiffness, structural integrity, and mechanical performance of the coating.[152,160] Here, the extent of deformation of the coating is measured under controlled mechanical force. This property plays a role in determining the influence of the coating in the biological environment, including cells and proteins used for antifouling applications. **Figure 6a**(**i**) schematically describes the force–indentation curve corresponding to the three distinct indentation regimes (ranging from before indentation to the AFM tip creating an indent) at a constant temperature. In **Figure 6a**(**ii**), the Young's modulus for thermosensitive smart polymeric particle was calculated by utilizing the data from **Figure 6a**(**i**) and applying the Hertzian model.[152] The stiffness of the particles can be easily modulated during the synthesis phase, and it is correlated with the amount of cross-linker used; that is, the highest cross-linking density indicates the highest elastic modulus.[128] An easy method to ascertain the adhesion strength of the coating is the Scratch test, where a sharp tool (such as a blade) is used to scratch across the coated surface.[161] The critical point of failure is the load or pressure at which the coating starts to delaminate or detach or fracture from the material surface. Other techniques such as tensile strength and nanoindentation are commonly employed to evaluate the adhesion strength. Stability (mechanical stability and retention of the function) is another crucial aspect to consider before any biomedical or clinical application of the coatings.[102,164] In a study by Ghosh et al., the in vitro stability of NIPAM-*co*-APMA nGel-coated materials was analyzed by exposing them to biological fluids such as phosphate buffered saline (PBS) and fetal bovine serum (FBS). In addition, the in vivo stability is assessed by surgically implanting them in mice models.[102]

4.2.3. *Coating thickness*

This thickness can be controlled by manipulating several parameters, such as the hydrodynamic size and the cross-linking density of the particles in the coating, along with the concentration of the bulk solution used.[35,146] Presurface treatments such as surface activation by plasma treatment,[165,166] surface functionalization by chemical methods such as silanization,[161,167,168] or covalent grafting,[168] or application of an intermediate layer such as PEI to impart positive charge, improve anchorage, and enhance the stability of the coating.[102,115] The employed coating deposition technique determines the coating thickness [105,112] (as described in Section 4.1), while the degree of cross-linking density influences the swelling behavior and the mechanical properties of the particles within the coating, which also contribute to its thickness.[124,152] The coating thickness is measured by various techniques, which include (i) nonoptical methods — AFM[167] and quartz crystal microbalance (QCM), (ii) noncontact (nondestructive) optical methods such as ellipsometry and optical interferometry, and (iii) contact optical methods such as focused ion beam (FIB) milling and SEM. QCM detects change in the resonant frequency upon mass addition or removal from the crystal surface and is widely applicable in the field of biomedical coatings, in antifouling surfaces/membranes, or as biosensors for detection of protein adsorption and other biomolecules.[169,170] Ellipsometry determines the thickness and optical properties (refractive index) of the coating by analyzing the change in polarization of the reflected beam from the coated surface.[171–173] Xia et al. fabricated the PNIPAAmSt layer by the spin-coating method and plotted its temperature versus thickness curve by ellipsometry.[146] The results indicated that beyond the VPTT, the thickness of the adsorbed μGel layer declined while the refractive index increased, resulting in dehydration and expulsion of water from its interface (**Figure 6b**). The film was exposed to two full thermal cycles, and the thickness was measured at 22°C and 37°C. Optical interferometry utilizes the interference patterns of light waves and variations in the optical path length to measure thickness.[174,175] These noncontact optical techniques offer practical advantages over the destructive characterization methods.[174] For instance, in SEM, a conductive layer of Au or platinum is sputter coated on the μ/nGel-coated surface to avoid charging artifacts and significantly enhance the image quality. This alters the substrate, rendering it unsuitable for further use. Similarly, in FIB milling, the thickness is analyzed by milling the coated layers with FIBs.[176]

4.2.4. *Imaging of the coating*

Visualizing the presence and uniformity of the coating is crucial as inconsistent coverage hinders biomedical and clinical applications. Imaging at the nanoscale level requires employing high-resolution visualization techniques, such as SEM, AFM, in vitro imaging system (IVIS), and fluorescence microscopy.

For imaging by SEM (briefly discussed in Sections 4.2.1 and 4.2.3 under Chemical analysis and Coating thickness), the coated material is mounted in the high-vacuum chamber. Subsequently, an electron beam is focused on the material surface. This interaction emits electrons, which are collected by the detectors. The signals produce highly detailed images to assess the micro/nanostructure, uniformity, and integrity of the functional coating.[177,178] The AFM probe rasters over the region of interest to capture its topography[162,179] (discussed in detail in Sections 4.4.2 and 4.2.3, Mechanical properties and Coating thickness, respectively). The IVIS is a noninvasive, bio-optical imaging system that is based on the principles of fluorescence and bioluminescence imaging. This sensitive technique is highly applicable for detection and real-time imaging of specific labeled biological interfaces/processes (such as coated surface, cell tracking, or efficiency of drug release at a diseased site).[180,181]

In a study by Ghosh et al., AFM was used to visualize the NIPAM-*co*-APMA nGel-coated material surfaces that were distinctly different from their corresponding bare (noncoated) surfaces. The coating was rendered traceable and detectable by incubating it in fluorescein isothiocyanate (FITC) dye and subsequently imaged by IVIS. The captured images by IVIS along with the fluorescence signals and AFM indicated the homogenous presence of the coating on Teflon surface (briefly discussed in Section 4.2.2., under Mechanical properties) even on exposure to PBS and FBS after 21 days, as shown in **Figure 6c.**[102] Fluorescence microscopy plays an instrumental role in tracking the release profiles and distribution of bioactive moieties (drugs, dyes, or markers),[182,183] and visualizing the fluorescent-labeled coatings[94,102,165] by utilizing the excitation and emission wavelength of the corresponding moieties.

5. Microgels and nanogels as platforms for biomedical applications

The unique attributes of the soft, dynamic polymeric particles play an instrumental role for a plethora of biomedical applications and make them indispensable in therapeutic and theranostic fields. On postmodification, multiple functionalities, including bioactive agents, imaging modalities, biolubricants, and biosensors can be introduced within the same particles to form a highly applicable suspension based[182,184,185] or as a surface coating[102,128,186] on the biomaterial/medical device without altering the overall (bulk) material.[102,187,188] The surface coating acts as an interface between the intrinsic properties of the bulk material and the surrounding biological system, which further aims to augment the biocompatibility, longevity, and durability with improved functions.[102] **Tables 3 and 4** present the list of μ/nGel suspension and coating-based biomedical applications along with their significant outcomes.

5.1. *Suspension-based biomedical applications*

5.1.1. *Antimicrobial delivery*

The porous architecture and high swelling property of μ and nGels facilitate the incorporation of antimicrobial agents such as

Table 3. List of μ/nGel suspensions for different biomedical applications along with their significant outcomes.

Suspension based on	**Biomedical applications**	**Significant outcomes**	**Ref**
p(NIPAM-co-DMAPMA) nGel	(a) Drug delivery (b) Antibacterial	(a) Delivery of hydrophobic antibacterial agent, Triclosan. (b) Minimum inhibitory concentration and minimal bactericidal concentration decreased by three orders of magnitude against *S. aureus* and *S. epidermidis*, in comparison to free. Triclosan	189
pNIPMAM-*co*-3-acrylamidophenylboronic acid (APBA) with dyes fluorescein-acrylate and Nile blue acrylamide nGel	Intracellular tracking and delivery	Intrinsic difference in reactivity of NIPMAM-*co*-APBA with ester and amide conjugation. FL-Ac allows for controlled delivery, while NB-AAm allows for intracellular tracking with the confined cargo.	182
p(2-(diethylamino)ethyl methacrylate)PDEAEMA nGel	Antitumor drug delivery	pH-controlled drug release with higher release of DOX in an acidic environment (unhealthy cells).	190
P(MEA-*co*-PEGA-*co*-AAEM) μGel encapsulated with γ-Fe_2O_3 magnetic NP (MNP)	Drug delivery	Drug delivery system under thermal and magnetic stimuli for a linear drug release profile.	191
p(NIPAM-*co*-DMAPMA nGel quaternized with 1-bromododecane chain	Gene delivery	(a) Form complex structures with pDNA and prevent it from serum degradation. (b) Mediates gene transfer efficiently by endosomal release of genetic material into the intracellular space. (c) Provides rational design for the improvement of nonviral **gene therapy for diseases by delivering gene vectors.**	192

(Continued)

Table 3. (*Continued*)

Suspension based on	Biomedical applications	Significant outcomes	Ref
Drug release	Gelatin hybrid μGel encapsulated with p(NIPAM-co-Aam) and iron oxide MNPs	(a) Release of bovine serum albumin (model drug) under thermo-magnetic stimuli. (b) Controlled release of the drug adjusted by parameters including quantity of p(NIPAM-*co*-AAm) and iron oxide, duration of cross-linking and alternating magnetic frequency factors such as frequency, field intensity, and exposure time.	193
Image-guided and light-triggered therapy for cancer treatment	Silica hybrid sphere functionalized by chelated Gd3+, Ce6, carbon dot, and p(NIPAM-co-MAA)	(a) Multimodal theranostic platform with chemotherapy, photothermal therapy, and photodynamic therapy. (b) Controlled release of DOX under temperature/pH stimuli along with improved signals for the image-guided therapy. (c) Biocompatible and enhanced therapeutic efficiency.	194
Drug delivery	P(NIPAm-*co*-AA) nGel	Dual-responsive (temperature and pH) nGels indicated maximum cumulative release of ibuprofen at pH 7.4 (above the pKa) and temperature 37°C.	195
(a) Sustained drug release for bactericidal properties (b) Sensing platform	P(N-isopropylacrylamide-*co*-vinylferrocene) P(NIPAM-*co*-VFc) μGels	Electrochemical switching of the particle polarity (hydrophobic/hydrophilic) enabling the uptake and release of cargo.	79

Table 4. List of μ/nGel surface coatings for different biomedical applications along with their significant outcomes.

Coating based on	Biomedical applications	Significant outcomes	Ref
P(NIPMAM) μGel	Antifouling	(a) 98% reduction of bacterial adhesion. (b) Mechanical modulation of stiffness of nGel particles and coating thickness.	128
p(sulfobetaine methacrylate-*co*-2-aminoethyl methacrylate) μGel/ polydopamine/sulfobetaine-modified polyethylene-imine μGel	(a) Antifouling (b) Anticoagulant (c) Antithrombogenic	(a) Structural integrity of coating maintained. (b) Improved hydrophilicity in PBS whole blood.	164
p(NIPAM-*co*-DMAPMA) nGel quaternized with 1-bromododecane chain	(a) Antifouling (b) Antimicrobial (c) Drug carrier for hydrophobic molecule, Triclosan	(a) One of the first to be reported, antimicrobial drug encapsulated in hydrophobic cavity of nGel network for application as a multifunctional coating; (b) 99.99% bactericidal efficiency reported, with the release of Triclosan and disruption of the bacterial cell membrane.	165
p(NIPAM) μGel	Cell manipulation tool	(a) Noninvasive cell sorting by shear force assay. (b) Tuning of cell attachment and detachment on the thermoresponsive coating. (c) Cell patterning to improve wound healing assay. All these properties were employed to modulate the cell behavior.	94

(*Continued*)

Table 3. (*Continued*)

Coating based on	**Biomedical applications**	**Significant outcomes**	**Ref**
p(NIPMAM) nGel	Immunomodulation	(a) Coating improves the biological performance of the PVDF hernia mesh. (b) Reduced immunogenic responses in the acute phase.	102
P(NIPAM) cross-linked with PEG diacrylate μGel	(a) Immunomodulation (b) Biofouling	(a) Applied as conformal coatings indicating significant reduction of in vitro fibrinogen adsorption, leukocyte adhesion, and primary human monocyte/macrophage adhesion and spreading. (b) Suppress **the expression of proinflammatory cytokines such as TNF-α, IL-1β, and MCP-1 in response to the implants. (c) Behaves as a protective coating to** prevent biofouling, leukocyte adhesion, and activation and any adverse host responses to foreign materials.	196
p(NIPAM) μGel	Tissue engineering	(a) Exploiting the temperature responsive property of NIPAM film to study the adhesion forces involved in cell attachment and spreading. (b) Noninvasive cell detachment procedure. (c) Future study includes understanding cell attachment patterns based on the uneven distribution of stiffness of the nGel particles.	197
p(NIPAM) nGel	(a) Tissue engineering (b) Wound healing	Tailored biointerface for directional migration of cells and enhanced adhesion on the underlying biomaterial.	198

(CMC-Ag) nGel-immobilized on cotton fabric	(a) Antibacterial fabrics for wound dressings. (b) Prevent chronic infections	(a) Almost >99% reduction of bacterial colonies observed at a 300 ppm Ag content. (b) The nGel-coated fabric displayed excellent antiadhesive behavior against *S. aureus* and *E. coli*.	199
Fe_3O_4-PNA μGel	Biolubrication	(a) Surface friction of nGel coating can be reversibly adjusted using NIR photothermal light. (b) Potential application as friction-modulating coatings and lubricating additives.	154
PDA and MoS_2@chitosan@ p(NIPAM) (MCN)	(a) Biolubrication (b) Bioadhesive (c) Potential applications as smart microfluidics device and robots	(a) Lubricating coating on titanium alloy used for replacement of Ti_6Al_4V artificial joint heads. (b) Fast adjustments in friction by NIR light using the photothermal core, MoS_2, and thermosensitive shell (hence, degree of hydration), NIPAM.	13
pNIPAm-co-AA nGels with MPB-capable moieties (PD-1 antigen and biocytin)	Biosensors	(a) Quick assessment of the MPB and its kinetic parameters without using any labels. (b) Comprehensive knowledge of the protein–protein interactions for treatment of diseases and understanding biochemical pathways of the body.	4

drugs,[200,201] antimicrobial peptides (AMPs)[202,203] or NPs[204,205] into their polymeric network and are ideal candidates for their controlled release profile, thereby minimizing any systemic side effects.[3,206] The hydrophobic antibacterial drug Triclosan has poor solubility and stability in an aqueous environment.[207] Zu et al. addressed this challenge by quaternizing p(NIPAM-*co*-N-[3(dimethylamino)propyl] methacrylamide) (p(NIPAM-*co*-DMAPMA)) nGel particles with 1-bromododecane, resulting in hydrophobic domains and Triclosan encapsulated within the particles.[189] The positively charged nGel particles attached to the bacterial cell wall by electrostatic interactions and disrupted the membrane, causing the release of Triclosan at the bacterial site and enhancing the antimicrobial effectiveness by 1018-fold against *Staphylococcus aureus*. These engineered particles possessed nano-injection capabilities, enabling the transport and delivery of the drug at specific sites. Incorporation of AMPs into polymeric networks is becoming a popular alternative to commercial antibiotics due to their increased stability and antimicrobial effect while concurrently reducing their cytotoxicity.[35,189] The research by Nordström et al. highlighted the applicability of μGels as potential delivery systems for AMPs and their interactions with bacterial membranes for optimum delivery of the peptides.[204] Cationic AMPs, LL-37 and DPK-060, were loaded into anionic-based p(ethyl acrylate-*co*-methacrylic acid) μGels, and the peptides were protected against enzymatic (protease) degradation even at high charge densities. The net charge of the particle remained negative even at high peptide loads, which suppressed their binding onto negatively charged bacterial membranes due to like-charge interactions. Instead, several factors such as a decrease in the peptide length and a decrease in the charge density of the μGels influenced the release of peptide for lysis of the bacterial membrane and antimicrobial efficiency.

5.1.2. *Nanocarriers for drug delivery*

Besides facilitating the delivery of hydrophilic drugs, the μ and nGels also act as vehicles for hydrophobic drugs by different mechanisms such as developing compartments such as the core–shell structures, smart particles with switchable hydrophobic properties, incorporation of hydrophobic domains into the network, and host–guest interactions.[208]

The poor water solubility, absorption, and bioavailability of the hydrophobic drug, cyclosporine A, was mitigated by using hyaluronic acid-g-PNIPAAm μGels as a potential carrier for the topical ocular delivery. The formulation was tested on rabbit eyes and demonstrated as safe and nonirritant.[209] As discussed in Section 3, polymeric gels are exploited for their stimuli-responsive property and thus applied as vehicles for drug delivery.[210] Multi-responsive β-lapachone (LP)-loaded PNIPAM-*co*-AA nGel particles triggered by temperature and pH were synthesized by Kim et al. for the controlled release of the anticancer drug β-LP.[211] The LCST shifted to a higher temperature with the addition of AA due to the acting repulsive forces between the COOH groups of AA and hydrophobic interactions with NIPAM. This property was exploited to modulate the release kinetics of the drug (by avoiding burst release) while still encapsulating the drug. The cumulative release profile indicated maximum release at pH 8 and an almost similar release profile between acidic and neutral pH. This double-controlled nGel platform showed potential for targeted drug delivery to the intestine at physiological pH and temperature via oral administration without losing the encapsulated drug along the gastrointestinal tract.

5.1.3. *Imaging agents for image-guided therapy*

Magnetic μ and nGel particles are a hybrid of stimuli-responsive polymeric gels and magnetic NP. In a study by Sun et al., the synergistic effect of magnetic resonance (MR) and computed tomography (CT) imaging was represented by nGel platforms with higher than 85% cytocompatibility.[212] Here, AuNPs bound with PEI and imaging agent gadolinium (Gd) chelates PEI–Au–Gd NPs were loaded into alginate (AG) nGels to form AG/PEI–Au–Gd NPs. The nGel-based PEI–Au–Gd NPs displayed a 12-fold increase in r_1 relaxivity, reduction in rapid clearance, and enhanced cellular uptake than the NP itself. The dual mode of MR and CT images indicated an enhanced contrast, prolonged time for tumor imaging, and good X-ray attenuation properties, respectively. In another example, Yang et al. fabricated a theranostic nanoplatform to deliver a highly efficient image-guided therapy for cancer along with chemotherapy, photothermal therapy, and photodynamic therapy.[194] The mesoporous silica was functionalized with combined chelated Gd^{3+} ions, carbon dots as a photothermal agent, chlorin e6 (Ce6) as a photosensitizer, and p(NIPAm-*co*-(methacrylic acid)) (MAA) as thermo/pH-sensitive coating on the particles. On irradiation with NIR light, the carbon dots and Ce6 were photoactivated to exterminate the tumor cells. The polymeric coating acted as a gatekeeper for the controlled release of DOX, while the chelated Gd^{3+} improved the magnetic resonance imaging (MRI) signal.

5.2. *Coating-based biomedical applications*

5.2.1. *Antifouling surface coatings*

Antifouling-based nGel coatings are noncytotoxic by nature as they utilize the inherent hydration layer and form a physical barrier. This facilitates the repulsion (bouncing off) of the incoming bacteria from the surface (without inactivating/killing them) and hinders nonspecific adsorption of proteins to the surface; consequently producing an antifouling system, one such example is nGel coatings based on NIPAM, NIPMAM, PEG, or PEO. Recently, our group has reported a study based on mechanically modulating the stiffness and thickness of p(NIPMAM) μGel coating thickness as a tool to prevent bacterial adhesion. The flow chamber experiments indicated maximum antifouling, that is, 99.7% reduction in the condition with the lowest cross-linking density and thickest μGel surface coating.[128] In this method, a simple, straightforward, and cost-effective spray-coating method was implemented to achieve/form a promising antifouling platform. Oliu et al. established an impedimetric biosensing platform [3D interdigitated electrode arrays

(3D-IDEAs)] for measuring the real-time responses of both resistant and nonresistant strains of *E. coli* bacteria to ampicillin.[213] The 3D-IDEA device was coated with a polycation-based anchoring layer, PEI, while the spatial immobilization of bacteria selectively at the bottom of the interdigitated barriers was achieved by deposition of PEI followed by microcontact printing of antifouling particles, p(NIPMAM) μGels, on top of the barriers, preventing bacteria from adhering to those regions and thereby drastically increasing the sensitivity of the sensor. Consequently, the changes in the response of the two strains of *E. coli* bacterial medium were easily detected through the impedimetric measurements. The insights from the study can open new avenues in developing bioassays for toxicity studies and the possibility to tailor antibiotic treatment for different bacterial responses for precision medicine.

5.2.2. *Bioactive surface coatings*

Antibacterial-based nGel surface coatings are perfect models to combat complications caused by biofilm formation, infection, and inflammation associated with implants and medical devices. This is mainly achieved by two strategies — contact-killing,[5,214] wherein bacteria are killed by contact with the implant surface consisting of antibacterial moieties or by the pattern of the surface and by release-based mechanisms as antimicrobial agents,[213,215] where the antibacterial molecule is preloaded into the coating and it gets activated (released) when in proximity to the bacteria.

Using the same quaternized (p(NIPAM-*co*-DMAPMA) nGel particles loaded with Triclosan) as in the work of Zu et al.,[189] Keskin et al. applied them as surface coatings and demonstrated exceptional bacterial killing efficiency against *S. aureus* up to 99.99%, thus proving an excellent approach to tackle biomaterial-associated infections (BAI).[165] In another example, Liu et al. developed a facile multifunctional antibacterial surface by one-step co-deposition of p(NIPAM-*co*-sulfobetaine methacrylate) (p(NIPAM-*co*-SBMA)) μGel combined with tannic acid (TA).[188] Here, the antifouling property of NIPAM and the zwitterion, SBMA, along with the bactericidal character of TA, was exploited to serve three goals simultaneously, namely, antifouling, contact-killing, and release mechanism of the bacteria, which increases the longevity, killing efficiency against both Gram-positive (*S. aureus*) and gram-negative (*E. coli*) bacteria, and surface regeneration properties, and hence improves the quality of the material. To elucidate, the hydration layer created by the monomers (NIPAM and SBMA) in the μGel prevented the initial bacterial adhesion to the substrate (antifouling) by over ~72 hours, while TA was responsible for killing ~93% of the "already attached" bacteria on the substrate. Furthermore, the temperature-responsive behavior of the μGel also enabled almost 90% removal of debris (dead bacteria) from the surface.

6. Conclusion and future directions

μ/nGels are robust particles that encompass the characteristics of both hydrogel and μ/NPs with the unique chemical functionality, tunable properties, and responsiveness to different stimuli. We discussed the trajectory of the smart particles spanning from its synthesis phase to applicability and provided in-depth discussion on smart particles, physicochemical characterization techniques for the particles, and coatings along with the deposition methods. Finally, we explored the paradigm shift brought about by smart polymeric particles in suspension and surface coatings in basic science, new research avenues, and enhanced applications in the biomedical domain.

Smart μ/nGel particles can be formed during the synthesis phase by modulating the polymer composition and/or by incorporating active moieties, such as dyes, organic components, bioactive and biological molecules. These particles are responsive to external cues, including chemical (pH and redox), physical (temperature and light), and biological (enzymes) stimuli. Recent scientific breakthroughs include the formation of multi-responsive smart particles by combining attributes of the μ/NPs along with dual or multiple stimuli-responsive moieties, thereby yielding even "smarter" and "sophisticated" μ/nGels. Scientists can skillfully integrate the smart polymeric particles with the inherent properties of the bulk material to form a biologically compatible interface to produce functional surface coatings. As there are no coating deposition techniques that serve as a "one size, fits all" strategy, it is imperative to determine the method only after careful consideration of the desired outcome (such as physicochemical properties of the substrate, substrate-coating combination, and coating thickness). Therefore, leveraging these functionalized smart polymeric particles empowers μ/nGel-based suspensions and surface coatings and is implemented to serve an array of biomedical applications. This encompasses protecting the biomaterials from bacterial invasion, controlled delivery of drugs in the targeted microenvironment, and rendering theranostic treatment methods such as imaging, engineering of cells and tissue, smart biosensors, self-cleaning ability, and facile tools for cell manipulation (noninvasive cell adherent/nonadherent mechanism).

Despite the excellent properties of smart μ/nGels systems, several key challenges need to be addressed to ensure their smooth transition from bench research to practical clinical applications.[3,35] One crucial aspect includes the biosafety and in vivo degradation of μ/nGel platforms.[216] At present, the majority of investigated μ/nGel systems are predominantly based on synthetic polymeric materials, which often face the challenge of lack of biocompatibility and biodegradation properties. Therefore, it is necessary to investigate μ/nGels derived from natural or bio-based polymers such as lignin, cellulose, and alginate.[217,218] It remains a popular research area that combines bio-based materials and synthetic polymers to create innovative, biodegradable, and functional μ/nGel systems with outstanding physicochemical and biocompatibility characteristics.[219] Particularly, in the field of combating biofilm-related infections, further exploration involves diversifying the antimicrobial components loaded in the μ/nGels carriers as required.[220] Some studies have found that loading antimicrobial NPs, such as iron oxide NPs,[221] silver (Ag) NPs,[222] and copper sulfide NPs,[223] into the polymeric carriers result in NP/Gel systems with remarkable antibacterial properties and biofilm elimination ability via mechanisms distinct from conventional antibiotics. Hence, the co-delivery of antibacterial NPs and antibiotics within the μ/nGel carriers could be a potential way to harness the synergistic effects of the delivered agents, achieving better therapeutic effects against biofilm infections. In addition, the targeting capability of μ/nGels to the specific in vivo areas or biofilms needs to be further explored.[199,224] Currently, the knowledge about overexpressed proteins in biofilms or diseased tissues is insufficient. Therefore, it is essential to search for and investigate novel targeted ligands to improve the targeting efficiency of μ/nGels and minimize the impact on healthy cells.

References

1. Plamper FA, Richtering W. Functional microgels and microgel systems. *Acc Chem Res* [Internet]. 2017 Feb 21;50(2):131–140. [cited 2020 Aug 2]. https://pubs.acs.org/sharingguidelines.
2. Sung B, Kim MH, Abelmann L. Magnetic microgels and nanogels: Physical mechanisms and biomedical applications. *Bioeng Transl Med* [Internet]. 2021 Jan 1;6(1):e10190. [cited 2023 Jun 30]. https://onlinelibrary.wiley.com/doi/full/10.1002/btm2.10190.
3. Kittel Y, Kuehne AJC, De Laporte L. Translating therapeutic microgels into clinical applications. *Adv Healthc Mater* [Internet]. 2022 Mar 1;11(6):2101989. [cited 2022 Jun 28]. https://onlinelibrary.wiley.com/doi/full/10.1002/adhm.202101989.
4. Yang HM, Teoh JY, Yim GH, Park Y, Kim YG, Kim J, et al. Label-free analysis of multivalent protein binding using bioresponsive nanogels and surface plasmon resonance (SPR). *ACS Appl Mater Interfaces* [Internet]. 2020 Feb 5;12(5):5413–5419. [cited 2023 Oct 6]. https://pubs.acs.org/doi/full/10.1021/acsami.9b17328.
5. Sójka O, Keskin D, van der Mei HC, van Rijn P, Gagliano MC. Nanogel-based coating as an alternative strategy for biofilm control in drinking water distribution systems. *Biofouling* [Internet]. 2023;39:121–134. [cited 2023 May 1]. https://www.tandfonline.com/doi/abs/10.1080/08927014.2023.2190023.

6. Soni KS, Desale SS, Bronich TK. Nanogels: An overview of properties, biomedical applications and obstacles to clinical translation. *J Control Release* [Internet]. 2016 Oct 10;240:109. [cited 2023 Jun 21]. /pmc/articles/PMC4862943/.
7. Bilardo R, Traldi F, Brennan CH, Resmini M. The role of crosslinker content of positively charged NIPAM nanogels on the in vivo toxicity in zebrafish. *Pharmaceutics* [Internet]. 2023 Jul 1;15(7):1900. [cited 2024 May 16]. https://www.mdpi.com/1999-4923/15/7/1900/htm.
8. Das SS, Bharadwaj P, Bilal M, Barani M, Rahdar A, Taboada P, et al. Stimuli-responsive polymeric nanocarriers for drug delivery, imaging, and theragnosis. *Polymers* [Internet]. 2020 Jun 22;12(6):1397. [cited 2024 May 16]. https://www.mdpi.com/2073-4360/12/6/1397/htm.
9. de Lima CSA, Balogh TS, Varca JPRO, Varca GHC, Lugão AB, Camacho-Cruz LA, et al. An updated review of macro, micro, and nanostructured hydrogels for biomedical and pharmaceutical applications. *Pharmaceutics* [Internet]. 2020 Oct 15;12(10):970. [cited 2024 Feb 23]. https://www.mdpi.com/1999-4923/12/10/970/htm.
10. Yin Y, Hu B, Yuan X, Cai L, Gao H, Yang Q. Nanogel: A versatile nano-delivery system for biomedical applications. *Pharmaceutics* [Internet]. 2020 Mar 1;12(3):290. [cited 2022 Jun 28]. /pmc/articles/PMC7151186/.
11. Stevens CA, Kaur K, Klok HA. Self-assembly of protein-polymer conjugates for drug delivery. *Adv Drug Deliv Rev.* 2021 Jul 1; 174:447–460.
12. Bergman MJ, Gnan N, Obiols-Rabasa M, Meijer JM, Rovigatti L, Zaccarelli E, et al. A new look at effective interactions between microgel particles. *Nat Commun* [Internet]. 2018 Dec 1;9(1):5039. [cited 2022 Oct 14]. /pmc/articles/PMC6262015/.
13. Chen Z, Liu Y, Sun Y, Liu G, Zhou F, Liu W. Mussel-inspired adhesive friction-controlled lubricating coating for titanium alloy used in artificial joint replacement. *Appl Surf Sci.* 2023 Jul 30; 626:157265.
14. Wu W, Zhou S. Hybrid micro-/nanogels for optical sensing and intracellular imaging. *Nano Rev* [Internet]. 2010 Jan 1;1(1):5730. [cited 2021 Jan 13]. https://www.tandfonline.com/doi/full/10.3402/nano.v1i0.5730.

15. Mergel O, Schneider S, Tiwari R, Kühn PT, Keskin D, Stuart MCA, et al. Cargo shuttling by electrochemical switching of core-shell microgels obtained by a facile one-shot polymerization. *Chem Sci* [Internet]. 2019 Feb 6;10(6):1844–1856. [cited 2021 Mar 23]. https://hdl.handle.net/21.11102/b794f67f-caf8-11e8-a934-e41f1366df48.
16. Razavi A, Rutsch M, Wismath S, Kupnik M, von Klitzing R, Rahimzadeh A. Frequency-dependent ultrasonic stimulation of poly(N-isopropylacrylamide) microgels in water. *Gels* [Internet]. 2022 Oct 1;8(10):628. [cited 2024 May 16]. https://www.mdpi.com/2310-2861/8/10/628/htm.
17. Qi H, Yang L, Shan P, Zhu S, Ding H, Xue S, et al. The stability maintenance of protein drugs in organic coatings based on nanogels. *Pharmaceutics* [Internet]. 2020 Feb 1;12(2):115. [cited 2024 Apr 4]. https://www.mdpi.com/1999-4923/12/2/115/htm.
18. Chen Z, Zhao Y, Zhu X, Möller M. Formation of monodisperse polymer@SiO_2 core–shell nanoparticles via polymerization in emulsions stabilized by amphiphilic silica precursor polymers: HLB dictates the reaction mechanism and particle size. *Macromolecules* [Internet]. 2019 Aug 13;52(15):5670–5678. [cited 2024 Apr 4]. https://pubs.acs.org/doi/full/10.1021/acs.macromol.9b00841.
19. Takeuchi S, Cesari A, Soma S, Suzuki Y, Casulli MA, Sato K, et al. Preparation of ultrasmall cyclodextrin nanogels by an inverse emulsion method using a cationic surfactant. *Chem Commun (Camb)* [Internet]. 2023 Mar 30;59(27):4071–4074. [cited 2024 Apr 4]. https://pubs.rsc.org/en/content/articlehtml/2023/cc/d3cc00523b.
20. Jans A, Lölsberg J, Omidinia-Anarkoli A, Viermann R, Möller M, De Laporte L, et al. High-throughput production of micrometer sized double emulsions and microgel capsules in parallelized 3D printed microfluidic devices. *Polymers* [Internet]. 2019 Nov 15;11(11):1887. [cited 2024 Apr 4]. https://www.mdpi.com/2073-4360/11/11/1887/htm.
21. Zhang B, Wang X, Zhu A, Ma K, Lv Y, Wang X, et al. Enzyme-initiated reversible addition-fragmentation chain transfer polymerization. *Macromolecules* [Internet]. 2015 Nov 10;48(21):7792–802. [cited 2024 Apr 4]. https://pubs.acs.org/doi/full/10.1021/acs.macromol.5b01893.

22. Hamzah YB, Hashim S, Rahman WAWA. Synthesis of polymeric nano/microgels: A review. *J Polym Res* [Internet]. 2017 Aug 3; 24(9):1–19. [cited 2023 Jun 15]. https://link.springer.com/article/10.1007/s10965-017-1281-9.
23. Ahmad Z, Salman S, Khan SA, Amin A, Rahman ZU, Al-Ghamdi YO, et al. Versatility of hydrogels: From synthetic strategies, classification, and properties to biomedical applications. *Gels* [Internet]. 2022 Mar 7;8(3):167. [cited 2024 Jun 20]. https://www.mdpi.com/2310-2861/8/3/167/htm.
24. Pinelli F, Ferracin F, Perale G, Rossi F. Synthesis and applications of nanogels via covalent cross-linking strategies. *Adv Chem Eng.* 2023 Jan 1;62:35–58.
25. Zhang M, Choi W, Kim M, Choi J, Zang X, Ren Y, et al. Recent advances in environmentally friendly dual-crosslinking polymer networks. *Angew Chemie Int Ed* [Internet]. 2024 Jun 10;63(24):e202318035. [cited 2024 Jun 20]. https://onlinelibrary.wiley.com/doi/full/10.1002/anie.202318035.
26. Yang J, Chen Y, Zhao L, Zhang J, Luo H. Constructions and properties of physically cross-linked hydrogels based on natural polymers. *Polym Rev* [Internet]. 2023;63(3):574–612. [cited 2024 Jun 20]. https://www.tandfonline.com/doi/abs/10.1080/15583724.2022.2137525.
27. Mavila S, Eivgi O, Berkovich I, Lemcoff NG. Intramolecular cross-linking methodologies for the synthesis of polymer nanoparticles. *Chem Rev* [Internet]. 2016 Feb 10;116(3):878–961. [cited 2024 Jun 20]. https://pubs.acs.org/doi/full/10.1021/acs.chemrev.5b00290.
28. Hu W, Wang Z, Xiao Y, Zhang S, Wang J. Advances in crosslinking strategies of biomedical hydrogels. *Biomater Sci* [Internet]. 2019 Feb 26;7(3):843–855. [cited 2024 Jun 20]. https://pubs.rsc.org/en/content/articlehtml/2019/bm/c8bm01246f.
29. Vancoillie G, Frank D, Hoogenboom R. Thermoresponsive poly(oligo ethylene glycol acrylates). *Prog Polym Sci.* 2014 Jun 1;39(6):1074–1095.
30. Liu P, Pearce CM, Anastasiadi RM, Resmini M, Castilla AM. Covalently crosslinked nanogels: An NMR study of the effect of monomer reactivity on composition and structure. *Polymers* [Internet]. 2019 Feb 18;11(2):353. [cited 2024 Apr 4]. https://www.mdpi.com/2073-4360/11/2/353/htm.

31. Kala S, Agarwal A, Sogan N, Naik SN, Nagpal BN, Patanjali PK, et al. Chitosan-acrylate nanogel for durable anti mosquito finishing of cotton fabric and its dermal toxicity profiling on Swiss albino mice. *Colloids Surf B Biointerfaces.* 2019 Sep 1;181:789–797.
32. Radwan RR, Ali HE. Radiation-synthesis of chitosan/poly (acrylic acid) nanogel for improving the antitumor potential of Rutin in hepatocellular carcinoma. *Drug Deliv Transl Res* [Internet]. 2021 Feb 1;11(1):261–278. [cited 2024 Apr 4]. https://link.springer.com/article/10.1007/s13346-020-00792-7.
33. Duan QY, Zhu YX, Jia HR, Wang SH, Wu FG. Nanogels: Synthesis, properties, and recent biomedical applications. *Prog Mater Sci.* 2023 Oct 1;139:101167.
34. Salek P, Trousil J, Novackova J, Hromadkova J, Mahun A, Kobera L. Poly[2-(dimethylamino)ethyl methacrylate-co-ethylene dimethacrylate]nanogel by dispersion polymerization for inhibition of pathogenic bacteria. *RSC Adv* [Internet]. 2021 Oct 10;11(53):33461. [cited 2024 Apr 4]. /pmc/articles/PMC9042309/.
35. Keskin D, Zu G, Forson AM, Tromp L, Sjollema J, van Rijn P. Nanogels: A novel approach in antimicrobial delivery systems and antimicrobial coatings. *Bioact Mater* [Internet]. 2021 Oct 1;6(10):3634–3657. [cited 2021 Apr 8]. https://linkinghub.elsevier.com/retrieve/pii/S2452199X21001055.
36. Allouche J. Synthesis of organic and bioorganic nanoparticles: An overview of the preparation methods. In: Fadeel B, Pietroiusti A, Shvedova A, editors. *Nanomater A Danger or a Promise? A Chem Biol Perspect.* London: Springer; 2013 Nov 1; 27–74. [cited 2024 Apr 4]. https://link.springer.com/chapter/10.1007/978-1-4471-4213-3-2.
37. Capek I. Preparation of polymer-based nanomaterials. *Nanocomposite Struct Dispersions.* New York: Wiley; 2019. p. 175–265.
38. Liu G, An Z. Frontiers in the design and synthesis of advanced nanogels for nanomedicine. *Polym Chem* [Internet]. 2014 Feb 4;5(5):1559–1565. [cited 2023 Jun 15]. https://pubs.rsc.org/en/content/articlehtml/2014/py/c3py01502e.
39. Mauri E, Giannitelli SM, Trombetta M, Rainer A. Synthesis of nanogels: Current trends and future outlook. *Gels* [Internet]. 2021 Mar 29;7(2):36. [cited 2023 Jun 15]. https://www.mdpi.com/2310-2861/7/2/36/htm.

40. Vignali VS, Miranda B, Lodoso-Torrecilla I, Van Nisselroy CAJ, Hoogenberg BJ, Dantuma S, et al. Biocatalytically induced surface modification of the tobacco mosaic virus and the bacteriophage M13. *Chem Commun* [Internet]. 2018 Dec 18;55(1):51–54. [cited 2024 May 16]. https://pubs.rsc.org/en/content/articlehtml/2019/cc/c8cc08042a.
41. Matyjaszewski K. Atom transfer radical polymerization (ATRP): Current status and future perspectives. *Macromolecules* [Internet]. 2012 May 22;45(10):4015–4039. [cited 2024 Apr 4]. https://pubs.acs.org/doi/full/10.1021/ma3001719.
42. Jung KO, Siegwart DJ, Lee H-il, Sherwood G, Peteanu L, Hollinger JO, et al. Biodegradable nanogels prepared by atom transfer radical polymerization as potential drug delivery carriers: Synthesis, biodegradation, in vitro release, and bioconjugation. *J Am Chem Soc* [Internet]. 2007 May 9;129(18):5939–5945. [cited 2023 Oct 9]. https://pubmed.ncbi.nlm.nih.gov/17439215/.
43. Wang X, Guo X, Cohen Stuart MA, Wang J, Ding P. Cationic nanogels enable gold nanoparticle immobilization and regulated catalytic activity. *Polymers* [Internet]. 2023 Apr 19;15(8):1935. [cited 2024 Apr 4]. https://www.mdpi.com/2073-4360/15/8/1935/htm.
44. Giannitelli SM, Limiti E, Mozetic P, Pinelli F, Han X, Abbruzzese F, et al. Droplet-based microfluidic synthesis of nanogels for controlled drug delivery: Tailoring nanomaterial properties via pneumatically actuated flow-focusing junction. *Nanoscale* [Internet]. 2022 Jul 29;14(31):11415–11428. [cited 2023 Jun 20]. https://pubmed.ncbi.nlm.nih.gov/35903969/.
45. Moad G. RAFT (reversible addition–fragmentation chain transfer) crosslinking (co)polymerization of multi-olefinic monomers to form polymer networks. *Polym Int* [Internet]. 2015 Jan 1;64(1):15–24. [cited 2024 Apr 4]. https://onlinelibrary.wiley.com/doi/full/10.1002/pi.4767.
46. Liu Y, Jiang X. Why microfluidics? Merits and trends in chemical synthesis. *Lab Chip* [Internet]. 2017 Nov 21;17(23):3960–3978. [cited 2024 Apr 4]. https://pubs.rsc.org/en/content/articlehtml/2017/lc/c7lc00627f.
47. Liu D, Cito S, Zhang Y, Wang C-F, Sikanen TM, Santos HA, et al. A versatile and robust microfluidic platform toward high throughput synthesis of homogeneous nanoparticles with tunable

properties. *Adv Mater* [Internet]. 2015 Apr 1;27(14):2298–2304. [cited 2024 Apr 4]. https://onlinelibrary.wiley.com/doi/full/10.1002/adma.201405408.

48. Araújo F, Shrestha N, Gomes MJ, Herranz-Blanco B, Liu D, Hirvonen JJ, et al. In vivo dual-delivery of glucagon like peptide-1 (GLP-1) and dipeptidyl peptidase-4 (DPP4) inhibitor through composites prepared by microfluidics for diabetes therapy. *Nanoscale* [Internet]. 2016 May 19;8(20):10706–10713. [cited 2024 Apr 4]. https://pubs.rsc.org/en/content/articlehtml/2016/nr/c6nr00294c.
49. Saravanakumar SM, Cicek PV. Microfluidic mixing: A physics-oriented review. *Micromachines* [Internet]. 2023 Sep 25;14(10):1827. [cited 2024 Apr 4]. https://www.mdpi.com/2072-666X/14/10/1827/htm.
50. Niculescu AG, Mihaiescu DE, Grumezescu AM. A review of microfluidic experimental designs for nanoparticle synthesis. *Int J Mol Sci* [Internet]. 2022 Aug 1;23(15):8293. [cited 2024 Apr 8]. /pmc/articles/PMC9368202/.
51. B Guerzoni LP, Rose JC, Gehlen DB, Jans A, Haraszti T, Wessling M, et al. Cell encapsulation in soft, anisometric poly(ethylene) glycol microgels using a novel radical-free microfluidic system. *Small* [Internet]. 2019 May 1;15(20):1900692. [cited 2024 May 16]. https://onlinelibrary.wiley.com/doi/full/10.1002/smll.201900692.
52. De Crozals G, Bonnet R, Farre C, Chaix C. Nanoparticles with multiple properties for biomedical applications: A strategic guide. *Nano Today*. 2016 Aug 1;11(4):435–463.
53. Kumar P, Wasim L, Chopra M, Chhikara A. Co-delivery of vorinostat and etoposide via disulfide cross-linked biodegradable polymeric nanogels: Synthesis, characterization, biodegradation, and anticancer activity. *AAPS PharmSciTech* [Internet]. 2018 Feb 1;19(2):634–647. [cited 2024 May 16]. https://link.springer.com/article/10.1208/s12249-017-0863-5.
54. Muir VG, Burdick JA. Chemically modified biopolymers for the formation of biomedical hydrogels. *Chem Rev* [Internet]. 2021 Sep 22;121(18):10908–10949. [cited 2024 May 16]. https://pubs.acs.org/doi/full/10.1021/acs.chemrev.0c00923.
55. Fu GD, Jiang H, Yao F, Xu LQ, Ling J, Kang ET. Preparation of fluorescent organometallic porphyrin complex nanogels of

controlled molecular structure via reverse-emulsion click chemistry. *Macromol Rapid Commun* [Internet]. 2012 Sep 26;33(18):1523–1527. [cited 2024 May 16]. https://onlinelibrary.wiley.com/doi/full/10.1002/marc.201200337.

56. Singh N, Lyon LA. Synthesis of multifunctional nanogels using a protected macromonomer approach. *Colloid Polym Sci* [Internet]. 2008 Aug;286(8–9):1061–1069. [cited 2024 May 16]. https://pubmed.ncbi.nlm.nih.gov/19079744/.
57. Saraswathy M, Stansbury JW, Nair DP. Thiol-functionalized nanogels as reactive plasticizers for crosslinked polymer networks. *J Mech Behav Biomed Mater* [Internet]. 2017 Oct 1;74:296–303. [cited 2024 May 16]. https://pubmed.ncbi.nlm.nih.gov/28648989/.
58. Hatton FL, Lovett JR, Armes SP. Synthesis of well-defined epoxy-functional spherical nanoparticles by RAFT aqueous emulsion polymerization. *Polym Chem* [Internet]. 2017 Aug 22;8(33):4856–4868. [cited 2024 May 16]. https://pubs.rsc.org/en/content/articlehtml/2017/py/c7py01107e.
59. Chen Y, Tezcan O, Li D, Beztsinna N, Lou B, Etrych T, et al. Overcoming multidrug resistance using folate receptor-targeted and pH-responsive polymeric nanogels containing covalently entrapped doxorubicin. *Nanoscale* [Internet]. 2017 Jul 27;9(29):10404–10419. [cited 2024 May 16]. https://pubs.rsc.org/en/content/articlehtml/2017/nr/c7nr03592f.
60. Wei X, Xiong H, He S, Wang Y, Zhou D, Jing X, et al. A facile way to prepare functionalized dextran nanogels for conjugation of hemoglobin. *Colloids Surf B Biointerfaces*. 2017 Jul 1;155:440–448.
61. Mast J, Verleysen E, Hodoroaba VD, Kaegi R. Characterization of nanomaterials by transmission electron microscopy: Measurement procedures. In: Friedrichs S, Linsinger T, editors. *Characterization of Nanoparticles: Measurement and Processing of Nanopartices*. 1st ed. Amsterdam: Elsevier; 2020 Jan 1; p. 29–48.
62. Zu G, Meijer M, Mergel O, Zhang H, van Rijn P. 3D-Printable hierarchical nanogel-GelMA composite hydrogel system. *Polym* [Internet]. 2021 Jul 29;13(15):2508. [cited 2022 Apr 21]. https://www.mdpi.com/2073-4360/13/15/2508/htm.
63. Nizardo NM, Alimin DF, Lestari MLAD. Synthesis and characterization of dual-responsive poly(N-vinylcaprolactam-co-N-methylolacrylamide) nanogels. *Des Monomers Polym* [Internet].

2022 Dec 31;25(1):155–164. [cited 2024 Apr 8]. https://www.tandfonline.com/doi/abs/10.1080/15685551.2022.2086412.

64. Varghese S, Chaudhary JP, Ghoroi C. One-step dry synthesis of an iron based nano-biocomposite for controlled release of drugs. *RSC Adv* [Internet]. 2020 Apr 1;10(23):13394–133404. [cited 2024 Apr 8]. https://pubs.rsc.org/en/content/articlehtml/2020/ra/d0ra01133a.
65. Danaei M, Dehghankhold M, Ataei S, Hasanzadeh Davarani F, Javanmard R, Dokhani A, et al. Impact of particle size and polydispersity index on the clinical applications of lipidic nanocarrier systems. *Pharmaceutics* [Internet]. 2018 May 18;10(2):57. [cited 2024 Apr 8]. /pmc/articles/PMC6027495/.
66. Masarudin MJ, Cutts SM, Evison BJ, Phillips DR, Pigram PJ. Factors determining the stability, size distribution, and cellular accumulation of small, monodisperse chitosan nanoparticles as candidate vectors for anticancer drug delivery: Application to the passive encapsulation of [14C]-doxorubicin. *Nanotechnol Sci Appl* [Internet]. 2015 Dec 11;8:67. [cited 2024 Apr 8]. /pmc/articles/PMC4686320/.
67. Varenne F, Coty JB, Botton J, Legrand FX, Hillaireau H, Barratt G, et al. Evaluation of zeta potential of nanomaterials by electrophoretic light scattering: Fast field reversal versus slow field reversal modes. *Talanta*. 2019 Dec 1;205:120062.
68. Bhaladhare S, Bhattacharjee S. Chemical, physical, and biological stimuli-responsive nanogels for biomedical applications (mechanisms, concepts, and advancements): A review. *Int J Biol Macromol.* 2023 Jan 31;226:535–553.
69. Cao Z, Zhou X, Wang G. Selective release of hydrophobic and hydrophilic cargos from multi-stimuli-responsive nanogels. *ACS Appl Mater Interfaces* [Internet]. 2016 Oct 26;8(42):28888–288896. [cited 2024 Apr 4]. https://pubs.acs.org/doi/full/10.1021/acsami.6b10360.
70. Chen S, Yuan S, Bian Q, Wu B. NIR light, pH, and redox-triple responsive nanogels for controlled release. *Soft Matter [Internet].* 2023 Sep 13;19(35):6754–6760. [cited 2024 Apr 4]. https://pubs.rsc.org/en/content/articlehtml/2023/sm/d3sm00667k.
71. Lekjinda K, Sunintaboon P, Jakaew P, Jearanaiwitayakul T, Ubol S, Jenjob R, et al. Facile synthesis of multi-stimuli responsive

biocompatible trimethyl chitosan-based nanogel as potential nanocarrier for photothermal chemotherapy. *Sustain Chem Pharm.* 2023 Dec 1;36:101290.

72. Lee M-R, Baek K-H, Jin HJ, Jung Y-G, Shin I, Lee M-R, et al. Targeted enzyme-responsive drug carriers: Studies on the delivery of a combination of drugs. *Angew Chemie Int Ed* [Internet]. 2004 Mar 19;43(13):1675–1678. [cited 2024 Apr 4]. https://onlinelibrary.wiley.com/doi/full/10.1002/anie.200353204.
73. Zhang X, Li D, Huang J, Ou K, Yan B, Shi F, et al. Screening of pH-responsive long-circulating polysaccharide–drug conjugate nanocarriers for antitumor applications. *J Mater Chem B* [Internet]. 2019 Jan 2;7(2):251–264. [cited 2024 Apr 4]. https://pubs.rsc.org/en/content/articlehtml/2019/tb/c8tb02474j.
74. Rizwan M, Yahya R, Hassan A, Yar M, Azzahari AD, Selvanathan V, et al. pH Sensitive hydrogels in drug delivery: Brief history, properties, swelling, and release mechanism, material selection and applications. *Polymers (Basel)* [Internet]. 2017 Apr 12;9(4):137. [cited 2024 Apr 4]. /pmc/articles/PMC6432076/.
75. Pikabea A, Ramos J, Forcada J. Production of cationic nanogels with potential use in controlled drug delivery. *Part Part Syst Charact* [Internet]. 2014 Jan 1;31(1):101–109. [cited 2024 Apr 4]. https://onlinelibrary.wiley.com/doi/full/10.1002/ppsc.201300265.
76. Chen S, Bian Q, Wang P, Zheng X, Lv L, Dang Z, et al. Photo, pH and redox multi-responsive nanogels for drug delivery and fluorescence cell imaging. *Polym Chem* [Internet]. 2017 Oct 10;8(39):6150–6157. [cited 2024 Apr 4]. https://pubs.rsc.org/en/content/articlehtml/2017/py/c7py01424d.
77. Zhou L, Wang H, Li Y. Stimuli-responsive nanomedicines for overcoming cancer multidrug resistance. *Theranostics* [Internet]. 2018;8(4):1059. [cited 2024 Apr 4]. /pmc/articles/PMC5817110/.
78. Qian H, Wang X, Yuan K, Xie C, Wu W, Jiang X, et al. Delivery of doxorubicin in vitro and in vivo using bio-reductive cellulose nanogels. *Biomater Sci* [Internet]. 2013 Dec 23;2(2):220–232. [cited 2024 Apr 4]. https://pubs.rsc.org/en/content/articlehtml/2014/bm/c3bm60176e.
79. Mergel O, Schneider S, Tiwari R, Kühn PT, Keskin D, Stuart MCA, et al. Cargo shuttling by electrochemical switching of core–shell microgels obtained by a facile one-shot polymerization. *Chem Sci.* 2019;10(6):1844–1856.

80. Jung SH, Schneider S, Plamper F, Pich A. Responsive supramolecular microgels with redox-triggered cleavable crosslinks. *Macromolecules* [Internet]. 2020 Feb 11;53(3):1043–1053. [cited 2024 May 23]. https://pubs.acs.org/doi/full/10.1021/acs.macromol.9b01292.
81. Bae YH, Okano T, Kim SW. "On–Off" thermocontrol of solute transport. I. Temperature dependence of swelling of n-isopropylacrylamide networks modified with hydrophobic components in water. *Pharm Res An Off J Am Assoc Pharm Sci* [Internet]. 1991;8(4):531–537. [cited 2024 Apr 4]. https://link.springer.com/article/10.1023/A:1015871732706.
82. Zhang XZ, Zhuo RX, Cui JZ, Zhang JT. A novel thermo-responsive drug delivery system with positive controlled release. *Int J Pharm.* 2002 Mar 20;235(1–2):43–50.
83. Luckanagul JA, Bhuket PRN, Muangnoi C, Rojsitthisak P, Wang Q, Rojsitthisak P. Self-assembled thermoresponsive nanogel from grafted hyaluronic acid as a biocompatible delivery platform for curcumin with enhanced drug loading and biological activities. *Polymers (Basel)* [Internet]. 2021 Jan 2;13(2):1–14. [cited 2024 Apr 4]. /pmc/articles/PMC7825653/.
84. Patnaik S, Sharma AK, Garg BS, Gandhi RP, Gupta KC. Photoregulation of drug release in azo-dextran nanogels. *Int J Pharm.* 2007 Sep 5;342(1–2):184–193.
85. Abedi F, Ghandforoushan P, Adeli F, Yousefnezhad M, Mohammadi A, Moghaddam SV, et al. Development of stimuli-responsive nanogels as drug carriers and their biomedical application in 3D printing. *Mater Today Chem.* 2023 Apr 1;29:101372.
86. Schlaich C, Li M, Cheng C, Donskyi IS, Yu L, Song G, et al. Mussel-inspired polymer-based universal spray coating for surface modification: Fast fabrication of antibacterial and superhydrophobic surface coatings. *Adv Mater Interfaces* [Internet]. 2018 Mar 1;5(5):1701254. [cited 2023 Feb 6]. https://onlinelibrary.wiley.com/doi/full/10.1002/admi.201701254.
87. Aslzad S, Heydari P, Abdolahinia ED, Amiryaghoubi N, Safary A, Fathi M, et al. Chitosan/gelatin hybrid nanogel containing doxorubicin as enzyme-responsive drug delivery system for breast cancer treatment. *Colloid Polym Sci* [Internet]. 2023 Mar 1; 301(3):273–281. [cited 2024 Apr 4]. https://link.springer.com/article/10.1007/s00396-023-05066-5.

88. Li M, Zhao G, Su WK, Shuai Q. Enzyme-responsive nanoparticles for anti-tumor drug delivery. *Front Chem.* 2020 Jul 30;8:556896.
89. Zangabad PS, Mirkiani S, Shahsavari S, Masoudi B, Masroor M, Hamed H, et al. Stimulus-responsive liposomes as smart nanoplatforms for drug delivery applications. *Nanotechnol Rev* [Internet]. 2018 Feb 23;7(1):95. [cited 2024 Apr 4]. /pmc/articles/PMC5796673/.
90. Sun H, Meng F, Cheng R, Deng C, Zhong Z. Reduction-sensitive degradable micellar nanoparticles as smart and intuitive delivery systems for cancer chemotherapy. *Expert Opin Drug Deliv* [Internet]. 2013 Aug;10(8):1109–1122. [cited 2024 Apr 4]. https://www.tandfonline.com/doi/abs/10.1517/17425247.2013.783009.
91. Li H, Mergel O, Jain P, Li X, Peng H, Rahimi K, et al. Electroactive and degradable supramolecular microgels. *Soft Matter* [Internet]. 2019 Oct 30;15(42):8589–8602. [cited 2024 May 23]. https://pubs.rsc.org/en/content/articlehtml/2019/sm/c9sm01390c.
92. Rabiee N, Hajebi S, Bagherzadeh M, Ahmadi S, Rabiee M, Roghani-Mamaqani H, et al. Stimulus-responsive polymeric nanogels as smart drug delivery systems. *Acta Biomater* [Internet]. 2019 Jul 1;92:1. [cited 2021 Nov 1]. /pmc/articles/PMC6661071/.
93. Özlem Nazli K, Pester CW, Konradi A, Böker A, van Rijn P. Cross-linking density and temperature effects on the self-assembly of SiO_2 — PNIPAAm core–shell particles at interfaces. *Chem — A Eur J* [Internet]. 2013 Apr 26;19(18):5498–5498. [cited 2024 May 16]. https://onlinelibrary.wiley.com/doi/full/10.1002/chem.201390062.
94. Uhlig K, Wegener T, Hertle Y, Bookhold J, Jaeger M, Hellweg T, et al. Thermoresponsive microgel coatings as versatile functional compounds for novel cell manipulation tools. *Polymers (Basel)* [Internet]. 2018 Jun 12;10(6). [cited 2020 Aug 9]. /pmc/articles/PMC6404125/?report=abstract.
95. Jochum FD, Theato P. Temperature- and light-responsive smart polymer materials. *Chem Soc Rev* [Internet]. 2013 Aug 5;42(17):7468–7483. [cited 2024 May 16]. https://pubs.rsc.org/en/content/articlehtml/2013/cs/c2cs35191a.
96. Mauro M. Gel-based soft actuators driven by light. *J Mater Chem B* [Internet]. 2019 Jul 10;7(27):4234–4242. [cited 2024 May 16]. https://pubs.rsc.org/en/content/articlehtml/2019/tb/c8tb01893f.

97. Han K, Go D, Hoenders D, Kuehne AJC, Walther A. Switchable supracolloidal coassembly of microgels mediated by host/guest interactions. *ACS Macro Lett* [Internet]. 2017 Mar 21;6(3):310–314. [cited 2024 May 16]. https://pubs.acs.org/doi/full/10.1021/acsmacrolett.7b00053.
98. Wang Y, Luo Y, Zhao Q, Wang Z, Xu Z, Jia X. An enzyme-responsive nanogel carrier based on PAMAM dendrimers for drug delivery. *ACS Appl Mater Interfaces* [Internet]. 2016 Aug 10;8(31):19899–19906. [cited 2024 Apr 4]. https://pubs.acs.org/doi/full/10.1021/acsami.6b05567.
99. Kumar N, Singh S, Sharma P, Kumar B, Kumar A. Single-, dual-, and multi-stimuli-responsive nanogels for biomedical applications. *Gels* [Internet]. 2024 Jan 14;10(1):61. [cited 2024 Apr 4]. https://www.mdpi.com/2310-2861/10/1/61/htm.
100. Basak S. The age of multistimuli-responsive nanogels: The finest evolved nano delivery system in biomedical sciences. *Biotechnol Bioprocess Eng* [Internet]. 2020 Sep 1;25(5):655–669. [cited 2024 Apr 4]. https://link.springer.com/article/10.1007/s12257-020-0152-0.
101. Agrawal G, Schürings MP, Van Rijn P, Pich A. Formation of catalytically active gold–polymer microgel hybrids via a controlled in situ reductive process. *J Mater Chem A* [Internet]. 2013 Oct 8;1(42):13244–13251. [cited 2024 May 23]. https://pubs.rsc.org/en/content/articlehtml/2013/ta/c3ta12370g.
102. Ghosh D, Keskin D, Forson AM, Rosman CWK, Bron R, Siebenmorgen CM, et al. A universal nanogel-based coating approach for medical implant materials. *Adv NanoBiomed Res* [Internet]. 2023 Apr 19;3(7):2200141. [cited 2023 Apr 20]. https://onlinelibrary.wiley.com/doi/full/10.1002/anbr.202200141.
103. Keskin D, Mergel O, Van Der Mei HC, Busscher HJ, Van Rijn P. Inhibiting bacterial adhesion by mechanically modulated microgel coatings. *Biomacromolecules.* 2019 Jan 14;20(1):243–253.
104. Struchkov NS, Alexandrov EV, Romashkin AV, Silakov GO, Rabchinskii MK. Uniform graphene oxide films fabrication via spray-coating for sensing application. *Fullerenes, Nanotubes Carbon Nanostruct* [Internet]. 2020 Mar 3;28(3):214–220. [cited 2023 Nov 29]. https://www.tandfonline.com/doi/abs/10.1080/1536383X.2019.1686623.

105. Ahmad R, Wolfbeis OS, Hahn YB, Alshareef HN, Torsi L, Salama KN. Deposition of nanomaterials: A crucial step in biosensor fabrication. *Mater Today Commun.* 2018 Dec 1;17:289–321.
106. Shim E. Coating and laminating processes and techniques for textiles. In: Müssig J, editor. *Smart Textile Coatings and Laminates.* Woodhead Publishing Series in Textiles. Cambridge: Woodhead Publishing; 2010. p. 10–41.
107. Wu Z, Chen Y, Liu H, Hua W, Duan J, Kong L. A Review of the developments of the characteristics and mechanisms of airless spraying on complex surfaces. *Coatings* [Internet]. 2023 Dec 17;13(12):2095. [cited 2024 Apr 30]. https://www.mdpi.com/2079-6412/13/12/2095/htm.
108. Bauer M, Duerkop A, Baeumner AJ. Critical review of polymer and hydrogel deposition methods for optical and electrochemical bioanalytical sensors correlated to the sensor's applicability in real samples. *Anal Bioanal Chem* [Internet]. 2023 Jan 1;415(1):83–95. [cited 2023 Oct 16]. https://link.springer.com/article/10.1007/s00216-022-04363-2.
109. Müller V, Jobbagy M, Djurado E. Coupling sol-gel with electrospray deposition: Towards nanotextured bioactive glass coatings. *J Eur Ceram Soc.* 2021 Nov 1;41(14):7288–7300.
110. Li X, Wang C. Engineering nanostructured anodes via electrostatic spray deposition for high performance lithium ion battery application. *J Mater Chem A* [Internet]. 2012 Dec 4;1(2):165–182. [cited 2024 Apr 30]. https://pubs.rsc.org/en/content/articlehtml/2013/ta/c2ta00437b.
111. Park SH, Lei L, D'Souza D, Zipkin R, DiMartini ET, Atzampou M, et al. Efficient electrospray deposition of surfaces smaller than the spray plume. *Nat Commun* [Internet]. 2023 Dec 1;14(1):4896. [cited 2024 Apr 30]. /pmc/articles/PMC10425365/.
112. Ulyashin AG, Hadjadj A, Butt MA. Thin-film coating methods: A successful marriage of high-quality and cost-effectiveness — A brief exploration. *Coatings* [Internet]. 2022 Aug 4;12(8):1115. [cited 2024 May 1]. https://www.mdpi.com/2079-6412/12/8/1115/htm.
113. Scherino L, Giaquinto M, Micco A, Aliberti A, Bobeico E, La Ferrara V, et al. A time-efficient dip coating technique for the deposition of microgels onto the optical fiber tip. *Fibers* [Internet]. 2018 Sep 28;6(4):72. [cited 2024 May 1]. https://www.mdpi.com/2079-6439/6/4/72/htm.

114. Lončarević D, Čupić Ž. The perspective of using nanocatalysts in the environmental requirements and energy needs of industry. In: Kuila T, editor. *Ind Appl Nanomater*. Cambridge: Elsevier; 2019. p. 91–122.
115. Schmidt S, Hellweg T, von Klitzing R. Packing density control in P(NIPAM-co-AAc) microgel monolayers: Effect of surface charge, pH, and preparation technique. *Langmuir*. 2008 Nov;24(21): 12595–12602.
116. Kaur H, Bulasara VK, Gupta RK. Influence of pH and temperature of dip-coating solution on the properties of cellulose acetate-ceramic composite membrane for ultrafiltration. *Carbohydr Polym*. 2018 Sep 1;195:613–621.
117. Iqbal MH, Kerdjoudj H, Boulmedais F. Protein-based layer-by-layer films for biomedical applications. *Chem Sci* [Internet]. 2024;15(25):9408–9437. [cited 2024 Jun 3]. https://pubs.rsc.org/en/content/articlehtml/2024/sc/d3sc06549a.
118. Sigolaeva L, Pergushov D, Oelmann M, Schwarz S, Brugnoni M, Kurochkin I, et al. Surface functionalization by stimuli-sensitive microgels for effective enzyme uptake and rational design of biosensor setups. *Polymers* (*Basel*) [Internet]. 2018 Jul 19;10(7):791. [cited 2020 May 30]. http://www.mdpi.com/2073-4360/10/7/791.
119. Cutright CC, Harris JL, Ramesh S, Khan SA, Genzer J, Menegatti S, et al. Surface-bound microgels for separation, sensing, and biomedical applications. *Adv Funct Mater* [Internet]. 2021 Nov 1;31(47):2104164. [cited 2023 Jul 17]. https://onlinelibrary.wiley.com/doi/full/10.1002/adfm.202104164.
120. Karg M, Pich A, Hellweg T, Hoare T, Lyon LA, Crassous JJ, et al. Nanogels and microgels: From model colloids to applications, recent developments, and future trends. *Langmuir*. 2019 May 14;35(19):6231–6255.
121. Takizawa M, Sazuka Y, Horigome K, Sakurai Y, Matsui S, Minato H, et al. Self-Organization of soft hydrogel microspheres during the evaporation of aqueous droplets. *Langmuir* [Internet]. 2018 Apr 17;34(15):4515–4525. [cited 2024 May 22]. https://pubs.acs.org/doi/full/10.1021/acs.langmuir.8b00230.
122. Kaliyaraj Selva Kumar A, Zhang Y, Li D, Compton RG. A mini-review: How reliable is the drop casting technique? *Electrochem Commun*. 2020 Dec 1;121:106867.

123. Nandy S, Chae KH. Chemical synthesis of ferrite thin films. *Ferrite Nanostructured Magn Mater Technol Appl*; 2023. p. 309–334.
124. Mavukkandy MO, McBride SA, Warsinger DM, Dizge N, Hasan SW, Arafat HA. Thin film deposition techniques for polymeric membranes – A review. *J Memb Sci.* 2020;610:118258.
125. Le Roux JD, Paul DR. Preparation of composite membranes by a spin coating process. *J Memb Sci.* 1992 Nov 4;74(3):233–252.
126. Sunitha K, Vasudev H. A short note on the various thermal spray coating processes and effect of post-treatment on Ni-based coatings. *Mater Today Proc.* 2022 Jan 1;50:1452–1457.
127. Schlenoff JB. Zwitteration: Coating surfaces with zwitterionic functionality to reduce nonspecific adsorption. *Langmuir* [Internet]. 2014 Aug 19;30(32):9625–9636. [cited 2024 Apr 11]. https://pubs.acs.org/doi/full/10.1021/la500057j.
128. Keskin D, Mergel O, van der Mei HC, Busscher HJ, van Rijn P. Inhibiting bacterial adhesion by mechanically modulated microgel coatings. *Biomacromolecules* [Internet]. 2019 Jan 14;20(1):243–253. [cited 2019 Nov 17]. https://pubs.acs.org/doi/10.1021/acs.biomac.8b01378.
129. Nishinaka M, Jinno H, Jimbo Y, Lee S, Wang J, Lee W, et al. High-transconductance organic electrochemical transistor fabricated on ultrathin films using spray coating. *Small Struct* [Internet]. 2021 Mar 1;2(3):2000088. [cited 2023 Nov 29]. https://onlinelibrary.wiley.com/doi/full/10.1002/sstr.202000088.
130. Drevet R, Benhayoune H. Advanced biomaterials and coatings. *Coatings* [Internet]. 2022 Jul 7;12(7):965. [cited 2023 Oct 2]. https://www.mdpi.com/2079-6412/12/7/965/htm.
131. Boda SK, Li X, Xie J. Electrospraying an enabling technology for pharmaceutical and biomedical applications: A review. *J Aerosol Sci.* 2018 Nov 1;125:164–181.
132. Ciarleglio G, Russo T, Toto E, Santonicola MG. Fabrication of alginate/ozoile gel microspheres by electrospray process. *Gels* [Internet]. 2024 Jan 1;10(1). [cited 2024 Jun 17]. /pmc/articles/PMC10815707/.
133. Yule AJ, Shrimpton JS, Watkins AP, Balachandran W, Hu D. Electrostatically atomized hydrocarbon sprays. *Fuel.* 1995 Jul 1;74(7):1094–1103.

134. Maski D, Durairaj D. Effects of charging voltage, application speed, target height, and orientation upon charged spray deposition on leaf abaxial and adaxial surfaces. *Crop Prot.* 2010 Feb 1;29(2):134–141.
135. Zhou H, Ou M, Dong X, Zhou W, Dai S, Jia W. Spraying performance and deposition characteristics of an improved air-assisted nozzle with induction charging. *Front Plant Sci* [Internet]. 2024;15:1309088. [cited 2024 Jun 6]. /pmc/articles/PMC11010639/.
136. Zhu C, Fu Y, Yu Y. Designed nanoarchitectures by electrostatic spray deposition for energy storage. *Adv Mater* [Internet]. 2019 Jan 1;31(1):1803408. [cited 2024 Apr 30]. https://onlinelibrary.wiley.com/doi/full/10.1002/adma.201803408.
137. Grzenda MJ, Atzampou M, Samateh A, Jitianu A, Zahn JD, Singer JP. Microscale templating of materials across electrospray deposition regimes. *Coatings* [Internet]. 2023 Mar 1;13(3):599. [cited 2024 Jun 17]. https://www.mdpi.com/2079-6412/13/3/599/htm.
138. Sahoo SK, Manoharan B, Sivakumar N. Introduction: Why perovskite and perovskite solar cells? In: Wong L, Snaith H, editors. *Perovskite Photovoltaics Basic to Advanced Concepts Implement.* Amsterdam: Elsevier; 2018. p. 1–24.
139. ten Elshof JE. Chemical solution deposition techniques for epitaxial growth of complex oxides. *Ep Growth Complex Met Oxides.* 2015 Jan 1;69–93.
140. Bakhtiary-Noodeh M, Moradian S, Ranjbar Z. Improvement of the edge protection of an automotive electrocoating in presence of a prepared epoxy-amine microgel. *Prog Org Coatings.* 2017;317: 134–147.
141. Zarubica A. Modified nanostructured titania based thin films in photocatalysis: Kinetic and mechanistic approach. *React Kinet Mech Catal* [Internet]. 2015 Jun 1;115(1):159–174. [cited 2024 May 1]. https://link.springer.com/article/10.1007/s11144-014-0830-z.
142. Orthopaedic Research UK. Antibacterial coating via hybrid layer-by-layer deposition system for fighting infections in orthopaedic implants — Orthopaedic Research UK [Internet]. [cited 2023 Sep 29]. https://oruk.org/case-study/antibacterial-coating-via-hybrid-layer-by-layer-deposition-system-for-fighting-infections-in-orthopaedic-implants-2/?cn-reloaded=1.

143. Vatankhah-Varnosfaderani M, Hu X, Li Q, Adelnia H, Ina M, Sheiko SS. Universal coatings based on zwitterionic–dopamine copolymer microgels. *ACS Appl Mater Interfaces.* 2018 Jun;10(24):20869–20875.
144. Quint SB, Pacholski C. Extraordinary long range order in self-healing non-close packed 2D arrays. *Soft Matter* [Internet]. 2011 Apr 5;7(8):3735–3738. [cited 2024 May 22]. https://pubs.rsc.org/en/content/articlehtml/2011/sm/c1sm05058c.
145. Minato H, Takizawa M, Hiroshige S, Suzuki D. Effect of charge groups immobilized in hydrogel microspheres during the evaporation of aqueous sessile droplets. *Langmuir* [Internet]. 2019 Aug 13;35(32):10412–10423. [cited 2024 May 22]. https://pubs.acs.org/doi/full/10.1021/acs.langmuir.9b01933.
146. Xia Y, He X, Cao M, Chen C, Xu H, Pan F, et al. Thermoresponsive microgel films for harvesting cells and cell sheets. *Biomacromolecules* [Internet]. 2013 Oct 14;14(10):3615–3625. [cited 2024 May 3]. https://pubs.acs.org/doi/full/10.1021/bm4009765.
147. Moreira J, Vale AC, Alves NM. Spin-coated freestanding films for biomedical applications. *J Mater Chem B* [Internet]. 2021 May 12;9(18):3778–3799. [cited 2023 Dec 4]. https://pubs.rsc.org/en/content/articlehtml/2021/tb/d1tb00233c.
148. Chen BT. Investigation of the solvent-evaporation effect on spin coating of thin films. *Polym Eng Sci* [Internet]. 1983 May 1;23(7):399–403. [cited 2024 May 3]. https://onlinelibrary.wiley.com/doi/full/10.1002/pen.760230706.
149. Yilbas BS, Al-Sharafi A, Ali H. Surfaces for self-cleaning. In: Yilbas BS, Al-Sharai A, Ali H, editors. *Self-Cleaning Surfaces Water Droplet Mobil.* London: Elsevier; 2019. p. 45–98.
150. Mishra A, Bhatt N, Bajpai AK. Nanostructured superhydrophobic coatings for solar panel applications. *Nanomater Coatings Fundam Appl*; 2019. 397–424.
151. Giaquinto M, Ricciardi A, Aliberti A, Micco A, Bobeico E, Ruvo M, et al. Light-microgel interaction in resonant nanostructures. *Sci Rep* [Internet]. 2018 Jun 19;8(1):1–13. [cited 2024 May 28]. https://www.nature.com/articles/s41598-018-27197-4.
152. Wellert S, Richter M, Hellweg T, Von Klitzing R, Hertle Y. Responsive microgels at surfaces and interfaces. *Z Phys Chem* [Internet]. 2015 Aug 28;229(7–8):1225–1250. [cited 2024 Apr 15].

https://www.degruyter.com/document/doi/10.1515/zpch-2014-0568/html.
153. Walzak MJ, Davidson R, Biesinger M. The use of XPS, FTIR, SEM/EDX, contact angle, and AFM in the characterization of coatings. *J Mater Eng Perform* [Internet]. 1998;7(3):317–23. [cited 2024 Jan 26]. https://link.springer.com/article/10.1361/105994998770347747.
154. Chen Z, Feng Y, Zhao N, Liu Y, Liu G, Zhou F, et al. Near-infrared-light-modulated lubricating coating enabled by photothermal microgels. *ACS Appl Mater Interfaces* [Internet]. 2021 Oct 20;13(41):49322–49330. [cited 2023 Jul 11]. https://pubs.acs.org/doi/full/10.1021/acsami.1c14646.
155. Dias FG d. A, Veiga AG, Gomes APA d. CP, da Costa MF, Rocco MLM. Using XPS and FTIR spectroscopies to investigate polyamide 11 degradation on aging flexible risers. *Polym Degrad Stab*. 2022 Jan 1;195:109787.
156. Krishna DNG, Philip J. Review on surface-characterization applications of X-ray photoelectron spectroscopy (XPS): Recent developments and challenges. *Appl Surf Sci Adv*. 2022 Dec 1;12:100332.
157. Greczynski G, Hultman L. X-ray photoelectron spectroscopy: Towards reliable binding energy referencing. *Prog Mater Sci*. 2020 Jan 1;107:100591.
158. Sharma A, Garg T, Aman A, Panchal K, Sharma R, Kumar S, et al. Nanogel--An advanced drug delivery tool: Current and future. *Artif Cells, Nanomedicine, Biotechnol* [Internet]. 2016 Jan 1;44(1):165–177. [cited 2022 Jun 30]. https://pubmed.ncbi.nlm.nih.gov/25053442/.
159. Rahimi M, Yousef M, Cheng Y, Meletis EI, Eberhart RC, Nguyen K. Formulation and characterization of a covalently coated magnetic nanogel. *J Nanosci Nanotechnol* [Internet]. 2009 Jul;9(7):4128. [cited 2024 Jun 7]. /pmc/articles/PMC2878735/.
160. Keskin D, Mokabbar T, Pei Y, van Rijn P. The relationship between bulk silicone and benzophenone-initiated hydrogel coating properties. *Polymers* (*Basel*). 2018 May 16;10(5):534.
161. Buratti E, Sanzari I, Dinelli F, Prodromakis T, Bertoldo M. Formation and stability of smooth thin films with soft microgels made of poly(N-isopropylacrylamide) and poly(acrylic acid). *Polym* [Internet]. 2020 Nov 10;12(11):2638. [cited 2024 Apr 23]. https://www.mdpi.com/2073-4360/12/11/2638/htm.

162. Nguyen-Tri P, Ghassemi P, Carriere P, Nanda S, Assadi AA, Nguyen DD. Recent Applications of advanced atomic force microscopy in polymer science: A review. *Polymers* (*Basel*) [Internet]. 2020 May 1;12(5):1142. [cited 2024 Feb 23]. /pmc/articles/PMC7284686/.
163. Lazzeri L, Cascone MG, Narducci P, Vitiello N, D'Acunto M, Giusti P. Atomic force microscopy wear characterization of biomedical polymer coatings. *Tribotest* [Internet]. 2006 Jul 1;12(3):257–265. [cited 2024 Feb 23]. https://onlinelibrary.wiley.com/doi/full/10.1002/tt.24.
164. Yao M, Sun X, Guo Z, Zhao Z, Yan Z, Yao F, et al. Bioinspired zwitterionic microgel-based coating: Controllable microstructure, high stability, and anticoagulant properties. *Acta Biomater.* 2022 Oct 1;151:290–303.
165. Keskin D, Tromp L, Mergel O, Zu G, Warszawik E, van der Mei HC, et al. Highly efficient antimicrobial and antifouling surface coatings with triclosan-loaded nanogels. *ACS Appl Mater Interfaces* [Internet]. 2020 Dec 30;12(52):57721–57731. [cited 2021 Jan 4]. https://pubs.acs.org/doi/10.1021/acsami.0c18172.
166. Jakšić Z, Jakšić O. Biomimetic nanomembranes: An overview. *Biomimetics* [Internet]. 2020 May 29;5(2):24. [cited 2024 Jun 5]. https://www.mdpi.com/2313-7673/5/2/24/htm.
167. Sanzari I, Buratti E, Huang R, Tusan CG, Dinelli F, Evans ND, et al. Poly(N-isopropylacrylamide) based thin microgel films for use in cell culture applications. *Sci Rep* [Internet]. 2020 Dec 1;10(1):6126. [cited 2020 Dec 16]. https://pubmed.ncbi.nlm.nih.gov/32273560/.
168. Hu X, Wang T, Li F, Mao X. Surface modifications of biomaterials in different applied fields. *RSC Adv* [Internet]. 2023 Jul 7;13(30):20495–20511. [cited 2024 Apr 11]. https://pubs.rsc.org/en/content/articlehtml/2023/ra/d3ra02248j.
169. Clegg JR, Ludolph CM, Peppas NA. QCM-D assay for quantifying the swelling, biodegradation, and protein adsorption of intelligent nanogels. *J Appl Polym Sci* [Internet]. 2020 Jul 7;137(25):48655. [cited 2024 Feb 24]. /pmc/articles/PMC8562820/.
170. Santi M, Saha P, Walkowiak JJ, Rubner J, Wessling M, Pich A. In-line characterization of the temperature-responsive behavior of surface-bound microgel coatings by QCM-D: A novel strategy for protein repellence evaluation. ACS *Appl Mater Interfaces* [Internet].

2022 Mar 2;14(8):10907–10916. [cited 2024 Feb 24]. https://pubs.acs.org/doi/full/10.1021/acsami.1c21814.

171. Collins RW. Measurement technique of ellipsometry. *Springer Ser Opt Sci* [Internet]. 2018;212:19–58. [cited 2024 Feb 23]. https://link.springer.com/chapter/10.1007/978-3-319-75377-5-2.
172. Zavgorodnya O, Carmona-Moran CA, Kozlovskaya V, Liu F, Wick TM, Kharlampieva E. Temperature-responsive nanogel multilayers of poly(N-vinylcaprolactam) for topical drug delivery. *J Colloid Interface Sci.* 2017 Nov 15;506:589–602.
173. Li B, Kappl M, Han L, Cui J, Zhou F, Del Campo A, et al. Goosebumps-inspired microgel patterns with switchable adhesion and friction. *Small* [Internet]. 2019 Aug 1;15(35):1902376. [cited 2024 Apr 8]. https://onlinelibrary.wiley.com/doi/full/10.1002/smll.201902376.
174. Park J, Kim JA, Ahn H, Bae J, Jin J. A review of thickness measurements of thick transparent layers using optical interferometry [Internet]. *Int J Precis Eng Manuf.* 2019 Mar 1;20(3):463–477. [cited 2024 Feb 24]. https://link.springer.com/article/10.1007/s12541-019-00105-0.
175. Zhang P, Ma X, Zhang C, Balasuriya N, Serpe MJ, Chen H, et al. X-ray triggered color-tunable microgel-based interferometers for radiation dose sensing. *Chem Eng J.* 2023 May 15;464:142519.
176. Honda K, Takaki T, Kang D. Recent advances in electron microscopy for the diagnosis and research of glomerular diseases [Internet]. *Kidney Res Clin Pract.* 2023 Mar 1;42(2):155. [cited 2024 May 29]. /pmc/articles/PMC10085717/.
177. Winnersbach P, Hosseinnejad A, Breuer T, Fechter T, Jakob F, Schwaneberg U, et al. Endogenous nitric oxide-releasing microgel coating prevents clot formation on oxygenator fibers exposed to in vitro blood flow [Internet]. *Membranes (Basel).* 2022 Jan 1;12(1):73. [cited 2024 Apr 3]. /pmc/articles/PMC8779597/.
178. Xia Y, Tang D, Wu H, Wang X, Cao M, He H, et al. Cell attachment/detachment behavior on poly(N-isopropylacrylamide)-based microgel films: The effect of microgel structure and swelling ratio. *J Mater Sci.* 2018;53(12):8795–8806.
179. Schulte MF, Bochenek S, Brugnoni M, Scotti A, Mourran A, Richtering W. Stiffness tomography of ultra-soft nanogels by atomic force microscopy. *Angew Chemie Int Ed* [Internet]. 2021 Feb

1;60(5):2280–2287. [cited 2024 Apr 4]. https://onlinelibrary.wiley.com/doi/full/10.1002/anie.202011615.

180. Ramachandran S, Thiyagarajan S, Dhinakar Raj G, Uma A. Non-invasive in vivo imaging of fluorescence-labeled bacterial distributions in aquatic species. *Int J Vet Sci Med.* 2017 Dec 1;5(2): 187–195.
181. Karimi S, Ahl D, Vågesjö E, Holm L, Phillipson M, Jonsson H, et al. In Vivo and in vitro detection of luminescent and fluorescent *Lactobacillus reuteri* and application of red fluorescent mCherry for assessing plasmid persistence [Internet]. *PLoS One.* 2016 Mar 1; 11(3):151969. [cited 2020 Aug 5]. /pmc/articles/PMC4803345/?report=abstract.
182. Zu G, Mergel O, Ribovski L, Bron R, Zuhorn IS, Rijn P. Nanogels with selective intracellular reactivity for intracellular tracking and delivery [Internet]. *Chem – A Eur J.* 2020 Nov 26;26(66):15084–15088. [cited 2021 Feb 25]. https://onlinelibrary.wiley.com/doi/10.1002/chem.202001802.
183. González-Urías A, Manzanares-Guevara LA, Licea-Claveríe Á, Ochoa-Terán A, Licea-Navarro AF, Bernaldez-Sarabia J, et al. Stimuli responsive nanogels with intrinsic fluorescence: Promising nanovehicles for controlled drug delivery and cell internalization detection in diverse cancer cell lines. *Eur Polym J.* 2021 Feb 5;144:110200.
184. Ribovski L, de Jong E, Mergel O, Zu G, Keskin D, van Rijn P, et al. Low nanogel stiffness favors nanogel transcytosis across an in vitro blood–brain barrier. *Nanomedicine.* 2021 Jun 1;34:102377.
185. Zu G, Cao Y, Dong J, Zhou Q, van Rijn P, Liu M, et al. Development of an Aptamer-conjugated polyrotaxane-based biodegradable magnetic resonance contrast agent for tumor-targeted imaging. *ACS Appl Bio Mater.* 2019 Jan;2(1):406–416.
186. Rodriguez-Abetxuko A, Sánchez-deAlcázar D, Cortajarena AL, Beloqui A, Rodriguez-Abetxuko A, Beloqui A, et al. A versatile approach for the assembly of highly tunable biocatalytic thin films [Internet]. *Adv Mater Interfaces.* 2019 Sep 1;6(18):1900598 [cited 2023 Oct 18]. https://onlinelibrary.wiley.com/doi/full/10.1002/admi.201900598.
187. Ghosh D, Peterson BW, de Waal C, de Vries J, Kaper H, Zu G, et al. Effects of sterilization on nanogel-based universal coatings:

An essential step for clinical translation [Internet]. *Mater Des.* 2024 Feb 1;238:112689 [cited 2024 Jan 31]. https://linkinghub.elsevier.com/retrieve/pii/S0264127524000613.

188. Liu Y, Mao S, Zhu L, Chen S, Wu C. Based on tannic acid and thermoresponsive microgels design a simple and high-efficiency multifunctional antibacterial coating. *Eur Polym J.* 2021 Jun 15; 153:110498.
189. Zu G, Steinmüller M, Keskin D, van der Mei HC, Mergel O, van Rijn P. Antimicrobial nanogels with nanoinjection capabilities for delivery of the hydrophobic antibacterial agent triclosan [Internet]. *ACS Appl Polym Mater.* 2020 Dec 11;2(12):5779–5789. [cited 2021 Jan 12]. doi:10.1021/acsapm.0c01031.
190. Pikabea A, Villar-Álvarez E, Forcada J, Taboada P. pH-controlled doxorubicin delivery from PDEAEMA-based nanogels. *J Mol Liq.* 2018 Sep 15;266:321–329.
191. Yang Y, Ren Z, Li X, Yan Y, Liu J, Lian M, et al. Preparation and characterization of magnetic microgels with linear thermosensitivity over a wide temperature range [Internet]. *E-Polymers.* 2023 Jan 1;23(1):20230161 [cited 2024 May 20]. https://www.degruyter.com/document/doi/10.1515/epoly-2023-0161/html?lang=en.
192. Zhang H, Keskin D, Zuhorn IS, de Haan-Visser WH, Zu G, van Rijn P. Aliphatic quaternary ammonium functionalized nanogels for gene delivery [Internet]. *Pharmaceutics.* 2021 Nov 1;13(11), 1964 [cited 2023 May 9]. /pmc/articles/PMC8618000/.
193. Sung B, Shaffer S, Sittek M, Alboslemy T, Kim C, Kim MH. Alternating magnetic field-responsive hybrid gelatin microgels for controlled drug release [Internet]. *J Vis Exp.* 2016 Feb 13;2016(108):53680 [cited 2024 May 20]. /pmc/articles/PMC4828155/.
194. Yang D, Yang G, Gai S, He F, Lv R, Dai Y, et al. Imaging-guided and light-triggered chemo-/photodynamic/photothermal therapy based on Gd (III) chelated mesoporous silica hybrid spheres [Internet]. *ACS Biomater Sci Eng.* 2016 Nov 14;2(11):2058–2071 [cited 2024 Mar 12]. https://pubmed.ncbi.nlm.nih.gov/33440542/.
195. Jin X, Wang Q, Sun J, Panezail H, Wu X, Bai S. Dual temperature- and pH-responsive ibuprofen delivery from poly(N-isopropylacrylamide-co-acrylic acid) nanoparticles and their fractal features [Internet]. *Polym Bull.* 2017 Sep 1;74(9):3619–3638

[cited 2024 May 15]. https://link.springer.com/article/10.1007/s00289-017-1915-4.

196. Bridges AW, Singh N, Burns KL, Babensee JE, Andrew Lyon L, García AJ. Reduced acute inflammatory responses to microgel conformal coatings. *Biomaterials.* 2008 Dec 1;29(35):4605–4615.
197. Schmidt S, Zeiser M, Hellweg T, Duschl C, Fery A, Möhwald H. Adhesion and mechanical properties of PNIPAM microgel films and their potential use as switchable cell culture substrates [Internet]. *Adv Funct Mater.* 2010 Oct 8;20(19):3235–3243 [cited 2021 Mar 25]. http://doi.wiley.com/10.1002/adfm.201000730.
198. Sechi A, G Freitas JM, Wünnemann P, Töpel A, Paschoalin RT, Ullmann S, et al. Surface-grafted nanogel arrays direct cell adhesion and motility [Internet]. *Adv Mater Interfaces.* 2016 Oct 1;3(20):1600455 [cited 2023 Oct 18]. https://onlinelibrary.wiley.com/doi/full/10.1002/admi.201600455.
199. Sharma A, Verma C, Mukhopadhyay S, Gupta A, Gupta B. Silver nanoparticle-embedded nanogels for infection-resistant surfaces [Internet]. *ACS Appl Nano Mater.* 2022 Jun 24;5(6):8546–8556 [cited 2023 Oct 2]. https://pubs.acs.org/doi/full/10.1021/acsanm.2c01959.
200. Kim PH, Yim HG, Choi YJ, Kang BJ, Kim J, Kwon SM, et al. Injectable multifunctional microgel encapsulating outgrowth endothelial cells and growth factors for enhanced neovascularization [Internet]. *J Control Release.* 2014 Aug 10;187:1–13 [cited 2023 Jun 22]. https://pubmed.ncbi.nlm.nih.gov/24852096/.
201. Zhao Y, Sun W, Saldaña MDA. Nanogels of poly-N-isopropylacrylamide, poly-N,N-diethylacrylamide and acrylic acid for controlled release of thymol [Internet]. *J Polym Res.* 2018 Dec 1;25(12):1–12 [cited 2024 Mar 5]. https://link.springer.com/article/10.1007/s10965-018-1644-x.
202. Aminu N, Chan SY, Yam MF, Toh SM. A dual-action chitosan-based nanogel system of triclosan and flurbiprofen for localised treatment of periodontitis. *Int J Pharm.* 2019 Oct 30;570:118659.
203. Singh S, Datta A, Borro BC, Davoudi M, Schmidtchen A, Bhunia A, et al. Conformational aspects of high content packing of antimicrobial peptides in polymer microgels [Internet]. *ACS Appl Mater Interfaces.* 2017 Nov 22;9(46):40094–400106 [cited 2024 Mar 5]. https://pubs.acs.org/doi/full/10.1021/acsami.7b13714.

204. Nordström R, Nyström L, Andrén OCJ, Malkoch M, Umerska A, Davoudi M, et al. Membrane interactions of microgels as carriers of antimicrobial peptides [Internet]. *J Colloid Interface Sci.* 2018 Mar 1;513:141–150 [cited 2023 Jun 2]. https://pubmed.ncbi.nlm.nih.gov/29145017/.
205. Khan A, Khan TH, El-Toni AM, Aldalbahi A, Alam J, Ahamad T. In situ formation and immobilization of silver nanoparticles using thermo-responsive microgel particles and their cytotoxicity evaluation. *Mater Lett.* 2019 Jan 15;235:197–201.
206. Ballesteros CAS, Correa DS, Zucolotto V. Polycaprolactone nanofiber mats decorated with photoresponsive nanogels and silver nanoparticles: Slow release for antibacterial control. *Mater Sci Eng C.* 2020 Feb 1;107:110334.
207. Wechsler ME, Stephenson RE, Murphy AC, Oldenkamp HF, Singh A, Peppas NA. Engineered microscale hydrogels for drug delivery, cell therapy, and sequencing [Internet]. *Biomed Microdevices.* 2019 Mar 3;21(2):31 [cited 2024 Mar 5]. /pmc/articles/PMC6615733/.
208. Borro BC, Nordström R, Malmsten M. Microgels and hydrogels as delivery systems for antimicrobial peptides [Internet]. *Colloids Surf B Biointerfaces.* 2020 Mar 1;187:110835 [cited 2023 Jun 2]. https://pubmed.ncbi.nlm.nih.gov/32033885/.
209. Dave R, Randhawa G, Kim D, Simpson M, Hoare T. Microgels and nanogels for the delivery of poorly water-soluble drugs. *Mol Pharm.* 2022;19(6):1704–1721 Jun 6 [cited 2024 Mar 13]. https://pubs.acs.org/doi/full/10.1021/acs.molpharmaceut.1c00967.
210. Zhao Q, Zhang S, Wu F, Li D, Zhang X, Chen W, et al. Rational design of nanogels for overcoming the biological barriers in various administration routes. *Angew Chemie Int Ed.* 2021 Jun 25;60(27):14760–14778 [cited 2022 Jun 23]. https://onlinelibrary.wiley.com/doi/full/10.1002/anie.201911048.
211. Kim YK, Kim EJ, Lim JH, Cho HK, Hong WJ, Jeon HH, Chung BG. Dual Stimuli-Triggered Nanogels in Response to Temperature and pH Changes for Controlled Drug Release. *Nanoscale Res Lett.* 2019 Jun 5;14(1):392.
212. Sun W, Zhang J, Zhang C, Zhou Y, Zhu J, Peng C, et al. A unique nanogel-based platform for enhanced dual mode tumor MR/CT imaging [Internet]. *J Mater Chem B.* 2018 Jul 25;6(29):4835–4842

[cited 2024 Mar 12]. https://pubs.rsc.org/en/content/articlehtml/2018/tb/c8tb01044g.

213. Brosel-Oliu S, Mergel O, Uria N, Abramova N, Van Rijn P, Bratov A. 3D impedimetric sensors as a tool for monitoring bacterial response to antibiotics [Internet]. *Lab Chip*. 2019 Apr 9;19(8):1436–1447 [cited 2024 Jan 18]. https://pubs.rsc.org/en/content/articlehtml/2019/lc/c8lc01220b.
214. Keskin D, Zu G, Forson AM, Tromp L, Sjollema J, van Rijn P. Nanogels: A novel approach in antimicrobial delivery systems and antimicrobial coatings. *Bioact Mater*. 2021;6(8):2764–2775. doi:10.1016/j.bioactmat.2021.04.029.
215. Alvarez-Lorenzo C, Garcia-Gonzalez CA, Bucio E, Concheiro A. Stimuli-responsive polymers for antimicrobial therapy: Drug targeting, contact-killing surfaces and competitive release [Internet]. *Expert Opin Drug Delivery*. 2016 Aug 2;13(8):1109–1119 [cited 2023 Jul 3]. https://www.tandfonline.com/doi/abs/10.1080/17425247.2016.1178719.
216. Manimaran V, Nivetha RP, Tamilanban T, Narayanan J, Vetriselvan S, Fuloria NK, et al. Nanogels as novel drug nanocarriers for CNS drug delivery. *Front Mol Biosci*. 2023 Aug 8;10: 1232109.
217. Altuntaş E, Özkan B, Güngör S, Özsoy Y. Biopolymer-based nanogel approach in drug delivery: Basic concept and current developments [Internet]. *Pharm*. 2023 Jun 2;15(6):1644 [cited 2024 May 23]. https://www.mdpi.com/1999-4923/15/6/1644/htm.
218. Yiamsawas D, Kangwansupamonkon W, Kiatkamjornwong S. Lignin-based nanogels for the release of payloads in alkaline conditions. *Eur Polym J*. 2021 Feb 15;145:110241.
219. Mougin NC, Van Rijn P, Park H, Müller AHE, Böker A. Hybrid capsules via self-assembly of thermoresponsive and interfacially active bionanoparticle–polymer conjugates [Internet]. *Adv Funct Mater*. 2011 Jul 8;21(13):2470–2476 [cited 2024 May 23]. https://onlinelibrary.wiley.com/doi/full/10.1002/adfm.201002315.
220. Ali AA, Al Bostami RD, Al-Othman A. Nanogel-based composites for bacterial antibiofilm activity: Advances, challenges, and prospects [Internet]. *RSC Adv*. 2024 Mar 3;14(15):10546 [cited 2024 May 23]. /pmc/articles/PMC10985586/.

221. Taylor EN, Kummer KM, Durmus NG, Leuba K, Tarquinio KM, Webster TJ. Superparamagnetic iron oxide nanoparticles (SPION) for the treatment of antibiotic-resistant biofilms [Internet]. *Small.* 2012 Oct 8;8(19):3016–3027 [cited 2024 May 23]. https://onlinelibrary.wiley.com/doi/full/10.1002/smll.201200575.
222. Kalishwaralal K, BarathManiKanth S, Pandian SRK, Deepak V, Gurunathan S. Silver nanoparticles impede the biofilm formation by Pseudomonas aeruginosa and Staphylococcus epidermidis. *Colloids Surf B Biointerfaces.* 2010 Sep 1;79(2):340–344.
223. Al-Hakkani MF. Biogenic copper nanoparticles and their applications: A review [Internet]. *SN Appl Sci.* 2020 Mar 1;2(3):1–20 [cited 2024 May 23]. https://link.springer.com/article/10.1007/s42452-020-2279-1.
224. Attama AA, Nnamani PO, Onokala OB, Ugwu AA, Onugwu AL. Nanogels as target drug delivery systems in cancer therapy: A review of the last decade. *Front Pharmacol.* 2022 Sep 8;13:874510.

https://doi.org/10.1142/9789819806782_0012

Chapter 12

Engineering Approaches to Control Morphogen Gradients in Tissue Models

Sebastian Kühn, Valentina Magno, Uwe Freudenberg, Carsten Werner*

Leibniz Institute of Polymer Research Dresden, Max Bergmann Center of Biomaterials, 01069 Dresden, Germany
**Corresponding author: werner@ipfdd.de*

Controlling morphogen gradients is critical for directing tissue development and regeneration. This chapter provides a comprehensive overview of the engineering approaches applied to achieve precise regulation of these gradients in vitro. The key requirements for establishing and manipulating morphogen gradients in cell cultures and tissue models will be described, highlighting the need for spatiotemporal control. Strategies for engineering the extracellular matrix (ECM) and designing artificial signaling centers will be presented, with a special focus on the principles, applications, and technological innovations of biomaterial-based approaches. The chapter concludes with an outlook on emerging technologies and future challenges in the field, emphasizing the potential for integrating engineering tools with biological systems to better replicate developmental processes and improve tissue engineering applications.

1. Introduction

The term "morphogen" was first introduced by Alan Turing in 1952[1] and describes long-range soluble signaling molecules (proteins or small molecules) that form concentration gradients across developing tissues, guiding cellular differentiation, pattern formation, and morphogenesis. These gradients are established through the controlled production, transport, and removal of morphogens, and shape the development of organisms by providing spatial information that cells use to determine their positional identity within a developing tissue.[2–4]

A simple *source-sink* model describing the generation of morphogen gradients was presented by Francis Crick in 1970.[5] According to this model, the source — or signaling center — is typically a group of cells that produce the signaling molecules, which then diffuse through the tissue across distances that go from a few to several dozen cell diameters.[2] Sink cells distant from the source are responsible for the depletion of the morphogens, which will then form a concentration gradient that decreases with distance as the morphogens are removed at the sink. Cells at different positions across this restricted region sense different concentrations of the signaling molecules, leading to the activation of distinct genetic programs depending on the local morphogen concentration, and thus defining multiple cell fates from a single gradient.

While the source-sink model remains valid and widely used as a framework in developmental biology, research over the years has expanded our understanding of how morphogen gradients are controlled and interpreted, underscoring their complexity and dynamic nature.[6–8]

The cellular interpretation of morphogen gradients is a multifaceted process that extends beyond mere concentration levels. It encompasses the duration of exposure to the morphogen and the specific cellular context at the time of exposure. Morphogen gradients are dynamic entities, subject to modulation by a variety of factors. These include the rate at which morphogens are produced, the efficiency with which they diffuse through the cellular environment, and the role of extracellular matrix (ECM) components that have the capacity to bind and immobilize morphogens, thereby influencing their distribution.

The characteristics and behavior of these gradients can be mathematically described using several quantitative models. These include the diffusion and degradation model, which considers the spread and breakdown of morphogens; the reaction–diffusion model, accounting for interactions between morphogens and other substances; and the source-sink model, focusing on the balance between morphogen production and elimination.[6–8] Additionally, contemporary computational models play a crucial role in simulating and understanding morphogen dynamics.[9]

To quantitatively describe and compare morphogen gradients, key parameters such as the maximum concentration, the gradient's slope, and the temporal dynamics are essential. These parameters are critical in defining the gradient's shape, which in turn influences the precision of developmental processes.[2,4]

The cellular response to morphogen gradients involves a complex interplay of signaling pathways and gene regulatory networks. This intricate integration ensures that cells respond accurately to morphogen cues, thereby enabling the spatial and temporal regulation of gene expression that is vital for the correct formation of tissues and organs. The efficacy of morphogen gradients in guiding developmental processes is underpinned by feedback mechanisms that contribute to the robustness of the gradient, as well as by the inherent physical properties of the morphogens themselves, such as their stability and diffusion potential. These aspects are crucial in ensuring that the gradients maintain their functional integrity and continue to provide reliable developmental cues.[4]

A classic examples of morphogen gradient function are the Dorsal (Dl) and the Bicoid (Bcd) proteins in Drosophila melanogaster (the latter being the first molecularly identified morphogen).[10] These proteins respectively establish the dorsal–ventral and anterior–posterior axes of the embryo, culminating in the segmented body plan of the fruit fly[4] (**Figure 1a**). Another well-studied example is the vertebrate

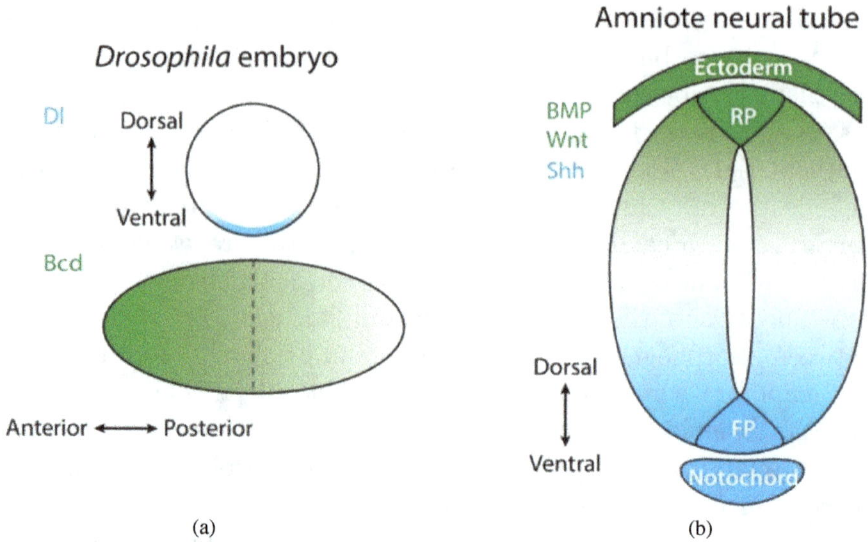

Figure 1. Examples of morphogen gradients in (a) Drosophila and (b) vertebrate neural tube model systems. RP, roof plate. FP, floor plate. Figure modified and reproduced under a Creative Commons Attribution 4.0 International License from Kircheva, Briscoe.[4]

neural tube, where opposing morphogen gradients control the pattern formation along the dorsal–ventral axis (**Figure 1b**). Here, the ectoderm and the roof plate secrete signaling molecules from the bone morphogenetic protein (BMP) and Wnt families that form a dorsal-to-ventral gradient. The notochord and the floor plate on the opposite side produce Sonic Hedgehog (Shh), which forms a ventral-to-dorsal gradient. The result is the specification of different neural progenitor domains along the dorsal–ventral axis of the neural tube.[4]

The control of morphogen gradients in cell, organoid, and tissue cultures is pivotal for understanding developmental processes and advancing tissue engineering toward bigger, more complex, and more realistic tissue and disease models. Owing to the plethora of cell and tissue culture methods and the intricate requirements of the biological systems, many different approaches have been developed to gain control over morphogen gradient formation in vitro. These include biomaterials-based approaches, microfluidic techniques (microfluidics), and engineered cells. Each approach comes with specific advantages and drawbacks, and the intricate engineering requirements can often only be met through the integration of multiple systems. The aim of this chapter is to provide a comprehensive overview of these methodologies, starting from some fundamental requirements and providing a checklist for better evaluation and comparison.

2. Gaining gradient control in vitro — engineering requirements

Engineering methods to generate and control morphogen gradients in cell and tissue cultures must fulfill several critical requirements to effectively mimic in vivo conditions and support advanced biological research.[11] First, they need to produce biologically relevant gradient profiles and concentration ranges. This means that the gradients should reflect the natural levels and spatial distributions of morphogens observed within developing tissues, ensuring that experimental outcomes are physiologically meaningful and applicable to real biological systems. Second, these methods should have the capability to establish complex gradients, encompassing the creation of nonlinear gradients, gradients along various axes, and those involving multiple morphogens. Such complexity is essential for replicating the intricate signaling environments found in vivo, where multiple factors often interact in a highly coordinated manner to influence cell fate and tissue patterning. For example, the dorsoventral patterning of the neural tube is orchestrated by opposing gradients of Shh and BMP together with members of the transforming growth factor beta (TGFβ) superfamily, while the anteroposterior patterning occurs simultaneously through opposing gradients of retinoic acid (RA) and fibroblast growth factor (FGF)/Wnt.[12] The generation of such complex gradients requires high spatiotemporal control. Spatial control entails precise positioning, high resolution, and an appropriate range to ensure that the morphogen gradients can be accurately applied to specific regions within the cell culture. High spatial resolution is particularly important for studying fine-scale developmental processes and for targeting specific cell populations within heterogeneous cultures. Temporal control is equally important, as developing tissues are highly dynamic systems.

Beyond these technical aspects, the required know-how and equipment will determine the possible applications of the different methods. Here, user-friendly interfaces, compatibility with existing laboratory equipment, and straightforward protocols that do not require extensive technical expertise can facilitate a wider adoption in various research settings.[11] Moreover, the methods for gradient control should allow for analytical access to monitor the biological response. Versatility, modularity, and tunability of the approaches further foster their utility, allowing the adaption for different experimental conditions, morphogens, and research objectives. A broader application of the different methods, especially with respect to potential commercial/medical applications, goes hand in hand with the requirement for reproducibility and robustness of the formed gradients to produce stable and repeatable outcomes across different experiments and laboratory settings. Finally, scalability and throughput are essential considerations. Following the long-cherished yet distant dream of growing whole organs in the lab, methods that can be scaled up for larger-sized tissue cultures are essential to engineer structures beyond the millimeter scale. On the other hand, miniaturization supports high-throughput approaches for applications such as drug screening and large-scale developmental biology research.

3. Morphogen gradient formation in the ECM or in polymer networks that mimic ECM functionality

3.1. *How does nature control morphogen gradients?*

In the tissue, the ECM provides mechanical, topographical, and chemical cues that mediate key signaling processes during morphogenesis.[13,14] Although the exact composition of the ECM is highly tissue specific, the main components are collagens, various non-collagenous proteins (e.g., fibronectin, elastin, and laminin), proteoglycans (e.g., perlecan, aggrecan, and syndecan), and the conjugated glycosaminoglycans (GAGs, e.g., hyaluronic acid, chondroitin/dermatan sulfate, keratin sulfate, and heparan sulfate/heparin). The proteoglycans, consisting of a central core protein and several associated side chain GAGs, form big macromolecular complexes within the ECM, or the glycocalyx of a cell surface that determine the interplay with soluble morphogens/signaling molecules, and thus gradient formation.[8,15,16] The ability of these proteoglycans to interact with these signaling proteins accordingly plays a crucial role during morphogenesis, as it not only regulates morphogen presentation and activity (e.g., receptor binding) and increases protein stability, but also mediates gradient formation by restricting diffusion.[17] In this context, especially sulfated GAGs (sGAGs) have been identified as key players to control morphogen distribution.[18] The high negative charge of heparan sulfate or heparin, introduced by their sulfate groups, facilitates electrostatic interactions with positively charged domains on the protein surface.[19,20] Although hydrophobic interactions contribute to the binding as well, the morphogen affinity depends predominantly on the density and pattern of the sulfate groups. This strong complexation in turn reduces the effective diffusion of morphogens and enables the formation of localized and stable gradients. One example is the regulation of vascular development through the different isoforms of vascular endothelial growth factor (VEGF) with adjusted affinity for sulfated sGAGs.[21] While the strong binding of the VEGF165-isoform to heparan sulfate proteoglycans (HSPGs) in the ECM contributes to the formation of steeper and shorter gradients, the absence of the heparin-binding domain in the VEGF121-isoform, and thus weaker binding, facilitates signaling further away from the source.

3.2. *Polymer networks mimicking ECM functionality*

In tissue engineering and regenerative medicine, highly hydrated polymer networks — hydrogels — are widely applied as scaffold materials to mimic key ECM functionalities for embedded cells and tissues.[22] They can be composed of isolated biopolymers, including ECM-derived components such as collagen, gelatin, laminin, or

GAGs; multi-component mixtures such as decellularized ECM (e.g., Matrigel®), synthetic polymers such as poly(hydroxyethyl methacrylate) (PHEMA), poly-(ethylene glycol) (PEG), poly(vinyl alcohol) (PVA), polyacrylamide (PAm), and polyoxazolines (POx) or combinations thereof.[23–26] The chemical versatility of hydrogels, especially of biohybrid and fully synthetic polymer networks, provides an excellent opportunity for spatiotemporal control of the extracellular environment with respect to key physicochemical signaling cues, including their mechanical properties, their capability for enzymatic remodeling, provision of cell adhesion ligands, and administration of soluble morphogens.[27] For the last point, different strategies to achieve graded morphogen distributions that can guide tissue development have been introduced. These strategies can be grouped into two main categories: (a) imprinting approaches where the morphogen gradients are covalently or non-covalently immobilized within the hydrogels, and (b) diffusion-based approaches that utilize non-covalent protein–polymer network interactions to shape the morphogen distribution as it diffuses from a source.

3.2.1. *Engineering gradients by imprinting*

Over the years, different approaches have been developed to fabricate hydrogels with chemical, mechanical, or structural gradients that are also utilized to imprint graded morphogen distributions within hydrogel scaffolds.[11] For example, controlled fluid mixing and subsequent crosslinking with biocompatible (click) chemistries can be utilized to form and lock the gradients in place upon hydrogel formation.[28,29] Moreover, light-mediated micro-patterning can be used to tether morphogens into the polymer networks after their formation.[30,31] For example, in order to study endothelial cell (EC) migration and tubule formation, agarose hydrogels were differentially exposed by a multiphoton confocal laser to immobilize VEGF gradients. Thiols on the agarose, protected by photolabile groups, were uncaged with the light exposure and allowed the Michael-type addition with maleimide-functionalized VEGF.[30] While laser-based approaches offer high spatial resolution,[32] covalently linking the protein to the polymer network or to a secondary tethering agent might alter the protein structure and thus the presentation to cell receptors. This can affect morphogen activity and therefore needs to be precisely optimized for a specific system.

Using the same fabrication techniques as above, hydrogels can be functionalized with gradients of morphogen-binding moieties that allow for subsequent non-covalent immobilization of the signaling proteins.[33] Using molecular components (or their derivatives) that mediate morphogen signaling in the ECM in vivo ensures a more natural presentation of proteins. While immobilized gradients are well defined and stable, their temporal control remains limited. Here, the use of stimuli-responsive tethering agents, such as photolabile linkers[34] or protease-cleavable peptides,[35] would allow to manipulate the gradient and introduce a dynamic component. Since the encapsulation of cells with these methods still remains challenging, they are often used to study migration[29] or invasion,[31] but are less suited to guide more complex tissue organization in 3D.

As an alternative to gradient formation directly during the hydrogel fabrication or through post-modification of the matrix, gradients can also be imprinted through the assembly of different hydrogel building blocks. For example, hydrogel particles with different nerve growth factor (NGF) loadings were arranged into a scaffold to form an NGF gradient that would promote axonal outgrowth and bridging when implanted in peripheral nerve defects.[36] Decoupling the gradient integration from the hydrogel synthesis/post-modification significantly increases the degree of freedom with respect to the scaffold design and simplifies the embedding of cells within these scaffolds.[37] Overall, a major advantage of imprinted gradients is that they are not restricted by morphogen diffusion and therefore are particularly suited to establish long-range gradients.[36] However, these gradients lack a morphogen source in the classical sense and thus are very limited with respect to temporal control.

3.2.2. *Engineering gradients by tuning diffusion*

Akin to the developing tissue, morphogen gradients in cell and tissue cultures are commonly formed through passive diffusion, where molecules move from a region of high concentration to one of lower concentration, exposing the cells along the way to this gradient.[1] While the formation of these diffusion-based gradients begins at the morphogen source (see Section 4), the polymer networks used to culture the cells play a significant role in determining how morphogens diffuse. Due to the volume exclusion effects caused by the polymer components and the presence of cells, diffusion in hydrated polymer networks (i.e., hydrogels) and cell aggregates is slower compared to diffusion in aqueous solutions,[38,39] with the rate inversely related to the density of the network and cells. Additionally, effective diffusion depends on the interaction of the morphogens with the polymer network. By adjusting these protein–polymer interactions, key parameters of the gradient — such as maximum concentration, slope, and time dynamics — can be controlled. As a somewhat evident solution, ECM constituents and their derivatives with specific morphogen binding sites are often incorporated into the hydrogel networks to complex morphogens and control their distribution. For example, sGAGs with high affinity for various morphogens can be used to stabilize and shape signaling gradients.[27,40] Chemical modifications, such as selective removal of the highly charged sulfate groups,[41,42] or varying the sGAG chain length,[43,44] allows to modify the affinity and further modulate the gradient formation. Alternatively, proteins or peptides with high morphogen affinity can fulfill this function.[45,46] Since the cells are in direct contact with the materials, the entirety of their physico- and biochemical properties will determine the response of the embedded cells, and material solutions are needed with independent control over these features. To address this challenge, we have developed a biohybrid hydrogel platform based on PEG and highly negatively charged sGAGs, where the affinity for soluble signaling molecules/ morphogens can be controlled independently from the other network properties (e.g., stiffness, degradability, cell adhesiveness, structure) following a rational holistic material design.[40,41,47,48] The protein-network affinity in these hydrogels is predominantly governed by the electrostatic interactions between the negatively charged sulfate groups of the sGAGs and positively charged domains on the morphogen surface originating from lysin or arginine amino acid side chains, and can be described by the integral charge density, which depends on the concentration of sGAGs and the local charge density on the sGAG derivatives with different sulfation patterns.

Going beyond ECM-derived biopolymers such as sGAGs, compounds that can emulate the molecular interactions with morphogens are suitable polymer network building blocks. For example, synthetic sulfated/sulfonated polymers can display similar morphogen binding like sGAGs through electrostatic and hydrophobic interactions and function in a biological context.[49,50] Owing to their higher chemical versatility compared to biopolymers, such synthetic polymers could further potentiate morphogen gradient control in hydrogels.

Over the years, a large variety of options have emerged for adapting the composition, crosslinking chemistry, and shape/format of hydrogel materials to various cell and tissue culture setups, including standard well plate-based methods or specialized systems, such as microfluidic devices. Overall, following nature's lead by gaining control over morphogen diffusion through molecular interactions with polymer matrices is essential to establish functional gradients in cell and tissue cultures. Seeking further inspiration in biology, but also exploring existing and completely new material systems with the help of tools from data science and artificial intelligence (AI), will help to foster our understanding of morphogen gradient formation and open up new possibilities to direct tissue development in vitro.

4. Engineering the morphogen source — artificial signaling centers

In traditional cell culture, morphogens are supplied through the culture medium and fail to establish functional gradients as they are homogeneously distributed. Therefore, it requires engineering solutions to achieve a directed morphogen supply from a source and emulate the function of localized morphogen-secreting cell clusters in developing tissues. This includes cellular approaches, where cells are genetically engineered to secrete specific morphogens, microfluidic systems based on reservoir compartments and perfusable microchannels, biomaterials for controlled morphogen release, or combinations thereof.[51,52]

4.1. *Cell-based approaches*

Engineering of cells to modify their genetic program offers control over morphogen secretion and renders them into artificial signaling centers, also referred to as artificial "organizer cells." Of great interest is the pioneering research conducted by the group of Sergiu Pasca, who introduced the concept of *assembloids.*[53,54] Developed in the field of human brain organogenesis, assembloids are defined as complex, 3D cell culture systems that mimic the structure and function of human tissues or organs by combining different types of organoids or tissue-specific cell types.[54] Researchers can manipulate the cellular composition and spatial arrangement of different cell types to establish controlled environments where morphogen gradients naturally form or are experimentally induced by using genetically engineered cell lines that produce specific proteins in response to an external stimulus.[55–57] These gradients influence the behavior of cells within the assembloid, directing them to form a distinct structure or to differentiate into specific cell types in a manner that mirrors natural developmental processes. In a recent example, cell engineering was applied to produce an inducible Shh-expressing human pluripotent stem cell (hPSC) line that could introduce an Shh signaling gradient into developing forebrain organoids and specify positional identity in a distance-dependent manner.[57] The inducible cells were embedded as a small cluster at one pole of a bigger hPSC spheroid, and the aggregate was differentiated into the forebrain in the presence of doxycycline to induce an asymmetric Shh signal. Immunofluorescence staining successfully demonstrated the formation of an Shh gradient, with the fluorescence intensity diminishing with increasing distance from the artificial organizer. The generated morphogen gradient was key in providing positional information for the recapitulation of in vivo-like topographies, enabling the forebrain organoids to pattern along the dorso-ventral and anterior–posterior positional axes.

Challenges of assembloid-based approaches include the limited control over the positioning of morphogen-producing cells and the lack of fine-tunability in morphogen production. Optogenetic methods represent a powerful cell-based alternative for generating morphogen gradients. Such approaches use light as an external stimulus to locally switch on specific signaling pathways, offering both high spatial precision and temporal control. For instance, the use of optogenetic Cre recombinase systems to regulate the spatial production of Shh has proven effective in creating signaling centers and guiding pattern formation in stem cell differentiation models.[58,59] While providing spatiotemporal control over morphogen release, these systems are irreversible. A more elegant approach was recently reported,[60] which showed fine regulation over both activation and inactivation of Shh production, demonstrating how concentration and duration of morphogen exposure are critical to determine and maintain the ventral identity of neural progenitors.

Nevertheless, optogenetics comes with disadvantages. First, it requires highly specialized knowledge and equipment, which can make the experimental setups more complex, expensive, and technically challenging. Second, due to the limited penetration depth of light, challenges might arise when trying to establish consistent gradients in 3D cultures. Third, both the expression of optogenetic proteins and the exposure to light can potentially alter the cell state. Considering these factors, while optogenetics provides valuable precision, it may require careful optimization and troubleshooting to effectively generate morphogen gradients in vitro.

Genetic engineering further allows to create synthetic morphogen systems via the synthetic Notch (synNotch) receptor.[61] The extracellular recognition domain of the transmembrane receptor can be swapped to bind specific ligands, whereas synthetic transcription factors can be used as the intracellular transcriptional domain to establish user-defined signaling pathways. This allows to render arbitrary molecules, such as green fluorescent protein (GFP), into morphogens and establish complex signalling gradients and feedback circuits.[62] While cell-secreted morphogen gradients from aforementioned methods can be distorted through the simultaneous expression of signaling antagonists, the choice of "inert" molecules ensures the orthogonality of the system.

Overall, the desire to define and control the genetic program of a cell has generated versatile toolsets to turn cells into customizable signaling centers. Current challenges that limit their broad application to date are the high technical expertise and experience required to generate the transgenic cell lines and the inherent variability of cell-based systems that make it difficult to establish robust gradients.

4.2. *Microfluidics*

Microfluidic tools have gained immense popularity in the fields of developmental biology, tissue engineering, and regenerative medicine for various applications ranging from cell culture, high-throughput phenotypic and genetic assays, drug screening, and scaffold synthesis.[63] In fact, many methods to establish morphogen gradients in cell and tissue culture also rely on microfluidic devices, where cells and morphogen solutions are arranged in microfabricated setups of reservoir compartments and/or perfusable channels in the size range of tens to hundreds of micrometers.[64,65]

Commonly, these devices/chips are fabricated from transparent, flexible polydimethylsiloxane (PDMS) using photolithography or soft lithography approaches.[66] As an alternative, 3D printing is used to achieve the microscale structures of the device but further enables the precise deposition of cells and polymer matrices (hydrogels) to print tissue-like structures.[65] Together with other biofabrication techniques, such as molding,[67] it allows to implement perfusable microchannels within the tissue construct and create luminal structures in an attempt to resemble the vasculature[68] or the intestinal tube.[69]

The process of gradient formation in microfluidic devices can be categorized into diffusion-based and flow-based methods.[64] In the latter, cells in the flow chamber are exposed to parallel fluid streams with different solute concentrations, resulting in a gradient that runs orthogonally to the flow direction. Due to the predictable laminar (non-turbulent) flow, mixing of the parallel streams is prevented (except for slow molecular diffusion), which ensures gradient stability. The laminar flow is achieved at low Reynolds numbers, which, for a given fluid, depend on the microchannel diameter and the velocity.[52] Continuous gradients and further modulation of the gradient shape can be achieved by pre-mixing the different streams.[70] For example, a gradient maker device was used to expose human embryonic stem cells (ESCs) to Wnt-activating gradients and induce regional differentiation that resembles the rostro-caudal organization of the neural tube development[71] (**Figure 2a**).

The fluid flow in microfluidic systems is a double-sided sword, as it allows to introduce fluid shear stress as a signaling cue that is crucial for some morphogenic processes such as vascular development, but might also lead to undesired alterations in cell behavior.[64,65]

Alternatively, gradients can be formed through diffusion from a source reservoir on one side through the incubation chamber containing the cells into the sink compartment on the other side. Source and sink can be simple reservoir compartments[72] or perfused channels that are connected to external reservoirs and pumps[12,73] (**Figure 2b**). In these diffusion-based approaches, hydrated polymer networks can be utilized to mimic the ECM, shield cells from undesired shear stress,[73] or further modulate gradient formation,[40] as also highlighted in Section 4.2. Compared to flow-based methods, where cells are exposed to the same stable gradient as long as the flow is maintained, cells experience more dynamic concentrations if the gradient is formed via diffusion.[64]

A major advantage of microfluidic approaches is the morphogen supply from an external source. This ensures a high degree of temporal control, where signals can be easily switched on and off during the culture.[74,75] Furthermore, researchers can easily switch to

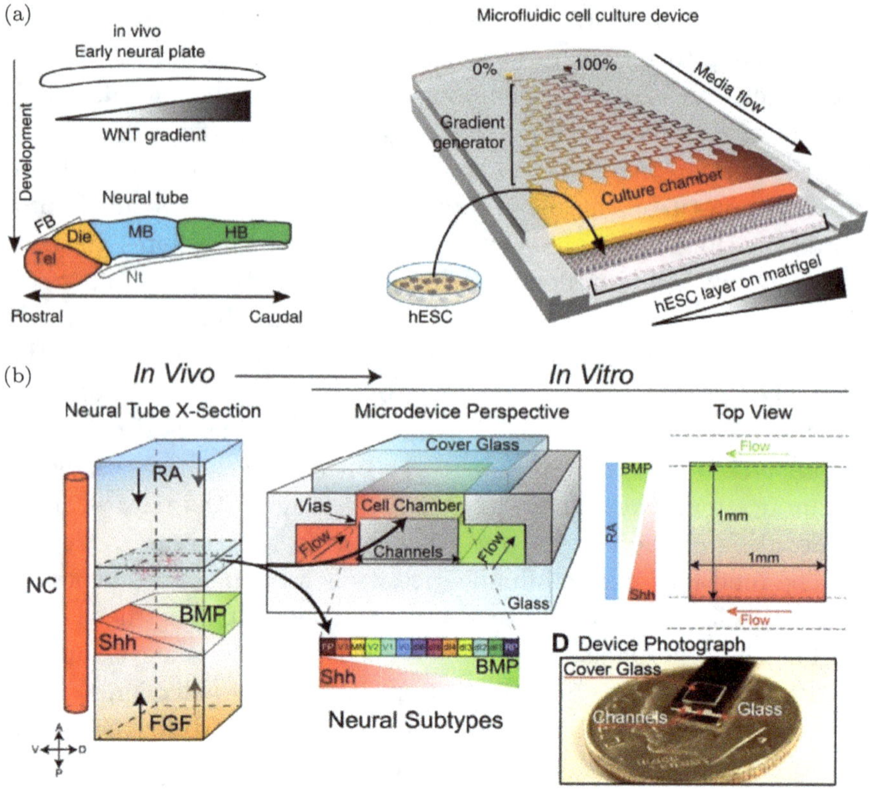

Figure 2. Examples of (a) flow-based and (b) diffusion-based microfluidic methods for morphogen gradient control. FB, forebrain. Tel, telencephalon. Die, diencephalon. MB, midbrain. HB, hindbrain. Nt/NC, notochord. Figures modified and reproduced under license from Rifes et al.[71] (License Number 5885350786100) and Demers et al.[12] (License Number 1534565).

different morphogens in the culture medium in order to create more complex gradients, such as dual,[76] opposing, or orthogonal gradients.[12] For example, this was used to recapitulate neural tube patterning through diffusion-based complex gradients of RA, FGF, BMP, and purmorphamine, which activates Shh signaling[12] (**Figure 2b**). The external supply of soluble molecules through the culture medium further simplifies the adaptation to different biological systems or enables to explore chemical gradients beyond morphogens, such as oxygen.[77]

While the versatility, precision, and robustness of microfluidic devices render them powerful tools to study the influence of asymmetric morphogen distribution on tissue development, their small size (e.g., required to maintain a laminar flow) inherently imposes several challenges that limit the upscaling of the technology to grow bigger structures. With tens to hundreds of micrometers, the channel size and thus the size of the cell and tissue cultures are limited.[65] Furthermore, fast-growing cultures such as organoids display high spatial and architectural dynamics,[78] which might be difficult to match with rigidly designed microfluidic systems. In most microfluidic setups, the whole culture or only large regions are exposed to the gradient from the outside, which restricts the development of bigger structures due to limited diffusion and also often does not resemble natural morphogenesis, limits spatial resolution, and can hamper a more complex spatial organization.[65,52]

However, the aforementioned biofabrication techniques to implement perfusable channels directly within tissue constructs offer a promising solution to overcome diffusion limits and achieve larger-scale and more functional structures.[79]

Another drawback of microfluidic approaches is the complexity of the technology.[64,65] The design and fabrication of microfluidic devices require expert knowledge, complex equipment, and can be time-consuming. Although more and more commercial solutions are developed, this currently still limits the accessibility of the technology. While the complexity and accessories, such as pumps and tubing, can impose a significant challenge on automatization and add extra costs compared to standard culture dishes and plates, the small size of the chambers facilitates parallelization and high-throughput applications.[64,65,80]

Altogether, microfluidic approaches offer powerful tools to study tissue development due to their precision and versatility, especially for small-scale structures such as cells and aggregates. Advances in biofabrication techniques, such as the printing of cells and hydrogels, offer the possibility to implement perfusable networks and overcome the diffusion limits that currently hinder the formation of large-scale and more functional tissue constructs.

4.3. *Biomaterial-based approaches*

Another method for directing morphogen supply involves morphogen-loaded biomaterials, which can be placed in various formats relative to or within the cell and tissue cultures. Over time, this field has advanced from simple techniques like morphogen-soaked paper carriers and polymer beads[81] to more sophisticated hydrogel materials.[82–84] These hydrogels offer greater versatility and provide a better interface with developing tissues in terms of mechanical properties and composition. Control over morphogen supply and distribution can be achieved at the molecular level by modulating the release and at the micro- or macroscopic level by arranging the hydrogel units. In order to control the morphogen release, the same molecular design principles as previously discussed in Section 3 for the hydrogel functionalization with morphogens can also be employed here, and will therefore only be summarized shortly. Morphogens can be covalently attached to the polymer network via degradable linker molecules that release the morphogen, for example, after photolytic[84] or proteolytic cleavage.[85] Such stimuli-responsive release introduces temporal control and the possibilities for dynamic gradient modulation; however, with the disadvantage of chemically modifying the morphogen and potentially altering the activity.

Alternatively, hydrogel functionalization with morphogen-binding moieties to control the morphogen diffusion (see Section 3.2) facilitates sustained supply from the source. Being able to tune the morphogen affinity through the polymer network design, the release kinetics can be adjusted to program the gradient formation. For example, the same sGAG-based hydrogel platform that was introduced in Section 3.2 as a tunable cell and tissue culture matrix to shape gradient formation can be loaded with morphogens and serve as a source with highly tunable morphogen release.[83,86,87] Since many morphogens show sGAG affinity, this material platform can be easily adapted to a wide range of biological systems and offers high modularity as well as the ability to create complex gradients, for example, comprising different morphogens.[88] However, biomaterial-based sources offer only a finite morphogen supply, and establishing stable gradients can be challenging due to the constantly depleting morphogen source.

Moving to the micro-/macroscopic level, the advantage of hydrogel-based morphogen sources lies in the malleability of the material, allowing researchers to process it into various shapes and formats with biofabrication techniques such as printing, molding, injection/pipetting, microfluidics, and so on.[37,89,90] This ensures the compatibility with different culture setups, such as well plates[86] and microfabricated devices,[72,91] and facilitates the broad application of hydrogel-based morphogen sources. For example, through embedding or assembly, hydrogel reservoirs can be combined with other hydrogels serving as culture matrix to form multiphasic composite materials.[37] These so-called *gel-in-gel* systems offer the possibility to control the gradient on the level of the morphogen source and through the diffusion in the cell-laden ECM-mimicking hydrogel. Printing of hydrogels and cells offers exciting possibilities to realize complex tissue structures and precisely integrate morphogen sources.[37,90] The ability to pipette and layer small volumes of hydrogels in well plates, or to produce microgels, ensures the necessary downscaling that facilitates high-throughput applications.[92] While stimuli-responsive release and programmed release kinetics govern the gradient formation once the materials are integrated, the possibilities to easily apply or remove them further enhance the temporal control.

Processing hydrogels into microgels, that is, small hydrogel microparticles/beads in the size range of natural signaling centers (e.g., single cells or cell aggregates), further expands the application of hydrogel materials as morphogen sources. For example, their small size offers high spatial resolution and allows for seamless integration into the tissue constructs through pipetting, micromanipulation, microinjection, or printing.[89] Furthermore, rendering the microgels magnetic offers remote control over their positioning.[93] This high degree of positional control could pave the way to achieve complex dual, opposing, and orthogonal gradients in tissue constructs by using multiple microgel sources. This is particularly exciting given the recent development of granular hydrogels consisting exclusively of microgels, also termed microporous annealed particles (MAPs).[89] To this day, morphogen-loaded beads/microgels are extensively used as morphogen sources,[52] not least because of the relatively little know-how that is required for their application compared to more complex microfluidic setups or the use of engineered cells, as they can be simply manually positioned in the culture setup.[83,94,95] Nevertheless, while bulk hydrogels offer a high source capacity due to their large volume, a challenge for bead-based approaches is to ensure sufficient morphogen supply from a small volume in the range of pL to nL. Overcoming this problem by using populations of microgels[96,97] comes at the cost of lower resolution and spatial precision. If single microgels are used, the material design must not only ensure high morphogen loading and tunable release, but also enable reproducible synthesis to minimize the variability between individual units with respect to size and composition/loading, the main parameters determining the morphogen release. One option we recently developed to overcome these challenges is detailed in Section 5.

In summary, morphogen-loaded hydrogels offer a straightforward, robust, and highly customizable means to integrate morphogen sources into cell and tissue cultures. Their value and performance increase with the ability to control the release through the underlying material design. Since hydrogels already serve as scaffold materials, and given the recent advances in biofabrication techniques, hydrogel-based morphogen sources emerge as powerful building blocks to guide development in tissue engineering applications.

5. Microgel units to instruct development (μGUIDEs)

Current microgel materials provide only limited control over morphogen release and gradient formation,[82,84,94,96–98] which hinders their effectiveness in modulating tissue development signals. For instance, crosslinked agarose microbeads with covalently conjugated protein-affine dyes and initially designed for protein purification, have been widely applied for generating morphogen gradients in embryos and organoids grown in vitro.[94,97,99] Such microbeads offer no affinity tuning and are heterogeneous in size and morphology, thereby greatly reducing precision and resolution. More recently, photo-cleavable DNA microbeads have been proposed for the UV-triggered release of covalently immobilized morphogens.[84] Despite offering spatial control and on-demand release, these microbeads provide only burst release, limiting sustained delivery over time. Furthermore, the covalent tethering of the morphogens to the DNA network may alter their natural conformation and affect how they interact with cell receptors, thus reducing their biological activity. Leveraging the electrostatic interactions between positively charged domains of morphogens and negatively charged sGAGs of proteoglycans, which typically occur in the ECM, microgels using covalently conjugated heparin have been developed to provide sustained morphogen release,[82] but remained limited in controlling the gradient profiles.

In order to address these challenges and the intricate signalling requirements of developing tissues, we have introduced a library of μGUIDEs (i.e., μGel Units to Instruct Development) with unprecedented modularity, source capacity, and computationally guided gradient control[83] (**Figure 3**).

Based on star-shaped PEG (starPEG)-sGAG hydrogels and on a rational material design concept,[48] our μGUIDEs allow a multiparametric adjustment of their charge characteristics that is key to programmable morphogen release. Specifically, both the integral volume charge density (P1) and the local charge density (P2) can be customized by varying the volume concentration and sulfation patterns of the sGAGs present in the hydrogel network (**Figure 3a**). With this effective and decoupled fine-tuning, single μGUIDEs, serving as point sources, can program precisely localized short-term (hours) and long-term (days) morphogen gradients. Thanks to the rational material design of the μGUIDEs, mathematical modeling can be applied to obtain a quantitative prediction and customization of the dynamic molecular signals. Considering the critical importance of vascularization for in vitro tissue functionality and scalability, we have validated our system with a focus on vascular

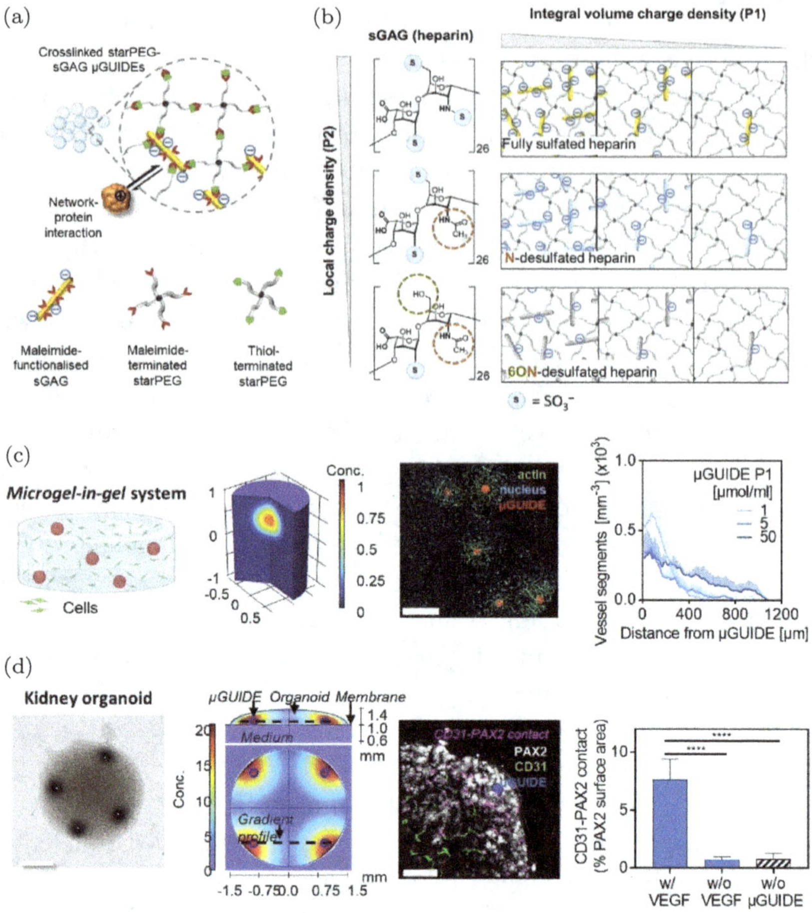

Figure 3. Microgel units to instruct development (μGUIDEs) (**a**) Schematic of the μGUIDE composition. (**b**) Control of μGUIDEs charge characteristics through the systematic variation of integral charge density (P1, sGAG volume concentration) and local charge density (P2, sGAG sulfation pattern). (**c**) μGUIDE-mediated generation of VEGF gradients to direct endothelial cell network formation in a microgel-in-gel system. From left to right: schematic of the microgel-in-gel system, VEGF gradient simulation using a reaction–diffusion model, representative confocal image showing local network assembly around VEGF-loaded μGUIDEs, control of network formation with varying P1. Scale bar, 800 μm. (**d**) VEGF-loaded μGUIDEs to stimulate the local vascularization of kidney organoids. From left to right: phase contrast image of a kidney organoid

Figure 3. (*Continued on facing page*) with four μGUIDEs deposited at the periphery of the cell pellet, VEGF gradient simulation within the kidney organoid transwell culture using a reaction–diffusion model, representative confocal image showing the interaction of CD31+ endothelial cells with PAX2+ renal vesicles, quantitative analysis of endothelial cells–renal vesicles interactions. Scale bars, 1 mm and 500 μm. Figure modified and reproduced with permission from Kühn et al.[83] (License Number 5957041316610).

development. First, we demonstrated the effectiveness of VEGF-loaded μGUIDEs in a microgel-in-gel vasculogenesis model, demonstrating the formation of tailored morphogen concentration gradients that control developmental processes with cellular resolution (**Figure 3b**). Further, we have extended the application of μGUIDEs to human-induced PSC (hiPSC)-derived kidney organoid cultures, marking a pioneering step in the implementation of localized VEGF gradients within organoids. Computationally predicted gradients from μGUIDEs enhanced vascular development in targeted organoid regions, leading to a more physiologically relevant organization (**Figure 3c**). While presented for the administration of VEGF, μGUIDEs could be easily applied for the administration of numerous GAG-binding morphogens (e.g., Shh, BMPs, TGF-β, FGFs),[27,100] also in combination and in staged sequences. Furthermore, thanks to the independent modulation of biomolecular and physical characteristics, the microgels could be adaptable to different in vitro tissue and disease models.

6. Outlook

In summary, morphogen gradients are pivotal in orchestrating the spatial and temporal aspects of development. They provide a versatile and robust mechanism for cells to acquire positional information, leading to the organized formation of tissues and organs. The ongoing efforts to manipulate these gradients in in vitro settings continue to unravel the intricate details of how these gradients are established, maintained, and interpreted, providing deeper insights into the fundamental processes that underlie tissue and disease development. Despite the impressive progress in the field, a long list of questions regarding fundamental aspects remains, such as the robustness of morphogen gradients against fluctuations in secretion or diffusion, the scaling of the gradients with tissue growth, or the patterning across different length scales.[4] Progress with respect to analysis techniques, such as quantitative imaging or single-cell transcriptomics, will provide more quantitative and complex data that, together with new theoretical frameworks, will further improve our understanding. Therefore, methods that provide access to such analytical tools become increasingly important. This could be microfluidic chips with the possibility for on-chip imaging or microgel-based morphogen sources, such as μGUIDEs, which are highly compatible with different culture setups and analysis techniques.

Furthermore, controlling morphogen gradients in vitro enables the engineering of more complex, realistic, functional, and bigger tissue constructs and disease models for medical applications such as drug screening or transplantation/therapy. To date, most stem cell-derived tissue models only capture the premature embryonic stages of development and fail to resemble adult tissue with respect to cell populations, organization, and functionality.[78] Here, methods for gradient control will be assessed by their compatibility with advanced biofabrication technologies, such as bioprinting that continue to show increasing ability to assemble larger and more complex tissue constructs.[69] Our efforts to create vascularized (perfusable) tissues not only benefit from the know-how of previous microfluidic approaches but will also enable microfluidic-based morphogen signaling. Organizer cells and more robust microgel-based signaling centers can establish the localized gradients that are needed to shape tissue organization and complexity.

On the other hand, morphogen gradient methods that facilitate miniaturization of the tissue models for combinatorial and high-throughput applications, such as drug screening or systematic studies of morphogen signaling, will be able to take full advantage of the new technological era that is just beginning to emerge around machine learning and AI. When applied in this context, large, parameterized systematics like the μGUIDE platform will be able to cover big parameter spaces and explore conditions that could even go beyond nature's example.

In addition to embryonic development and morphogenesis, the methods for gradient control can have interesting applications with respect to regeneration and disease of adult tissue. Adult tissue, for example, the intestinal epithelial lining,[101] is being constantly remodeled, facing the same problems as embryonic tissues to maintain its spatial organization. While the role of morphogen gradients for developmental diseases is well appreciated, the consideration of soluble signaling gradients, for example in cancer biology, is relatively little discussed.[101,102] Understanding the role of soluble signaling gradients in adult tissue regeneration and disease can lead to new therapeutic concepts that reach past the molecular and cellular level, which currently dominates medical research. In this context, an inherent benefit of biomaterial-based approaches, and to a lesser extent also cell-based approaches due to higher safety concerns, is their dual use. The same material applied in vitro as a morphogen source could fulfill the same function in vivo, either in a therapeutic context or to verify in vitro data.

References

1. Turing AM. The chemical basis of morphogenesis. *Philos Trans R Soc Lond B Biol Sci.* 1952;237(641):37.
2. Rogers KW, Schier AF. Morphogen gradients: From generation to interpretation. *Annu Rev Cell Dev Biol.* 2011;27:377.
3. Simsek MF, Özbudak EM. Patterning principles of morphogen gradients. *Open Biol.* 2022;12(10):220224.
4. Kicheva A, Briscoe, J. Control of tissue development by morphogens. *Annu Rev Cell Dev Biol.* 2023;39(1):91–121.
5. Crick F. Diffusion in embryogenesis. *Nature.* 1970;225:420.
6. Lander AD. Morpheus unbound: Reimagining the morphogen gradient. *Cell.* 2007;128(2):245.
7. Kicheva A, Pantazis P, Bollenbach T, Kalaidzidis Y, Bittig T, Jülicher F, González-Gaitán M. Kinetics of morphogen gradient formation. *Science (1979).* 2007;315(5811):521.

8. Wartlick O, Kicheva A, González-Gaitán M. Morphogen gradient formation. *Cold Spring Harb Perspect Biol.* 2009;1(3).
9. Niu B, Bach TN, Chen X, Chandratre KR, Isaac Murray J, Zhao Z, et al. Computational modeling and analysis of the morphogenetic domain signaling networks regulating *C. elegans* embryogenesis. *Comput Struct Biotechnol J.* 2022;20:3653.
10. Driever W, Nsslein-volhard C. The Bicoid protein determines position in the drosophila embryo in a concentration-dependent manner. *Cell.* 1988;54:95.
11. Li C, Ouyang L, Armstrong JPK, Stevens MM. Advances in the fabrication of biomaterials for gradient tissue engineering. *Trends Biotechnol.* 2021;39(2):150.
12. Demers CJ, Soundararajan P, Chennampally P, Cox GA, Briscoe J, Collins SD, et al. Development-on-chip: In vitro neural tube patterning with a microfluidic device. *Development.* 1884;143(11):1884.
13. Kleinman HK, Philp D, Hoffman MP. Role of the extracellular matrix in morphogenesis. *Curr Opin Biotechnol.* 2003;14(5):526.
14. Cruz Walma DA, Yamada KM. The extracellular matrix in development. *Development (Cambridge).* 2020;147(10):dev175596.
15. Hayashi Y, Kobayashi S, Nakato H. Drosophila glypicans regulate the germline stem cell niche. *J Cell Biol.* 2009;187(4):473–480.
16. Gude F, Froese J, Manikowski D, Di Iorio D, Grad JN, Wegner S, et al. Hedgehog is relayed through dynamic heparan sulfate interactions to shape its gradient. *Nat Commun.* 2023;14(1):1.
17. Stapornwongkul KS, Vincent JP. Generation of extracellular morphogen gradients: The case for diffusion. *Nat Rev Genet.* 2021; 22(6):393.
18. Raman R, Sasisekharan V, Sasisekharan R. Structural insights into biological roles of protein–glycosaminoglycan interactions. *Chem Biol.* 2005;12(3):267.
19. Capila I, Linhardt RJ. Heparin–protein interactions. *Angew Chem Int Ed Engl.* 2002;41(3):390.
20. Meneghetti MC, Hughes AJ, Rudd TR, Nader HB, Powell AK, Yates EA, Lima MA. Heparan sulfate and heparin interactions with proteins. *J R Soc Interface.* 2015;12(110):20150589.
21. Vempati P, Popel AS, MacGabham F. Extracellular regulation of VEGF: Isoforms, proteolysis, and vascular patterning. *Cytokine Growth Factor Rev.* 2008;25(1):1.

22. Magno V, Meinhardt A, Werner C. Polymer hydrogels to guide organotypic and organoid cultures. *Adv Funct Mater.* 2020;30(48): 2000097.
23. Peppas NA, Hilt JZ, Khademhosseini A, Langer R. Hydrogels in biology and medicine: From molecular principles to bionanotechnology. *Adv Mater.* 2006;18(11):1345.
24. Prince E, Kumacheva E. Design and applications of man-made biomimetic fibrillar hydrogels. *Nat Rev Mater.* 2019;4(2):99.
25. Hussey GS, Dziki JL, Badylak SF. Extracellular matrix-based materials for regenerative medicine. *Nat Rev Mater.* 2018;3(7):159.
26. Lutolf MP, Gilbert PM, Blau HM. Designing materials to direct stem-cell fate. *Nature.* 2009;462(7272):433.
27. Freudenberg U, Liang Y, Kiick KL, Werner C. Glycosaminoglycan-based biohybrid hydrogels: A sweet and smart choice for multifunctional biomaterials. *Adv Mater.* 2016;28(40):8861.
28. Moore K, Macsween M, Shoichet M. Immobilized concentration gradients of neurotrophic factors guide neurite outgrowth of primary neurons in macroporous scaffolds. *Tissue Eng.* 2006;12(2): 060303124145002.
29. DeLong SA, Moon JJ, West JL. Covalently immobilized gradients of bFGF on hydrogel scaffolds for directed cell migration. *Biomaterials.* 2005;26(16):3227.
30. Aizawa Y, Wylie R, Shoichet M. Endothelial cell guidance in 3D patterned scaffolds. *Adv Mater.* 2010;22(43):4831.
31. Fisher SA, Tam RY, Fokina A, Mahmoodi MM, Distefano MD, Shoichet MS. Photo-immobilized EGF chemical gradients differentially impact breast cancer cell invasion and drug response in defined 3D hydrogels. *Biomaterials.* 2018;178(1):751.
32. Ruskowitz ER, Deforest CA. Photoresponsive biomaterials for targeted drug delivery and 4D cell culture. *Nat Rev Mater.* 2018;3.
33. Hubka KM, Carson DD, Harrington DA, Farach-Carson MC. Perlecan domain I gradients establish stable biomimetic heparin binding growth factor gradients for cell migration in hydrogels. *Acta Biomater.* 2019;97:385.
34. Kloxin AM, Kasko AM, Salinas CN, Anseth KS. Photodegradable hydrogels for dynamic tuning of physical and chemical properties. *Science (1979).* 2009;324(5923):59.
35. Tsurkan MV, Chwalek K, Prokoph S, Zieris A, Levental KR, Freudenberg U, et al. Defined polymer-peptide conjugates to form

cell-instructive starPEG-heparin matrices in situ. *Adv Mater.* 2013;25(18):2606.

36. Hsu RS, Chen PY, Fang JH, Chen YY, Chang CW, Lu YJ, et al. Adaptable microporous hydrogels of propagating NGF-gradient by injectable building blocks for accelerated axonal outgrowth. *Adv Sci.* 2019;6(16):1900520.
37. Kühn S, Sievers J, Stoppa A, Träber N, Zimmermann R, Welzel PB, et al. Cell-Instructive multiphasic gel-in-gel materials. *Adv Funct Mater.* 2020;30(26):1908857.
38. el-Kareh AW, Braunstein SL, Secomb TW. Effect of cell arrangement and interstitial volume fraction on the diffusivity of monoclonal antibodies in tissue. *Biophys J.* 1993;64(5):1638.
39. Amsden B. Solute diffusion in hydrogels. An examination of the retardation effect. *Polymer Gels and Netw.* 1998;6(1):13.
40. Limasale YDP, Atallah P, Werner C, Freudenberg U, Zimmermann R. Tuning the local availability of VEGF within glycosaminoglycan-based hydrogels to modulate vascular endothelial cell morphogenesis. *Adv Funct Mater.* 2020;30(44):2000068.
41. Zieris A, Dockhorn R, Röhrich A, Zimmermann R, Müller M, Welzel PB, et al. Biohybrid networks of selectively desulfated glycosaminoglycans for tunable growth factor delivery. *Biomacromolecules.* 2014;15(12):4439.
42. Ono K, Hattori H, Takeshita S, Kurita A, Ishihara M. Structural features in heparin that interact with VEGF165 and modulate its biological activity. *Glycobiology.* 1999;9(7):705.
43. Kühn S, Freyse J, Atallah P, Rademann J, Freudenberg U, Werner C. Tuning the network charge of biohybrid hydrogel matrices to modulate the release of SDF-1. *Biol Chem.* 2021;402(11):1453.
44. Köhling S, Blaszkiewicz J, Ruiz-Gómez G, Fernández-Bachiller MI, Lemmnitzer K, Panitz N, et al. Syntheses of defined sulfated oligohyaluronans reveal structural effects, diversity and thermodynamics of GAG-protein binding. *Chem Sci.* 2019;10(3):866.
45. Martino MM, Tortelli F, Mochizuki M, Traub S, Ben-David D, Kuhn GA, et al. Engineering the growth factor microenvironment with fibronectin domains to promote wound and bone tissue healing. *Sci Transl Med.* 2011;3(100):100ra89.
46. Belair DG, Miller MJ, Wang S, Darjatmoko SR, Binder BYK, Sheibani N, et al. Differential regulation of angiogenesis using degradable VEGF-binding microspheres. *Biomaterials.* 2016;93:27.

47. Freudenberg U, Sommer JU, Levental KR, Welzel PB, Zieris A, Chwalek K, et al. Using mean field theory to guide biofunctional materials design. *Adv Funct Mater.* 2012;22(7):1391.
48. Freudenberg U, Atallah P, Limasale YDP, Werner C. Charge-tuning of glycosaminoglycan-based hydrogels to program cytokine sequestration. *Faraday Discuss.* 2019;219:244.
49. Paluck SJ, Nguyen TH, Maynard HD. Heparin-mimicking polymers: Synthesis and biological applications. *Biomacromolecules.* 2016;17(11):3417.
50. Neves MI, Araújo M, Moroni L, da Silva RMP, Barrias CC. Glycosaminoglycan-inspired biomaterials for the development of bioactive hydrogel networks. *Molecules.* 2020;25(4):978.
51. Zheng SL, Loh KM. Creating artificial signaling gradients to spatially pattern engineered tissues. *Curr Opin Biotechnol.* 2022;78:102810.
52. Samal P, van Blitterswijk C, Truckenmüller R, Giselbrecht S. Grow with the flow: When morphogenesis meets microfluidics. *Adv Mater.* 2019;31(17).
53. Pasca SP. Assembling human brain organoids. *Science.* 2019; 363(6423):126–127.
54. Schmidt C. The rise of the assembloid. *Nature.* 2021;597(7878):S22.
55. Gossen M, Bujard H. Tight control of gene expression in mammalian cells by tetracycline-responsive promoters. *Proc Natl Acad Sci U S A.* 1992;89(12):5547.
56. Welman A, Barraclough J, Dive C. Generation of cells expressing improved doxycycline-regulated reverse transcriptional transactivator rtTA2S-M2. *Nat Protoc.* 2006;1(2):803.
57. Cederquist GY, Asciolla JJ, Tchieu J, Walsh RM, Cornacchia D, Resh MD, et al. Specification of positional identity in forebrain organoids. *Nat Biotechnol.* 2019;37(4):436.
58. De Santis R, Etoc F, Rosado-Olivieri EA, Brivanlou AH. Self-organization of human dorsal-ventral forebrain structures by light induced SHH. *Nat Commun.* 2021;12(1):1.
59. Legnini I, Emmenegger L, Zappulo A, Rybak-Wolf A, Wurmus R, Martinez AO, et al. Spatiotemporal, optogenetic control of gene expression in organoids. *Nat Methods.* 2023;20(10):1544.
60. Benzinger D, Briscoe J. Illuminating morphogen and patterning dynamics with optogenetic control of morphogen production. *bioRxiv.* 2024;2024.06.11.598403.

61. Morsut L, Roybal KT, Xiong X, Gordley RM, Coyle SM, Thomson M, et al. Engineering customized cell sensing and response behaviors using synthetic notch receptors. *Cell.* 2016;164(4):780.
62. Toda S, McKeithan WL, Hakkinen TJ, Lopez P, Klein OD, Lim WA. Engineering synthetic morphogen systems that can program multicellular patterning. *Science (1979).* 2020;370(6514):327.
63. Hamblin MR, Karimi M, editors. *Biomedical applications of microfluidic devices.* 1st ed. London: Academic Press; 2020.
64. Sonnen KF, Merten CA. Microfluidics as an emerging precision tool in developmental biology. *Dev Cell.* 2019;48(3):293.
65. Bonner MG, Gudapati H, Mou X, Musah S. Microfluidic systems for modeling human development. *Development.* 2022; 149(3):dev199463.
66. Seiffert S, Thiele J. Microfluidics: Theory and Practice for Beginners. Berlin, Boston: De Gruyter; 2020.
67. O'Grady B, Balikov DA, Wang JX, Neal EK, Ou YC, Bardhan R, et al. Spatiotemporal control and modeling of morphogen delivery to induce gradient patterning of stem cell differentiation using fluidic channels. *Biomater Sci.* 2019;7(4):1358.
68. Miller JS, Stevens KR, Yang MT, Baker BM, Nguyen DHT, Cohen DM, et al. Rapid casting of patterned vascular networks for perfusable engineered three-dimensional tissues. *Nat Mater.* 2012;11(9):768.
69. Brassard JA, Nikolaev M, Hübscher T, Hofer M, Lutolf MP. Recapitulating macro-scale tissue self-organization through organoid bioprinting. *Nat Mater.* 2021;20(1):22.
70. Li Jeon N, Baskaran H, Dertinger SKW, Whitesides GM, Van De Water L, Toner M. Neutrophil chemotaxis in linear and complex gradients of interleukin-8 formed in a microfabricated device. *Nat Biotechnol.* 2002;20(8):826.
71. Rifes P, Isaksson M, Rathore GS, Aldrin-Kirk P, Møller OK, Barzaghi G, et al. Modeling neural tube development by differentiation of human embryonic stem cells in a microfluidic WNT gradient. *Nat Biotechnol.* 2020;38(11):1265.
72. Regier MC, Tokar JJ, Warrick JW, Pabon L, Berthier E, Beebe DJ, et al. User-defined morphogen patterning for directing human cell fate stratification. *Sci Rep.* 2019;9(1):1.
73. Manfrin A, Tabata Y, Paquet ER, Vuaridel AR, Rivest FR, Naef F, et al. Engineered signaling centers for the spatially

controlled patterning of human pluripotent stem cells. *Nat Methods.* 2019;16(7):640.

74. Sorre B, Warmflash A, Brivanlou AH, Siggia ED. Encoding of temporal signals by the TGF-β pathway and implications for embryonic patterning. *Dev Cell.* 2014;30(3):334.
75. Sonnen KF, Lauschke VM, Uraji J, Falk HJ, Petersen Y, Funk MC, et al. Modulation of phase shift between Wnt and notch signaling oscillations controls mesoderm segmentation. *Cell.* 2018; 172(5):1079.
76. Attayek PJ, Ahmad AA, Wang Y, Williamson I, Sims CE, Magness ST, et al. In vitro polarization of colonoids to create an intestinal stem cell compartment. *PLoS One.* 2016;11(4):1.
77. Shih HC, Lee TA, Wu HM, Ko PL, Liao WH, Tung YC. Microfluidic collective cell migration assay for study of endothelial cell proliferation and migration under combinations of oxygen gradients, tensions, and drug treatments. *Sci Rep.* 2019;9(1):1.
78. Brassard JA, Lutolf MP. Engineering stem cell self-organization to build better organoids. *Cell Stem Cell.* 2019;24(6):860.
79. Hofer M, Lutolf MP. Engineering organoids. *Nat Rev Mater.* 2021;6(5):402.
80. Bhattacharjee N, Folch A. Large-scale microfluidic gradient arrays reveal axon guidance behaviors in hippocampal neurons. *Microsyst Nanoeng.* 2017;3(1):17003.
81. Eichele G, Tickle C, Alberts BM. Microcontrolled release of biologically active compounds in chick embryos: Beads of 200-μm diameter for the local release of retinoids. *Anal Biochem.* 1984;142(2):542.
82. Kent RN, Said M, Busch ME, Poupard ER, Tsai A, Xia J, et al. Physical and soluble cues enhance tendon progenitor cell invasion into injectable synthetic hydrogels. *Adv Funct Mater.* 2022;32(48): 2207556.
83. Kühn S, Magno V, Zimmermann R, Limasale YDP, Atallah P, Stoppa A, Männel MJ, Thiele J, Friedrichs J, Freudenberg U, Werner C. Microgels With Electrostatically Controlled Molecular Affinity to Direct Morphogenesis. *Adv Mater.* 2024;37:2409731.
84. Afting C, Walther T, Drozdowski OM, Schlagheck C, Schwarz US, Wittbrodt J, et al. DNA microbeads for spatio-temporally controlled morphogen release within organoids. *Nat Nanotechnol.* 2024;19:1849–1857.

85. Foster GA, Headen DM, González-García C, Salmerón-Sánchez M, Shirwan H, García AJ. Protease-degradable microgels for protein delivery for vascularization. *Biomaterials.* 2017;113:170.
86. Atallah P, Schirmer L, Tsurkan M, Putra Limasale YD, Zimmermann R, Werner C, et al. In situ-forming, cell-instructive hydrogels based on glycosaminoglycans with varied sulfation patterns. *Biomaterials.* 2018;181:227.
87. Freudenberg U, Zieris A, Chwalek K, Tsurkan MV, Maitz MF, Atallah P, et al. Heparin desulfation modulates VEGF release and angiogenesis in diabetic wounds. *J Control Release.* 2015;220:79.
88. Zieris A, Chwalek K, Prokoph S, Levental KR, Welzel PB, Freudenberg U, et al. Dual independent delivery of pro-angiogenic growth factors from starPEG-heparin hydrogels. *J Control Release.* 2011;156(1):28.
89. Daly AC, Riley L, Segura T, Burdick JA. Hydrogel microparticles for biomedical applications. *Nat Rev Mater.* 2020;5(1):20.
90. Moroni L, Burdick JA, Highley C, Lee SJ, Morimoto Y, Takeuchi S, et al. Biofabrication strategies for 3D in vitro models and regenerative medicine. *Nat Rev Mater.* 2018;3(5):21.
91. Koh I, Hagiwara M. Gradient to sectioning CUBE workflow for the generation and imaging of organoids with localized differentiation. *Commun Biol.* 2023;6:299.
92. Dennison NR, Fusenig M, Grönnert L, Maitz MF, Martinez MAR, Wobus M, et al. Precision culture scaling to establish high-throughput vasculogenesis models. *Adv Healthc Mater.* 2024;13: e2400388.
93. Rose JC, Gehlen DB, Haraszti T, Köhler J, Licht CJ, De Laporte L. Biofunctionalized aligned microgels provide 3D cell guidance to mimic complex tissue matrices. *Biomaterials.* 2018;163:128.
94. Ben-Reuven L, Reiner O. Toward spatial identities in human brain organoids-on-chip induced by morphogen-soaked beads. *Bioengineering.* 2020;7(4):1.
95. Träber N, Uhlmann K, Girardo S, Kesavan G, Wagner K, Friedrichs J, et al. Polyacrylamide bead sensors for in vivo quantification of cell-scale stress in zebrafish development. *Sci Rep.* 2019; 9(1):1.
96. Bratt-Leal AM, Nguyen AH, Hammersmith KA, Singh A, McDevitt TC. A microparticle approach to morphogen delivery

within pluripotent stem cell aggregates. *Biomaterials.* 2013;34(30): 7227.

97. Vanslambrouck JM, Wilson SB, Tan KS, Groenewegen E, Rudraraju R, Neil J, et al. Enhanced metanephric specification to functional proximal tubule enables toxicity screening and infectious disease modelling in kidney organoids. *Nat Commun.* 2022; 13(1):5943.
98. Wang Y, Irvine DJ. Engineering chemoattractant gradients using chemokine-releasing polysaccharide microspheres. *Biomaterials.* 2011;32(21):4903.
99. Lee HC, Hastings C, Oliveira NMM, Pérez-Carrasco R, Page KM, Wolpert L, et al. 'Neighbourhood watch' model: Embryonic epiblast cells assess positional information in relation to their neighbours. *Development.* 2022;149(10):dev200295.
100. Häcker U, Nybakken K, Perrimon N. Heparan sulphate proteoglycans: The sweet side of development. *Nat Rev Mol Cell Biol.* 2005; 6(7):530.
101. Muthuswamy SK. Self-organization in cancer: Implications for histopathology, cancer cell biology, and metastasis. *Cancer Cell.* 2021; 39(4):443.
102. Potter JD. Morphogens, morphostats, microarchitecture and malignancy. *Nat Rev Cancer.* 2007;7(6):464.

https://doi.org/10.1142/9789819806782_0013

Chapter 13

In Vitro Models of Vasculogenesis for High-Throughput Applications

Jens Friedrichs[1#], Yanuar Dwi Putra Limasale[1#], Maximilian Fusenig[1#], Nicholas R. Dennison[1#], Kathryn A. Rosowski[2], Peter D. Westenskow[2] and Carsten Werner[1*]

[1]*Leibniz Institute of Polymer Research Dresden, Max Bergmann Center of Biomaterials, 01060 Dresden, Germany*
[2]*Roche Pharma Research and Early Development, Roche Innovation Center Basel, F. Hoffmann-La Roche Ltd., Basel, 4070, Switzerland*
**Corresponding author: werner@ipfdd.de*
#*These authors contributed equally.*

This chapter examines the historical development and recent innovations in hydrogel technology for in vitro vasculogenesis models. It explores their incorporation into high-throughput culture formats and vasculogenesis-compatible platforms, emphasizing their utility in screening angiogenesis inhibitors, especially in retinal disease research. This chapter confronts the challenges and constraints of existing high-throughput screening models and discusses strategies to enhance their efficiency and reliability, thereby propelling progress in drug discovery and disease modeling.

1. The vascular system in health and disease

The human body contains an extensive network of blood vessels that, if stretched end to end, would span 60,000 miles — enough to circle the globe more than twice.[1] This complex system, composed of arteries, veins, and capillaries, forms the backbone of the circulatory system, enabling the continuous blood flow essential for life.

Robust and elastic arteries transport oxygen-rich blood from the heart to the body's tissues. Veins, equipped with valves and thinner walls, carry oxygen-depleted blood back to the heart for reoxygenation. Capillaries, the thinnest and most numerous blood vessels, are the primary locations for the exchange of gases, nutrients, and waste products. They form networks that reach nearly every cell in the body, effectively linking arteries to veins and facilitating efficient exchanges between blood and tissues.[2] This circulatory infrastructure is not only critical for meeting the metabolic needs of tissues — delivering essential nutrients and removing waste — but also plays a key role in the structural organization and developmental processes of tissues and organs. Blood vessels act as selective barriers, regulating the transport of macromolecules, therapeutic drugs, and toxins, thus protecting tissues from harmful substances while allowing the absorption of beneficial compounds.[3]

Key processes in vascular development include vasculogenesis and angiogenesis. Vasculogenesis, the formation of new vessels from progenitor cells, not only occurs primarily during embryonic development but also contributes to adult tissue repair. Angiogenesis, on the other hand, is initiated by conditions such as hypoxia or tissue damage and entails the development of new vessels from preexisting ones. These processes are regulated by endothelial cells that line the circulatory system and respond to various physical and chemical stimuli. Growth factors such as vascular endothelial growth factor (VEGF) and fibroblast growth factors (FGFs) are instrumental in these processes, driving the development and refinement of vascular structures.[2,4,5] The interactions between endothelial cells and other vascular cells, such as smooth muscle cells and pericytes, are fundamental to vascular stability and function. Supported by a complex network of signaling molecules and extracellular matrix (ECM) components, these interactions ensure effective cell adhesion and communication. This is crucial not only for maintaining the structural integrity of blood vessels but also for their adaptive response to various physiological and pathological conditions.[3]

Disturbed angiogenesis and vasculogenesis are central to a wide array of pathological conditions, affecting everything from cancer to cardiovascular health. In oncology, uncontrolled angiogenesis drives tumor growth and metastasis. Tumors secrete angiogenic factors such as VEGF to promote endothelial cell proliferation, resulting in the formation of new, often dysfunctional blood vessels. This abnormal vasculature not only supports tumor expansion but also creates hypoxic areas within tumors that further stimulate angiogenic signaling and perpetuate a self-sustaining cycle of growth.[6] In cardiovascular diseases, an imbalance in angiogenic processes results in insufficient vessel formation under ischemic conditions, resulting in reduced blood supply to cardiac or peripheral tissues. In addition, endothelial dysfunction impedes the proper vascular response to hypoxia, exacerbating tissue damage and accelerating disease progression.[7] Similarly, chronic inflammatory conditions such as rheumatoid arthritis and psoriasis are adversely affected by disrupted angiogenic activity. In these diseases, aberrant new vessel formation in inflamed tissues sustains and intensifies inflammation, driven by endothelial cells that, in response to inflammatory cytokines, proliferate and migrate excessively, intensifying inflammation and tissue damage.[8] Importantly, impaired angiogenesis and vasculogenesis also contribute to eye diseases such as diabetic retinopathy (DR) and age-related macular degeneration (AMR). These processes are further detailed in a subsequent section of this chapter (Section 3).

2. Vascular model systems

Various methodologies have been developed to replicate human vasculature in vitro, enhancing our understanding of vascular biology and pathology. Simple two-dimensional (2D) angiogenesis models use endothelial cells on a flat surface to form tube-like structures, allowing for fundamental mechanistic studies and drug screenings.[9,10] They lack the complex three-dimensional (3D) interactions and mechanical forces found in living tissues, limiting their physiological relevance.[3,11] Therefore, further discussion on 2D vasculogenesis models is deferred to Section 4.2, where they will be explored within the context of high-throughput experiments.

Several 3D culture models provide a more accurate physiological representation of human vasculature by allowing endothelial cells to self-assemble into vascular-like networks within a supportive matrix that simulates the ECM.[11]

Spheroid-based assays effectively replicate the angiogenesis process by enabling endothelial cells to form multicellular spheroids that sprout into the surrounding matrix under proangiogenic conditions. These assays are invaluable for studying cancer progression, the effects of hypoxia on vascular growth, and the impact of therapeutic agents.[12]

Bioprinting has transformed in vitro vasculogenesis with its precision in structuring cells and ECM components, crafting intricate tissue constructs with embedded vascular networks. This technique adeptly customizes vascular geometries to mirror the complexity of natural vessels and facilitates the integration of diverse cell types, reflecting the dynamic interplay within living tissues.[13]

Organ-on-a-chip systems represent a major leap forward, combining microfluidic technologies with biological modeling to simulate the complex functions of tissues and organs.[14] These systems precisely reproduce the dynamic flow and mechanical stresses that blood vessels experience, offering unparalleled insights into how these physical factors influence vascular behavior and disease processes.[15]

By bridging the gap between traditional cell culture and animal models, these systems offer a more ethical and cost-effective means of research, reducing reliance on animal testing and improving the translation of findings from the laboratory to clinical settings.

2.1. *Cells for vascular model systems*

In the field of in vitro vasculogenesis assays, various cell sources are employed to meet specific scientific objectives and experimental designs. Primary cells such as human umbilical vein endothelial cells (HUVECs) and human microvascular endothelial cells, directly extracted from tissues, are highly valued for their physiological accuracy as they retain many in vivo phenotypic and functional characteristics.[16] These cells are essential for studies requiring a high degree of biological relevance, offering a closer representation of the in vivo environment. However, challenges such as donor variability, limited availability, and cellular senescence after several cell culture passages can impact the reproducibility and scalability of research using primary cells.

Alternatively, immortalized cell lines provide a more consistent and sustainable option for large-scale genetic or pharmacological screenings, offering uniformity and extended culture longevity. They help standardize experimental conditions, minimizing the variability associated with primary cell usage. However, these cell lines may not perfectly mimic primary cell behavior due to altered genomic integrity from prolonged culture, potentially leading to discrepancies between in vitro results and actual biological outcomes.[17]

Induced pluripotent stem cells (iPSCs), due to their potential to transform into any cell type, including endothelial cells, have revolutionized vasculogenesis research. Their adaptability significantly amplifies their value in personalized medicine and disease modeling.[18] However, generating iPSC-derived endothelial cells involves intricate differentiation protocols that demand precise control over the cell culture environment, growth factors, and signaling pathways, introducing challenges in consistency and reliability in experimental results.[19]

Vasculogenesis assays often incorporate a variety of cell types in coculture systems to more accurately simulate in vivo cellular interactions and enhance network formation.[3] Pericytes and smooth muscle cells provide essential mechanical support and stability to emerging blood vessels, communicating directly with endothelial cells through gap junctions and paracrine signaling, exchanging growth factors such as platelet-derived growth factor (PDGF) and angiopoietin-1, crucial for vascular maturation and functionality. Fibroblasts contribute to the vascular niche by creating a dense ECM and secreting growth factors such as FGF and VEGF. These factors support endothelial cell adhesion and survival, while also promoting their proliferation and migration.[20] Mesenchymal stem cells (MSCs) release a wide array of angiogenic cytokines and growth factors, significantly enhancing endothelial tube formation and stability. They modify the local microenvironment through the release of transforming growth factor beta (TGF-β), VEGF, and stromal-derived factor 1 (SDF-1).[21] In addition, the inclusion of immune cells such as macrophages in vasculogenesis assays emphasizes their role in vascular development and repair. They secrete proangiogenic factors and remodel the vascular matrix, contributing to the dynamic restructuring necessary for the proper organization and function of newly formed blood vessels.[22]

2.2. *Materials for vascular model systems*

In the field of in vitro vasculogenesis, natural materials are widely utilized for their biocompatibility and ability to mimic the ECM. These materials promote cellular behaviors essential for vascular network formation, although each presents unique challenges.[23]

Matrigel, composed of laminin, collagen IV, entactin, and growth factors such as FGF and TGF-β, effectively replicates the natural basement membrane.[24,25] This composition makes Matrigel highly effective for promoting rapid and robust capillary formation in vitro, facilitating the study of angiogenesis and other cellular processes that require a well-defined basement membrane. Despite its utility, Matrigel faces challenges with batch variability and ethical issues due to its derivation from mouse tumors.[24] To address these challenges, efforts are underway to develop synthetic or recombinant alternatives that can replicate the beneficial properties of Matrigel without the associated variability and ethical concerns.[26]

Collagen serves as the primary structural protein in the ECM and supports endothelial cell functions critical to vasculogenesis. Collagen hydrogels enhance endothelial cell attachment and proliferation, facilitating vascular tube formation and maturation.[27] However, collagen's inherent variability sometimes complicates its utility in experimental settings. Variations in the fiber structure between batches can significantly affect the consistency and reproducibility of experimental results, presenting a substantial challenge in standardizing research protocols.[28]

Fibrin hydrogels are recognized for supporting capillary integrity and serving as a reservoir for growth factors essential to angiogenesis.[3] Fibrin gel scaffolds have shown superiority over collagen for long-term vascular stability due to their resistance to gel contraction, a factor that inhibits network maturation.[29] Fibrin gels have been crucial in forming perfusable networks within microfluidic devices, enhancing our understanding of how soluble signals, the viscoelastic properties of the ECM, and shear forces influence vascular morphology and function.[30–32] Moreover, these in vitro perfusable vessels have been instrumental in modeling various physiological and pathological conditions, such as immune cell trafficking,[33] cancer,[34] and DR.[35]

Gelatin, a form of denatured collagen, has been demonstrated to support the formation of extensive and functional vascular networks.[36] Gelatin offers thermoreversible properties that aid in physical manipulation but lacks the mechanical strength and biological signals of native collagen, often requiring chemical modification to improve functionality.[37] Gelatin functionalized with methacrylate groups forms photocrosslinkable gelatin methacrylate (GelMA) hydrogels.[38,39] The degree of methacrylation in GelMA hydrogels can modulate cellular behaviors and the extent of vascular network formation both in vitro and in vivo.[40,41] However, using ultraviolet (UV) light for cross-linking GelMA introduces challenges, as it can damage encapsulated cells and degrade sensitive biological molecules within the hydrogel.

To overcome the limitations inherent in natural materials for 3D culture systems, synthetic materials, particularly polyethylene glycol (PEG)-based hydrogels, are increasingly favored due to their adaptability and tunability.[42,43] PEG's water solubility, chain flexibility, and ability to be functionalized with reactive groups enable precise adjustment of mechanical properties to match a variety of cellular environments, from soft, brain-like tissues to stiff, bone-like structures.[44] However, these hydrogels often lack essential features such as cellular adhesion sites, adequate degradability, and the bioactivity needed for vascular network formation.[45,46] To address these shortcomings, researchers have developed biohybrid materials that merge synthetic polymers with ECM-derived biopolymers such as collagen,[47] fibrin,[48] and glycosaminoglycans (GAGs).[49]

Synthetic hydrogels are crucial in tissue engineering and regenerative medicine due to their versatility and tunability. However, they often lack robust growth factor-binding moieties, which are essential for establishing and maintaining effective signaling gradients crucial for vessel formation. This limitation necessitates continuous supplementation of proangiogenic growth factors. A potent solution to this problem is the functionalization of synthetic hydrogels with ECM components, particularly GAGs. These unbranched, high-molecular-weight polysaccharides, either covalently linked to protein cores to form proteoglycans or freely diffused, are integral to many essential ECM functions. Notably, GAGs aid in the localized delivery of various soluble signaling molecules, contributing to the formation of morphogen gradients. These gradients play a crucial role in key biological processes such as embryogenesis,[50] development,[51,52] homeostasis,[52] inflammation,[53] and tumor formation and progression.[54] Among naturally occurring GAGs, heparin is distinguished by its high sulfate content.[55,56] Heparin's unique sulfation pattern enables selective interactions with key angiogenic factors such as VEGF and FGF, crucial for promoting angiogenesis and vasculogenesis.[57,58] Its ability to bind and stabilize growth factors enhances their biological activities, significantly amplifying signaling pathways vital for new blood vessel formation.[59] This feature is especially critical in tissue engineering and regenerative medicine, where precise control over growth factor activity can significantly impact tissue repair and regeneration outcomes. Consequently, GAGs such as heparin are increasingly utilized to develop more effective biomaterials due to their multifunctional capabilities.[60]

Recently, a biohybrid GAG-based hydrogel platform has been developed, utilizing heparin and its derivatives alongside multi-arm PEG (starPEG).[49] This system forms covalently linked hydrogels under physiological conditions through a bioorthogonal Michael-type addition, enabling the direct embedding of cells during hydrogel formation.[49,56] The dual-modular starPEG–GAG design enables the decoupling of biophysical properties from biochemical functionalities by adjusting the molar ratio of the biologically inactive starPEG relative to the GAG. This adjustment regulates the hydrogel's stiffness while maintaining consistent biochemical interactions with embedded cells and the reversible electrostatic binding of growth factors to heparin.[49,61] In addition, peptide sequences can be included at the starPEG terminals as targets for matrix metalloproteinases (MMPs), creating cell-responsive hydrogels essential for cellular proliferation.[49] Finally, maleimide groups on the GAG component facilitate the covalent attachment of cell-adhesive peptides during network formation.[49] These features are essential for the hydrogel's ability to mimic the dynamic nature of the ECM, promoting cellular adhesion and mechanotransduction, as well as matrix remodeling by cell-secreted proteases. These properties have allowed for the implementation of these hydrogels in the development of several 3D cultures, including neuronal stem cells,[62] kidney cells,[63] and tumor cells.[64–68]

StarPEG–GAG hydrogels have also proven effective in promoting the development of complex endothelial cell networks, thereby facilitating vasculogenesis.[49,69–71] By adjusting the hydrogel's viscoelastic properties and optimizing growth factor concentrations, optimal matrix conditions for supporting HUVEC growth and vascular network formation were thoroughly explored. In addition, coculturing endothelial cells with MSCs has been shown to enhance network growth and stability for up to four weeks, with supporting cells thriving near developing endothelial filaments.[72] Complementary research assessed how variations in GAG concentration and sulfation patterns influence the availability of cell-instructive signaling proteins within the hydrogel, significantly impacting endothelial capillary network formation in the 3D model.[73]

3. Engineered retinal models for angiogenesis and vascular stability

Abnormal angiogenesis significantly influences numerous eye diseases, particularly retinopathies. The retina's health relies on the blood supply from two distinct vascular networks: the inner retinal microvasculature and the extraretinal choriocapillaris. The transport of nutrients, waste products, and other factors is tightly controlled in these vasculatures to maintain homeostasis and prevent invasion of pathogenic substances into the retina. This control is provided by two types of blood–retina barriers (BRBs): the inner BRB (iBRB) and the outer BRB (oBRB).[74] The iBRB is established by the tight junctions between endothelial cells of the retinal microvasculature and is supported by the astrocytes, pericytes, and Müller cells surrounding the vessels.[75] The oBRB is instead established mainly by the tight junctions of the retinal pigmented epithelium (RPE), located below the photoreceptors of the eye. The photoreceptors, RPE, and choriocapillaris form a single functional unit, and impairments of one component can lead to damage of the entire unit.[76]

Certain conditions can trigger the breakdown of the BRB, causing vascular instability and leakage. Resulting ischemia leads to the release of proangiogenic factors such as VEGF, initiating pathogenic angiogenesis. This process involves the sprouting of new vessels from existing ones, which are often unstable and leaky, causing an imbalance in fluid exchange in the retina and, ultimately, edema. Excessive neovascularization in the retina or subretinal space can impair visual acuity and, in severe cases, lead to fibrotic scarring and increased traction forces, escalating the risk of painful retinal detachment. These symptoms are characteristic of blinding diseases including AMD[77] and DR.[78]

AMD is a leading cause of vision loss, affecting approximately 200 million people worldwide.[79,80] Although there is a genetic component and risk is strongly increased with both age and smoking, the disease's underlying causes are multifactorial and not yet fully understood.[81] AMD primarily impacts the macula, a specialized region of the retina responsible for central vision and detail perception. Aging RPE accumulates stress and damage, potentially leading to macular atrophy and subsequently dry AMD (dAMD), a slow-progressing form of the disease with limited current treatment for advanced stages. The most severe vision loss in AMD is associated with the late-stage, rapidly progressing wet/neovascular form (nAMD). In nAMD, ischemia in RPE triggers pathological angiogenesis and fibrosis in the subretinal space.[76]

DR, a major complication of diabetes, is the leading cause of vision loss in the Western world.[82] It's projected that 1 in 15 diabetics will develop diabetic macular edema (DME), a severe outcome of DR.[83] In diabetic patients with chronic hyperglycemia, iBRB disruption triggers alterations in the retinal microvasculature, including the failure of cell junctions, loss of pericytes, and immune cell infiltration into the retina. Cumulatively, these changes lead to vasoregression and excessive vascular permeability. In cases of proliferative DR, VEGF overproduction within ischemic zones induces the growth of new unstable and leaky vessels. This can lead to neurovascular defects that cause vision loss.[78]

Historically, nAMD and DME were treated with photocoagulation, a laser-based process designed to eliminate pathological vessels and seal those that are leaky. In addition, nAMD could also be managed with photodynamic therapy, which involves the use of photosensitizers that, when activated by light, produce reactive oxygen species leading to the closure of new vessels.[78] DME, on the other hand, could be treated with anti-inflammatory corticosteroids.[78]

Anti-VEGF therapies, now the standard of care for these diseases, involve direct injections of anti-VEGF antibodies or antagonists into the retina. These therapies mitigate the proangiogenic response, preventing the formation of new leaky vessels and reducing vision loss. However, the requirement for repeated injections poses a significant burden for patients and healthcare providers. Furthermore, not all patients respond optimally, and anti-VEGF treatment slows but does not stop disease progression. This leaves a vast area of discovery for new therapies for vision loss due to retinal disease. A promising new approach combines anti-VEGF treatment with the modulation of other pathways, such as angiopoietin-2 (Ang-2),[84] which further stabilizes vessel, prevents pathological vascularization, and delays fibrosis.[85,86] Exploring treatments that restore the initial loss of barrier function and prevent early vascular leakiness is another interesting research direction. Identifying unknown factors controlling angiogenesis and vessel stability will be crucial for identifying new therapeutic targets. In addition, investigating the disease's earlier stages, including the causes of initial ischemia, the role of vascular support cells, and the involvement of inflammation and immune response, could provide new opportunities for discovery.

Besides the ethical concerns, high costs, and long times associated with animal studies, animal models offer limited predictability when translated to human systems. This is primarily due to the absence of certain tissue structures in some animal eyes that are present in humans, as well as an inability to induce disease phenotypes. In contrast, 3D human tissue models offer the flexibility to adjust cell types, modify the ECM, and optimize soluble factors to induce more physiological phenotypes. Consequently, in vitro human vasculature systems have emerged as a crucial tool for discovering new drugs for retinal diseases. These systems effectively replicate BRBs and enable researchers to investigate the pathways and mechanisms involved in maintaining proper barrier function.[87] Moreover, they offer the opportunity to model the intricate process of angiogenesis and simulate conditions, leading to the pathological vessel growth observed in retinal diseases.

Various technologies are currently in use to establish stable, self-assembled vessels. These vessels are capable of forming perfusable networks, demonstrating appropriate polarization, and establishing physiological connections to support cells and the surrounding ECM. However, it can be challenging to model chronic disease phenotypes, such as those found in DR, in short-term in vitro cultures. Organ-on-a-chip iBRB models allow for long-term culture of cells. The addition of a "diabetic cocktail" can mimic early DR phenotypes, including pericyte loss and vessel regression, enabling the study of early disease mechanisms.[35] Moreover, these long-term cultured systems can be used to examine the effects of existing drugs and new molecules on aged vessels. The inclusion of RPE cells into retinal vasculature has led to the development of various oBRB models for further mechanistic studies of AMD and fibrosis.[88–90] The incorporation of immune cells and inflammatory signals can also be explored to investigate the various pathways and signaling cascades involved in both healthy and diseased states. In these models, flow is often applied to study the role of shear stress in vascular health. Assays such as fluorescent imaging, vascular perfusion, RNA sequencing, and molecular profiling of secreted and expressed proteins provide precise measurement of cellular and molecular interactions.

While the vascular model systems discussed have greatly contributed to our understanding of vascular biology, they do come with several limitations. Many of these models require complex and time-consuming manual preparation and maintenance. In addition, they often lack the scalability necessary for extensive drug testing and can suffer from inconsistencies in reproducibility. These challenges can hinder the efficient execution of large-scale studies and limit their suitability for high-throughput screening (HTS) applications. HTS assays, however, offer the potential for rapid, large-scale testing with greater precision and consistency. This makes them invaluable tools for accelerating drug discovery and advancing vascular research. Therefore, there is a critical need to develop models that can overcome the limitations of current systems, providing a more practical and reliable approach to advancing vascular research.

4. High-throughput vascular model systems

4.1. *General aspects of high-throughput screenings*

HTS methodologies have become indispensable in biotechnology and pharmaceutical research, revolutionizing how researchers identify drug candidates from compound libraries.[91,92] HTS systems prioritize assay miniaturization, reducing the consumption of reagents and cells while enabling the simultaneous testing of numerous conditions within a single experiment. Technological advancements such as automated liquid handling and the development of novel assay platforms have streamlined the entire screening process from drug development to toxicity evaluation.

4.2. *Two-dimensional high-throughput vasculogenesis models*

2D in vitro assays have long been foundational in HTS studies, primarily involving cells cultured on flat surfaces in culture plates where interactions occur on a single plane between the cells, the culture vessels, and the supplied culture media. The simplicity of these cultures is crucial for producing rapid and reproducible results, making them indispensable in the early stages of drug discovery.[93]

The emergence of microarrays has greatly broadened the capabilities of traditional 2D assays. These arrays, made up of uniform spots of ECM proteins, growth factors, or a mix of natural and synthetic hydrogel coatings, form cellular islands of various shapes and sizes. This arrangement allows for exact control over the cellular microenvironment, facilitating systematic studies on how different ligands and environmental factors influence cell behaviors such as adhesion, growth, and differentiation. Protein microarrays, for example, are crucial in screening interactions between cells and various ECM components, yielding vital insights into cellular responses that are key in cancer research and tissue engineering.[94–97]

Microfluidic devices mark another significant progression in 2D HTS technology. Despite their initial development and validation complexity, these devices offer considerable benefits, including reduced volumes of reagents and cells. They allow for precise manipulation of experimental variables and can simulate blood or interstitial fluid flow, especially useful in studying diseases such as cancer, where cell migration and drug responses under fluidic conditions are critical.[98–100]

2D high-throughput vasculogenesis models are essential for identifying compounds that influence blood vessel formation, crucial for treating ischemic heart disease and peripheral arterial disease. These models have significantly enhanced tissue perfusion and functionality by improving vascular formation.[101,102] For example, studies using immortalized endothelial and mural cell lines have simplified the investigation of premature vessel-like structure formation upon exposure to chemical compounds.[103] High-content fluorescence microscopy, combined with screening in the 1,536-well plate format, has enabled an HTS campaign across a library of 2,816 drugs, identifying potent angiogenesis inhibitors linked to relevant pathways, providing a robust platform for identifying and preliminary validating hit compounds.[104]

In oncology, the focus has shifted toward using 2D HTS to discover antiangiogenic agents that inhibit vasculogenesis, depriving tumors of their necessary blood supply. Researchers simulate the tumor microenvironment using 2D endothelial monolayers treated with tumor-conditioned media, employing high-content screening assays to assess the efficacy of compounds in preventing capillary-like structures, a response typically induced and exaggerated by tumor-derived signals. This approach has expedited the screening of numerous potential antiangiogenic agents, accelerating anticancer therapy development.[105,106]

Beyond drug discovery, 2D vasculogenesis systems have also been used in genetic screenings to manipulate genes crucial for endothelial function and vessel formation. RNA interference[107] and CRISPR-Cas9[108] technologies have been applied to selectively knock down or edit genes. Large-scale RNAi screenings, for instance, have targeted extensive libraries of kinases and phosphatases to determine their roles in endothelial cell migration and network formation on a 2D matrix.[109]

Despite these advancements, traditional 2D assays often fail to mimic the complex 3D interactions observed in vivo, potentially leading to less predictive or nonphysiological results.[110] This limitation underscores the need for continued innovation and refinement in assay development to bridge the gap between 2D in vitro models and their in vivo counterparts, enhancing the translational relevance of these studies.

4.3. *Three-dimensional high-throughput vasculogenesis models*

Despite their advantages, current 3D cell culture preparation methods for HTS assays face substantial hurdles. These systems must maintain consistent mechanical and biochemical properties across matrices or scaffolds for reproducibility, a technically demanding task at scale. Uniform nutrient and gas exchange is vital in thicker 3D cultures, as gradients can cause cell death or abnormal behaviors, affecting the uniform treatment and exposure of cells to test compounds and reducing screening accuracy. Furthermore, the throughput of 3D culture technologies trails that of 2D systems due to increased manual handling, complexity, and longer setup times, hindering their use in high-throughput settings. These operational inefficiencies slow the adoption and integration of 3D cell culture technologies in HTS applications.[111–113]

Several methods have been developed to induce scaffold-free cell aggregations, known as spheroids, thereby addressing some of the mentioned technical challenges. These techniques include hanging droplet cultures,[114,115] cell assemblies in ultra-low attachment multiwell plates[116] or prefabricated microwell arrays,[117] and specially designed microfluidic setups.[118] The composition of these 3D spheroids can vary, including endothelial cells, mixtures of endothelial and mural cells, and even additional human stem cells for targeted HTS studies.[119] For example, spheroids composed of endothelial cells and MSCs, formed in microwell cavities, have been demonstrated to develop into vascularized microtissues, enabling highly parallelized studies of blood (stem) cell migration and interaction with preassembled 3D stromal cell spheroids in a robust and highly reproducible 3D culture setup.[120]

Besides cell aggregation, critical advances in automated methodologies help tackle challenges of scalability and uniformity through, for example, sophisticated cell seeding schemes on patterned surfaces, enhanced bioprinting techniques,[121] injection molding,[122–124] and soft lithography.[125,126] These innovative strategies aim to standardize HTS processes and increase physiological relevance without compromising efficiency.[127]

Microfluidic 3D culture systems involving ECM-mimicking scaffolds have allowed the development of screening platforms where 3D cell cultures are exposed to multiple soluble compounds supplied through individual microfluidic channels.[99,128] In an advanced example of its potential applications, a multiunit microfluidic array based on the 96-well plate format was developed to generate vascularized organ-on-chip cultures for drug screening.[129] Here, individual tissue chambers were filled with endothelial cell–fibrin mixtures to develop vascularized microtissues, after which the cultures were exposed to bioactive compounds. The addition of cancer cells to this culture system led to the creation of a vascular cancer niche model, which was effectively used to screen for anticancer and antivascular drug effects. In addition to drug screening, the detailed study and thorough understanding of the biologic interplay of cocultured cell populations in 3D microenvironments are of equal interest and importance. Along these lines, an injection-molded polystyrene-based microfluidic array was established to investigate the influence of MSCs on the vasculogenesis of co-embedded HUVECs in fibrin-based 3D hydrogel cultures.[126] In this study, high-content confocal microscopy was combined with statistical approaches and molecular biological techniques to cluster MSCs according to their vasculogenic bioactivity. In addition, this approach identified vasculogenic and matricellular genes in MSCs correlated to their vasculogenic bioactivity, paving the way for future studies streamlining complex vasculogenesis models for drug screening purposes and bioactivity assessments for regenerative, cell-based therapies.

4.4. *Statistics-driven improvement of high-throughput screening reproducibility*

Developing high-throughput cell culture systems still requires comprehensive strategies to ensure reliability and reproducibility, particularly for 3D assays. Achieving uniformity among conditions remains a significant hurdle that can directly impact the validity of the screening outcomes.[130,131]

To address these limitations, the integration of statistics-driven approaches such as design of experiments (DoE) and multiple linear regression (MLR) modeling has proven crucial, particularly in enhancing the efficiency and effectiveness of HTS methodologies. DoE provides a systematic method for planning and conducting experiments with multiple controllable parameters, limiting the overall number of conditions tested while maximizing the amount and quality of information extracted from each investigation.[132] This comprehensive analysis is achieved through MLR, which allows quantifying the effects of each parameter (including their combinations) on the measured outcome, providing predictive insights that inform subsequent testing phases.[133]

By applying DoE and MLR, significant factors and their interactions can be rapidly identified, thus optimizing the whole screening process from assay development to hit identification and target validation, reducing the number of required experiments, and focusing valuable and often limited resources only on the most promising conditions.[134]

4.5. *Precision culture scaling to establish high-throughput vasculogenesis models*

Implementing DoE and MLR, we have previously established the precision culture scaling (PCS-X) methodology, customizing hydrogel-based tissue and disease-specific high-throughput models under varied boundary conditions (**Figure 1a**).[135]

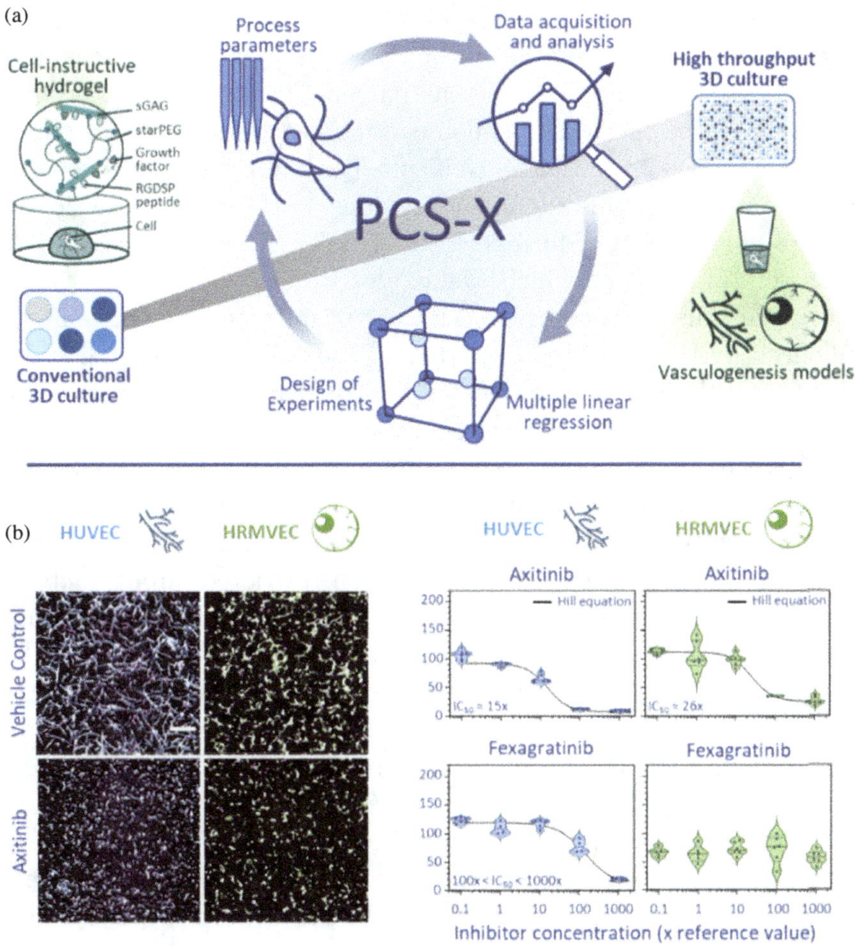

Figure 1. (**a**) Schematic summary of the precision culture scaling (PCS-X) methodology to customize high-throughput 3D tissue and disease models. PCS-X uses DoE methods to systematically alter the culture composition. Culture conditions are prepared in an automated and parallelized manner, with quantitative readouts collected and analyzed. MLR modeling of the data identifies individual and interactive effects of the investigated parameters, guiding further adjustments. To exemplifying the approach, high-throughput 3D vasculogenesis models were developed from HUVECs and human retinal microvascular endothelial cells (HRMVECs) in mono- and cocultures with mesenchymal stromal cells or retinal microvascular pericytes, respectively. (**b**) Exemplary dose-dependent effects of two bioactives on HUVEC and HRMVEC vasculogenesis. Figure modified and reproduced under license from Wiley (License Number 5812510111980).

PCS-X integrates DoE-assisted combinatorial experimental designs and automated and parallelized processing schemes for hydrogel-based 3D cultures, image-based quantification of key morphological readouts, and MLR modeling to analyze the effects of independently varied hydrogel and culture parameters on cell growth. This methodology builds on our previous work with conventional low-throughput culture formats (i.e., droplet 3D hydrogel cultures in 24-well plates), in which we explored how vasculogenesis of endothelial cells, in both mono- and coculture with stromal cells, is influenced by the material properties of the cell-encapsulating starPEG–GAG 3D hydrogel matrix.[69,70,136] Adapting these protocols to high-throughput 3D vasculogenesis models in a low-volume 384-well format required significant changes in the volume ratio of hydrogel to cell culture medium, impacting diffusion conditions and thus the supply of hydrogel-embedded cells with nutrients and metabolites. We systematically varied three diffusion-restricted input factors known to influence vasculogenesis to assess their individual and combined effects on HUVECs and retinal endothelial cells (HRMVECs). These adjustments helped in studying the effects of these factors on endothelial and stromal cell vasculogenesis, emphasizing the importance of nutrient competition and growth factor dynamics in these cell-dense 3D microenvironments.

To prepare parallelized 3D hydrogel cultures, we developed an automated liquid handling routine to form the gels in a low-volume 384-well plate. This routine allowed for the parallelized preparation of multiple combinations of exogenous conditions, including the biochemical and physical properties of the cell-embedding gels and their growth factor supplementation. High-content confocal microscopy and advanced 3D image analysis were employed to quantify vasculogenic structures, facilitating high-throughput hydrogel formulations and culture conditions screening. The DoE framework enabled systematic variation in key factors such as cell density and growth factor concentrations (of VEGF, basic fibroblast growth factor (bFGF), and SDF-1α), which were analyzed through MLR to model and predict optimal culture conditions and understand biologically relevant mechanisms regulating vasculogenesis of gel-embedded endothelial cells.

The PCS-X methodology was also extended to drug screening, testing established vasculogenesis inhibitors across various concentrations, which faithfully assessed and scored compound- and dose-dependent responses in a highly parallelized and precise manner in single assay plates (**Figure 1b**). This confirmed PCS-X's robustness in drug discovery, enhancing reliability and reducing variability, which is crucial for advancing personalized medicine. Future prospects include integrating biochemical secretome analyses to increase the descriptive power of PCS-X and expanding the range of modeled tissues and diseases, aiming to enhance the predictiveness of in vitro biomedical research and reducing reliance on animal testing.

5. Conclusions and outlook

The vascular system is critical in health and disease, with abnormal angiogenesis and vasculogenesis contributing to conditions such as cancer, cardiovascular diseases, and inflammatory disorders.[2,4–8] A variety of vascular model systems have been developed, ranging from simple 2D endothelial cell cultures to complex 3D microfluidic platforms and organoids.[9–15] These models use different cell sources, each with their own unique benefits and drawbacks, and coculture systems for a more physiologically relevant environment.[16,18–21,137] Various materials, including natural, synthetic, and biohybrid, are used to create 3D culture systems that aim to mimic the ECM, supporting angiogenesis and vasculogenesis.[22,25,41–43,45] GAG-based hydrogels in particular have shown great promise in supporting vascularization.[66–73] HTS methodologies have revolutionized drug discovery in 2D formats, but their application to 3D cell culture systems presents challenges. These issues arise from the complexity of 3D cultures, which, while providing more physiologically relevant data, are harder to control and standardize. Factors such as cell seeding density, nutrient diffusion, and matrix composition can greatly influence the behavior of cells in 3D cultures, leading to variability in experimental outcomes. To address these challenges, new technologies and protocols are being developed, such as the PCS-X methodology that integrates DoE-assisted combinatorial designs, automated processing, image-based quantification, and MLR modeling to customize hydrogel-based tissue-specific high-throughput in vitro models.[135] This approach has been successfully applied to 3D vasculogenesis models, demonstrating its robustness and potential in advancing high-throughput drug testing.

In addition, the application of artificial intelligence in high-throughput analysis of vascular network morphology can automate the quantification of vascular structures, enhancing the efficiency of complex 3D network analysis.[138–143] When combined with genomics, proteomics, and secretomics, a comprehensive understanding of vasculogenesis mechanisms can be achieved, leading to the identification of novel targets for vascular diseases.[144–147] Future models may incorporate multiple tissues or organoids with perfusable vasculature, integrating fluid flow, shear stress, and gradient formation, enabling more predictive vascular models.[27,29–34] Further advancements in biomaterials, such as dynamic and responsive hydrogels, can better mimic the complex and dynamic nature of the ECM,[41–44,148] enhancing the physiological relevance of 3D vasculogenesis models and providing a more accurate representation of the in vivo environment.[148–150]

References

1. Wolfson J. Your vascular system: The unsung hero of human longevity. *Integr Med (Encinitas).* 2024;22:52–53.
2. Trimm E, Red-Horse K. Vascular endothelial cell development and diversity. *Nat Rev Cardiol.* 2023;20:197–210.
3. O'Connor C, Brady E, Zheng Y, Moore E, Stevens KR. Engineering the multiscale complexity of vascular networks. *Nat Rev Mater.* 2022;7:702–716.
4. Risau W. Mechanisms of angiogenesis. *Nature.* 1997;386:671–674.
5. Zhang C. The role of inflammatory cytokines in endothelial dysfunction. *Basic Res Cardiol.* 2008;103:398–406.
6. De Palma M, Biziato D, Petrova TV. Microenvironmental regulation of tumour angiogenesis. *Nat Rev Cancer.* 2017;17:457–474.

7. Simons M, Ware JA. Therapeutic angiogenesis in cardiovascular disease. *Nat Rev Drug Discov.* 2003;2:863–871.
8. Aguilar-Cazares D, Chavez-Dominguez R, Carlos-Reyes A, Lopez-Camarillo C, Hernandez de la Cruz ON, Lopez-Gonzalez JS. Contribution of angiogenesis to inflammation and cancer. *Front Oncol.* 2019;9:492467.
9. Goodwin AM. In vitro assays of angiogenesis for assessment of angiogenic and anti-angiogenic agents. *Microvasc Res.* 2007;74: 172–183.
10. Arnaoutova I, George J, Kleinman HK, Benton G. The endothelial cell tube formation assay on basement membrane turns 20: State of the science and the art. *Angiogenesis.* 2009;12:267–274.
11. Zucchelli E, Majid QA, Foldes G. New artery of knowledge: 3D models of angiogenesis. *Vasc Biol.* 2020;1:H135–H143.
12. Vakhrushev IV, Nezhurina EK, Karalkin PA, Tsvetkova AV, Sergeeva NS, Majouga AG, et al. Heterotypic multicellular spheroids as experimental and preclinical models of sprouting angiogenesis. *Biology (Basel).* 2022;11(1):18.
13. Mota C, Camarero-Espinosa S, Baker MB, Wieringa P, Moroni L. Bioprinting: From tissue and organ development to in vitro models. *Chem Rev.* 2020;120(9):10547–10607.
14. Shirure VS, Hughes CCW, George SC. Engineering vascularized organoid-on-a-chip models. *Annu Rev Biomed Eng.* 2021;23: 141–167.
15. Ingber DE. Human organs-on-chips for disease modelling, drug development and personalized medicine. *Nat Rev Genet.* 2022; 23:467–491.
16. McCloskey MC, Zhang VZ, Ahmad SD, Walker S, Romanick SS, Awad HA, et al. Sourcing cells for in vitro models of human vascular barriers of inflammation. *Front Med Technol.* 2022;4:979768.
17. Voloshin N, Tyurin-Kuzmin P, Karagyaur M, Akopyan Z, Kulebyakin K. Practical use of immortalized cells in medicine: Current advances and future perspectives. *Int J Mol Sci.* 2023;24(16):12716.
18. Klein D. iPSCs-based generation of vascular cells: Reprogramming approaches and applications. *Cell Mol Life Sci.* 2018;75(8): 1411–1433.

19. Zhao Z, Chen X, Dowbaj AM, Sljukic A, Bratlie K, Lin L, et al. Organoids. *Nat Rev Methods Primers.* 2022;2:1–21.
20. Warren E, Gerecht S. Beyond the endothelium: The role of mural cells in vascular biology: In vitro systems to study endothelial/pericyte cell interactions. *Vasc Biol.* 2023;5(1):e220021.
21. Melchiorri AJ, Nguyen BNB, Fisher JP. Mesenchymal stem cells: Roles and relationships in vascularization. *Tissue Eng Part B Rev.* 2014;20(3):218–228.
22. Nucera S, Biziato D, de Palma M. The interplay between macrophages and angiogenesis in development, tissue injury and regeneration. *Int J Dev Biol.* 2011;55:495–503.
23. Xing H, Lee H, Luo L, Kyriakides TR. Extracellular matrix-derived biomaterials in engineering cell function. *Biotechnol Adv.* 2020;42:107421.
24. Hughes CS, Postovit LM, Lajoie GA. Matrigel: A complex protein mixture required for optimal growth of cell culture. *Proteomics.* 2010;10(9):1886–1890.
25. Kleinman HK, Martin GR. Matrigel: Basement membrane matrix with biological activity. *Semin Cancer Biol.* 2005;15:378–386.
26. Aisenbrey EA, Murphy WL. Synthetic alternatives to Matrigel. *Nat Rev Mater.* 2020;5:539–551.
27. Devillard CD, Marquette CA. Vascular tissue engineering: Challenges and requirements for an ideal large scale blood vessel. *Front Bioeng Biotechnol.* 2021;9:721843.
28. Antoine EE, Vlachos PP, Rylander MN. Review of collagen I hydrogels for bioengineered tissue microenvironments: Characterization of mechanics, structure, and transport. *Tissue Eng Part B Rev.* 2014;20(6):683–696.
29. Park YK, Tu TY, Lim SH, Clement IJM, Yang SY, Kamm RD. In vitro microvessel growth and remodeling within a three-dimensional microfluidic environment. *Cell Mol Bioeng.* 2014;7(1):15–25.
30. Whisler JA, Chen MB, Kamm RD. Control of perfusable microvascular network morphology using a multiculture microfluidic system. *Tissue Eng Part C Methods.* 2014;20:543–552.
31. Wan Z, Zhang S, Zhong AX, Xu L, Coughlin MF, Pavlou G, et al. Transmural flow upregulates PD-L1 expression in microvascular networks. *Adv Sci.* 2024;11(26):e2400921. doi:10.1002/advs.202400921.

32. Zhang S, Wan Z, Pavlou G, Zhong AX, Xu L, Kamm RD, et al. Interstitial flow promotes the formation of functional microvascular networks in vitro through upregulation of matrix metalloproteinase-2. *Adv Funct Mater.* 2022;32:2206767.
33. Hirth E, Cao W, Peltonen M, Kapetanovic E, Dietsche C, Svanberg S, et al. Self-assembled and perfusable microvasculature-on-chip for modeling leukocyte trafficking. *Lab Chip.* 2023;24: 292–304.
34. Haase K, Offeddu GS, Gillrie MR. Kamm RD. Endothelial regulation of drug transport in a 3D vascularized tumor model. *Adv Funct Mater.* 2020;30:2002444.
35. Maurissen TL, Spielmann AJ, Schellenberg G, Bickle M, Vieira JR, Lai SY, et al. Modeling early pathophysiological phenotypes of diabetic retinopathy in a human inner blood-retinal barrier-on-a-chip. *Nat Commun.* 2024;15:1372.
36. Chen YC, Lin RZ, Qi H, Yang Y, Bae H, Melero-Martin JM, et al. Functional human vascular network generated in photocrosslinkable gelatin methacrylate hydrogels. *Adv Funct Mater.* 2012;22: 2027–2039.
37. Lukin I, Erezuma I, Maeso L, Zarate J, Desimone MF, Al-Tel TH, et al. Progress in gelatin as biomaterial for tissue engineering. *Pharmaceutics.* 2022;14:1177.
38. Klotz BJ, Gawlitta D, Rosenberg AJWP, Malda J, Melchels FPW. Gelatin-methacryloyl hydrogels: Towards biofabrication-based tissue repair. *Trends Biotechnol.* 2016;34:394–407.
39. Van Den Bulcke AI, Bogdanov B, de Rooze N, Schacht EH, Cornelissen M, Berghmans H. Structural and rheological properties of methacrylamide modified gelatin hydrogels. *Biomacromolecules.* 2000;1:31–38.
40. Bupphathong S, Lim J, Fang HW, Tao HY, Yeh CE, Ku TA, et al. Enhanced vascular-like network formation of encapsulated HUVECs and ADSCs coculture in growth factors conjugated GelMA hydrogels. *ACS Biomater Sci Eng.* 2024;10(5):3306–3315.
41. Nichol JW, Koshy ST, Bae H, Hwang CM, Yamanlar S, Khademhosseini A, et al. Cell-laden microengineered gelatin methacrylate hydrogels. *Biomaterials.* 2010;31:5536–5544.
42. Allen ME, Hindley JW, Baxani DK, Ces O, Elani Y. Hydrogels as functional components in artificial cell systems. *Nat Rev Chem.* 2022;6:562–578.

43. Gyles DA, Castro LD, Silva JOC, Ribeiro-Costa RM. A review of the designs and prominent biomedical advances of natural and synthetic hydrogel formulations. *Eur Polym J.* 2017;88:373–392.
44. Zhu J. Bioactive modification of poly(ethylene glycol) hydrogels for tissue engineering. *Biomaterials.* 2010;31:4639–4656.
45. Caliari SR, Burdick JA. Practical guide to hydrogels for cell culture. *Nat Methods.* 2016;13:405–414.
46. Liu J, Long H, Zeuschner D, Räder AFB, Polacheck WJ, Kessler H, et al. Synthetic extracellular matrices with tailored adhesiveness and degradability support lumen formation during angiogenic sprouting. *Nat Commun.* 2021;12:1–12.
47. Singh RK, Seliktar D, Putnam AJ. Capillary morphogenesis in PEG-collagen hydrogels. *Biomaterials.* 2013;34:9331–9340.
48. Benavides OM, Quinn JP, Pok S, Petsche Connell J, Ruano R, Jacot JG. Capillary-like network formation by human amniotic fluid-derived stem cells within fibrin/poly(ethylene glycol) hydrogels. *Tissue Eng Part A.* 2015;21:1185–1194.
49. Tsurkan MV, Chwalek K, Prokoph S, Zieris A, Levental KR, Freudenberg U, et al. Defined polymer-peptide conjugates to form cell-instructive starpeg-heparin matrices in situ. *Adv Mater.* 2013; 25:2606–2610.
50. Kramer KL. Specific sides to multifaceted glycosaminoglycans are observed in embryonic development. *Semin Cell Dev Biol.* 2010;21:631–637.
51. Häcker U, Nybakken K, Perrimon N. Heparan sulphate proteoglycans: The sweet side of development. *Nat Rev Mol Cell Biol.* 2005; 6:530–541.
52. Bishop JR, Schuksz M, Esko JD. Heparan sulphate proteoglycans fine-tune mammalian physiology. *Nature.* 2007;446:1030–1037.
53. Parish CR. The role of heparan sulphate in inflammation. *Nat Rev Immunol.* 2006;6:633–643.
54. Sasisekharan R, Shriver Z, Venkataraman G, Narayanasami U. Roles of heparan-sulphate glycosaminoglycans in cancer. *Nat Rev Cancer.* 2002;2:521–528.
55. Capila I, Linhardt RJ. Heparin–protein interactions. *Angew Chem Int Ed Engl.* 2002;41:390–412.
56. Atallah P, Schirmer L, Tsurkan M, Limasale YDP, Zimmermann R, Werner C, et al. In situ-forming, cell-instructive hydrogels based

on glycosaminoglycans with varied sulfation patterns. *Biomaterials.* 2018;181:227–239.

57. Pike DB, Cai S, Pomraning KR, Firpo MA, Fisher RJ, Shu XZ, et al. Heparin-regulated release of growth factors in vitro and angiogenic response in vivo to implanted hyaluronan hydrogels containing VEGF and bFGF. *Biomaterials.* 2006;27:5242–5251.
58. Riley CM, Fuegy PW, Firpo MA, Shu XZ, Prestwich GD, Peattie RA. Stimulation of in vivo angiogenesis using dual growth factor-loaded crosslinked glycosaminoglycan hydrogels. *Biomaterials.* 2006;27:5935–5943.
59. Taylor S, Folkman J. Protamine is an inhibitor of angiogenesis. *Nature.* 1982;297:307–312.
60. Freudenberg U, Liang Y, Kiick KL, Werner C. Glycosaminoglycan-based biohybrid hydrogels: A sweet and smart choice for multi-functional biomaterials. *Adv Mater.* 2016;28:8861–8891.
61. Welzel PB, Prokoph S, Zieris A, Grimmer M, Zschoche S, Freudenberg U, et al. Modulating biofunctional starPEG heparin hydrogels by varying size and ratio of the constituents. *Polymers (Basel).* 2011;3:602–620.
62. Papadimitriou C, Celikkaya H, Cosacak MI, Mashkaryan V, Bray L, Bhattarai P, et al. 3D Culture method for Alzheimer's disease modeling reveals interleukin-4 rescues Aβ42-induced loss of human neural stem cell plasticity. *Dev Cell.* 2018;46;85–101.e8.
63. Weber HM, Tsurkan MV, Magno V, Freudenberg U, Werner C. Heparin-based hydrogels induce human renal tubulogenesis in vitro. *Acta Biomater.* 2017;57:59–69.
64. Bray LJ. Secker C, Murekatete B, Sievers J, Binner M, Welzel PB, et al. Three-dimensional in vitro hydro-and cryogel-based cell-culture models for the study of breast-cancer metastasis to bone. *Cancers (Basel).* 2018;10:292.
65. Dhawan A, von Bonin M, Bray LJ, Freudenberg U, Bejestani EP, Werner C, et al. Functional interference in the bone marrow micro-environment by disseminated breast cancer cells. *Stem Cells.* 2016;34:2224–2235.
66. Chwalek K, Bray LJ, Werner C. Tissue-engineered 3D tumor angiogenesis models: Potential technologies for anti-cancer drug discovery. *Adv Drug Deliv Rev.* 2014;79:30–39.

67. Taubenberger AV, Bray LJ, Haller B, Shaposhnykov A, Binner M, Freudenberg U, et al. 3D extracellular matrix interactions modulate tumour cell growth, invasion and angiogenesis in engineered tumour microenvironments. *Acta Biomater.* 2016;36:73–85.
68. Bray LJ, Binner M, Holzheu A, Friedrichs J, Freudenberg U, Hutmacher DW, et al. Multi-parametric hydrogels support 3D in vitro bioengineered microenvironment models of tumour angiogenesis. *Biomaterials.* 2015;53:609–620.
69. Freudenberg U, Zieris A, Chwalek K, Tsurkan MV, Maitz MF, Atallah P, et al. Heparin desulfation modulates VEGF release and angiogenesis in diabetic wounds. *J Control Release.* 2015;220: 79–88.
70. Zieris A, Prokoph S, Levental KR, Welzel PB, Grimmer M, Freudenberg U, et al. FGF-2 and VEGF functionalization of starPEG-heparin hydrogels to modulate biomolecular and physical cues of angiogenesis. *Biomaterials.* 2010;31:7985–7994.
71. Zieris A, Chwalek K, Prokoph S, Levental KR, Welzel PB, Freudenberg U, et al. Dual independent delivery of pro-angiogenic growth factors from starPEG-heparin hydrogels. *J Control Release.* 2011;156:28–36.
72. Chwalek K, Tsurkan MV, Freudenberg U, Werner C. Glycosaminoglycan-based hydrogels to modulate heterocellular communication in in vitro angiogenesis models. *Sci Rep.* 2014;4:4414.
73. Limasale YDP, Atallah P, Werner C, Freudenberg U, Zimmermann R. Tuning the local availability of VEGF within glycosaminoglycan-based hydrogels to modulate vascular endothelial cell morphogenesis. *Adv Funct Mater.* 2020;30:2000068.
74. Matthew C, Humphries P. The blood-retina barrier: Tight junctions and barrier modulation. *Adv Exp Med Biol.*2012;763:70–84.
75. Díaz-Coránguez M, Ramos C, Antonetti DA. The inner blood-retinal barrier: Cellular basis and development. *Vision Res.* 2017; 139:123.
76. Bird AC. Therapeutic targets in age-related macular disease. *J Clin Invest.* 2010;120:3033.
77. Chirco KR, Sohn EH, Stone EM, Tucker BA, Mullins RF. Structural and molecular changes in the aging choroid: Implications for age-related macular degeneration. *Eye.* 2017;31(1):10–25.

78. Kusuhara S, Fukushima Y, Ogura S, Inoue N, Uemura A. Pathophysiology of diabetic retinopathy: The old and the new. *Diabetes Metab J.* 2018;42:364–376.
79. Jiang B, Jiang C, Li J, Lu P. Trends and disparities in disease burden of age-related macular degeneration from 1990 to 2019: Results from the Global Burden of Disease Study 2019. *Front Public Health.* 2023;11:1138428.
80. Chaudhuri M, Hassan Y, Bakka V, Bellary Pattanashetty MS, Abdin ZU, Siddiqui HF. Age-related macular degeneration: An exponentially emerging imminent threat of visual impairment and irreversible blindness. *Cureus.* 2023;15:e39624.
81. Yates JRW, Moore AT. Genetic susceptibility to age related macular degeneration. *J Med Genet.* 2000;37:83–87.
82. Wong TY, Cheung CMG, Larsen M, Sharma S, Simó R. Diabetic retinopathy. *Nat Rev Dis Primers.* 2016;2:16012.
83. Chen YP, Wu AL, Chuang CC, Chen SN. Factors influencing clinical outcomes in patients with diabetic macular edema treated with intravitreal ranibizumab: Comparison between responder and non-responder cases. *Sci Rep.* 2019;9:10952.
84. Regula JT, Lundh von Leithner P, Foxton R, Barathi VA, Cheung CMG, Tun SBB, et al. Targeting key angiogenic pathways with a bispecific CrossMAb optimized for neovascular eye diseases. *EMBO Mol Med.* 2016;8:1265–1288.
85. Linder M, Bennink L, Foxton RH, Kirkness M, Westenskow PD. In vivo monitoring of active subretinal fibrosis in mice using collagen hybridizing peptides. *Lab Animal.* 2024;53:196–204.
86. Canonica J, Foxton R, Garcia Garrido M, Lin CM, Uhles S, Shanmugam S, et al. Delineating effects of angiopoietin-2 inhibition on vascular permeability and inflammation in models of retinal neovascularization and ischemia/reperfusion. *Front Cell Neurosci.* 2023;17:1192464.
87. Ragelle H, Dernick K, Westenskow PD, Kustermann S. Retinal microvasculature-on-a-chip for modeling VEGF-induced permeability. *Methods in Mol Biol.* 2022;2475:239–257.
88. Nam U, Lee S, Jeon JS. Generation of a 3D outer blood-retinal barrier with advanced choriocapillaris and its application in diabetic retinopathy in a microphysiological system. *ACS Biomater Sci Eng.* 2023;9:4929–4939.

89. Song MJ, Quinn R, Nguyen E, Hampton C, Sharma R, Park TS, et al. Bioprinted 3D outer retina barrier uncovers RPE-dependent choroidal phenotype in advanced macular degeneration. *Nature Methods.* 2022;20(1):149–161.
90. Chung M, Lee S, Lee BJ, Son K, Jeon NL, Kim JH, et al. Wet-AMD on a chip: Modeling outer blood-retinal barrier in vitro. *Adv Healthc Mater.* 2018;7(2):4359–4368.
91. MacArron R, Banks MN, Bojanic D, Burns DJ, Cirovic DA, Garyantes T, et al. Impact of high-throughput screening in biomedical research. *Nat Rev Drug Discov.* 2011;10:188–195.
92. An WF, Tolliday N. Cell-based assays for high-throughput screening. *Mol Biotechnol.* 2010;45:180–186.
93. Wei F, Wang S Gou X. A review for cell-based screening methods in drug discovery. *Biophys Rep.* 2021;7:504–516.
94. Hong HJ, Koom WS, Koh WG. Cell microarray technologies for high-throughput cell-based biosensors. *Sensors (Switzerland).* 2017;17:1293.
95. Berthuy OI, Muldur SK, Rossi F, Colpo P, Blum LJ, Marquette CA. Multiplex cell microarrays for high-throughput screening. *Lab Chip.* 2016;16:4248–4262.
96. Fernandes TG, Diogo MM, Clark DS, Dordick JS, Cabral JMS. High-throughput cellular microarray platforms: Applications in drug discovery, toxicology and stem cell research. *Trends Biotechnol.* 2009;27:342–349.
97. Nguyen EH, Daly WT, Le NNT, Farnoodian M, Belair DG, Schwartz MP, et al. Versatile synthetic alternatives to Matrigel for vascular toxicity screening and stem cell expansion. *Nat Biomed Eng.* 2017;1:1–14.
98. Ali M, Kim W, Park J. Droplet microfluidic technologies for next-generation high-throughput screening. *Front Lab Chip Technol.* 2023;2:1230791.
99. Du G, Fang Q, den Toonder JMJ. Microfluidics for cell-based high throughput screening platforms — A review. *Anal Chim Acta.* 2016;903:36–50.
100. De Stefano P, Bianchi E, Dubini G. The impact of microfluidics in high-throughput drug-screening applications. *Biomicrofluidics.* 2022;16:31501.

101. Huang H, Lindgren A, Wu X, Liu NA. Lin S. High-throughput screening for bioactive molecules using primary cell culture of transgenic zebrafish embryos. *Cell Rep.* 2012;2:695–704.
102. Ryoo H, Kimmel H, Rondo E, Underhill GH. Advances in high throughput cell culture technologies for therapeutic screening and biological discovery applications. *Bioeng Transl Med.* 2023;9(3): e10627. doi:10.1002/btm2.10627.
103. Iwata Y, Klaren WD, Lebakken CS, Grimm FA, Rusyn I. High-content assay multiplexing for vascular toxicity screening in induced pluripotent stem cell-derived endothelial cells and human umbilical vein endothelial cells. *Assay Drug Dev Technol.* 2017;15:267–279.
104. Li S, Hsu C-W, Sakamuru S, Zou C, Huang R, Xia M. Identification of angiogenesis inhibitors using a co-culture cell model in a high-content and high-throughput screening platform. *SLAS Technol.* 2018;23:217–225.
105. Evensen L, Link W, Lorens, JB. Imaged-based high-throughput screening for anti-angiogenic drug discovery. *Curr Pharm Des.* 2011;16:3958–3963.
106. Evensen L, Micklem DR, Link W, Lorens JB. A novel imaging-based high-throughput screening approach to anti-angiogenic drug discovery. *Cytometry A.* 2010;77:41–51.
107. Mittal V, Nolan DJ. Genomics and proteomics approaches in understanding tumor angiogenesis. *Expert Rev Mol Diagn.* 2007;7: 133–147.
108. Gopal Hariprabu KN, Sathya M, Vimalraj S. CRISPR/Cas9 in cancer therapy: A review with a special focus on tumor angiogenesis. *Int J Biol Macromol.* 2021;192:913–930.
109. Williams SP, Gould CM, Nowell CJ, Karnezis T, Achen MGA, Simpson K, et al. Systematic high-content genome-wide RNAi screens of endothelial cell migration and morphology. *Sci Data.* 2017;4:1–11.
110. Horvath P, Aulner N, Bickle M, Davies AM, Del Nery E, Ebner D, et al. Screening out irrelevant cell-based models of disease. *Nat Rev Drug Discov.* 2016;15:751–769.
111. Bersini S, Gilardi M, Arrigoni C, Talò G, Zamai M, Zagra L, et al. Human in vitro 3D co-culture model to engineer vascularized

bone-mimicking tissues combining computational tools and statistical experimental approach. *Biomaterials.* 2016;76:157–172.

112. Carragher N, Piccinini F, Tesei A, Trask OJ Jr, Bickle M, Horvat P. Concerns, challenges and promises of high-content analysis of 3D cellular models. *Nat Rev Drug Discov.* 2018;17:606–606.
113. Wang Y, Jeon H. 3D cell cultures toward quantitative high-throughput drug screening. *Trends Pharmacol Sci.* 2022;43: 569–581.
114. Yang X, Pan R, Ning K, Xie Y, Chen F, Sun W, et al. High throughput generating stable spheroids with tip-refill wafer. *Biotechnol J.* 2024;19:2300427.
115. Tung YC, Hsiao AY, Allen SG, Torisawa Y, Ho M, Takayama S, et al. High-throughput 3D spheroid culture and drug testing using a 384 hanging drop array. *Analyst.* 2011;136:473–478.
116. Shi W, Kwon J, Huang Y, Tan J, Uhl CG, He R, Zhou C, et al. Facile tumor spheroids formation in large quantity with controllable size and high uniformity. *Sci Rep.* 2018;8:1–9.
117. Mosaad EO, Chambers KF, Futrega K, Clements JA, Doran MR. The Microwell-mesh: A high-throughput 3D prostate cancer spheroid and drug-testing platform. *Sci Rep.* 2018;8(1):253.
118. Li L, Chen Y, Wang H, An G, Wu H, Huang W. A high-throughput, open-space and reusable microfluidic chip for combinational drug screening on tumor spheroids. *Lab Chip.* 2021;21:3924–3932.
119. De Moor L, Merovci I, Baetens S, Verstraeten J, Kowalska P, Krysko DV, et al. High-throughput fabrication of vascularized spheroids for bioprinting. *Biofabrication.* 2018;10:035009.
120. Ryoo H, Kimmel H, Rondo E, Underhill GH. Advances in high throughput cell culture technologies for therapeutic screening and biological discovery applications. *Bioeng Transl Med.* 2023;9:e10627.
121. Meng F, Meyer CM, Joung D, Vallera DA, McAlpine MC, Panoskaltsis-Mortari A. 3D bioprinted in vitro metastatic models via reconstruction of tumor microenvironments. *Adv Mater.* 2019; 31:e1806899.
122. Ko J, Ahn J, Kim S, Lee Y, Lee J, Park D, et al. Tumor spheroid-on-a-chip: A standardized microfluidic culture platform for investigating tumor angiogenesis. *Lab Chip.* 2019;19:2822–2833.
123. Kim S, Ko J, Lee SR, Park D, Park S, Jeon NL. Anchor-IMPACT: A standardized microfluidic platform for high-throughput antiangiogenic drug screening. *Biotechnol Bioeng.* 2021;118:2524–2535.

124. Yu J, Lee S, Song J, Lee SR, Kim S, Choi H, et al. Perfusable micro-vascularized 3D tissue array for high-throughput vascular phenotypic screening. *Nano Converg.* 2022;9:16.
125. Sobrino A, Phan DTT, Datta R, Wang X, Hachey SJ, Romero-López M, et al. 3D microtumors in vitro supported by perfused vascular networks. *Sci Rep.* 2016;6:31589.
126. Lam J, Lee B, Yu J, Kwee BJ, Kim Y, Kim J, et al. A microphysiological system-based potency bioassay for the functional quality assessment of mesenchymal stromal cells targeting vasculogenesis. *Biomaterials.* 2022;290:121826.
127. Ryoo H, Kimmel H, Rondo E, Underhill GH. Advances in high throughput cell culture technologies for therapeutic screening and biological discovery applications. *Bioeng Transl Med.* 2023:9:e10627.
128. Toh YC, Lim TC, Tai D, Xiao G, van Noort D, Yu H, et al. A microfluidic 3D hepatocyte chip for drug toxicity testing. *Lab Chip.* 2009;9:2026–2035.
129. Phan DTT, Wang X, Craver BM, Sobrino A, Zhao D, Chen JC, et al. A vascularized and perfused organ-on-a-chip platform for large-scale drug screening applications. *Lab Chip.* 2017;17:511–520.
130. Blay V, Tolani B, Ho SP, Arkin MR. High-throughput screening: Today's biochemical and cell-based approaches. *Drug Discov Today.* 2020;25:1807–1821.
131. Mennen SM, Alhambra C, Allen L, Barberis M, Berritt S, Brandt TA, et al. The evolution of high-throughput experimentation in pharmaceutical development and perspectives on the future. *Org Process Res Dev.* 2019;23:1213–1242.
132. Politis SN, Colombo P, Colombo G, Rekkas DM. Design of experiments (DoE) in pharmaceutical development. *Drug Dev Ind Pharm.* 2017;43:889–901.
133. Slinker BK, Glantz SA. Multiple linear regression: Accounting for multiple simultaneous determinants of a continuous dependent variable. *Circulation.* 2008;117:1732–1737.
134. Potyrailo R, Rajan K, Stoewe K, Takeuchi I, Chisholm B, Lam H. Combinatorial and high-throughput screening of materials libraries: Review of state of the art. *ACS Comb Sci.* 2011;13:579–633.
135. Dennison NR, Fusenig M, Grönnert L, Maitz MF, Ramirez Martinez MA, Wobus M, et al. Precision culture scaling to establish high-throughput vasculogenesis models. *Adv Healthc Mater.* 2024;13(18):e2400388. doi:10.1002/adhm.202400388.

136. Freudenberg U, Atallah P, Limasale YDP, Werner, C. Charge-tuning of glycosaminoglycan-based hydrogels to program cytokine sequestration. *Faraday Discuss.* 2019;219:244–251.
137. Voloshin N, Tyurin-Kuzmin P, Karagyaur M, Akopyan Z, Kulebyakin K. Practical use of immortalized cells in medicine: Current advances and future perspectives. *Int J Mol Sci.* 2023; 24:12716.
138. Kim S, Lee J, Ko J, Park S, Lee SR, Kim Y, et al. Angio-Net: Deep learning-based label-free detection and morphometric analysis of in vitro angiogenesis. *Lab Chip.* 2024;24:751–763.
139. Tronolone JJ, Mathur T, Chaftari CP, Jain A. Evaluation of the morphological and biological functions of vascularized microphysiological systems with supervised machine learning. *Ann Biomed Eng.* 2023;51:1723–1737.
140. Mathur T, Tronolone JJ, Jain A. AngioMT: An in silico platform for digital sensing of oxygen transport through heterogenous microvascular networks. *BioRxiv.* 2023; doi:10.1101/2023.01.09.523275.
141. Tronolone JJ, Mathur T, Chaftari CP, Sun Y, Jain A. Machine learning chained neural network analysis of oxygen transport amplifies the physiological relevance of vascularized microphysiological systems. *Bioeng Transl Med.* 2023;8:e10582.
142. Rappez L, Akinbote A, Cherubini M, Uhlmann V, Haase K. Label-free phenotyping of human microvessel networks. *BioRxiv.* 2024;7823–7830. doi:10.1101/2024.02.20.581133.
143. Pandian NKR, Farell A, Davis E, Li Chang Teo J, Sundaram S, Eykmans J, et al. Three-dimensional flow assessment of microvascular beds with interstitial space. *bioRxiv.* 2024; 2024.02.28.582152. doi:10.1101/2024.02.28.582152.
144. Hemmen K, Reinl T, Buttler K, Behler F, Dieken H, Jänsch L, et al. High-resolution mass spectrometric analysis of the secretome from mouse lung endothelial progenitor cells. *Angiogenesis.* 2011;14:163–172.
145. Rafii S, Butler JM, Ding BS. Angiocrine functions of organ-specific endothelial cells. *Nature.* 2016;529:316–325.
146. Gurung S, Restrepo NK, Chestnut B, Klimkaite L, Sumanas S. Single-cell transcriptomic analysis of vascular endothelial cells in zebrafish embryos. *Sci Rep.* 2022;12:13065.

147. Hernandez-Fernaud JR, Reid SE, Neilson LJ, Zanivan S. Quantitative mass spectrometry-based proteomics in angiogenesis. *Proteomics Clin Appl.* 2013;7:464–476.
148. Chaudhuri O, Gu L, Darnell M, Klumpers D, Bencherif SA, Weaver JC, et al. Substrate stress relaxation regulates cell spreading. *Nat Commun.* 2015;6:1–7.
149. Wei Z, Lei M, Wang Y, Xie Y, Xie X, Lan D, et al. Hydrogels with tunable mechanical plasticity regulate endothelial cell outgrowth in vasculogenesis and angiogenesis. *Nat Commun.* 2023;14:1–16.
150. Wei Z, Schnellmann R, Pruitt HC, Gerecht S. Hydrogel network dynamics regulate vascular orphogenesis. *Cell Stem Cell.* 2020;27:798–812.e6.

https://doi.org/10.1142/9789819806782_0014

Chapter 14

Polymeric Biomaterials for Studying Neurodegeneration and Facilitating Brain Repair

Ben Newland

School of Pharmacy and Pharmaceutical Sciences,
Cardiff University, King Edward VII Avenue,
CF10 3NB Cardiff, United Kingdom
Leibniz Institute of Polymer Research Dresden,
Max Bergmann Center of Biomaterials, 01060 Dresden, Germany
NewlandB@cardiff.ac.uk

Biomaterials consisting of polymer materials have been investigated intensively for applications in neural tissue engineering and brain repair. More recently, they are also being used to develop models of diseases, typically in vitro or ex vivo, but also in vivo, which better recapitulate certain aspects of the pathology of the human condition. Herein, an overview of the use of biomaterials in neural cells, tissue, and the brain is provided in relation to the strive for better models and therapeutics for common neurological conditions of the nervous system.

1. Introduction

From ancient times to the present day, our understanding of the brain and its function has been shaped by the dominant technological and scientific paradigms of the era.[1] Analogies, such as those drawn from hydraulics, electricity, and computers, have served as conceptual bridges, allowing us to grasp the unfathomable within the framework of the familiar.

In antiquity, the brain was often likened to a hydraulic system, with the flow of fluids thought to govern its functions. During the Renaissance, this analogy gained prominence, thanks to the work of figures like René Descartes, who envisioned the brain as a hydraulic machine, with animal spirits coursing through its ventricles to convey sensory information and commands.[2]

The advent of electricity in the 18th century brought about a new wave of analogical thinking, with the brain now being compared to a complex electrical circuit. Scientists such as Luigi Galvani and Alessandro Volta experimented with the electrical properties of nerve tissue,[3] furthered later by Hermann von Helmholtz who coined the phrase "action potential," laying the groundwork for our modern understanding of neural communication.[2,4]

The transition to a more rigorous and empirically grounded understanding in the late 19th and early 20th centuries marked the advent of modern neuroscience. Santiago Ramón y Cajal's meticulous investigations into the microstructure of the nervous system, facilitated by Golgi's staining technique, fundamentally altered the landscape of neurological inquiry.[5–7] From the refinement of microscopy techniques enabling the visualization of individual neurons to the advent of noninvasive imaging modalities like functional magnetic resonance imaging (fMRI), each leap in instrumentation has afforded further insights into the brain's structure and function.[8]

As our exploration of the brain continues into the 21st century, our knowledge of its structure and function expands, but with an ever-growing list of research questions. Many fundamental principles, such as the nature of consciousness itself, remain elusive. Advancements in technologies have played a considerable role in landmark breakthroughs, such as the location of the causal gene in Huntington's disease.[9] However, much research is still needed to unravel the complex mechanistic interplay underlying the pathology of neurodegenerative diseases and for the development of new therapeutics.

In this chapter, the use of polymeric biomaterials in our recent efforts toward understanding brain function in health and disease will be discussed. Whilst clearly not on the level of profoundness of other technological breakthroughs such as the invention of microscopy, DNA sequencing, gene editing, and so on, polymer-based biomaterials are nevertheless becoming important tools for basic neuroscience research. They are also playing an ever-increasing role in the delivery of therapeutics and neural tissue engineering. In 2009, David Williams set out to define biomaterials[10]; however, for the purposes herein, biomaterials can broadly be classed as substances created for use with or within cells or tissues and organisms. These can often be based on repeating structural units, or polymers, which can arise through natural or synthetic origins, and the uses of these materials can vary widely.

The experimental use of polymeric biomaterials in the brain is now a broad topic, thus forcing a degree of selectivity in the topics covered in this chapter. For example, electrically conductive polymers are paving the way for high-resolution mapping of neural activity and advanced neuroprosthetic devices, but this burgeoning research field can be considered a topic worthy of in-depth review by itself.[11,12]

Herein, the focus will be upon polymeric biomaterials and their emerging use for modeling neurological disorders as well as their development for regenerative medicine, which includes the local delivery of therapeutics to the brain.

2. Polymeric biomaterials for neuroscience applications

2.1. *General considerations*

It would be useful to have a set of key parameters or requirements that must be met when designing biomaterials for use in the brain. However, many of the physical and chemical requirements will vary depending on the intended use, and thus cannot be reduced to a simple tabulated formula. Furthermore, certain design necessities

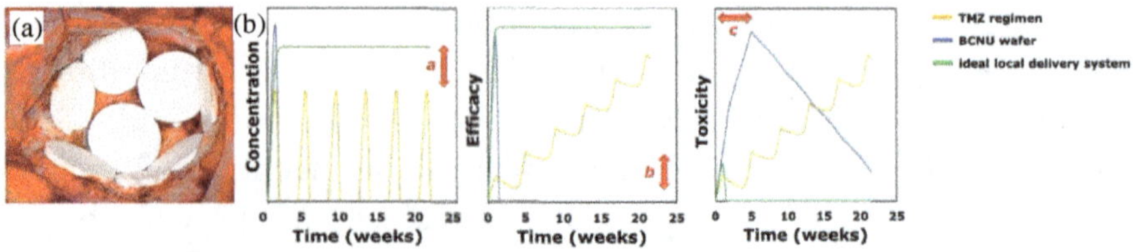

Figure 1. (**a**) Gliadel wafers containing carmustine (BCNU) implanted into a brain tumor resection cavity (reprinted with permission from Wolinsky et al.[16]). (**b**) Graphical illustration of concentration, efficacy, and toxicity profiles of an idealized drug delivery system against systemic temozolomide (TMZ) adjuvant therapy and stiff Gliadel wafers. (a) Limitations in systemic toxicity of TMZ are overcome with local delivery. An ideal delivery system will possess a mitigated bolus and 0th- or 1st-order release kinetics. (b) Achievement of a high local concentration of TMZ in local delivery systems sustained over many weeks produces a reduction in tumor burden and an increase in efficacy in the idealized local delivery systems. (c) Gliadel wafers' stiffness mismatch with tissue contributes to toxicity and negative side effects after >80% of drug release. A soft idealized local drug delivery system will have minimal acute toxicity and no negative chronic immune response. Figure adapted with permission from Tabet et al.[13]

may lose importance with a change in the overall dimensions of the biomaterials. For example, it is becoming increasingly clear that polymeric implants should ideally be soft, to reduce any mechanical mismatch between the implant and the soft brain tissue.[13] Gliadel wafers, the only FDA-approved local drug delivery system for glioblastoma (**Figure 1**), cause adverse events,[14] which include cerebral edema, convulsion, impaired healing, and infection,[15] which can continue long after the drug has been released.[13]

These have been attributed to the stiff nature of the wafer causing damage to the soft surrounding tissue.[13] However, such design constraints may no longer hold true as the dimensions of the biomaterial are reduced to the micro or nanoscale, allowing for better accommodation by the host tissue. Several microscale biomaterials, such as poly(D,L-lactic) acid (PLA) and poly(lactic-co-glycolic acid) (PLGA) microspheres, typically with orders of magnitude higher stiffness than that of the brain, are often shown to be biocompatible.[17–19] The coupling of size and stiffness-dependent adverse reactions may also be further complicated by aspect ratio, where despite their nano- or microscale dimensions, the toxicity of high-aspect ratio materials such as nanotubes may be reduced by reducing the mechanical stiffness or integrity.[20–23]

For the vast majority of intended applications, with the possible exception of immunostimulatory adjuvants, for example, for cancer therapeutics, the biomaterial should be biocompatible. Biocompatibility is a term often used loosely, especially when describing results from cytotoxicity experiments in vitro. However, the term should encompass the ability of the material to perform its functions and interact with living systems without the risk of injury, toxicity, rejection by the immune system, or inappropriate local or systemic effects (paraphrased from Ghasemi-Mobarakeh et al.[24]). It should be borne in mind that an elevated immune activity often accompanies neurodegenerative disorders,[25–27] so any intervention should seek to minimize any additional exacerbation or burden. Any implantation procedure, for example, even the introduction of a small cannula, will cause a host inflammatory response, typically in the form of a scar formed by reactive astrocytes[28,29] accompanied by the invasion of microglia and monocytes from the periphery.[30] The key point here is that an immune response is inevitable during any local intervention in the brain tissue; therefore, the biomaterial in use should ideally not cause further elevation beyond that of the surgical procedure.

For macroscale biomaterial implants (e.g., typically 0.5 mm or larger), the author hypothesizes that an endeavor should be made to create the material as close as is practically possible to the stiffness of the brain tissue. Brain tissue varies in stiffness depending on the region, age, and disease state of the patient.[31–33] White matter has shown to be slightly softer than gray matter, with storage moduli of 2.7 kPa and 3.1 kPa reported via magnetic resonance imaging (MRI) of healthy volunteers for white and gray matter respectively.[31] The elastic moduli recorded directly on rat brain tissue by atomic force microscopy also show a difference at 294 Pa and 454 Pa for white and gray matter respectively.[34] Others have shown the stiffness of brain tissue decreases with age[32] (approx. 2.5 kPa), though brain tissue is extremely soft during development in the embryo (15–35 Pa).[35] In general, the Young's modulus of brain tissue has been shown to range from a few hundred pascals up to 10 kPa, a range that could be considered suitable for biomaterials designed for use in the brain.

Another aspect of biomaterial design that must be considered is the surface charge. Research in the field of nonviral gene vectors has highlighted how charge affects the function. For example, these nanoscale particles, consisting of nucleic acid often condensed onto a polymer or lipid material, would generally be cationic to allow cell uptake and high transfection levels,[36,37] though this can come at the price of cellular toxicity.[29,36] However, cationic nonviral vectors that work well in vitro do not necessarily work well in the brain.[38] In fact, research by Mastorakos et al. showed vectors close to being neutrally charged showed far higher transfection levels in the brain.[39,40] They showed that the nanoparticles injected into the striatum showed much better distribution in the brain parenchyma if they were neutral rather than if they were cationic.[39] Seeing that the extracellular matrix (ECM) has a mix of negatively charged proteoglycans and other components,[35,41] the lack of an opposite charge improves mobility. More recent research has also shown how brain tissue will rapidly (within 20 minutes) adhere and grow into cationic and neutral scaffolds but is prevented from doing so in negatively charged scaffolds.[42] This may have implications for the future design of removable devices, where low tissue adherence is required.

Although there are many factors to consider for biomaterials used in neuroscience, one that the author feels is "generally accepted" by the field is that implantable or injectable systems must be biodegradable. As stated earlier, the side effects of Gliadel wafers (biodegradable) persist after all the drug is released and are considered to be due to the material's mechanical presence rather than the breakdown products of its composition.[13] Lewitus et al. showed that fibers that degraded within a week showed less of an astrocyte or microglia response after intracranial implantation than slow-degrading fibers.[43] Interestingly, the host response was not analyzed at the point of equal degradation of the various materials used; therefore, the polymer that degraded in a few days would of course have less of an astrocyte scar than one still present four weeks later (longest time analyzed).

For numerous applications, such as sustained drug release to the brain, much longer persistence times of the delivery system would be required. A simple, but thorough and highly revealing study by Bjugstad et al. showed that both degradable and nondegradable poly(ethylene) glycol (PEG)-based hydrogel implants were well tolerated, over a 56-day period.[44] The microglia in the constantly changing environment adjacent to the biodegradable materials seemed more activated than those in proximity to nondegradable hydrogels,[44] showing that biodegradation may not in fact be an automatic prerequisite for biocompatibility.

The effects of biomaterial design characteristics are clearly sometimes at odds, and a lack of large-scale comparative studies of various materials in the brain makes general criteria impossible to define. Much will depend on the intended use, but the already inflamed, compromised, or degenerated condition of the diseased brain must be considered to ensure minimal exacerbation when the goal is neural protection or repair.

2.2. *A broad overview of biomaterial types*

Biomaterial designs, and the methods for which polymeric biomaterials are synthesized, have advanced substantially over the past 30 years, vastly expanding the possible range, complexity, and function of delivery systems, scaffolds, and so on. Commonplace methods such as peptide synthesis allow the creation of libraries of peptides that can be used to functionalize biomaterials,[45] or even be used to create the biomaterial itself.[46–48] Likewise, living polymerization methods allow tailoring of the polymer components with high fidelity and the creation of varying structures from linear polymer chains to highly branched polymers.[49] The scope for creating fully

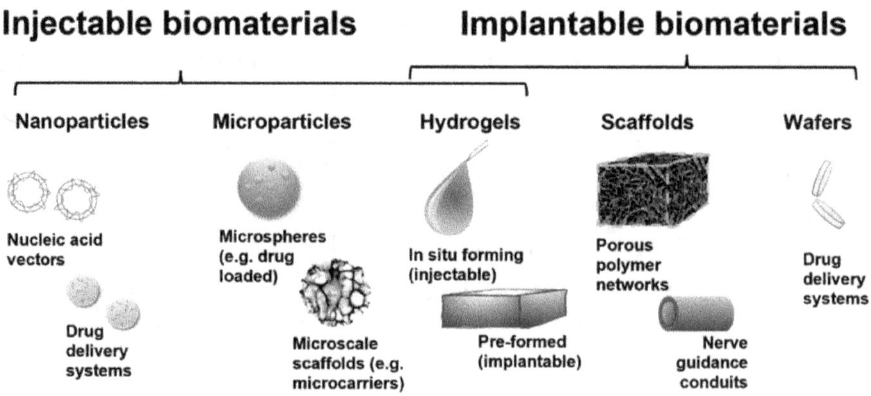

Figure 2. A general overview of the different categories of biomaterials that are being investigated for applications in neural tissue engineering and brain repair.

synthetic polymeric biomaterials is vast, and when combined with biopolymers now being extracted from a wide range of living organisms, the possibility for tailoring semisynthetic materials seems only limited by the creator's imagination. In an attempt to bring simplicity to the subject of biomaterials for use in neuroscience applications, **Figure 2** illustrates some broad categories into which these materials may be classified.

An obvious starting point in characterizing biomaterials for use in the brain could be the consideration of whether the material is to be injected (i.e., through a needle or cannula) or implanted (i.e., placed into a lesion or surgical cavity). Nanoscale materials, such as those of size 1–1,000 nm, will of course be injectable. These can include drug delivery systems such as a drug contained within a polymer matrix, either embedded in the network or encapsulated within a shell or a lipid membrane. Alternatively, nonviral gene vectors typically consist of a cationic moiety (either polymer or lipid) that condenses nucleic acids via electrostatic interactions, which together become a nanoparticle. As stated above, the overall charge of the nanoparticle affects cell uptake and distribution in the brain, but in general, achieving widespread and efficient transfection in the brain can be difficult.

Microscale materials such as microspheres, microparticles, and microcarriers can also be used for drugs, biologics, or cell delivery. With a lower surface area-to-volume ratio than nanoparticles, microparticles designed to release their cargo through biodegradation can potentially achieve long release times (several months[19,50]) though many still release the majority of their cargo within a few days.[51] Due to their comparatively large size, movement of microscale materials through the brain parenchyma cannot be expected, so should therefore be considered for applications where a focal effect of the intervention is required. These materials have therefore been extensively studied for use in brain cancers, Parkinson's disease, and stroke, where there is at least some degree of delineation between affected tissue and the otherwise functioning brain. Highly porous microparticles, which are scaled-down versions of implantable scaffolds, are often produced using an emulsion to create a droplet of the material and are sometimes referred to as microcarriers.[52,53] Since these are highly compressible, and theoretically the host brain cells can grow throughout the structure, these biomaterials create less "dead space" than solid microspheres.

As mentioned above, tailoring the mechanical properties of the biomaterials to be close to those of brain tissue is highly favorable. Soft hydrogels are therefore an attractive strategy for mediating brain repair,[54] and many researchers allow the gelation to be performed in situ. In this way, the precursor solution for the hydrogel is injected into the region of interest and allowed to gel in the host tissue. Arbitrary shapes such as a stroke lesion or tumor resection cavity can theoretically be evenly filled, as the initial liquid fills the form prior to gelation. This is relatively straightforward to perform, though the author's own previous experience with intracranial injection of temperature-based gelating hydrogels[55] showed care is needed in the setup to avoid premature gelation. Preformed hydrogels retain the advantage of soft mechanical properties but require implantation and thus seem not so heavily investigated for applications in the brain.

The term "scaffold" is rather broad, as hydrogels can also be considered a type of scaffold, but herein they are referred to as a macroporous or a fibrous structure, thus differing from the traditional hydrogel. Such scaffolds can be formed by a variety of methods such as electrospinning, three-dimensional (3D) printing, cryogelation, and salt templating to name a few,[56] but many may effectively have dense hydrogel struts, with large open pores, making the scaffold somewhat sponge-like in appearance.[57] Scaffolds of a variety of shapes and sizes have been created to suit the intended application such as bridging certain brain regions, though elongated biomaterials such as nerve guidance conduits have been extensively analyzed, particularly for the spinal cord[58–61] and for peripheral nerve regeneration.[62,63]

Finally, as mentioned previously, stiff wafers, created from biodegradable polymers, have been used to deliver chemotherapeutics to the tumor resection cavity, though with the aforementioned problems. It should be noted that many of these materials can be used in combination, such as embedding microspheres or nanoparticles, within a hydrogel.[64,65] Furthermore, although **Figure 2** depicts these as individual categories for the purpose of this text, in reality, these lines of characterization can be blurred. For instance, lipid nanocapsules loaded with a chemotherapeutic have been crosslinked together to form a hydrogel.[66] Similarly, researchers have used microscale beads of hydrogel, which when injected together form a slurry, which effectively fills a stroke cavity and anneals to form a macroscale hydrogel.[67–69]

3. Biomaterials for modeling neurodegeneration

Biomaterial scaffolds and delivery systems have been investigated for their potential therapeutic uses in the brain for decades[70–72] but have only more recently been employed in modeling neurological disorders. Our ability to develop new therapeutics is heavily reliant on model systems in which we can test therapeutic safety and efficacy. It is therefore important that these models recapitulate key features of the disease and are appropriate for answering the specific research questions. Whilst animal models remain vital for drug development, there has been a recent push to improve in vitro cell models of diseases, both through improving cell sources (such as iPSC-derived neurons) and through improving the extracellular environment to better reflect that of the human brain. For example, the heavily used two-dimensional (2D), flat tissue culture plastic substrate is now being superseded by 3D models, to better recapitulate the cell–cell and cell–ECM crosstalk.[73–76] These are typically in the form of hydrogels, porous scaffolds, or a combination of both, seeded with neurons and other glia.[77]

Matrigel, a mixture of extracellular components derived from mouse sarcomas, has been used as a substrate to embed neurons. For example, human neural progenitor cells genetically altered to produce the amyloid-B and tau aggregates characteristic of Alzheimer's disease have been seeded in Matrigel disks.[78] These showed better expression of mature human four-repeat tau isoforms than the same cells in 2D culture.[79] However, Matrigel requires large-scale animal use in its production and the composition is poorly defined, leading researchers to develop alternative biomaterials for tissue culture models. There is now a plethora of hydrogel systems for use in cell culture,[77] though of note for disease modeling, a semisynthetic, heparin and PEG hydrogel successfully sustained the 3D culture of human astrocytes or neural stem cells.[80,81] Administration of Alzheimer's disease-associated amyloid-β42 peptide caused neuropathology that mimicked much of the human condition, including impaired neuronal network formation, dystrophic axonal ends, and loss of synapses, serving as a test bed for further understanding neurodegeneration and developing therapeutics.[80,82] Self-assembling peptides[83] and fully synthetic hydrogels[84] have been used for dopaminergic neuron generation, which, although focused on therapeutic use, could potentially be combined with neurotoxins to form an in vitro model of dopaminergic neuron loss in Parkinson's disease. Other biomaterials, such as macroporous microcarriers, provide an alternative to traditional 2D culture and 3D neurosphere culture, allowing month-long culture of large numbers of neural progenitor cells without the hypoxic or necrotic core normally associated with neurospheres.[53]

To bring in a higher level of complexity to model systems, researchers have cultured brain tissue, often in slices at the air–liquid interface.[85–88] This method comes at the cost of lower throughput, but now includes multiple cell types, and a preexisting neural network, even if somewhat damaged due to the slicing or processing. However, interventions in ex vivo slice culture would typically affect the whole slice and not allow focal manipulation of the tissue. For instance, the addition of a demyelinating agent (lysolecithin) to mimic demyelination in multiple sclerosis (MS) created global demyelination across the whole slice.[85] Inflammatory demyelination has also been caused by transfecting brain slice cultures with interferon gamma and tumor necrosis factor genes either by the addition of the gene delivery system to the medium or by adding them on top of the slice.[89] This still results in a global effect, though it should be noted higher transfection was achieved by direct administration of the gene vector on top of the slice than into the medium.

To better mimic regions of demyelinated lesions surrounded by healthy brain tissue, small cylinders of cryogel biomaterials have been soaked in lysolecithin and added next to rodent brain slices.[90] This created focal lesions of demyelination in the cortical gray matter (**Figure 3**) surrounded by otherwise healthy tissue, allowing observation of the remyelination process by the oligodendrocyte precursor cells. Spontaneous remyelination of the lesioned area could be observed after 14 days, creating a model to test therapeutics aimed at boosting remyelination.

It was observed with human brain tissue that these PEG cryogels allowed ingrowth of the tissue into the cryogel scaffold, causing damage when the cryogels was removed.[42] To overcome this, the addition of sulfonated groups to give a negative charge prevented this adhesion and allowed repeat addition and removal of the cryogel directly on top of the slices in the culture without causing damage. One can therefore envisage focal white matter lesions could be possible via the addition of a sulfonated cryogel soaked in lysolecithin directly onto regions of white matter in a brain slice, furthering the applicability of the model.

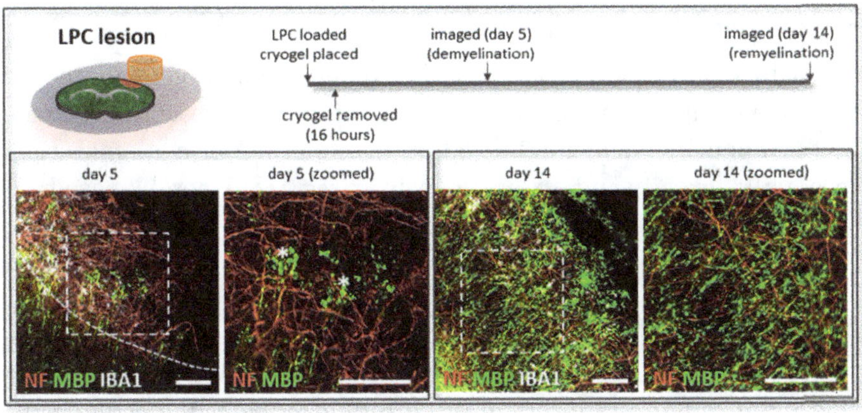

Figure 3. Biomaterials allow focal demyelination in brain tissue. A cryogel scaffold (depicted in orange) can be loaded with lysophosphatidylcholine (LPC) and placed next to mouse cortical tissue grown on a well-plate insert. This creates a region of demyelination by day 5 (absence of myelin basic protein [MBP]) with just debris marked with the asterisks between the neurofilament (NF) staining, with remyelination occurring by day 14. Scale bars represent 100 μm. Reproduced with permission from Eigel et al.[90]

The same principle of using cryogels to create focal models of demyelination has also been applied in vivo.[91] Focal demyelination of the cortical gray matter, which was surrounded by healthy tissue, was achieved. A hole was carefully bored in the skull, leaving a region of the cortex exposed onto which a 2-mm diameter and 0.5-mm thick PEG-based cryogel could be placed. Control animals received cryogels hydrated with phosphate-buffered saline (PBS), which caused no damage and no signs of toxicity. However, lysolecithin-soaked cryogels caused a small focal region of demyelination after two weeks.[91] Interestingly, this led to specific inhibitory neuron degeneration and synapse loss a week later, even in the absence of inflammation giving a new perspective for the development of neuroprotective therapies for MS.

Whilst these examples show that polymeric biomaterials can be used to refine models of neurodegenerative disorders, the vast majority of work using these materials in the brain has focused on the delivery of therapeutics and regenerative medicine.

4. Polymers for brain repair

As outlined in Section 2.2, there is a diverse range of biomaterials that have been used in the brain, predominantly focused on delivering therapeutics. The intended applications of these biomaterials are equally broad, and so are loosely categorized herein based on the payload rather than the clinical application.

4.1. *Nucleic acid delivery*

Taking Parkinson's disease as an example, the ability to manipulate genetic information in the brain could lead to better control of symptoms or to reducing the progression of the disease.[92,93] For example, enzyme replacement therapies aimed to improve patient quality of life via a more consistent level of dopamine replacement than can be achieved via oral medication with levodopa.[94] This strategy involves the overexpression of enzymes involved in the synthesis of dopamine, though it does not address the underlying disease pathology and progression. However, studies have also investigated whether the overexpression of neuroprotective proteins such as glial-cell line-derived neurotrophic factor (GDNF) or neurturin can slow disease progression.[95] Unfortunately, these trials have met with limited success to date, but gene therapies remain a promising approach for tackling neurodegenerative disorders.

Although the delivery of nucleic acids from nonviral vectors such as polymer-based delivery systems has typically lacked the efficacy shown by their viral counterparts, interest in polymeric biomaterials for nucleic acid delivery has grown substantially.[96–99] These typically involve a soluble polymer with a cationic moiety, which can condense negatively charged genetic material (i.e., plasmid DNA, short interfering RNA, micro-RNA, messenger RNA, or CRISPR/Cas9 components, etc.). Early animal studies of direct brain infusion of plasmid reporter genes (i.e., nontherapeutic genes that allow determination of transfection) used liposomes[97,100] before the polymer polyethyleneimine (PEI) was used in 1995.[101] Many of these studies used newborn rodents and proved the concept that transfection was possible. The author's experience of transfection in the *adult* rodent brain highlighted minimal reporter gene activity after intraventricular delivery with a cationic polymer.[38] This was disappointing, especially since robust overexpression of GDNF could be achieved many years previously with viral vectors.[102] However, as mentioned in Section 2.1, vectors that work well in vitro do not necessarily perform well in vivo, and studies have shown that reducing the overall surface charge of the polymer–plasmid complex (called polyplex) can vastly improve transfection efficacy in the rodent brain.[39,40] Vectors such as these, when administered via convection-enhanced delivery, are able to transfect similar volumes of animal brain tissue to an adeno-associated virus (**Figure 4**[103]).

Numerous nonviral gene delivery systems are being developed that cross the blood–brain barrier (BBB) with various efficacies. Off-target transfection is likely to remain a hurdle for systemic gene therapy, though several peptides for mediating the translocation of nanoparticles across the BBB have been developed.[104] Focused ultrasound can also aid translocation by transiently compromising the BBB.[99,104] Nose-to-brain delivery, via the olfactory system, is another potential way of allowing polymeric nanoparticles to enter

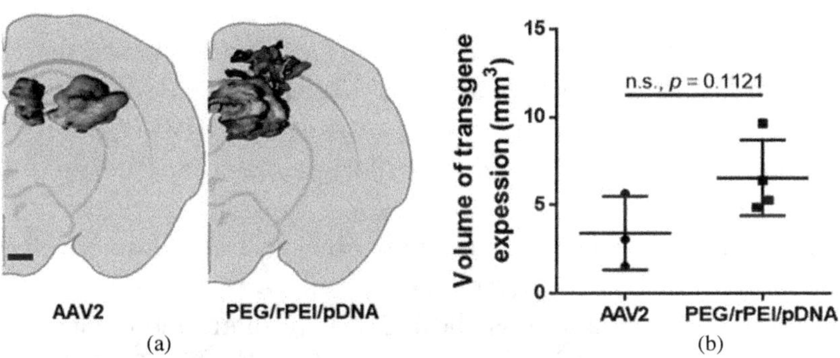

Figure 4. (**a**) Compact DNA nanoparticles (PEG/rPEI/pDNA) with near-neutral charge exhibit widespread transgene expression (**b**) that is similar in volume to an adeno-associated virus (AAV2). Scale bar represents 500 μm. Reproduced with permission from Rao et al.[103]

the brain, and intrathecal or intraventricular delivery via the cerebrospinal fluid also offers the potential to affect large areas of the brain.[99,104] The decision to focus on brain-wide transfection, versus regional transfection will likely influence the administration method. Local delivery may potentially be preferred for disorders with a focal pathology, such as Huntington's disease, Parkinson's disease, and stroke, for its specificity to the affected region without off-target side effects. In contrast, any system-wide effect for disorders such as MS and Alzheimer's disease will likely require systemic or at least intrathecal or intraventricular delivery.

4.2. *Controlled drug delivery*

Herein, the term "drug" will refer to small molecules, as proteins will be dealt with separately later. Whilst many drugs in routine use cross the BBB, those with a larger size or high surface charge fail to reach the target site in sufficient quantities for therapeutic effect.[105] This has become particularly problematic for developing brain cancer therapeutics.[106] Where other cancers have seen large improvements in prognosis over the past 30 years, brain cancers such as glioblastoma have seen little shift in the median overall survival rates.[107] Repositioning or repurposing chemotherapeutics amongst cancers could accelerate improvements in prognosis[108,109] but the BBB represents an obstacle, negating any chance of using many existing chemotherapeutics for brain cancers. Indeed, temozolomide, an alkylating agent that crosses the BBB, remains the single mainstay drug in common use for glioblastoma,[110] despite low potency and somewhat limited efficacy.[111]

Local drug delivery via an injectable or implantable delivery system opens the possibility of repurposing a wide range of drugs, which otherwise cannot pass the BBB.[112] Implantable devices have been extensively investigated in the field of glioblastoma research (clinical trial numbers: NCT00004028, NCT02433392, NCT00479765, NCT02858895, NCT01526837, NCT03596086). Many attempts have focused on the use of a drug encapsulated in a polymer wafer in a manner similar to the Gliadel wafer.[64] However, as previously alluded to, these cause side effects due to the mechanical mismatch between the stiff wafer and the soft brain tissue. To address this biocompatibility issue, many groups have focused on developing hydrogels as a means of sustaining drug release in the brain.[65,113]

For glioblastoma, drugs such as camptothecin, doxorubicin, gemcitabine, and paclitaxel have all been encapsulated in hydrogels, though release times vary.[51] Indeed, many use the hydrogel as a means of encapsulating another nanoscale drug delivery system[114,115] to improve the drug release profile. This begs the question of whether such nano- and microscale delivery systems can be used without a hydrogel. In the case of a resection cavity, even presentation of the drug at all regions of the perimeter becomes an issue, leading to the use of paste[116] or spray formulations,[117] or polymer meshes[118] for more even coverage and ease of application. However, researchers have also investigated injectable systems that hold the potential for use in nonresectable tumors,[51] though edema associated with local tumor destruction may present problems.

Locally injectable drug delivery systems to date have predominantly used PLGA as the encapsulating polymer.[119–121] These have the advantages of being well characterized and likely lead to a relatively straightforward path through regulatory approval. Such polymers allow the direct encapsulation of hydrophobic drugs and can be tailored to give a sustained drug release profile. However, a breadth of other delivery systems have been used,[122–124] though long-term release of a drug, or preferably multiple drugs, to give high therapeutic efficacy remains a difficult challenge.[51] Systemically administered drug–nanomaterial delivery systems will necessitate translocation of the BBB as dealt with above for nucleic acids, where a range of BBB penetrating strategies can be considered,[125] including the promising nose-to-brain delivery.[126,127]

4.3. *Sustained delivery of proteins*

Neurotrophic factors play an essential role in modulating the growth, repair, and survival of neurons in the brain.[128] Consequently, they may be able to slow or stop neurodegeneration in the brain.

They are under investigation for use as potential disease-modifying therapies for Parkinson's disease[129,130] and other neurodegenerative conditions such as Alzheimer's disease[131] and spinal cord injury.[132]

Many growth factors and their receptors have been identified in the central nervous system (CNS) and can be classified into four families as follows[133]: (1) neurotrophins such as brain-derived neurotrophic factor (BDNF) and nerve growth factor (NGF), (2) GDNF and their family ligands (GFLs), (3) neuropoietic cytokines, also known as the interleukin-6 family (IL-6), and (4) a family formed by cerebral dopamine neurotrophic factor (CDNF) and mesencephalic astrocyte-derived neurotrophic factor (MANF).

Growth factors are proteins with a relatively large molecular weight, making them unable to cross the BBB and reach the target area in the brain. Direct infusion into the brain offers a means of bypassing the BBB, and techniques such as convection-enhanced delivery allow greater spread of the protein throughout the brain parenchyma.[134,135] However, it should be noted that many proteins and trophic factors have a short half-life in vivo. For example, the half-life of GDNF is approximately 34 hours in cerebrospinal fluid[136] and 3–4 days in the brain.[137,138]

In an attempt to negate this short duration of action, repeat administrations have been considered. For example, one clinical trial, which focused on the direct infusion of GDNF to the striatum of patients with Parkinson's disease, utilized a port attached to the skull.[139,140] This port was connected to the cannulae embedded into the striatum, so that patients could be hooked up for repeat administrations every month. Whilst ultimately the clinical trial was not defined as a success based on its clinical endpoint,[139,140] many patients saw a remarkable improvement in daily function. However, it should be borne in mind that with a half-life of days and not months, patients actually received therapeutic doses for very little of the trial duration. Indeed, an in vivo study using a bolus injection of GDNF into the rat brain showed no immunohistochemical detection of the protein (above baseline) after just seven days.[141] This corroborates previous data showing GDNF to be largely absent from the brain by seven days, whereas neurturin was still detected at 21 days postinfusion into the rat striatum.[142] Such data, both the therapeutic potential and the shortcomings of delivery, provide a strong rationale for the development of a delivery system that protects the protein from degradation and releases it slowly to the surrounding tissue.

Early studies in local delivery used NGF stabilized with bovine serum albumin dispersed through a polymer disk.[143] These were 1 mm thick and 2 mm in diameter and highlighted the potential to deliver the protein to localized regions of the brain, even if with a limited diffusion distance of 1–2 mm from the implant site.[143] Progression to microparticles, predominantly formed from PLGA or other synthetic degradable polymers, allowed the encapsulation of proteins inside the polymer carrier, which could then be injected directly into the brain parenchyma,[19,144–146] or embedded in hydrogels for implantation.[147] One downside to this encapsulation strategy is that the release of the neurotrophic factor is typically short (e.g., up to one week),[50] though coatings added to microparticles, for example, poly(sebacic acid), can extend the release to several weeks.[148]

Another means of delivering proteins with more control and over an extended duration is to use electrostatic interactions between a charged carrier and charged regions of a protein. Interestingly, PLGA nanoparticles, with a negative surface charge, could sustain the release of the positively charged proteins neurotrophin-3 (NT3), stromal-derived growth factor 1α (SDF-1), and BDNF simply by mixing empty nanoparticles with these proteins.[149] The release rate and duration (~30 days) were no different to PLGA nanoparticles that had the proteins encapsulated within the polymer matrix.[149] Another protein, erythropoietin, with an overall negative charge, could not be delivered in a sustained manner, as this nanoparticle–protein combination would fail to show electrostatic interactions. The same group has used PLGA nanoparticles to effectively deliver other neuroprotective and charged proteins to the CNS, such as the pituitary adenylate cyclase-activating polypeptide (PACAP).[150]

This strategy of using affinity-based (electrostatic interaction) delivery systems has gained much traction for the sustained and controlled nature of protein delivery.[151] Nanoparticles, formed from the highly anionic biopolymer heparin, have been used to control the delivery of proteins in this way, with both SDF-1 and vascular endothelial growth factor (VEGF) being used to improve functional recovery postischemic stroke.[69,152] Heparin-based hydrogels have been extensively used to deliver proteins in a slow and sustained manner.[153–155] Cryogels containing heparin have also shown controlled release of growth factors[156–158] and the immunomodulatory cytokine IL-13[30] over multiple weeks. It should be noted that affinity-based control over therapeutic delivery is not restricted to proteins but can also be used for charged small molecules.[20,159–161] The clear limitation of this strategy is the restriction of molecules and proteins that can be delivered, as only those with sufficiently charged moieties can be delivered in this way.[162] Nevertheless, heparin-mimicking cryogels, designed with sulfonate groups, have been shown to deliver both GDNF and BDNF, with durations ranging up to 28 weeks.[141] Furthermore, since the proteins are not encapsulated inside the polymer matrix, but rather loaded to it, these could be refilled to give sustained release of GDNF for nine months (the longest time analyzed).[141] Such systems open the door to refilling by repeat administration, which may radically enhance the therapeutic efficacy of protein-based therapeutics in the future.

4.4. *Polymeric scaffolds for cell delivery and brain repair*

Cell therapies hold significant promise for the treatment of neurodegenerative disorders, offering diverse mechanisms that could address the underlying pathology of these conditions. Direct cell replacement therapy involves transplanting functional neurons or, more typically, neural progenitor cells, directly to the lesion area to replace those lost to disease, offering a direct means to restore neural circuits and improve neurological function. This strategy holds potential for disorders such as Parkinson's disease[163–165] and Huntington's disease,[166,167] where the brain regions of neuron loss are relatively focal. Alternatively, cell-based delivery of neuroprotective agents leverages genetically modified cells or stem cells engineered to secrete neurotrophic factors,[168,169] providing a sustained source of neuroprotection and supporting the survival of existing neurons. Again, such a strategy would be best suited to local administration to focal areas of neuron loss, which could of course include stroke[170] and some traumatic brain injuries.[171]

For more diffuse neurodegeneration, such as in amyotrophic lateral sclerosis (ALS) and MS, systematic administration of mesenchymal stem cells (MSCs) has been investigated.[172,173] The immunomodulatory properties of MSCs and their ability to promote repair through their secretome, which includes anti-inflammatory cytokines and growth factors, have led researchers to focus on secretome delivery rather than cell delivery.[153,174,175]

Polymeric scaffolds such as electrospun fibers, hydrogels, cryogels, and micro- or nanoscale delivery systems have been used in a variety of ways to improve the efficacy of cell transplantation. Of particular note, many have been developed to improve the survival and maturation of neuron progenitors posttransplantation to the brain. Although numerous animal studies show good graft survival, the percentage of surviving dopaminergic neuron progenitors can be very low indeed, with a recent systematic review showing an average of only 3% survival.[176] Cell death and poor maturation posttransplantation are likely to be multifaceted, and biomaterial strategies to improve survival have aimed at a number of potential mechanisms.[96] Encapsulating the transplanted cells in an in situ-forming hydrogel has been tried to ameliorate the host response to the graft.[54,55] Furthermore, this strategy can be enhanced by also using the hydrogel to release growth factors to the cells to provide neuroprotection and/or enhance differentiation of the progenitors to their mature phenotype.[177,178] Others have looked at encapsulating immunosuppressants in a polymer matrix to reduce the rejection of grafted cells.[179] Our group has proposed that a lack of nutrients, particularly glucose, could be a major contributor to cell death, particularly in the early stages before graft vascularization has occurred.[180,181] The clear region of hypoxia shown in the graft core soon after transplantation to the brain shows just how quickly the transplanted cells are using up their initial resources.[182] However, glucose, being such a small, hydrophilic, and uncharged molecule, is difficult to deliver in a sustained manner, though macroscale attempts too large for neural grafting are encouraging.[183,184]

Biomaterials could play a particularly interesting role in avoiding anoikis (cell death due to a lack of adhesion to a substrate).[156] With the advent of stem cell-derived neurons, the use of fetal tissue to replace neural networks is being increasingly seen as unfeasible, not least for the logistical challenges of obtaining donor tissue immediately prior to transplantation.[185] Instead, neurons can be derived from embryonic or induced pluripotent stem cell sources.[186–188] Transplantation of mature neurons leads to poor graft survival in in vivo models, and neuron progenitors fare better despite not being fully differentiated or functional.[189,190] This may be due to axotomy or severing of the dendrites, which occurs upon lifting the cells from their substrate prior to transplantation.

Indeed, Adil et al. showed that simply lifting neuron progenitor cells and replating them causes over 50% cell death (**Figure 5**),[191] which has been corroborated by Filippova et al.[192] These research teams showed that prior culture of the cells to a substrate that can be cotransplanted (i.e., no lifting of cell from a substrate) can vastly improve cell survival posttransplantation.[191–193] Others have used cells cultured on microcarriers to aid transplantation.[194,195] One can envisage that for cell replacement therapies, using the microcarrier as a substrate on which the cells can be differentiated and then transplanted may maximize the reduction of anoikis.

Together, it seems that biomaterials may be able to nourish grafted cells, helping differentiation and integration into the host brain and potentially provide a substrate for the transplantation process. Another interesting development, this time focused on systemic cell therapies, is the attachment of polymeric "backpacks" to monocyte cells prior to administration to an animal model of MS.[196] Whilst monocytes can be switched toward more proregenerative

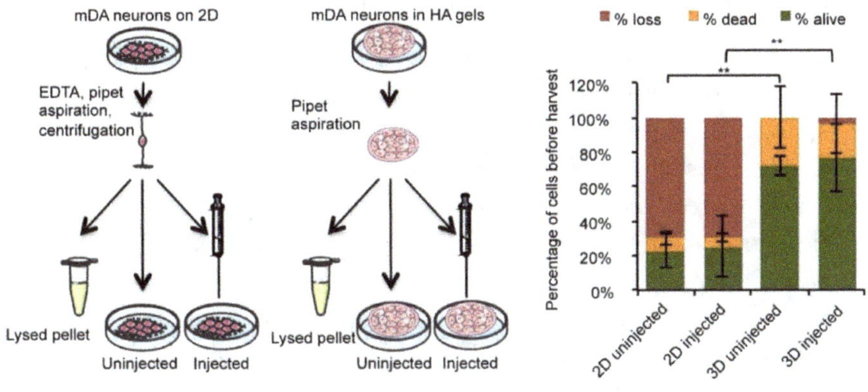

Figure 5. (**a**) Midbrain dopaminergic (mDA) neurons harvested from 2D or 3D cultures were either lysed, re-seeded, or re-seeded after injection through a 26-gauge needle. (**b**) Lactose dehydrogenase (LDH) levels were measured at different stages for cells harvested from 2D or 3D platforms and presented as a percentage of total cells prior to harvest. The percentage of live cells (green), dead cells (orange), and lost cells (red) were calculated for each condition. The majority of neurons are lost simply through lifting and replating the cells, whereas those remaining in hyaluronic acid (HA)-based hydrogels have high viability as no disruption to their neurites takes place. Reproduced with permission from Adil et al.[191]

phenotypes,[197] this change is dynamic, reversible, and is dependent on microenvironmental cues. Kapate et al. designed polymer disks ("backpacks") loaded with IL-4 and dexamethasone, which in combination stimulate anti-inflammatory and regulatory functions while dampening pro-inflammatory cytokine secretion.[196] Heparin was included to increase IL-4 loading, and a F(ab′) anti-CD45 antibody fragment was added, allowing approximately 60% of monocytes to bind to a backpack. This strategy gave ~5 days of release of the IL-4 and dexamethasone, helping the transplanted monocytes to retain their proregenerative phenotypes. One could envisage longer drug retention times and a variation of payloads opening previously unexplored avenues in systemic cell transplantation therapies.

5. Conclusions and thoughts for the future

Polymeric biomaterials have been used for a wide range of applications in neural tissue and the brain. As our understanding of brain physiology increases, biomaterial design will need to adapt to keep pace. This has thus far been marked by the shift from using stiff materials to soft biomaterial implants such as hydrogels. The issue of biodegradation needs to be carefully considered, as a slowly changing microenvironment caused by a degrading delivery system may potentially exacerbate the host response to the implant. This is still very much unclear, and further study is needed here, with closely comparable biomaterials to elucidate this point.

Significant progress has been made using biomaterial systems to deliver nucleic acids, small molecule drugs, proteins, and cells to the brain. As a field in general, particularly in the academic sector, we should perhaps shift our focus from early publication of our results to protection of intellectual property, particularly for promising therapeutic strategies. With each publication comes an erosion of the next inventive step, so we must be mindful of commercialization and clinical translation for the predominantly preclinical progress highlighted herein to evolve into the therapeutics of tomorrow.

References

1. Cobb M. A brief history of wires in the brain. *Front Ecol Evol.* 2021;9:760269.
2. Cobb M. *The Idea of the Brain: The Past and Future of Neuroscience.* Paris: Hachette UK; 2020.
3. Piccolino M. Animal electricity and the birth of electrophysiology: The legacy of Luigi Galvani. *Brain Res Bull.* 1998;46(5):381–407.
4. Schuetze SM. The discovery of the action potential. *Trends Neurosci.* 1983;6:164–168.
5. del Río-Hortega Bereciartu J. Pío del Río-Hortega: The revolution of glia. *Anat Rec.* 2020;303(5):1232–1241.
6. García-Marín V, García-López P, Freire M. Cajal's contributions to glia research. *Trends Neurosci.* 2007;30(9):479–487.

7. y Cajal R. Contribución al conocimiento de la neuroglia del cerebro humano. *Trab Lab Invest Biol.* 1913;11:255.
8. Bandettini PA. Twenty years of functional MRI: The science and the stories. *Neuroimage.* 2012;62(2):575–588.
9. O'Donovan MC. A novel gene containing a trinucleotide repeat that is expanded and unstable on Huntington's disease chromosomes. *Cell.* 1993;72(6):971–983.
10. Williams DF. On the nature of biomaterials. *Biomaterials.* 2009; 30(30):5897–5909.
11. McGlynn E, Nabaei V, Ren E, Galeote-Checa G, Das R, Curia G, et al. The future of neuroscience: Flexible and wireless implantable neural electronics. *Adv Sci.* 2021;8(10):2002693.
12. Minev IR. Electronic tissue technologies for seamless biointerfaces. *J Polym Sci.* 2023;61(16):1707–1712.
13. Tabet A, Jensen MP, Parkins CC, Patil PG, Watts C, Scherman OA. Designing next-generation local drug delivery vehicles for glioblastoma adjuvant chemotherapy: Lessons from the clinic. *Adv Healthc Mater.* 2019;8(3):1801391.
14. De Bonis P, Anile C, Pompucci A, Fiorentino A, Balducci M, Chiesa S, et al. Safety and efficacy of Gliadel wafers for newly diagnosed and recurrent glioblastoma. *Acta Neurochir.* 2012;154: 1371–1378.
15. Nishikawa R, Iwata H, Sakata Y, Muramoto K, Matsuoka T. Safety of Gliadel implant for malignant glioma: Report of postmarketing surveillance in Japan. *Neurol Med Chir.* 2021;61(9): 536–548.
16. Wolinsky JB, Colson YL, Grinstaff MW. Local drug delivery strategies for cancer treatment: Gels, nanoparticles, polymeric films, rods, and wafers. *J Control Release.* 2012;159(1):14–26.
17. Menei P, Daniel V, Montero-Menei C, Brouillard M, Pouplard-Barthelaix A, Benoit JP. Biodegradation and brain tissue reaction to poly(D,L-lactide-co-glycolide) microspheres. *Biomaterials.* 1993;14(6):470–478.
18. Chen W, He J, Olson JJ, Lu DR. Carboplatin-loaded PLGA microspheres for intracerebral implantation: In vivo characterization. *Drug Deliv.* 1997;4(4):301–311.
19. Garbayo E, Montero-Menei CN, Ansorena E, Lanciego JL, Aymerich MS, Blanco-Prieto MJ. Effective GDNF brain delivery

using microspheres—A promising strategy for Parkinson's disease. *J Control Release.* 2009;135(2):119–126.

20. Alghamdi M, Chierchini F, Eigel D, Taplan C, Miles T, Pette D, et al. Poly (ethylene glycol) based nanotubes for tuneable drug delivery to glioblastoma multiforme. *Nanoscale Adv.* 2020;2(10): 4498–4509.
21. Newland B, Taplan C, Pette D, Friedrichs J, Steinhart M, Wang W, et al. Soft and flexible poly(ethylene glycol) nanotubes for local drug delivery. *Nanoscale.* 2018;10(18):8413–8421.
22. Costa PM, Bourgognon M, Wang JTW, Al-Jamal KT. Functionalised carbon nanotubes: From intracellular uptake and cell-related toxicity to systemic brain delivery. *J Control Release.* 2016;241:200–219.
23. Nunes A, Bussy C, Gherardini L, Meneghetti M, Herrero MA, Bianco A, et al. In vivo degradation of functionalized carbon nanotubes after stereotactic administration in the brain cortex. *Nanomedicine.* 2012;7(10):1485–1494.
24. Ghasemi-Mobarakeh L, Kolahreez D, Ramakrishna S, Williams D. Key terminology in biomaterials and biocompatibility. *Curr Opin Biomed Eng.* 2019;10:45–50.
25. Ciccocioppo F, Bologna G, Ercolino E, Pierdomenico L, Simeone P, Lanuti P, et al. Neurodegenerative diseases as proteinopathies-driven immune disorders. *Neural Regen Res.* 2020;15(5):850–856.
26. Doty KR, Guillot-Sestier M-V, Town T. The role of the immune system in neurodegenerative disorders: Adaptive or maladaptive? *Brain Res.* 2015;1617:155–173.
27. Heneka MT, Kummer MP, Latz E. Innate immune activation in neurodegenerative disease. *Nat Rev Immunol.* 2014;14(7):463–477.
28. Moloney TC, Dockery P, Windebank AJ, Barry FP, Howard L, Dowd E. Survival and immunogenicity of mesenchymal stem cells from the green fluorescent protein transgenic rat in the adult rat brain. *Neurorehabil Neural Repair.* 2010;24(7):645–656.
29. Newland B, Moloney TC, Fontana G, Browne S, Abu-Rub MT, Dowd E, et al. The neurotoxicity of gene vectors and its amelioration by packaging with collagen hollow spheres. *Biomaterials.* 2013;34(8):2130–2141.
30. Schirmer L, Hoornaert C, Le Blon D, Eigel D, Neto C, Gumbleton M, et al. Heparin-based, injectable microcarriers for controlled

delivery of interleukin-13 to the brain. *Biomater Sci.* 2020;8(18): 4997–5004.

31. Green MA, Bilston LE, Sinkus R. In vivo brain viscoelastic properties measured by magnetic resonance elastography. *NMR Biomed.* 2008;21(7):755–764.
32. Takamura T, Motosugi U, Sasaki Y, Kakegawa T, Sato K, Glaser KJ, et al. Influence of age on global and regional brain stiffness in young and middle-aged adults. *J Magn Reson Imaging.* 2020; 51(3):727–733.
33. Schregel K, Wuerfel E, Garteiser P, Gemeinhardt I, Prozorovski T, Aktas O, et al. Demyelination reduces brain parenchymal stiffness quantified in vivo by magnetic resonance elastography. *Proc Natl Acad Sci U S A.* 2012;109(17):6650–6655.
34. Christ AF, Franze K, Gautier H, Moshayedi P, Fawcett J, Franklin RJM, et al. Mechanical difference between white and gray matter in the rat cerebellum measured by scanning force microscopy. *J Biomech.* 2010;43(15):2986–2992.
35. Long KR, Newland B, Florio M, Kalebic N, Langen B, Kolterer A, et al. Extracellular matrix components HAPLN1, lumican, and collagen I cause hyaluronic acid-dependent folding of the developing human neocortex. *Neuron.* 2018;99(4):702–719.e6.
36. Breunig M, Lungwitz U, Liebl R, Goepferich A. Breaking up the correlation between efficacy and toxicity for nonviral gene delivery. *Proc Natl Acad Sci U S A.* 2007;104(36):14454–14459.
37. Newland B, Abu-Rub M, Naughton M, Zheng Y, Pinoncely AV, Collin E, et al. GDNF gene delivery via a 2-(dimethylamino)ethyl methacrylate based cyclized knot polymer for neuronal cell applications. *ACS Chem Neurosci.* 2013;4:540–546.
38. Newland B, Aied A, Pinoncely AV, Zheng Y, Zhao T, Zhang H, et al. Untying a nanoscale knotted polymer structure to linear chains for efficient gene delivery in vitro and to the brain. *Nanoscale.* 2014;6(13):7526–7533.
39. Mastorakos P, Zhang C, Berry S, Oh Y, Lee S, Eberhart CG, et al. Highly PEGylated DNA nanoparticles provide uniform and widespread gene transfer in the brain. *Adv Healthc Mater.* 2015;4(7): 1023–1033.
40. Mastorakos P, Song E, Zhang C, Berry S, Park HW, Kim YE, et al. Biodegradable DNA nanoparticles that provide widespread gene delivery in the brain. *Small.* 2016;12(5):678–685.

41. Soles A, Selimovic A, Sbrocco K, Ghannoum F, Hamel K, Moncada EL, et al. Extracellular matrix regulation in physiology and in brain disease. *Int J Mol Sci.* 2023;24(8):7049.
42. Eigel D, Schuster R, Männel MJ, Thiele J, Panasiuk MJ, Andreae LC, et al. Sulfonated cryogel scaffolds for focal delivery in ex-vivo brain tissue cultures. *Biomaterials.* 2021;271:120712.
43. Lewitus DY, Smith KL, Shain W, Bolikal D, Kohn J. The fate of ultrafast degrading polymeric implants in the brain. *Biomaterials.* 2011;32(24):5543–5550.
44. Bjugstad KB, Lampe K, Kern DS, Mahoney M. Biocompatibility of poly(ethylene glycol)-based hydrogels in the brain: An analysis of the glial response across space and time. *J Biomed Mater Res A.* 2010;95A(1):79–91.
45. Quadir MA, Martin M, Hammond PT. Clickable synthetic polypeptides—routes to new highly adaptive biomaterials. *Chem Mater.* 2014;26(1):461–476.
46. Rodriguez AL, Bruggeman KF, Wang Y, Wang T-Y, Williams RJ, Parish CL, et al. Using minimalist self-assembling peptides as hierarchical scaffolds to stabilise growth factors and promote stem cell integration in the injured brain. *J Tissue Eng Regen Med.* 2018;12(3):e1571–e1579.
47. Wang T-Y, Bruggeman KF, Kauhausen JA, Rodriguez AL, Nisbet DR, Parish CL. Functionalized composite scaffolds improve the engraftment of transplanted dopaminergic progenitors in a mouse model of Parkinson's disease. *Biomaterials.* 2016;74:89–98.
48. Rodriguez A, Wang T-Y, Bruggeman K, Horgan C, Li R, Williams RJ, et al. In vivo assessment of grafted cortical neural progenitor cells and host response to functionalized self-assembling peptide hydrogels and the implications for tissue repair. *J Mater Chem B.* 2014;2(44):7771–7778.
49. Gao Y, Zhou D, Lyu J, Sigen A, Xu Q, Newland B, et al. Complex polymer architectures through free-radical polymerization of multivinyl monomers. *Nat Rev Chem.* 2020;4(4):194–212.
50. Garbayo E, Ansorena E, Lanciego JL, Aymerich MS, Blanco-Prieto MJ. Sustained release of bioactive glycosylated glial cell-line derived neurotrophic factor from biodegradable polymeric microspheres. *Eur J Pharm Biopharm.* 2008;69(3):844–851.

51. Wang Y, Bastiancich C, Newland B. Injectable local drug delivery systems for glioblastoma: A systematic review and meta-analysis of progress to date. *Biomater Sci.* 2023;11(5):1553–1566.
52. Feng L, Li D, Tian Y, Zhao C, Sun Y, Kou X, et al. One-step cell biomanufacturing platform: Porous gelatin microcarrier beads promote human embryonic stem cell-derived midbrain dopaminergic progenitor cell differentiation in vitro and survival after transplantation in vivo. *Neural Regen Res.* 2024;19(2):458–464.
53. Newland B, Ehret F, Hoppe F, Eigel D, Pette D, Newland H, et al. Macroporous heparin-based microcarriers allow long-term 3D culture and differentiation of neural precursor cells *Biomaterials.* 2020;230:119540.
54. Nih LR, Carmichael ST, Segura T. Hydrogels for brain repair after stroke: An emerging treatment option. *Curr Opin Biotechnol.* 2016;40:155–163.
55. Hoban DB, Newland B, Moloney TC, Howard L, Pandit A, Dowd E. The reduction in immunogenicity of neurotrophin overexpressing stem cells after intra-striatal transplantation by encapsulation in an in situ gelling collagen hydrogel. *Biomaterials.* 2013;34(37):9420–9429.
56. De France KJ, Xu F, Hoare T. Structured macroporous hydrogels: Progress, challenges, and opportunities. *Adv Healthc Mater.* 2018;7(1):1700927.
57. Bencherif SA, Sands RW, Bhatta D, Arany P, Verbeke CS, Edwards DA, et al. Injectable preformed scaffolds with shape-memory properties. *Proc Natl Acad Sci U S A.* 2012;109(48):19590–19595.
58. Abu-Rub MT, Billiar KL, van Es MH, Knight A, Rodriguez BJ, Zeugolis DI, et al. Nano-textured self-assembled aligned collagen hydrogels promote directional neurite guidance and overcome inhibition by myelin associated glycoprotein. *Soft Matter.* 2011;7(6):2770–2781.
59. Führmann T, Anandakumaran PN, Shoichet MS. Combinatorial therapies after spinal cord injury: How can biomaterials help? *Adv Healthc Mater.* 2017;6(10):1601130.
60. Suzuki H, Imajo Y, Funaba M, Ikeda H, Nishida N, Sakai T. Current concepts of biomaterial scaffolds and regenerative therapy for spinal cord injury. *Int J Mol Sci.* 2023;24(3):2528.

61. Yao L, Daly W, Newland B, Yao S, Wang W, Chen BKK, et al. Improved axonal regeneration of transected spinal cord mediated by multichannel collagen conduits functionalized with neurotrophin-3 gene. *Gene Ther.* 2013;20(12):1149.
62. Gonzalez-Perez F, Cobianchi S, Heimann C, Phillips JB, Udina E, Navarro X. Stabilization, rolling, and addition of other extracellular matrix proteins to collagen hydrogels improve regeneration in chitosan guides for long peripheral nerve gaps in rats. *Neurosurgery.* 2017;80(3):465–474.
63. Vijayavenkataraman S. Nerve guide conduits for peripheral nerve injury repair: A review on design, materials and fabrication methods. *Acta Biomater.* 2020;106:54–69.
64. Bastiancich C, Bozzato E, Henley I, Newland B. Does local drug delivery still hold therapeutic promise for brain cancer? A systematic review. *J Control Release.* 2021;337:296–305.
65. Bastiancich C, Malfanti A, Préat V, Rahman R. Rationally designed drug delivery systems for the local treatment of resected glioblastoma. *Adv Drug Deliv Rev.* 2021;177:113951.
66. Wang M, Bergès R, Malfanti A, Préat V, Bastiancich C. Local delivery of doxorubicin prodrug via lipid nanocapsule–based hydrogel for the treatment of glioblastoma. *Deliv Transl Res.* 2023; 13(1):1–17.
67. Nih LR, Sideris E, Carmichael ST, Segura T. Injection of microporous annealing particle (MAP) hydrogels in the stroke cavity reduces gliosis and inflammation and promotes NPC migration to the lesion. *Adv Mater.* 2017;29(32):1606471.
68. Sideris E, Kioulaphides S, Wilson KL, Yu A, Chen J, Carmichael ST, et al. Particle hydrogels decrease cerebral atrophy and attenuate astrocyte and microglia/macrophage reactivity after stroke. *Adv Ther.* 2022;5(8):2200048.
69. Nih LR, Gojgini S, Carmichael ST, Segura T. Dual-function injectable angiogenic biomaterial for the repair of brain tissue following stroke. *Nat Mater.* 2018;17(7):642–651.
70. Saltzman WM, Mak MW, Mahoney MJ, Duenas ET, Cleland JL. Intracranial delivery of recombinant nerve growth factor: Release kinetics and protein distribution for three delivery systems. *Pharm Res.* 1999;16:232–240.

71. Delcroix GJ-R, Garbayo E, Sindji L, Thomas O, Vanpouille-Box C, Schiller PC, et al. The therapeutic potential of human multipotent mesenchymal stromal cells combined with pharmacologically active microcarriers transplanted in hemi-parkinsonian rats. *Biomaterials.* 2011;32(6):1560–1573.
72. Shoichet MS, Winn SR, Gentile FT, Athavale S, Milton Harris JM. Poly(ethylene oxide)-grafted thermoplastic membranes for use as cellular hybrid bio-artificial organs in the central nervous system. *Biotechnol Bioeng.* 1994;43(7):563–572.
73. Bray LJ, Binner M, Holzheu A, Friedrichs J, Freudenberg U, Hutmacher DW, et al. Multi-parametric hydrogels support 3D in vitro bioengineered microenvironment models of tumour angiogenesis. *Biomaterials.* 2015;53:609–620.
74. Whitehouse C, Corbett N, Brownlees J. 3D models of neurodegeneration: Implementation in drug discovery. *Trends Pharmacol Sci.* 2023;44(4):208–221.
75. Centeno EG, Cimarosti H, Bithell A. 2D versus 3D human induced pluripotent stem cell-derived cultures for neurodegenerative disease modelling. *Mol Neurodegener.* 2018;13:1–15.
76. Newland B, Ehret F, Hoppe F, Eigel D, Pette D, Newland H, et al. Static and dynamic 3D culture of neural precursor cells on macroporous cryogel microcarriers. *MethodsX.* 2020;7:100805.
77. Mullis AS, Kaplan DL. Functional bioengineered tissue models of neurodegenerative diseases. *Biomaterials.* 2023;298:122143.
78. Kim YH, Choi SH, D'avanzo C, Hebisch M, Sliwinski C, Bylykbashi E, et al. A 3D human neural cell culture system for modeling Alzheimer's disease. *Nat Protoc.* 2015;10(7):985–1006.
79. Choi SH, Kim YH, Hebisch M, Sliwinski C, Lee S, D'Avanzo C, et al. A three-dimensional human neural cell culture model of Alzheimer's disease. *Nature.* 2014;515(7526):274–278.
80. Papadimitriou C, Celikkaya H, Cosacak MI, Mashkaryan V, Bray L, Bhattarai P, et al. 3D culture method for Alzheimer's disease modeling reveals interleukin-4 rescues Aβ42-induced loss of human neural stem cell plasticity. *Dev Cell.* 2018;46(1):85–101.e8.
81. Siddiqui T, Celikkaya H, Atasavum ZT, Popova S, Freudenberg U, Werner C, et al. Three-dimensional biohybrid starPEG–heparin hydrogel cultures for modeling human neuronal development and

Alzheimer's disease pathology. In: Chun J, editor. *Alzheimer's Disease: Methods and Protocols*. New York: Springer US; 2023. p. 159–170.

82. Siddiqui T, Cosacak MI, Popova S, Bhattarai P, Yilmaz E, Lee AJ, et al. Nerve growth factor receptor (NGFR) induces neurogenic plasticity by suppressing reactive astroglial Lcn2/Slc22a17 signaling in Alzheimer's disease. *NPJ Regen Med*. 2023;8(1):33.
83. Ni N, Hu Y, Ren H, Luo C, Li P, Wan J-B, et al. Self-assembling peptide nanofiber scaffolds enhance dopaminergic differentiation of mouse pluripotent stem cells in 3-dimensional culture. *PLoS One*. 2013;8(12):e84504.
84. Adil MM, Rodrigues GMC, Kulkarni RU, Rao AT, Chernavsky NE, Miller EW, et al. Efficient generation of hPSC-derived midbrain dopaminergic neurons in a fully defined, scalable, 3D biomaterial platform. *Sci Rep*. 2017;7(1):40573.
85. Zhang H, Jarjour AA, Boyd A, Williams A. Central nervous system remyelination in culture—A tool for multiple sclerosis research. *Exp Neurol*. 2011;230(1):138–148.
86. Ravi VM, Joseph K, Wurm J, Behringer S, Garrelfs N, d'Errico P, et al. Human organotypic brain slice culture: A novel framework for environmental research in neuro-oncology. *Life Sci Alliance*. 2019;2(4):e201900305.
87. Zhang J, Straehle J, Joseph K, Neidert N, Behringer S, Göldner J, et al. Isolation and profiling of viable tumor cells from human ex vivo glioblastoma cultures through single-cell transcriptomics. *STAR Proto*. 2023;4(3):102383.
88. Plug BC, Revers IM, Breur M, González GM, Timmerman JA, Meijns NRC, et al. Human post-mortem organotypic brain slice cultures: A tool to study pathomechanisms and test therapies. *Acta Neuropathol Commun*. 2024;12(1):83.
89. Mathew A, Pakan JMP, Collin EC, Wang W, McDermott KW, Fitzgerald U, et al. An ex-vivo multiple sclerosis model of inflammatory demyelination using hyperbranched polymer. *Biomaterials*. 2013;34(23):5872–5882.
90. Eigel D, Zoupi L, Sekizar S, Welzel PB, Werner C, Williams A, et al. Cryogel scaffolds for regionally constrained delivery of lysophosphatidylcholine to central nervous system slice cultures: A model of focal demyelination for multiple sclerosis research. *Acta Biomater*. 2019;97:216–229.

91. Zoupi L, Booker SA, Eigel D, Werner C, Kind PC, Spires-Jones TL, et al. Selective vulnerability of inhibitory networks in multiple sclerosis. *Acta Neuropathol.* 2021;141:415–429.
92. Bjorklund T, Kordower JH. Gene therapy for Parkinson's disease. *Mov Disord.* 2010;25(S1):S161–S173.
93. Chu Y, Bartus RT, Manfredsson FP, Olanow CW, Kordower JH. Long-term post-mortem studies following neurturin gene therapy in patients with advanced Parkinson's disease. *Brain.* 2020;143(3): 960–975.
94. Christine CW, Richardson RM, Van Laar AD, Thompson ME, Fine EM, Khwaja OS, et al. Safety of AADC gene therapy for moderately advanced Parkinson disease. *Neurology.* 2022;98(1): e40–e50.
95. Axelsen TM, Woldbye DPD. Gene therapy for Parkinson's disease, an update. *J Parkinsons Dis.* 2018;8:195–215.
96. Newland B, Newland H, Werner C, Rosser A, Wang W. Prospects for polymer therapeutics in Parkinson's disease and other neurodegenerative disorders. *Prog Polym Sci.* 2015;44:79–112.
97. Newland B, Dowd E, Pandit A. Biomaterial approaches to gene therapies for neurodegenerative disorders of the CNS. *Biomater Sci.* 2013;1(6):556–576.
98. Abdallah B, Hassan A, Benoist C, Goula D, Behr JP, Demeneix BA. A powerful nonviral vector for in vivo gene transfer into the adult mammalian brain: Polyethylenimine. *Hum Gene Ther.* 1996;7(16):1947–1954.
99. Yang J, Luly KM, Green JJ. Nonviral nanoparticle gene delivery into the CNS for neurological disorders and brain cancer applications. *Wiley Interdiscip Rev Nanomed Nanobiotechnol.* 2023;15(2):e1853.
100. Ono T, Fujino Y, Tsuchiya T, Tsuda M. Plasmid DNAs directly injected into mouse brain with lipofectin can be incorporated and expressed by brain cells. *Neurosci Lett.* 1990;117(3):259–263.
101. Boussif O, Lezoualc'h F, Zanta MA, Mergny MD, Scherman D, Demeneix B, et al. A versatile vector for gene and oligonucleotide transfer into cells in culture and in vivo: Polyethylenimine. *Proc Natl Acad Sci U S A.* 1995;92(16):7297–7301.
102. Dowd E, Monville C, Torres EM, Wong L-F, Azzouz M, Mazarakis ND, et al. Lentivector-mediated delivery of GDNF protects complex motor functions relevant to human Parkinsonism in a rat lesion model. *Eur J Neurosci.* 2005;22(10):2587–2595.

103. Rao D, Kwak G, Wang H, Eberhart CG, Hanes J, Suk JS. Bioreducible gene delivery platform that promotes intracellular payload release and widespread brain dispersion. *ACS Biomater Sci Eng.* 2023;9(8):4567–4572.
104. Tan J-KY, Sellers DL, Pham B, Pun SH, Horner PJ. Non-viral nucleic acid delivery strategies to the central nervous system. *Front Mol Neurosci.* 2016;9:108.
105. Pardridge WM. The blood-brain barrier: Bottleneck in brain drug development. *NeuroRx.* 2005;2:3–14.
106. Hersh AM, Alomari S, Tyler BM. Crossing the blood-brain barrier: Advances in nanoparticle technology for drug delivery in neuro-oncology. *Int J Mol Sci.* 2022;23(8):4153.
107. Cantrell JN, Waddle MR, Rotman M, Peterson JL, Ruiz-Garcia H, Heckman MG, et al. Progress toward long-term survivors of glioblastoma. *Mayo Clin Proc.* 2019;94(7):1278–1286.
108. Siegelin MD, Schneider E, Westhoff M-A, Wirtz CR, Karpel-Massler G. Current state and future perspective of drug repurposing in malignant glioma. *Semin Cancer Biol.* 2021;68:91–104.
109. Bahmad HF, Elajami MK, El Zarif T, Bou-Gharios J, Abou-Antoun T, Abou-Kheir W. Drug repurposing towards targeting cancer stem cells in pediatric brain tumors. *Cancer Metastasis Rev.* 2020;39(1):127–148.
110. Lukas RV, Wainwright DA, Ladomersky E, Sachdev S, Sonabend AM, Stupp R. Newly diagnosed glioblastoma: A review on clinical management. *Oncology (Williston Park).* 2019;33(3):91–100.
111. Kessel S, Cribbes S, Déry O, Kuksin D, Sincoff E, Qiu J, et al. High-throughput 3D tumor spheroid screening method for cancer drug discovery using Celigo image cytometry. *SLAS Technol.* 2017;22(4):454–465.
112. Alghamdi M, Gumbleton M, Newland B. Local delivery to malignant brain tumors: Potential biomaterial-based therapeutic/adjuvant strategies. *Biomater Sci.* 2021;9(18):6037–6051.
113. Bastiancich C, Danhier P, Préat V, Danhier F. Anticancer drug-loaded hydrogels as drug delivery systems for the local treatment of glioblastoma. *J Control Release.* 2016;243:29–42.
114. Zhao M, Danhier F, Bastiancich C, Joudiou N, Ganipineni LP, Tsakiris N, et al. Post-resection treatment of glioblastoma with an injectable nanomedicine-loaded photopolymerizable hydrogel induces long-term survival. *Int J Pharm.* 2018;548(1):522–529.

115. Brachi G, Ruiz-Ramirez J, Dogra P, Wang Z, Cristini V, Ciardelli G, et al. Intratumoral injection of hydrogel-embedded nanoparticles enhances retention in glioblastoma. *Nanoscale.* 2020;12(46): 23838–23850.
116. Smith SJ, Tyler BM, Gould T, Veal GJ, Gorelick N, Rowlinson J, et al. Overall survival in malignant glioma is significantly prolonged by neurosurgical delivery of etoposide and temozolomide from a thermo-responsive biodegradable paste. *Clin Cancer Res.* 2019;25(16):5094–5106.
117. McCrorie P, Mistry J, Taresco V, Lovato T, Fay M, Ward I, et al. Etoposide and Olaparib polymer-coated nanoparticles within a bioadhesive sprayable hydrogel for post-surgical localised delivery to brain tumours. *Eur J Pharm Biopharm.* 2020;157:108–120.
118. Di Mascolo D, Palange AL, Primavera R, Macchi F, Catelani T, Piccardi F, et al. Conformable hierarchically engineered polymeric micromeshes enabling combinatorial therapies in brain tumours. *Nat Nanotechnol.* 2021;16(7):820–829.
119. Hashemi M, Abnous K, Balarastaghi S, Hedayati N, Salmasi Z, Yazdian-Robati R. Mitoxantrone-loaded PLGA nanoparticles for increased sensitivity of glioblastoma cancer cell to TRAIL-induced apoptosis. *J Pharm Innov.* 2022;17(1):207–214.
120. Zhang Y-H, Yue Z-J, Zhang H, Tang G-S, Wang Y, Liu J-M. Temozolomide/PLGA microparticles plus vatalanib inhibits tumor growth and angiogenesis in an orthotopic glioma model. *Eur J Pharm Biopharm.* 2010;76(3):371–375.
121. Saucier-Sawyer JK, Seo Y-E, Gaudin A, Quijano E, Song E, Sawyer AJ, et al. Distribution of polymer nanoparticles by convection-enhanced delivery to brain tumors. *J Control Release.* 2016;232:103–112.
122. Heravi Shargh V, Luckett J, Bouzinab K, Paisey S, Turyanska L, Singleton WGB, et al. Chemosensitization of temozolomide-resistant pediatric diffuse midline glioma using potent nanoencapsulated forms of a N(3)-propargyl analogue. ACS *Appl Mater Interfaces.* 2021;13(30):35266–35280.
123. Bernal GM, LaRiviere MJ, Mansour N, Pytel P, Cahill KE, Voce DJ, et al. Convection-enhanced delivery and in vivo imaging of polymeric nanoparticles for the treatment of malignant glioma. *Nanomedicine.* 2014;10(1):149–157.

124. Allard E, Passirani C, Benoit J-P. Convection-enhanced delivery of nanocarriers for the treatment of brain tumors. *Biomaterials.* 2009;30(12):2302–2318.
125. Nance E, Pun SH, Saigal R, Sellers DL. Drug delivery to the central nervous system. *Nat Rev Mater.* 2022;7(4):314–331.
126. Lu Y, Wang JT-W, Li N, Zhu X, Li Y, Bansal S, et al. Intranasal administration of edaravone nanoparticles improves its stability and brain bioavailability. *J Control Release.* 2023;359:257–267.
127. Han S, Wang JT-W, Yavuz E, Zam A, Rouatbi N, Utami RN, et al. Spatiotemporal tracking of gold nanorods after intranasal administration for brain targeting. *J Control Release.* 2023;357: 606–619.
128. Henderson CE. Role of neurotrophic factors in neuronal development. *Curr Opin Neurobiol.* 1996;6(1):64–70.
129. Chmielarz P, Saarma M. Neurotrophic factors for disease-modifying treatments of Parkinson's disease: Gaps between basic science and clinical studies. *Pharm Rep.* 2020;72:1195–1217.
130. Barker RA, Björklund A, Gash DM, Whone A, Van Laar A, Kordower JH, et al. GDNF and Parkinson's disease: Where next? A summary from a recent workshop. *J Parkinsons Dis.* 2020;10(3): 875–891.
131. Schindowski K, Belarbi K, Buee L. Neurotrophic factors in Alzheimer's disease: Role of axonal transport. *Genes, Brain Behav.* 2008;7:43–56.
132. Harvey AR, Lovett SJ, Majda BT, Yoon JH, Wheeler LP, Hodgetts SI. Neurotrophic factors for spinal cord repair: Which, where, how and when to apply, and for what period of time? *Brain Res.* 2015; 1619:36–71.
133. Castilla-Cortázar I, Iturrieta I, García-Magariño M, Puche JE, Martín-Estal I, Aguirre GA, et al. Neurotrophic factors and their receptors are altered by the mere partial IGF-1 deficiency. *Neuroscience.* 2019;404:445–458.
134. Lewis O, Woolley M, Johnson D, Rosser A, Barua NU, Bienemann AS, et al. Chronic, intermittent convection-enhanced delivery devices. *J Neurosci Methods.* 2016;259:47–56.
135. Hamilton JF, Morrison PF, Chen MY, Harvey-White J, Pernaute RS, Phillips H, et al. Heparin coinfusion during convection-enhanced delivery (CED) increases the distribution of the

glial-derived neurotrophic factor (GDNF) ligand family in rat striatum and enhances the pharmacological activity of neurturin. *Exp Neurol.* 2001;168(1):155–161.

136. Luz M, Mohr E, Fibiger HC. GDNF-induced cerebellar toxicity: A brief review. *Neurotoxicology.* 2016;52:46–56.
137. Granholm A-C, Reyland M, Albeck D, Sanders L, Gerhardt G, Hoernig G, et al. Glial cell line-derived neurotrophic factor is essential for postnatal survival of midbrain dopamine neurons. *J Neurosci.* 2000;20(9):3182–3190.
138. Lapchak PA, Jiao S, Collins F, Miller PJ. Glial cell line-derived neurotrophic factor: Distribution and pharmacology in the rat following a bolus intraventricular injection. *Brain Res.* 1997; 747(1):92–102.
139. Whone A, Luz M, Boca M, Woolley M, Mooney L, Dharia S, et al. Randomized trial of intermittent intraputamenal glial cell line-derived neurotrophic factor in Parkinson's disease. *Brain.* 2019;142(3):512–525.
140. Whone AL, Boca M, Luz M, Woolley M, Mooney L, Dharia S, et al. Extended treatment with glial cell line-derived neurotrophic factor in Parkinson's disease. *J Parkinsons Dis.* 2019;9:301–313.
141. Hakami A, Narasimhan K, Comini G, Thiele J, Werner C, Dowd E, et al. Cryogel microcarriers for sustained local delivery of growth factors to the brain. *J Control Release.* 2024;369:404–419.
142. Hadaczek P, Johnston L, Forsayeth J, Bankiewicz KS. Pharmacokinetics and bioactivity of glial cell line-derived factor (GDNF) and neurturin (NTN) infused into the rat brain. *Neuropharmacology.* 2010;58(7):1114–1121.
143. Mahoney MJ, Saltzman WM. Millimeter-scale positioning of a nerve-growth-factor source and biological activity in the brain. *Proc Natl Acad Sci U S A.* 1999;96(8):4536–4539.
144. Garbayo E, Ansorena E, Lana H, Carmona-Abellan MdM, Marcilla I, Lanciego JL, et al. Brain delivery of microencapsulated GDNF induces functional and structural recovery in parkinsonian monkeys. *Biomaterials.* 2016;110:11–23.
145. Herrán E, Ruiz-Ortega JÁ, Aristieta A, Igartua M, Requejo C, Lafuente JV, et al. In vivo administration of VEGF- and GDNF-releasing biodegradable polymeric microspheres in a severe lesion model of Parkinson's disease. *Eur J Pharm Biopharm.* 2013;85(3, Part B):1183–1190.

146. Jollivet C, Aubert-Pouessel A, Clavreul A, Venier-Julienne M-C, Montero-Menei CN, Benoit J-P, et al. Long-term effect of intra-striatal glial cell line-derived neurotrophic factor-releasing microspheres in a partial rat model of Parkinson's disease. *Neurosci Lett.* 2004;356(3):207–10.
147. Lampe KJ, Kern DS, Mahoney MJ, Bjugstad KB. The administration of BDNF and GDNF to the brain via PLGA microparticles patterned within a degradable PEG-based hydrogel: Protein distribution and the glial response. *J Biomed Mater Res A.* 2011; 96A(3):595–607.
148. Wang Y, Cooke MJ, Sachewsky N, Morshead CM, Shoichet MS. Bioengineered sequential growth factor delivery stimulates brain tissue regeneration after stroke. *J Control Release.* 2013;172(1): 1–11.
149. Pakulska MM, Elliott Donaghue I, Obermeyer JM, Tuladhar A, McLaughlin CK, Shendruk TN, et al. Encapsulation-free controlled release: Electrostatic adsorption eliminates the need for protein encapsulation in PLGA nanoparticles. *Sci Adv.* 2016;2(5):e1600519.
150. Ho E, Deng Y, Akbar D, Da K, Létourneau M, Morshead CM, et al. Tunable surface charge enables the electrostatic adsorption-controlled release of neuroprotective peptides from a hydrogel–nanoparticle drug delivery system. *ACS Appl Mater Interfaces.* 2023;15(1):91–105.
151. Teal CJ, Lu SP, Shoichet MS. Engineering hydrogels for affinity-based release of therapeutic proteins. *Chem Mater.* 2024;36(2): 614–641.
152. Wilson KL, Joseph NI, Onweller LA, Anderson AR, Darling NJ, David-Bercholz J, et al. SDF-1 bound heparin nanoparticles recruit progenitor cells for their differentiation and promotion of angiogenesis after stroke. *Adv Healthc Mater.* 2023;12(19):2302081.
153. Silva D, Schirmer L, Pinho TS, Atallah P, Cibrão JR, Lima R, et al. Sustained release of human adipose tissue stem cell secretome from star-shaped poly(ethylene glycol) glycosaminoglycan hydrogels promotes motor improvements after complete transection in spinal cord injury rat model. *Adv Healthc Mater.* 2023;12(17):2202803.
154. Xu H-L, Tian F-R, Xiao J, Chen P-P, Xu J, Fan Z-L, et al. Sustained-release of FGF-2 from a hybrid hydrogel of heparin-poloxamer and decellular matrix promotes the neuroprotective

effects of proteins after spinal injury. *Int J Nanomedicine.* 2018; 13:681–694.

155. Shen B, Wang J, He T, Chen S, Wang X, Chen J, et al. Ultrasoft bioadhesive hydrogel as a versatile platform for the delivery of basic fibroblast growth factor to repair traumatic brain injury. *Chem Eng J.* 2024;483:149017.
156. Newland B, Welzel PB, Newland H, Renneberg C, Kolar P, Tsurkan M, et al. Tackling cell transplantation anoikis: An injectable, shape memory cryogel microcarrier platform material for stem cell and neuronal cell growth. *Small.* 2015;11(38):5047–5053.
157. Newland B, Newland H, Lorenzi F, Eigel D, Welzel PB, Fischer D, et al. Injectable glycosaminoglycan-based cryogels from well-defined microscale templates for local growth factor delivery. *ACS Chem Neurosci.* 2021;12(7):1178–1188.
158. Sievers J, Zimmermann R, Friedrichs J, Pette D, Limasale YDP, Werner C, et al. Customizing biohybrid cryogels to serve as ready-to-use delivery systems of signaling proteins. *Biomaterials.* 2021;278:121170.
159. Oommen OP, Duehrkop C, Nilsson B, Hilborn J, Varghese OP. Multifunctional hyaluronic acid and chondroitin sulfate nanoparticles: Impact of glycosaminoglycan presentation on receptor mediated cellular uptake and immune activation. *ACS Appl Mater Interfaces.* 2016;8(32):20614–20624.
160. Newland B, Varricchio C, Körner Y, Hoppe F, Taplan C, Newland H, et al. Focal drug administration via heparin-containing cryogel microcarriers reduces cancer growth and metastasis. *Carbohydr Polym.* 2020;245:116504.
161. Li Z, Zhang S, Liu M, Zhong T, Li H, Wang J, et al. Antitumor activity of the zinc oxide nanoparticles coated with low-molecular-weight heparin and doxorubicin complex in vitro and in vivo. *Mol Pharm.* 2022;19(11):4179–4190.
162. Li J, Mooney DJ. Designing hydrogels for controlled drug delivery. *Nat Rev Mater.* 2016;1(12):16071.
163. Björklund A, Barker RA. The basal forebrain cholinergic system as target for cell replacement therapy in Parkinson's disease. *Brain.* 2024;147(6):1937–1952.
164. Wang F, Sun Z, Peng D, Gianchandani S, Le W, Boltze J, et al. Cell-therapy for Parkinson's disease: A systematic review and meta-analysis. *J Transl Med.* 2023;21(1):601.

165. Fan Y, Winanto, Ng S-Y. Replacing what's lost: A new era of stem cell therapy for Parkinson's disease. *Transl Neurodegener.* 2020;9:1–10.
166. Rosser A, Svendsen CN. Stem cells for cell replacement therapy: A therapeutic strategy for HD? *Mov Disord.* 2014;29(11):1446–1454.
167. Conner LT, Srinageshwar B, Bakke JL, Dunbar GL, Rossignol J. Advances in stem cell and other therapies for Huntington's disease: An update. *Brain Res Bull.* 2023;199:110673.
168. Moloney TC, Rooney GE, Barry FP, Howard L, Dowd E. Potential of rat bone marrow-derived mesenchymal stem cells as vehicles for delivery of neurotrophins to the Parkinsonian rat brain. *Brain Res.* 2010;1359:33–43.
169. Hoban DB, Howard L, Dowd E. GDNF-secreting mesenchymal stem cells provide localized neuroprotection in an inflammation-driven rat model of Parkinson's disease. *Neuroscience.* 2015; 303(Supplement C):402–411.
170. Noh J-E, Oh S-H, Park I-H, Song J. Intracerebral transplants of GMP-grade human umbilical cord-derived mesenchymal stromal cells effectively treat subacute-phase ischemic stroke in a rodent model. *Front Cell Neurosci.* 2020;14:546659.
171. Zhang Z-X, Guan L-X, Zhang K, Zhang Q, Dai L-J. A combined procedure to deliver autologous mesenchymal stromal cells to patients with traumatic brain injury. *Cytotherapy.* 2008;10(2): 134–139.
172. Oh SH, Kim HN, Park H-J, Shin JY, Lee PH. Mesenchymal stem cells increase hippocampal neurogenesis and neuronal differentiation by enhancing the Wnt signaling pathway in an Alzheimer's disease model. *Cell Transplant.* 2015;24(6):1097–1109.
173. Sykova E, Cizkova D, Kubinova S. Mesenchymal stem cells in treatment of spinal cord injury and amyotrophic lateral sclerosis. *Front Cell Dev Biol.* 2021;9:695900.
174. Teixeira FG, Carvalho MM, Panchalingam KM, Rodrigues AJ, Mendes-Pinheiro B, Anjo S, et al. Impact of the secretome of human mesenchymal stem cells on brain structure and animal behavior in a rat model of Parkinson's disease. *Stem Cells Transl Med.* 2017;6(2):634–646.
175. Gomes Salgado AJB, Reis RLG, Sousa NJC, Gimble JM. Adipose tissue derived stem cells secretome: Soluble factors and their roles

in regenerative medicine. *Curr Stem Cell Res Ther.* 2010;5(2): 103–110.

176. Comini G, Dowd E. Human stem cell transplantation for Parkinson's disease: A systematic review of in situ survival and maturation of progenitors derived from human embryonic or induced stem cells in Parkinsonian models. *BioRxiv.* 2024: 2024.03.28.587203.
177. Moriarty N, Cabré S, Alamilla V, Pandit A, Dowd E. Encapsulation of young donor age dopaminergic grafts in a GDNF-loaded collagen hydrogel further increases their survival, reinnervation, and functional efficacy after intrastriatal transplantation in hemi-Parkinsonian rats. *Eur J Neurosci.* 2019;49(4):487–496.
178. Moriarty N, Pandit A, Dowd E. Encapsulation of primary dopaminergic neurons in a GDNF-loaded collagen hydrogel increases their survival, re-innervation and function after intra-striatal transplantation. *Sci Rep.* 2017;7(1):16033.
179. Gregory HN, Guillemot-Legris O, Crouch D, Williams G, Phillips JB. Electrospun aligned tacrolimus-loaded polycaprolactone biomaterials for peripheral nerve repair. *Regen Med.* 2024;19(4): 171–187.
180. Rizzo SA, Bartley O, Rosser AE, Newland B. Oxygen-glucose deprivation in neurons: Implications for cell transplantation therapies. *Prog Neurobiol.* 2021;205:102126.
181. Newland H, Eigel D, Rosser AE, Werner C, Newland B. Oxygen producing microscale spheres affect cell survival in conditions of oxygen-glucose deprivation in a cell specific manner: Implications for cell transplantation. *Biomater Sci.* 2018; 6:2571–2577.
182. Praet J, Santermans E, Daans J, Le Blon D, Hoornaert C, Goossens H, et al. Early inflammatory responses following cell grafting in the CNS trigger activation of the subventricular zone: A proposed model of sequential cellular events. *Cell Transplant.* 2015;24(8):1481–1492.
183. Denoeud C, Luo G, Paquet J, Boisselier J, Wosinski P, Moya A, et al. Enzyme-controlled, nutritive hydrogel for mesenchymal stromal cell survival and paracrine functions. *Commun Biol.* 2023; 6(1):1266.
184. Deschepper M, Petite H, Avramoglou DL, Paquet J, Pauthe E, Bidault L, et al. Time-controlled glucose releasing hydrogels

and applications thereof. *Google Patents*. 2018. Patent No. US201802136474A1

185. Barker RA. Designing stem-cell-based dopamine cell replacement trials for Parkinson's disease. *Nat Med.* 2019;25(7):1045–1053.
186. Garcia A, Fjodorova M, Lelos MJ, Perrier AL, Michael M, Bugi A, et al. I21 functional assessment of grafted human embryonic stem cells-derived progenitors in a rat model of Huntington's disease. *J Neurol Neurosurg Psychiatry.* 2018;89:A96.
187. Fjodorova M, Li M. Robust induction of DARPP32-expressing GABAergic striatal neurons from human pluripotent stem cells. In: Precious SV, Rosser AE, Dunnett SB, editors. *Huntington's Disease.* New York: Springer; 2018. p. 585–605.
188. Nolbrant S, Heuer A, Parmar M, Kirkeby A. Generation of high-purity human ventral midbrain dopaminergic progenitors for in vitro maturation and intracerebral transplantation. *Nat Protoc.* 2017;12(9):1962–1979.
189. Hills R, Mossman JA, Bratt-Leal AM, Tran H, Williams RM, Stouffer DG, et al. Neurite outgrowth and gene expression profile correlate with efficacy of human induced pluripotent stem cell-derived dopamine neuron grafts. *Stem Cells Dev.* 2023; 32(13–14):387–397.
190. de Luzy IR, Pavan C, Moriarty N, Hunt CPJ, Vandenhoven Z, Khanna A, et al. Identifying the optimal developmental age of human pluripotent stem cell-derived midbrain dopaminergic progenitors for transplantation in a rodent model of Parkinson's disease. *Exp Neurol.* 2022;358:114219.
191. Adil MM, Vazin T, Ananthanarayanan B, Rodrigues GMC, Rao AT, Kulkarni RU, et al. Engineered hydrogels increase the post-transplantation survival of encapsulated hESC-derived midbrain dopaminergic neurons. *Biomaterials.* 2017;136:1–11.
192. Filippova A, Bonini F, Efremova L, Locatelli M, Preynat-Seauve O, Béduer A, et al. Neurothreads: Development of supportive carriers for mature dopaminergic neuron differentiation and implantation. *Biomaterials.* 2021;270:120707.
193. Adil MM, Rao AT, Ramadoss GN, Chernavsky NE, Kulkarni RU, Miller EW, et al. Dopaminergic neurons transplanted using cell-instructive biomaterials alleviate parkinsonism in rodents. *Adv Funct Mater.* 2018;28(41):1804144.

194. Quittet M-S, Touzani O, Sindji L, Cayon J, Fillesoye F, Toutain J, et al. Effects of mesenchymal stem cell therapy, in association with pharmacologically active microcarriers releasing VEGF, in an ischaemic stroke model in the rat. *Acta Biomater.* 2015;15 (Supplement C):77–88.
195. Daviaud N, Garbayo E, Sindji L, Martínez-Serrano A, Schiller PC, Montero-Menei CN. Survival, differentiation, and neuroprotective mechanisms of human stem cells complexed with neurotrophin-3-releasing pharmacologically active microcarriers in an ex vivo model of Parkinson's disease. *Stem Cells Transl Med.* 2015; 4(6):670–684.
196. Kapate N, Dunne M, Kumbhojkar N, Prakash S, Wang LL-W, Graveline A, et al. A backpack-based myeloid cell therapy for multiple sclerosis. *Proc Natl Acad Sci U S A.* 2023;120(17):e2221535120.
197. Miron VE, Boyd A, Zhao J-W, Yuen TJ, Ruckh JM, Shadrach JL, et al. M2 microglia and macrophages drive oligodendrocyte differentiation during CNS remyelination. *Nat Neurosci.* 2013; 16:1211–1218.

https://doi.org/10.1142/9789819806782_0015

Chapter 15

Tumor Tissue Engineering: Modeling Prostate Cancer with Biomaterial-based Platforms

Julien Clegg[1], Laura H. Porter[2], Renea A. Taylor[3,4,5,6] and Daniela Loessner[1,2,7,8*]

[1]*Department of Chemical and Biological Engineering, Faculty of Engineering, Monash University, 3800 Victoria, Australia*
[2]*Department of Anatomy and Developmental Biology, Biomedicine Discovery Institute, Faculty of Medicine, Nursing and Health Sciences, Monash University, 3800 Victoria, Australia*
[3]*Department of Physiology, Biomedicine Discovery Institute, Faculty of Medicine, Nursing and Health Sciences, Monash University, 3800 Victoria, Australia*
[4]*Cancer Immunology Program, Cancer Research Division, Peter MacCallum Cancer Centre, Melbourne, 3000 Victoria, Australia*
[5]*Sir Peter MacCallum Department of Oncology, The University of Melbourne, Parkville, 3010 Victoria, Australia*
[6]*Cabrini Institute, Cabrini Health, Malvern, 3144 Victoria, Australia*
[7]*Department of Materials Science and Engineering, Faculty of Engineering, Monash University, 3800 Victoria, Australia*
[8]*Leibniz Institute of Polymer Research Dresden, Max Bergmann Centre of Biomaterials, Hohe Straße 6, 01069 Dresden, Germany*
**Corresponding author: Daniela.Loessner@monash.edu*

The extracellular matrix has been recognized as a pivotal component of the tumor microenvironment in cancer tissues. In prostate cancer, the matrix influences the heterogeneous cell populations and vice versa. During cancer initiation and progression, the dynamics of the tumor microenvironment unfold, demonstrating how different cell types contribute to the remodeling of the matrix and how various molecules influence cell behavior, functions, and responses to treatment. Experimental prostate cancer models traditionally include cell monolayers grown on plastic culture dishes and animal models. Novel three-dimensional (3D) models, such as hydrogels, scaffolds, or microfluidic devices, have emerged as biomaterial-based platforms to accurately resemble cell–cell and cell–matrix interactions as seen in patient tissues. Hydrogels advance 3D tissue models of prostate cancer, providing an intact matrix component with tunable biophysical and biochemical characteristics to suit research needs and dissect complex cellular interactions. The use of hydrogel approaches in prostate cancer research offers to bridge the gap between preclinical studies and clinical applications, by enhancing our understanding of disease mechanisms, therapeutic responses, and drug resistance, and by applying them in drug discovery and drug screening. This chapter focuses on the pathobiology and treatment of prostate cancer, elements of the tumor microenvironment, and currently used biomaterial-based platforms.

1. Prostate cancer

1.1. *The normal prostate*

The prostate is a male endocrine gland regulated by signaling molecules and androgen hormones.[1] Anatomically, the prostate gland is surrounded by fibromuscular stroma and encompasses three distinct regions, being the peripheral, transition, and central glandular zones.[1] The prostate gland is primarily involved in reproduction functions, such as secretion of protective proteins for spermatozoa.[2] Androgens play a key role in the function of the prostate gland, and in turn are involved in the metabolism of testosterone to dihydrotestosterone, with systemic effects on adult men.[3,4] Testosterone and dihydrotestosterone signal through androgen receptors (ARs) that regulate the normal processes of the prostate gland.

At the cellular level, the prostate gland has distinct cell phenotypes. Basal epithelial cells are located below the luminal cells and create a barrier between the stroma and lumen, which express cytokeratin 5. Secretory luminal epithelial cells are columnar in morphology, identifiable by the expression of cytokeratin 8, and produce and secrete prostate-specific antigen (PSA) or kallikrein-3, kallikrein-2, and prostatic acid phosphatase.[5] The third epithelial subtype is the neuroendocrine cells that express chromogranin A and is positioned throughout the basal and luminal layers. Neuroendocrine cells arise from the stem cells that reside within the basal layer, are considered androgen-independent, and contribute to the development of the prostate. They express chromogranin A, serotonin, bombesin, and parathyroid hormone-related proteins.[5–7] The stroma supports the maintenance and function of the prostate and mostly contains smooth muscle cells, fibroblasts, endothelial cells, pericytes, and various immune cell populations.[8–10]

Epithelial cells that line the lumen of the prostate gland undergo transformation from normal to cancerous cells, through genetic alterations, such as mutations or deletions in the protein tyrosine phosphatase (PTEN),[11] tumor suppressor gene (p53),[12] and retinoblastoma susceptibility gene 1 (RB1).[13] In most cases, the luminal epithelial cells, which are androgen-dependent, turn into cancerous cells, and in rare cases, the androgen-independent basal epithelial and neuroendocrine cells may be the source of prostate cancer.

1.2. *Clinical burden, treatment, and management*

Prostate cancer accounts for 15% of all male cancers worldwide, with an estimated 1.4 million new cases diagnosed in 2020, and its incidence is expected to double globally by 2040.[14] Prostate cancer is detected in biopsy specimens and graded according to the International Society of Urological Pathology (ISUP) Grade Group system. The highest-grade group (ISUP grade 5) indicates a high likelihood of aggressive disease.[15] For patients with localized low-volume, low-grade disease (ISUP grade 1), active surveillance can be used for clinical management.[16] Active surveillance, which typically involves regular assessment of serum PSA, multiparametric magnetic resonance imaging (mpMRI), and repeat biopsies, is employed to prevent overtreatment of clinically insignificant disease and decrease comorbidities associated with definitive therapy, such as urinary incontinence and erectile dysfunction.[17] Due to the increased risk of progression to metastatic disease and cancer-specific mortality, definitive therapy is recommended for patients with localized intermediate or high-risk prostate cancer.

Standard-of-care for localized prostate cancer targets the whole prostate gland by surgical removal, termed radical prostatectomy, or radiotherapy approaches, such as low-dose brachytherapy, high-dose brachytherapy, and external beam radiotherapy with or without hormonal therapy. To reduce overtreatment in a subset of patients, a developing therapy is focal therapy that specifically targets prostate cancer within a specific region of the prostate gland. Methods for focal therapy include high-intensity focused ultrasound, irreversible electroporation, interstitial laser, and photodynamic therapy; however, long-term follow-up data for these approaches are currently unavailable, and clinical trials are warranted to assess efficacy and survival outcomes.[18]

Biochemical recurrence can occur following definitive treatment for prostate cancer, indicated by a rise in serum PSA. In cases where the cancer has not metastasized based on conventional imaging (M0 prostate cancer) but there are residual or recurrent cancer cells in the prostate area or local lymph nodes (N1), salvage adjuvant therapy can be employed, including hormonal therapy or radiotherapy.[19,20] If cancer is detected beyond the prostate area by imaging using bone scans and/or nuclear positron emission tomographic (PET) scans, then it is classified as metastatic prostate cancer (M1). Prostate cancer often metastasizes to the bone, but other common sites include lymph nodes, liver, lung, and other distant organs. The extent and location of metastases play a significant role in the clinical management of metastatic prostate cancer.

Since it was first used nearly 80 years ago, androgen deprivation therapy (ADT) remains the cornerstone treatment for metastatic prostate cancer.[21] ADT initially reduces tumor burden due to the reliance of most prostate cancers on AR signaling for growth and survival. Whilst ADT was originally achieved by surgical orchidectomy, chemical methods are now typically employed with the use of gonadotropin-releasing hormone agonists or antagonists to block testosterone production. In addition to standard ADT to reduce systemic levels of androgens, the recent introduction of potent AR signaling inhibitors (ARSIs) can be used to further inhibit androgen-stimulated tumor growth. ARSIs inhibit AR signaling by preventing receptor dimerization, transport to the nucleus and recycling back to the cell surface (enzalutamide, apalutamide, and darolutamide) or inhibiting androgen production (abiraterone).[22]

The clinical landscape for the treatment of hormone-sensitive metastatic prostate cancer has evolved considerably in recent years. Several clinical trials have demonstrated the efficacy of combination therapies given at the time of ADT to improve progression-free and overall survival outcomes. These include ADT plus the chemotherapy docetaxel, ADT plus enzalutamide or abiraterone, or various forms of ADTs, including intermittent ADT.[23–28] Whilst most patients initially respond to ADT, either alone or in combination with chemotherapy or ARSIs, patients eventually progress on treatment to castrate-resistant prostate cancer (CRPC). Currently, curative treatment options are not available for CRPC and the disease remains incurable. However, there has been significant advances in the clinical management of CRPC over the past decade, with many therapeutic options now available that slow disease progression and improve survival outcomes. In many cases, CRPC still remains reliant on AR signaling due to several resistance mechanisms, including intracrine androgen synthesis, genomic alterations to the *AR* gene (amplification, gain-of-function mutations, and rearrangements), expression of AR splice variants, and alterations to the expression or activity of AR cofactors.[22] Thus, ADT and ARSIs are still used for CRPC. Chemotherapies are also effective treatments for metastatic prostate cancer, including the taxane-based chemotherapies docetaxel and cabazitaxel, and the platinum-based chemotherapy carboplatin, depending on the disease setting.[29–31]

In addition to AR-targeted therapy and chemotherapy, targeted therapeutic approaches are now being employed for the treatment of prostate cancer based on specific genomic aberrations or the expression of tumor-associated antigens in individual tumors. Examples include the use of poly (ADP-ribose) polymerase (PARP) inhibitors for prostate cancers with germline or somatic alterations in DNA damage repair genes, such as *BRCA1*, *BRCA2*, and *ATM* mutations. PARP inhibitors, including olaparib and talazoparib, have shown remarkable improvements in overall survival for these patients and have been approved for clinical use in certain settings.[31–34] Prostate-specific membrane antigen (PSMA)-directed therapeutics are another example of targeted therapy that is redefining the clinical management of prostate cancer. Lutetium-177 (Lu)-labeled PSMA 617 radioligand therapy (^{177}Lu-PSMA) delivers beta radiation to tumor cells expressing PSMA, a type II transmembrane protein that is overexpressed on prostate cancer cells in up to 90% of CRPC. ^{177}Lu-PSMA has been approved for late-stage disease based on improved outcomes in several clinical trials, and clinical trials are currently exploring its use earlier in disease progression as well as other radioligand therapies.[35,36] Clinical trials are also exploring immunotherapies as potential treatment options for prostate cancer, either alone or in combination with standard-of-care therapies. Whilst prostate cancer has traditionally been considered less responsive to immunotherapy compared to other cancers, sipuleucel-T is approved for metastatic CRPC and checkpoint inhibitors, such as pembrolizumab and nivolumab, are being studied in clinical trials for advanced prostate cancer.[37–40] Emerging immunotherapies in preclinical or clinical development also include chimeric antigen receptor (CAR) T-cell immunotherapy, bispecific T-cell engager molecules (BiTEs), vaccine approaches, and oncolytic viruses.[41–43]

In response to long-term ADT, treatment-induced neuroendocrine prostate cancer (NEPC) can emerge.[44] In rare cases, NEPC can also arise de novo or early in disease progression.[45] NEPC is associated with more aggressive clinical behavior and worse survival outcomes since it is typically resistant to traditional prostate cancer treatments, such as hormonal therapies and chemotherapy.[46,47] Currently, there is no standard-of-care approach for NEPC, but a combination of chemotherapy, immunotherapy, and other targeted therapies may be used.[47]

1.3. *Development and cellular composition of prostate cancer*

The development of prostate cancer involves a complex interplay between different cell types and a dynamic tumor microenvironment (TME) that arises from the normal extracellular matrix (ECM). Prostate cancer stem cells (CSCs) are another cell population found in the TME, with unique characteristics that contribute to therapy resistance, recurrence, and metastasis.[48] They display self-renewal and pluripotency and contribute to the heterogeneity of the TME. CSCs promote therapy resistance through upregulating drug efflux pumps, such as ABCG2, and modulate their proliferative capabilities, such as displaying a quiescent behavior that results in partial or full resistance to anticancer therapies.[49,50] During cancer progression, CSCs contribute to the epithelial–mesenchymal transition (EMT), whereby epithelial cells differentiate into mesenchymal-like phenotypes that support cell motility and invasion.[51] This process allows CSCs and tumor cells to metastasize from the prostate gland to distal tissues or organs like the bone or bone marrow.

The TME comprise various stromal cells, including cancer-associated fibroblasts (CAFs), the most abundant stromal cell type. CAFs are heterogeneous spatially and include various subtypes. They serve different biological functions, and their overall composition changes during disease progression.[52,53] A study by Pan et al. detected eight different CAF subtypes in prostate cancer, with two subtypes, CAF-C0 and CAF-C1, being the most abundant. CAF-C0 express microvasculature genes, such as *RGS5*, *MCAM*, and *MYH1*, and support intratumoral angiogenic processes. In contrast, CAF-C1 is associated with poor prognosis, resistance to castration therapy, and secreted ECM molecules, such as collagen type I, thereby promoting matrix remodeling within the TME.[53] CAFs interact with other cell types, such as immune cells, thereby antagonizing T-cell function and creating an immunosuppressive TME.[53,54] A study by Yu et al. showed that CAFs, but not normal fibroblasts, promote the aberrant transformation of benign prostatic hyperplasia BPH-1 cells into a cancerous phenotype.[55] These cells also express ARs, although during prostate cancer progression, CAFs downregulate its expression, which is a prognostic indicator of cancer progression. In a study by Leach et al., the loss of AR expression in CAFs caused an upregulation of the ECM-degrading enzyme matrix metalloproteinase 1 (MMP-1), and the downregulation of the adhesion molecules FBXO32 and FBN1, resulting in a TME that enables prostate cancer cells to migrate and invade.[56,57]

Endothelial cells also play a crucial role in the development and progression of prostate cancer. They, alongside pericytes, are recruited to the TME to form a new microvasculature for the delivery of nutrients and oxygen.[58] MMPs facilitate tumor angiogenesis and endothelial cell migration. Vascular endothelial growth factor 1 (VEGF1) and VEGF receptor 2 (VEGFR2) are among the most important signaling molecules for the recruitment of endothelial cells, cell differentiation and growth, mediating tumor angiogenesis. CAFs secrete fibroblast growth factors (FGFs), causing an autocrine and paracrine regulation of other cell types present within the TME.[58,59] For instance, FGF1 and FGF2 act upon endothelial cells and also impact cell differentiation, invasion, and growth. FGFs are also secreted by pericytes, cancer cells, and immune cells, such as tumor-associated macrophages (TAMs).[60]

TAMs are identifiable by their M2-like phenotype and are involved in tumor angiogenesis, modulating the TME and immunosuppression.[60] Immune cells, such as regulatory T-cells that inhibit antitumor immune responses and dendritic cells that become reeducated to adopt a tolerance to tumor cells, are examples of how prostate cancers display immune evasion and leverage immune signaling for tumor growth and progression.[61,62] Immune cells infiltrate the TME guided by cytokines and ECM molecules, such as growth factors, differentiation and recruitment molecules, and alter the cellular and noncellular components found within the TME.[63]

1.4. *Matrix macromolecules*

The TME comprise different ECM molecules that influence tumor progression.[64] These include collagens, laminins, fibronectins, growth factors, cytokines, and chemokines, as well as different proteoglycans and glycosaminoglycans.[51,64] These molecules make up the noncellular components of prostate cancer and support tumor growth and cell–ECM interactions. ECM molecules are categorized according to their type being growth factors, cytokines and chemokines, and matrix-degrading and matrix proteins.[65,66] Growth factors, such as VEGF, FGF, and transforming growth factor beta (TGF-β), are pivotal in stromal cell signaling, cell differentiation, and growth.[65] TGF-β acts as a tumor suppressor as well as a tumor promotor, inducing EMT, tumor angiogenesis, and providing tumor cells the ability to go undetected from the immune system.

Cytokines and chemokines act predominantly upon immune cells that are already present or beyond the TME for cell recruitment and steering cell behavior.[66] Interleukin-6 (IL-6) is elevated in prostate cancer and acts as a pro-inflammatory cytokine that promotes cancer cell growth and survival through the Janus kinase (JAK) signal transducer and activator of transcription (JAK-STAT) signaling pathway.[67] During the transition of castration sensitivity to resistance, IL-6 impacts cancer and stromal cells by activating ARs, independent of other TME factors and systemic androgen levels.[68] The CXCL12–CXCR4 axis is also important for cell migration and invasion leading to the movement of tumor and stromal cells to organs throughout the body, promoting metastasis. CXCL12, also known as stromal cell-derived factor-1 (SDF-1), binds CXCR4 on cancer cells, facilitating cell migration to tissues that are rich in CXCL12, such as the bone marrow.[69] Cytokines and chemokines interact with other ECM proteins and cell surface molecules like integrins to influence cell decisions and indirectly modulate the TME.[70]

Matrix-degrading enzymes, such as MMPs and urokinase plasminogen activator (uPA), modulate the TME by degrading matrix proteins, changing the stiffness and density of the tissue and encouraging both cell infiltration and migration.[71,72] High levels of these enzymes are correlated with poor prognosis.[73] Matrix proteins provide both structural support and influence cell behavior. Collagen is an integral ECM protein found in prostate cancer, with different collagen types being spatially heterogeneous and differentially expressed, which can be related clinically to the Gleason grade group.[74,75] During tumor progression, stomal cells stimulate the production of matrix proteins, resulting in collagen being deposited and an increasing stiffness of the tissue.[75] The increase in tissue stiffness signals tumor cells through focal adhesion kinase (FAK), promoting cell motility, growth, and invasion, and collagens can then later be degraded by MMPs promoting cell invasion and metastasis.[76,77] Fibronectin, laminin, and perlecan are also important regulators of cell growth and migration, as well as tumor angiogenesis. Both the deposition of ECM proteins and increased stiffness influence the efficacy of anticancer treatment; targeting tissue stiffness and ECM molecules are promising options for novel therapeutic strategies.[77]

2. Hydrogel-based biomaterial platforms

2.1. *Hydrogels for studying cancer cell behavior*

Tumors are highly complex organs that are tissue- and patient-specific, having unique genetic and behavioral characteristics that dynamically evolve over the progression of the disease.[78] This makes tumor tissues particularly hard to study and develop effective anticancer therapeutics. Current experimental approaches that are based on cell monolayers grown on plastic culture dishes do not accurately recapitulate the complexity of tumor tissues and result in a high drug attrition rate.[79] Hydrogels have emerged as alternative cell culture models and truly recapitulate the characteristics of tumor tissues.[80] Hydrogel-based models have been used to study cancer initiation, development, and progression, as well as drug resistance.

Hydrogels are water-absorbent 3D polymer networks that consist of polymers derived from natural sources, such as alginate from seaweed, chitosan from crustacean shells, and gelatin from animal skin or bone.[81,82] Synthetic sources of hydrogel polymers include polyvinyl alcohol (PVA), polyethylene glycol (PEG), and polyacrylamide (PAA).[82] Matrigel, derived from murine Engelbreth-Holm-Swarm tumors, is the current gold standard for hydrogel-based models and is highly heterogeneous in its protein composition.

Cross-linking of hydrogels occurs in the presence of water through physical, chemical, or radiation-based approaches and is described by the swelling ratio of the polymer network.[80,81] The physical properties of hydrogels dictate cell behavior and can be customized to change the elasticity, cross-linking density, and porosity.[81,83] The stiffness of tumor tissues has been recognized as a crucial feature of the TME. Thus, tailoring the stiffness of a hydrogel to match the in vivo condition is key when recreating tumor tissues in the laboratory.[84–86] The elastic modulus describes the stiffness of a hydrogel and can be tuned to mimic less stiff tumor tissues, such as breast (2–25 kPa),[87] pancreatic (mean 5.46 kPa),[88] and colon (median 7.51 kPa)[89] tumor tissues, while also matching more stiff tumors, such as bone tumors (50–130 kPa).[90] However, reported measurements are highly dependent on the measurement technique employed and the application, such as direct measurement on patient-derived or biopsied tissue.

Hydrogel-based models are further customized through chemical modifications, such as Michael-type addition chemistry, and by using different linkers to attach macromolecules, growth factors, instructive DNA sequences, signaling molecules, or drugs.[91–94] The ability to further functionalize hydrogels is a desirable feature as it expands the applications of hydrogels for use not only in 3D cell cultures, but also pharmaceutical and preclinical purposes (**Table 1**).

Hydrogels have been used to dissect the pathobiology of tumors, including the roles of stromal cells in cancer development and resistance, as well as how ECM molecules influence cancer cell functions and drug resistance. PEG–heparin hydrogels have been used to study angiogenesis of breast tumors, for example, an article by Koch et al. reported how CAFs help to form a microvascular-like network and epithelial morphology.[95] The addition of CAFs increased the density of the microvascular networks formed by endothelial cells and highlights the plasticity of the TME during intratumoral angiogenesis. Functionalized gelatin methacryloyl (GelMA) hydrogels have been used to study angiogenesis and formation of vascular-like networks using human umbilical vein endothelial cells and adipose-derived stem cells.[96] Therefore, GelMA hydrogels were functionalized using EDC/NHS chemistry with VEGF and basic FGF. The inclusion of growth factors into the polymer network altered the physical and chemical properties of the hydrogel and improved cytocompatibility as indicated by the cell viability and growth of encapsulated cells. The superior vascular-like network formation in GelMA hydrogels functionalized with growth factors was also demonstrated as compared to GelMA alone.[96] In a different study, a dextran-based hydrogel was used to mimic angiogenic processes using multicellular cultures comprised of fibroblasts, cancer, immune, and endothelial cells. Using this 3D model, cell–cell interactions and angiogenesis were supported and underpinned the suitability of the model as a platform for studying anticancer and antivascular effects.[97]

The immune component of the TME is critical for the progression of many solid tumors. Immune cell infiltration has been linked to cancer progression and implicated in tumor-associated processes, such as angiogenesis through ECM secretion of signaling molecules.[62,117] Disintegrable supramolecular gelatin (DSG) hydrogels

Table 1. Hydrogel biomaterials for studying tumor cell biology.

Main purpose	Hydrogel materials	Cellular composition	Citation
Tumor angiogenesis	PEG-Hep, semisynthetic	Tri-culture, CAFs + MCF -10a epithelial cells + mammary endothelial cells	95
	GelMA and GelMA + GFs, semisynthetic	Coculture, HUVECs + hADSCs	96
	Dextran-CD, semisynthetic	Multicellular, macrophages + neutrophils + fibroblasts + HUVECs + H838 cancer cells	97
Inclusion of immune cells	DSG, natural	Cocultures, endometrial cancer cells + M0/M1/M2 macrophages	98
	GelMA, semisynthetic	Mono/cocultures, microglia + patient-derived glioblastoma cells	99
	Gelatin–hyaluronan, natural	Monoculture, macrophages	100
Drug delivery and therapeutic studies	Poly-lysine-PEG, synthetic	Monoculture, TAMs, xenograft model	101
	Polysaccharide, 3D-printed	Non-small cell lung cancer patient-derived xenografts	102
	Collagen–alginate, natural	Monoculture, MCF-7 breast cancer cells	103
	pH-sensitive starch/PVA/g-C_3N_4, semisynthetic	Monoculture, MCF-7 breast cancer cells	104
	PVA–chitosan, semisynthetic	L929, K150 cell lines, animal model	105
	PEG–SH, PEG–DA, synthetic	Coculture, B16F10, DC2.4 melanoma, and dendritic cells	106

(*Continued*)

Table 1. (*Continued*)

Main purpose	Hydrogel materials	Cellular composition	Citation
	HA–SH/PEG–DA	Prostate cancer-derived xenografts	107
	PEG-4MAL with MMP cleavable VPM peptides, Matrigel	Prostate cancer organoids, OWCM154, OWCM155, OWCM1358	108
Modeling mechanical properties	Alginate–gelatin, natural	Monocultures, MCF-7, MDA-MB-231 breast cancer cells	109
	GelGMA/HAGMA-MMP, semisynthetic	Monoculture MCF-7 breast cancer cells	110
	PeptiGels, polyacrylamide, synthetic	**Monoculture, Suit2-007 pancreatic cancer cells**	**111**
	HAMA, semisynthetic	**Monoculture MCF-7 breast cancer cells**	**112**
Modeling metastasis or advanced disease	TempEasy 3D (Azide-functionalized PIC polymers), synthetic	Prostate cancer organoids MSK-PCa1, MG63 bone cells	113
	GelMA, semisynthetic	Primary osteoprogenitor cells, hBM-MSCs, LNCaP prostate cancer cells	114
	Chitosan–chondroitin sulphate, natural	PC-3 and 22Rv1 prostate cancer cells	115
3D cell culture	GelMA, semisynthetic	LNCaP, MCF-7 prostate and breast cancer cells	116

Materials: DA: diacrylate; Dextran-CD: dextran-polyethylene glycol-matrix metalloprotease; DSG: disintegrable supramolecular gelatin; GFs: growth factors; GelGMA: gelatin glycidyl methacrylate; GelMA: gel-methacrylate; HAGMA: yaluronic acid GMA; HAMA: HA-methacrylate; HA-SH: HA-thiol; hep: heparin; MAL: maleimide; MMP: matrix metalloprotease; PEG: poly(ethylene glycol); PEG-SH: PEG-thiol; PIC: polyisocyanopeptide; PVA: polyvinyl alcohol; VPM: GCRDVPMSMRGGDRCG cleavable peptide.

Cells: CAFs: cancer-associated fibroblasts; hADSCs: human adipose-derived stem cells; hBM-MSCs: human bone marrow mesenchymal stem cells; HUVECs: human umbilical vein endothelial cells; TAMs: tumor-associated macrophages.

have been used to study TAMs and immune infiltration,[98] while GelMA hydrogels have modeled glioblastoma activation of resident microglia immune cells. Using RNA sequencing, significant molecular alterations were found in multicellular cultures but not in monocultures.[99] Gelatin–hyaluronan hydrogels were further modified with fibronectin-derived peptides and used to understand macrophage invasion. MMP and rho-associated protein kinase (ROCK) inhibitors confirmed that migration occurred predominantly through the secretion of MMPs as opposed to ROCK secretion.[100] Yang et al. evaluated a polyphyllin II/resiquimod shear-thinning hydrogel for potential application as an injectable hydrogel for localized delivery of immunotherapies for gastric cancer. This study demonstrated that the biocompatibility and mechanical performances of the hydrogel efficiently modulate the behavior and tumor suppression of TAMs.[101] These studies highlight the far-reaching applications of hydrogels in cancer research and as drug delivery platforms.

Hydrogels are also integral tools to identify potential therapeutic targets. Cancer cells from different origins form spheroids in hydrogel-based models, thereby creating gradients in nutrients, gases, hypoxic, and therapeutic-resistant intratumoral regions.[118] Spheroids grown in hydrogels demonstrate higher concentrations of drugs to reach 50% inhibition (IC_{50}) as compared to cell monolayers grown on plastic culture dishes.[102]

Matrix stiffness also plays a role in the efficacy of drug treatment, as demonstrated with a collagen–alginate hydrogel that modeled stiffness effects of doxorubicin-treated cells and revealed a correlation between stiffness and treatment resistance.[103] Less stiff hydrogels led to larger spheroids and a lower fraction of proliferative cells after seven days of 3D culture, with increased resistance to doxorubicin treatment. This finding confirms the crucial role of TME stiffness and its effects on therapeutic efficacy and potentially identifies novel approaches to overcoming treatment resistance.[109] Hydrogels have been previously functionalized with growth factors[96] to study cell–ECM interactions; however, they can also be used to deliver drugs to cells in vitro, or solid tumor sites in vivo through approaches of functionalization or drug encapsulation.[104–106]

In summary, hydrogels are 3D models that accurately recapitulate the cell–cell, cell–ECM, and cell–drug interactions required to understand the pathobiology of tumors. These bioengineered 3D models have potential for clinical applications as drug-loaded delivery platforms for cancer patients. They have been used across different types of tumors, such as breast,[104] lung,[102] melanoma,[106] and prostate cancer,[115,119] to explore mechanisms underlying cancer development, progression, drug resistance, and biomechanics.[110–112]

2.2. *Hydrogels for studying prostate cancer*

Mimicking the complex TME of prostate cancer is difficult, as the incorporation of the spatial heterogeneity, cellular composition, and cell–cell and cell–ECM interplay into hydrogel-based models has yet to be fully realized. There have been only a few hydrogel-based models for the study of prostate cancer to assess drug efficacy and drug–cell interactions.[107,108,120] Prostate cancer is typically studied in cell monolayers and preclinical models, such as patient-derived xenografts. They have provided valuable insights into the mechanisms of prostate cancer progression and tissue tropism during metastatic initiation. There still remains a disconnect between simplistic in vitro approaches and animal models for identifying and validating therapeutic targets and assessing drug efficacy. Similarly, understanding how and why drugs fail or resistance occurs is difficult to follow up in animal models, as there are high financial and ethical costs that limit animal testing cohorts and the timescale or duration of these studies. Hydrogel-based prostate cancer models mimic the cell composition, matrix complexity, drug efficacy or resistance, and deliver information for animal studies thereby, minimizing the disconnect of the transition from basic to preclinical research.

In 2014, prostate cancer cell lines and patient-derived xenograft-derived cells were encapsulated in a thiolated hyaluronic acid/polyethylene glycol–diacrylate (HA-SH/PEG-DA) hydrogel, which maintained cell viability and AR expression over a culture period of two weeks.[107] Despite the maintenance of cell viability and AR expression, prostate cancer cells showed limited growth while demonstrating resistance to drug treatment. In 2017, HA–alginate hydrogels facilitated an EMT behavior of prostate cancer cells, which was absent in cell monolayers, while also showing an increased resistance of cells to drug treatment when grown in hydrogels as opposed to cell monolayers.[119] This is in part due to the formation of spheroids that create nutrient and gas gradients, but also diffusion requirements and polymer interactions between drug molecules and the hydrogel. HA–alginate hydrogels provided more accurate predictability for in vivo drug efficacy as compared to cell monolayers grown on plastic culture dishes.[119] Both studies highlight the importance and suitability of 3D models and their potential application for prostate cancer research, providing maintenance of hormone receptor levels, facilitating spheroid formation of cells and cytocompatibility of patient-derived xenografts.

It is widely known that prostate cancers demonstrate tissue tropism toward bone among other distal tissue sites,[121] with metastatic bone cancer (<90%) representing the majority of distal site migration as opposed to distal lymph nodes (8.7%), thorax (5.7%), and liver (4.5%) not being mutually exclusive. Different hydrogels have been employed to study the mechanism of prostate cancer-induced bone metastasis. The TempEasy 3D hydrogel that uses polyisocyanopeptide polymers functionalized with anti-CD3 antibodies has been employed as a coculture model to investigate prostate cancer-induced bone metastasis.[113] Growing prostate cancer organoids, or mini-replica of cancer tissues, together with MG-63 bone cells, supported cell–cell interactions and upregulated genes related to neovascularization and metastasis. Bone cells had an increased growth and stemness but a reduced secretion of ECM molecules and collagen deposition.

The overreliance on cell lines for in vitro studies contributes to the disconnect seen within the medical research field. Some studies are based only on tumor cell lines,[115,116] which simplify the cell–cell and cell–ECM interactions that they are attempting to understand.

It has been demonstrated that the use of patient-derived cells and the generation of organoids from other cancers, including prostate cancer, allows more realistic interactions and informs drug studies more faithfully.[108,113,122,123] Pioneering work from the Hans Clevers group has shown that cancer organoids can accurately recapitulate the heterogeneous cell populations and genotype diversity found in vivo across prostate, breast, and colon cancers. Hereby, spheroids are comprised of different primary cells and have highlighted the potential for high-throughput drug screening.[123–125] However, Matrigel is typically used to generate cancer organoids and biobanks to perform drug treatment studies. As new hydrogel-based models are being developed and validated, the combination of organoid generation in more sophisticated and tailorable hydrogels will significantly aid prostate cancer research.

Other platforms, such as GelMA-based hydrogels, have been used to include complex cell compositions in vitro. Hereby, LNCaP prostate cancer cells were grown together with bone marrow-derived adipocytes in a coculture setting, supporting cell–cell interactions that led to morphological changes in the bone marrow-derived adipocytes, de-lipidation, and de-differentiation upon coculture.[114] Additionally, a humanized bone organ in rodents was established, accurately reproducing the biological and morphological features found within human bone. This GelMA-based model allows for the study of the impact of human adipocytes on prostate cancer and the formation of bone metastasis. GelMA-based hydrogels also support spheroid formation of prostate cancer cells,[116] while other platforms, such as chitosan–chondroitin and PEG-4MAL hydrogels, have been employed to study EMT processes prior to prostate cancer metastasis and organoid responses to drug treatment.[108,115]

Hydrogels represent an important stepping stone for prostate cancer research, offering cell culture models that enable the 3D reconstruction of the TME. These 3D models introduce tailored structural ECM polymers for accurate mimicry of the pathophysiology as seen in patients while facilitating cell–cell and cell–ECM interactions. Studying these interactions in an experimental system that provides an accurate resemblance of the tumor tissue is crucial for understanding the drivers of prostate cancer progression, metastasis, and overcoming therapeutic resistance. However, significant challenges remain as very few hydrogel-based models have been specifically designed and validated for prostate cancer (**Figure 1**).

Future research in the field of tumor tissue engineering and hydrogel materials to model prostate cancer is required, as the current lack of precise and robust models necessitates validation of novel 3D approaches against both the gold standard, Matrigel, and patient-derived tissue references. There are pitfalls in both the biological and biomaterial complexity, in terms of cell lines and Matrigel. When generating 3D cell culture and organoid models for fundamental research and applications for high-throughput drug screening and drug discovery, both the use of primary prostate cancer cells and hydrogel-based models that allow for tissue complexity and dynamics need to be employed. Accurately mimicking the cell–cell and cell–ECM interactions, such as the signaling processes seen in CAFs, and the immunosuppressive behavior is crucial for prostate cancer. This molecular array needs to be benchmarked within a mechanically tailored system that fits within the current mechanical thresholds of prostate cancer tissues.

Besides, a rationally designed and validated hydrogel-based model must demonstrate the genomic stability of the original prostate cancer cells or tissues while avoiding artificial genomic instability and drift. Maintaining the patient's genotype and phenotype is essential for biomaterial-based platform for developing and screening new therapeutics and studying therapy responses and resistance.

3. Conclusion

In order to advance our understanding of the cellular complexity of prostate cancer and design new therapeutic approaches, advanced laboratory models are required. Hydrogel-based biomaterials have emerged as promising tools for studying cancer, offering 3D cell

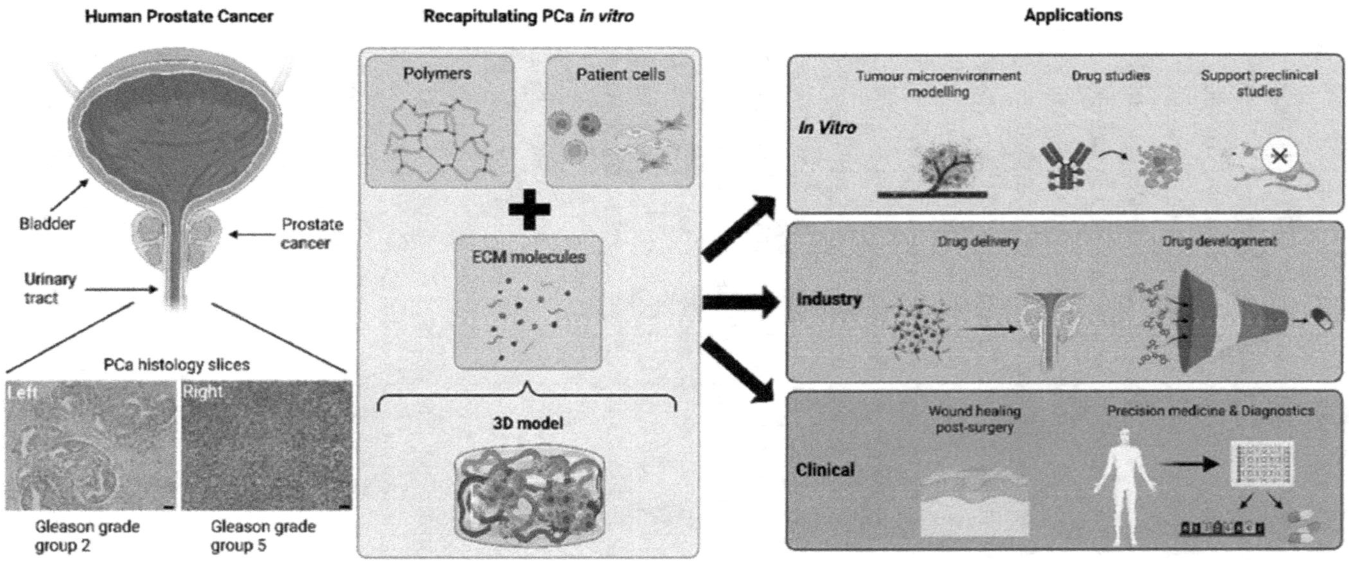

Figure 1. Hydrogel-based models for prostate cancer. Prostate cancer anatomy with two histology slices (scale bars 50 μm) stained with hematoxylin and eosin. (Left) Gleason grade group 2, (right) Gleason grade group 5. 3D models are useful cell culture tools that recapitulate important features of prostate cancer tissues. Illustration created in BioRender.com.

culture models that more accurately recapitulate the complex TME as compared to cell monolayers grown on plastic culture dishes. Hydrogels provide customization of the used polymers and tailorable mechanical, biological, and chemical features, enhancing their suitability for cancer research. Hydrogels have been employed for the study of prostate cancer and have high cytocompatibility for multicellular cultures that help to dissect either the primary TME characteristics or metastatic cascades. Despite this, only a few hydrogels have been robustly validated for prostate cancer, and future research requires benchmarking to Matrigel, the gold standard for 3D cancer models, and to prostate cancer tissue as a reference. High-fidelity hydrogel-based models have an immense value for preclinical research while contributing to drug discovery and novel therapeutic strategies that further improve clinical outcomes.

References

1. Aron L, Franco O, Hayward SW. Review of prostate anatomy and embryology and the etiology of benign prostatic hyperplasia. *Urol Clin North Am.* 2016;43(3):279–288.
2. Hayward SW, Cunha GR. The prostate: Development and physiology. *Radiol Clin North Am.* 2000;38(1):1–14.
3. Mantzoros CS, Georgiadis EI, Trichopoulos D. Contribution of dihydrotestosterone to male sexual behaviour. *BMJ.* 1995;310 (6990):1289–1291.
4. Marchetti PM, Barth JH. Clinical biochemistry of dihydrotestosterone. *Ann Clin Biochem.* 2013;50(2):95–107.
5. Long RM, Morrissey C, Fitzpatrick JM, Watson RWG. Prostate epithelial cell differentiation and its relevance to the understanding of prostate cancer therapies. *Clin Sci.* 2005;108(1):1–11.
6. Abdul M, Anezinis PE, Logothetis CJ, Hoosein NM. Growth inhibition of human prostatic carcinoma cell lines by serotonin antagonists. *Anticancer Res.* 1994;14(3A):1215–1220.
7. Aprikian AG, Tremblay L, Han K, Chevalier S. Bombesin stimulates the motility of human prostate-carcinoma cells through tyrosine phosphorylation of focal adhesion kinase and of integrin-associated proteins. *Int J Cancer.* 1997;72(3):498–504.

8. Henry GH, Malewska A, Joseph DB, Malladi VS, Lee J, Torrealba J, et al. A cellular anatomy of the normal adult human prostate and prostatic urethra. *Cell Rep.* 2018;25(12):3530–3542.e5.
9. Cunha GR, Hayward SW, Wang Y, Ricke WA. Role of the stromal microenvironment in carcinogenesis of the prostate. *Int J Cancer.* 2003;107(1):1–10.
10. Barron DA, Rowley DR. The reactive stroma microenvironment and prostate cancer progression. *Endocr Relat Cancer.* 2012;19(6): R187–R204.
11. Li J, Yen C, Liaw D, Podsypanina K, Bose S, Wang SI, et al. PTEN, a putative protein tyrosine phosphatase gene mutated in human brain, breast, and prostate cancer. *Science.* 1997;275 (5308):1943–1947.
12. Teroerde M, Nientiedt C, Duensing A, Hohenfellner M, Stenzinger A, Duensing S. Revisiting the role of p53 in prostate cancer. In: Bott SRJ, Ng KL, editors. *Prostate Cancer.* Brisbane: Exon Publications; 2021. p. 113–123.
13. Phillips SM, Barton CM, Lee SJ, Morton DG, Wallace DM, Lemoine NR, et al. Loss of the retinoblastoma susceptibility gene (RB1) is a frequent and early event in prostatic tumorigenesis. *Br J Cancer.* 1994;70(6):1252–1257.
14. James ND, Tannock I, N'Dow J, Feng F, Gillessen S, Ali SA, et al. The Lancet Commission on prostate cancer: Planning for the surge in cases. *Lancet.* 2024;403(10437):1683–1722.
15. Epstein JI, Egevad L, Amin MB, Delahunt B, Srigley JR, Humphrey PA, et al. The 2014 International Society of Urological Pathology (ISUP) Consensus Conference on Gleason Grading of Prostatic Carcinoma: Definition of grading patterns and proposal for a new grading system. *Am J Surg Pathol.* 2016;40(2):244–252.
16. Ischia JJ, Pang CY, Tay YK, Li Wai Suen CFD, Aw HC, Frydenberg M. Active surveillance for prostate cancer: An Australian experience. *BJU Int.* 2012;109(Suppl 3):40–43.
17. Donovan JL, Hamdy FC, Lane JA, Mason M, Metcalfe C, Walsh E, et al. Patient-reported outcomes after monitoring, surgery, or radiotherapy for prostate cancer. *N Engl J Med.* 2016;375(15): 1425–1437.
18. Hopstaken JS, Bomers JGR, Sedelaar MJP, Valerio M, Fütterer JJ, Rovers MM. An updated systematic review on focal therapy in

localized prostate cancer: What has changed over the past 5 years? *Eur Urol.* 2022;81(1):5–33.

19. Spratt DE, Dess RT, Zumsteg ZS, Lin DW, Tran PT, Morgan TM, et al. A systematic review and framework for the use of hormone therapy with salvage radiation therapy for recurrent prostate cancer. *Eur Urol.* 2018;73(2):156–165.
20. Yokomizo A, Wakabayashi M, Satoh T, Hashine K, Inoue T, Fujimoto K, et al. Salvage radiotherapy versus hormone therapy for prostate-specific antigen failure after radical prostatectomy: A randomised, multicentre, open-label, phase 3 trial (JCOG0401). *Eur Urol.* 2020;77(6):689–698.
21. Huggins C, Hodges CV. Studies on prostatic cancer: I. The effect of castration, of estrogen and of androgen injection on serum phosphatases in metastatic carcinoma of the prostate. *J Urol.* 2002;168(1):9–12.
22. Jacob A, Raj R, Allison DB, Myint ZW. Androgen receptor signaling in prostate cancer and therapeutic strategies. *Cancers.* 2021; 13(21):5417.
23. James ND, Sydes MR, Clarke NW, Mason MD, Dearnaley DP, Spears MR, et al. Addition of docetaxel, zoledronic acid, or both to first-line long-term hormone therapy in prostate cancer (STAMPEDE): Survival results from an adaptive, multiarm, multistage, platform randomised controlled trial. *Lancet.* 2016; 387(10024):1163–1177.
24. Attard G, Murphy L, Clarke NW, Cross W, Jones RJ, Parker CC, et al. Abiraterone acetate and prednisolone with or without enzalutamide for high-risk non-metastatic prostate cancer: A meta-analysis of primary results from two randomised controlled phase 3 trials of the STAMPEDE platform protocol. *Lancet.* 2022;399 (10323):447–460.
25. Kyriakopoulos CE, Chen Y-H, Carducci MA, Liu G, Jarrard DF, Hahn NM, et al. Chemohormonal therapy in metastatic hormone-sensitive prostate cancer: Long-term survival analysis of the randomized phase III E3805 CHAARTED trial. *J Clin Oncol.* 2018; 36(11):1080–1087.
26. Perera M, Roberts MJ, Klotz L, Higano CS, Papa N, Sengupta S, et al. Intermittent versus continuous androgen deprivation therapy for advanced prostate cancer. *Nat Rev Urol.* 2020;17(8):469–481.

27. Sweeney CJ, Martin AJ, Stockler MR, Begbie S, Cheung L, Chi KN, et al. Testosterone suppression plus enzalutamide versus testosterone suppression plus standard antiandrogen therapy for metastatic hormone-sensitive prostate cancer (ENZAMET): An international, open-label, randomised, phase 3 trial. *Lancet Oncol.* 2023;24(4):323–334.
28. Davis ID, Martin AJ, Stockler MR, Begbie S, Chi KN, Chowdhury S, et al. Enzalutamide with standard first-line therapy in metastatic prostate cancer. *N Engl J Med.* 2019;381(2):121–131.
29. Berthold DR, Pond GR, Soban F, de Wit R, Eisenberger M, Tannock IF. Docetaxel plus prednisone or mitoxantrone plus prednisone for advanced prostate cancer: Updated survival in the TAX 327 study. *J Clin Oncol.* 2008;26(2):242–245.
30. de Bono JS, Oudard S, Ozguroglu M, Hansen S, Machiels JP, Kocak I, et al. Prednisone plus cabazitaxel or mitoxantrone for metastatic castration-resistant prostate cancer progressing after docetaxel treatment: A randomised open-label trial. *Lancet.* 2010; 376(9747):1147–1154.
31. Corn PG, Heath EI, Zurita A, Ramesh N, Xiao L, Sei E, et al. Cabazitaxel plus carboplatin for the treatment of men with metastatic castration-resistant prostate cancers: A randomised, open-label, phase 1-2 trial. *Lancet Oncol.* 2019;20(10):1432–1443.
32. Hussain M, Mateo J, Fizazi K, Saad F, Shore N, Sandhu S, et al. Survival with olaparib in metastatic castration-resistant prostate cancer. *N Engl J Med.* 2020;383(24):2345–2357.
33. Mateo J, Carreira S, Sandhu S, Miranda S, Mossop H, Perez-Lopez R, et al. DNA-repair defects and olaparib in metastatic prostate cancer. *N Engl J Med.* 2015;373(18):1697–1708.
34. Agarwal N, Azad AA, Carles J, Fay AP, Matsubara N, Heinrich D, et al. Talazoparib plus enzalutamide in men with first-line metastatic castration-resistant prostate cancer (TALAPRO-2): A randomised, placebo-controlled, phase 3 trial. *Lancet.* 2023;402 (10398):291–303.
35. Sartor O, de Bono J, Chi KN, Fizazi K, Herrmann K, Rahbar K, et al. Lutetium-177-PSMA-617 for metastatic castration-resistant prostate cancer. *N Engl J Med.* 2021;385(12):1091–1103.
36. Hofman MS, Emmett L, Sandhu S, Iravani A, Joshua AM, Goh JC, et al. [^{177}Lu]Lu-PSMA-617 versus cabazitaxel in patients with metastatic castration-resistant prostate cancer (TheraP): A ran-

domised, open-label, phase 2 trial. *Lancet.* 2021;397(10276): 797–804.

37. Kantoff PW, Higano CS, Shore ND, Berger ER, Small EJ, Penson DF, et al. Sipuleucel-T immunotherapy for castration-resistant prostate cancer. *N Engl J Med.* 2010;363(5):411–422.
38. Fizazi K, Retz M, Petrylak DP, Goh JC, Perez-Gracia J, Lacombe L, et al. Nivolumab plus rucaparib for metastatic castration-resistant prostate cancer: Results from the phase 2 CheckMate 9KD trial. *J Immunother Cancer.* 2022;10(8):e004761.
39. Antonarakis ES, Piulats JM, Gross-Goupil M, Goh J, Ojamaa K, Hoimes CJ, et al. Pembrolizumab for treatment-refractory metastatic castration-resistant prostate cancer: Multicohort, open-label phase II KEYNOTE-199 study. *J Clin Oncol.* 2020; 38(5):395–405.
40. Yu EY, Piulats JM, Gravis G, Fong PCC, Todenhöfer T, Laguerre B, et al. Pembrolizumab plus olaparib in patients with metastatic castration-resistant prostate cancer: Long-term results from the phase 1b/2 KEYNOTE-365 cohort a study. *Eur Urol.* 2023; 83(1):15–26.
41. Porter LH, Zhu JJ, Lister NL, Harrison SG, Keerthikumar S, Goode DL, et al. Low-dose carboplatin modifies the tumor microenvironment to augment CAR T cell efficacy in human prostate cancer models. *Nat Commun.* 2023;14(1):5346.
42. Dorff TB, Narayan V, Forman SJ, Zang PD, Fraietta JA, June CH, et al. Novel redirected T-cell immunotherapies for advanced prostate cancer. *Clin Cancer Res.* 2022;28(4):576–584.
43. Das G, Ptacek J, Havlinova B, Nedvedova J, Barinka C, Novakova Z. Targeting prostate cancer using bispecific T-cell engagers against prostate-specific membrane antigen. *ACS Pharmacol Transl Sci.* 2023;6(11):1703–1714.
44. Beltran H, Rickman DS, Park K, Chae SS, Sboner A, MacDonald TY, et al. Molecular characterization of neuroendocrine prostate cancer and identification of new drug targets. *Cancer Discov.* 2011; 1(6):487–495.
45. Yamada Y, Beltran H. Clinical and biological features of neuroendocrine prostate cancer. *Curr Oncol Rep.* 2021;23(2):15.
46. Beltran H, Demichelis F. Therapy considerations in neuroendocrine prostate cancer: What next? *Endocr Relat Cancer.* 2021; 28(8):T67–T78.

47. de Kouchkovsky I, Chan E, Schloss C, Poehlein C, Aggarwal R. Diagnosis and management of neuroendocrine prostate cancer. *Prostate.* 2024;84(5):426–440.
48. Koukourakis IM, Platoni K, Kouloulias V, Arelaki S, Zygogianni A. Prostate cancer stem cells: Biology and treatment implications. *Int J Mol Sci.* 2023;24(19):14890.
49. Sabnis NG, Miller A, Titus MA, Huss WJ. The efflux transporter ABCG2 maintains prostate stem cells. *Mol Cancer Res.* 2017; 15(2):128–140.
50. Wang N, Docherty F, Brown HK, Reeves K, Fowles A, Lawson M, et al. Mitotic quiescence, but not unique "stemness," marks the phenotype of bone metastasis-initiating cells in prostate cancer. *FASEB J.* 2015;29(8):3141–3150.
51. Bahmad HF, Jalloul M, Azar J, Moubarak MM, Samad TA, Mukherji D, et al. Tumour microenvironment in prostate cancer: Toward identification of novel molecular biomarkers for diagnosis, prognosis, and therapy development. *Front Genet.* 2021; 12:652747.
52. Berglund E, Maaskola J, Schultz N, Friedrich S, Marklund M, Bergenstråhle J, et al. Spatial maps of prostate cancer transcriptomes reveal an unexplored landscape of heterogeneity. *Nat Commun.* 2018;9(1):2419.
53. Pan J, Ma Z, Liu B, Qian H, Shao X, Liu J, et al. Identification of cancer-associated fibroblasts subtypes in prostate cancer. *Front Immunol.* 2023;14:1133160.
54. Lakins MA, Ghorani E, Munir H, Martins CP, Shields JD. Cancer-associated fibroblasts induce antigen-specific deletion of CD8+ T cells to protect tumour cells. *Nat Commun.* 2018;9(1):948.
55. Yu S, Jiang Y, Wan F, Wu J, Gao Z, Liu D. Immortalized cancer-associated fibroblasts promote prostate cancer carcinogenesis, proliferation and invasion. *Anticancer Res.* 2017;37(8):4311–4318.
56. Leach DA, Need EF, Toivanen R, Trotta AP, Palenthorpe HM, Tamblyn DJ, et al. Stromal androgen receptor regulates the composition of the microenvironment to influence prostate cancer outcome. *Oncotarget.* 2015;6(18):16135–16150.
57. Leach DA, Buchanan G. Stromal androgen receptor in prostate cancer development and progression. *Cancers.* 2017;9(1):10.

58. Ioannidou E, Moschetta M, Shah S, Parker JS, Ozturk MA, Pappas-Gogos G, et al. Angiogenesis and anti-angiogenic treatment in prostate cancer: Mechanisms of action and molecular targets. *Int J Mol Sci.* 2021;22(18):9926.
59. Bedeschi M, Marino N, Cavassi E, Piccinini F, Tesei A. Cancer-associated fibroblast: Role in prostate cancer progression to metastatic disease and therapeutic·resistance. *Cells.* 2023;12(5):802.
60. Palano MT, Gallazzi M, Cucchiara M, Dehò F, Capogrosso P, Bruno A, et al. The tumor innate immune microenvironment in prostate cancer: An overview of soluble factors and cellular effectors. *Explor Target Anti-Tumor Ther.* 2022;3(5):694–718.
61. Stultz J, Fong L. How to turn up the heat on the cold immune microenvironment of metastatic prostate cancer. *Prostate Cancer Prostatic Dis.* 2021;24(3):697–717.
62. Messex JK, Liou G-Y. Impact of immune cells in the tumor microenvironment of prostate cancer metastasis. *Life.* 2023;13(2):333.
63. Huang J, Zhang L, Wan D, Zhou L, Zheng S, Lin S, et al. Extracellular matrix and its therapeutic potential for cancer treatment. *Signal Transduct Target Ther.* 2021;6(1):153.
64. Kang J, La Manna F, Bonollo F, Sampson N, Alberts IL, Mingels C, et al. Tumor microenvironment mechanisms and bone metastatic disease progression of prostate cancer. *Cancer Lett.* 2022; 530:156–169.
65. Corn PG, Wang F, McKeehan WL, Navone N. Targeting fibroblast growth factor pathways in prostate cancer. *Clin Cancer Res.* 2013;19(21):5856–5866.
66. Belli C, Trapani D, Viale G, D'Amico P, Duso BA, Della Vigna P, et al. Targeting the microenvironment in solid tumors. *Cancer Treat Rev.* 2018;65:22–32.
67. Nguyen DP, Li J, Tewari AK. Inflammation and prostate cancer: The role of interleukin 6 (IL-6). *BJU Int.* 2014;113(6):986–992.
68. Culig Z, Bartsch G, Hobisch A. Interleukin-6 regulates androgen receptor activity and prostate cancer cell growth. *Mol Cell Endocrinol.* 2002;197(1–2):231–238.
69. Sun YX, Wang J, Shelburne CE, Lopatin DE, Chinnaiyan AM, Rubin MA, et al. Expression of CXCR4 and CXCL12 (SDF-1) in human prostate cancers (PCa) in vivo. *J Cell Biochem.* 2003; 89(3):462–473.

70. Stewart DA, Cooper CR, Sikes RA. Changes in extracellular matrix (ECM) and ECM-associated proteins in the metastatic progression of prostate cancer. *Reprod Biol Endocrinol.* 2004;2:2.
71. Gong Y, Chippada-Venkata UD, Oh WK. Roles of matrix metalloproteinases and their natural inhibitors in prostate cancer progression. *Cancers.* 2014;6(3):1298–1327.
72. Dong Z, Saliganan AD, Meng H, Nabha SM, Sabbota AL, Sheng S, et al. Prostate cancer cell-derived urokinase-type plasminogen activator contributes to intraosseous tumor growth and bone turnover. *Neoplasia.* 2008;10(5):439–449.
73. Mekkawy AH, Pourgholami MH, Morris DL. Involvement of urokinase∆type plasminogen activator system in cancer: An overview. *Med Res Rev.* 2014;34(5):918–956.
74. Ojalill M, Rappu P, Siljamäki E, Taimen P, Boström P, Heino J. The composition of prostate core matrisome in vivo and in vitro unveiled by mass spectrometric analysis. *Prostate.* 2018;78(8): 583–594.
75. Luthold C, Hallal T, Labbé DP, Bordeleau F. The extracellular matrix stiffening: A trigger of prostate cancer progression and castration resistance? *Cancers.* 2022;14(12):2887.
76. Levental KR, Yu H, Kass L, Lakins JN, Egeblad M, Erler JT, et al. Matrix crosslinking forces tumor progression by enhancing integrin signaling. *Cell.* 2009;139(5):891–906.
77. Jiang Y, Zhang H, Wang J, Liu Y, Luo T, Hua H. Targeting extracellular matrix stiffness and mechanotransducers to improve cancer therapy. *J Hematol Oncol.* 2022;15(1):34.
78. Fisher R, Pusztai L, Swanton C. Cancer heterogeneity: Implications for targeted therapeutics. *Br J Cancer.* 2013;108(3):479–485.
79. Sun T, Jackson S, Haycock JW, MacNeil S. Culture of skin cells in 3D rather than 2D improves their ability to survive exposure to cytotoxic agents. *J Biotechnol.* 2006;122(3):372–381.
80. Lou J, Mooney DJ. Chemical strategies to engineer hydrogels for cell culture. *Nat Rev Chem.* 2022;6(10):726–744.
81. Richbourg N, Wechsler ME, Rodriguez-Cruz JJ, Peppas NA. Model-based modular hydrogel design. *Nat Rev Bioeng.* 2024;2(7): 1–13.
82. Bashir S, Hina M, Iqbal J, Rajpar A, Mujtaba M, Alghamdi N, et al. Fundamental concepts of hydrogels: Synthesis, properties, and their applications. *Polymers.* 2020;12(11):2702.

83. Lee J, Abdeen AA, Li Y, Goonetilleke S, Kilian KA. Gradient and dynamic hydrogel materials to probe dynamics in cancer stem cell phenotypes. *ACS Appl Bio Mater.* 2020;4(1):711–720.
84. Maskarinec G, Pagano IS, Little MA, Conroy SM, Park S-Y, Kolonel LN. Mammographic density as a predictor of breast cancer survival: The multiethnic cohort. *Breast Cancer Res.* 2013;15:1–8.
85. Bertero T, Oldham WM, Grasset EM, Bourget I, Boulter E, Pisano S, et al. Tumor-stroma mechanics coordinate amino acid availability to sustain tumor growth and malignancy. *Cell Metab.* 2019;29(1):124–140. e10.
86. Hupfer A, Brichkina A, Koeniger A, Keber C, Denkert C, Pfefferle P, et al. Matrix stiffness drives stromal autophagy and promotes formation of a protumorigenic niche. *Proc Natl Acad Sci U S A.* 2021;118(40):e2105367118.
87. Teixeira AM, Martins P. A review of bioengineering techniques applied to breast tissue: Mechanical properties, tissue engineering and finite element analysis. *Front Bioeng Biotechnol.* 2023;11:1161815.
88. Rubiano A, Delitto D, Han S, Gerber M, Galitz C, Trevino J, et al. Viscoelastic properties of human pancreatic tumors and in vitro constructs to mimic mechanical properties. *Acta Biomater.* 2018;67:331–340.
89. Kawano S, Kojima M, Higuchi Y, Sugimoto M, Ikeda K, Sakuyama N, et al. Assessment of elasticity of colorectal cancer tissue, clinical utility, pathological and phenotypical relevance. *Cancer Sci.* 2015; 106(9):1232–1239.
90. Lin Y, Yuan K, Yang Y, Yang S, Huang K, Yu Z, et al. Osteosarcoma progression in biomimetic matrix with different stiffness: Insights from a three-dimensional printed gelatin methacrylamide hydrogel. *Int J Biol Macromol.* 2023;252:126391.
91. Ahn W, Lee J-H, Kim SR, Lee J, Lee EJ. Designed protein- and peptide-based hydrogels for biomedical sciences. *J Mater Chem B.* 2021;9(8):1919–1940.
92. Kaur K, Wang Y, Matson JB. Linker-regulated H2S release from aromatic peptide amphiphile hydrogels. *Biomacromolecules.* 2020; 21(3):1171–1178.
93. Fan Y, Zhan M, Liang J, Yang X, Zhang B, Shi X, et al. Programming injectable DNA hydrogels yields tumor microenvironment-activatable and immune-instructive depots for augmented chemo-immunotherapy. *Adv Sci.* 2023;10(29):e2302119.

94. Voycheva C, Slavkova M, Popova T, Tzankova D, Stefanova D, Tzankova V, et al. Thermosensitive hydrogel-functionalized mesoporous silica nanoparticles for parenteral application of chemotherapeutics. *Gels*. 2023;9(9):769.
95. Koch MK, Jaeschke A, Murekatete B, Ravichandran A, Tsurkan M, Werner C, et al. Stromal fibroblasts regulate microvascular-like network architecture in a bioengineered breast tumour angiogenesis model. *Acta Biomater*. 2020;114:256–269.
96. Bupphathong S, Lim J, Fang H-W, Tao H-Y, Yeh C-E, Ku T-A, et al. Enhanced vascular-like network formation of encapsulated HUVECs and ADSCs coculture in growth factors conjugated GelMA hydrogels. *ACS Biomater Sci Eng*. 2024;10(5);3306–3315.
97. Kuehlbach C, Hensler S, Mueller MM. Recapitulating the angiogenic switch in a hydrogel-based 3D in vitro tumor-stroma model. *Bioengineering*. 2021;8(11):186.
98. Huang Y, Feng Q, Jiang H, Zhou W, Chen J, Gao J, et al. Mimicking the endometrial cancer tumor microenvironment to reprogram tumor-associated macrophages in disintegrable supramolecular gelatin hydrogel. *Int J Nanomedicine*. 2020;15: 4625–4637.
99. Chen J-WE, Lumibao J, Leary S, Sarkaria JN, Steelman AJ, Gaskins HR, et al. Crosstalk between microglia and patient-derived glioblastoma cells inhibit invasion in a three-dimensional gelatin hydrogel model. *J Neuroinflammation*. 2020;17(1):1–15.
100. Bahlmann LC, Baker AE, Xue C, Liu S, Meier-Merziger M, Karakas D, et al. Gelatin-hyaluronan click-crosslinked cryogels elucidate human macrophage invasion behavior. *Adv Funct Mater*. 2021;31(28):2008400.
101. Yang Y, Yang Y, Chen M, Chen J, Wang J, Ma Y, et al. Injectable shear-thinning polylysine hydrogels for localized immunotherapy of gastric cancer through repolarization of tumor-associated macrophages. *Biomaterials Sci*. 2021;9(19):6597–6608.
102. Gebeyehu A, Surapaneni SK, Huang J, Mondal A, Wang VZ, Haruna NF, et al. Polysaccharide hydrogel based 3D printed tumor models for chemotherapeutic drug screening. *Sci Rep*. 2021;11(1):372.
103. Li Y, Khuu N, Prince E, Tao H, Zhang N, Chen Z, et al. Matrix stiffness-regulated growth of breast tumor spheroids and their

response to chemotherapy. *Biomacromolecules.* 2020;22(2): 419–429.

104. Alipournazari P, Pourmadadi M, Abdouss M, Rahdar A, Pandey S. Enhanced delivery of doxorubicin for breast cancer treatment using pH-sensitive starch/PVA/g-C3N4 hydrogel. *Int J Biol Macromol.* 2024;265:130901.
105. Zhang Y, Zhang X, Zhang L, Yao J, Wang S, Zhao J, et al. A pH-responsive injectable hydrogel for enhanced chemo-chemodynamic synergistic therapy. *Colloids Surf A: Physicochem Eng Asp.* 2024; 690:133748.
106. Li J, Luo G, Zhang C, Long S, Guo L, Yang G, et al. In situ injectable hydrogel-loaded drugs induce anti-tumor immune responses in melanoma immunochemotherapy. *Mater Today Bio.* 2022;14:100238.
107. Fong EL, Martinez M, Yang J, Mikos AG, Navone NM, Harrington DA, et al. Hydrogel-based 3D model of patient-derived prostate xenograft tumors suitable for drug screening. *Mol Pharm.* 2014; 11(7):2040–2050.
108. Mosquera MJ, Kim S, Bareja R, Fang Z, Cai S, Pan H, et al. Extracellular matrix in synthetic hydrogel-based prostate cancer organoids regulate therapeutic response to EZH2 and DRD2 inhibitors. *Adv Mater.* 2022;34(2):e2100096.
109. Ertekin Ö, Monavari M, Krüger R, Fuentes-Chandía M, Parma B, Letort G, et al. 3D hydrogel-based microcapsules as an in vitro model to study tumorigenicity, cell migration and drug resistance. *Acta Biomater.* 2022;142:208–220.
110. Shou Y, Teo XY, Li X, Zhicheng L, Liu L, Sun X, et al. Dynamic magneto-softening of 3D hydrogel reverses malignant transformation of cancer cells and enhances drug efficacy. *ACS Nano.* 2023;17(3):2851–2867.
111. Lachowski D, Matellan C, Cortes E, Saiani A, Miller AF, del Río Hernández AE. Self-assembling polypeptide hydrogels as a platform to recapitulate the tumor microenvironment. *Cancers.* 2021;13(13):3286.
112. Wang J, Xu W, Qian J, Wang Y, Hou G, Suo A. Photo-crosslinked hyaluronic acid hydrogel as a biomimic extracellular matrix to recapitulate in vivo features of breast cancer cells. *Colloids Surf B: Biointerfaces.* 2022;209:112159.

113. Zhang Z, Chen W, Sun M, Aalders T, Verhaegh GW, Kouwer PH. TempEasy 3D hydrogel coculture system provides mechanistic insights into prostate cancer bone metastasis. *ACS Appl Mater Interfaces.* 2024;16(20):25773–25787.
114. Bessot A, Gunter J, Waugh D, Clements JA, Hutmacher DW, McGovern J, et al. GelMA and biomimetic culture allow the engineering of mineralized, adipose, and tumor tissue human microenvironments for the study of advanced prostate cancer in vitro and in vivo. *Adv Healthc Mater.* 2023;12(14):e2201701.
115. Xu K, Wang Z, Copland JA, Chakrabarti R, Florczyk SJ. 3D porous chitosan-chondroitin sulfate scaffolds promote epithelial to mesenchymal transition in prostate cancer cells. *Biomaterials.* 2020;254:120126.
116. Meinert C, Theodoropoulos C, Klein TJ, Hutmacher DW, Loessner D. A method for prostate and breast cancer cell spheroid cultures using gelatin methacryloyl-based hydrogels. *Methods Mol Biol.* 2018:175–194.
117. Stockmann C, Schadendorf D, Klose R, Helfrich I. The impact of the immune system on tumor: Angiogenesis and vascular remodeling. *Front Oncol.* 2014;4:69.
118. Fisher MF, Rao SS. Three-dimensional culture models to study drug resistance in breast cancer. *Biotechnol Bioeng.* 2020; 117(7):2262–2278.
119. Tang Y, Huang B, Dong Y, Wang W, Zheng X, Zhou W, et al. Three-dimensional prostate tumor model based on a hyaluronic acid-alginate hydrogel for evaluation of anti-cancer drug efficacy. *J Biomater Sci Polym Ed.* 2017;28(14):1603–1616.
120. Guimarães CF, Liu S, Wang J, Purcell E, Ozedirne T, Ren T, et al. Co-axial hydrogel spinning for facile biofabrication of prostate cancer-like 3D models. *Biofabrication.* 2024;16(2):025017.
121. Gandaglia G, Abdollah F, Schiffmann J, Trudeau V, Shariat SF, Kim SP, et al. Distribution of metastatic sites in patients with prostate cancer: A population-based analysis. *Prostate.* 2014; 74(2):210–216.
122. Bock C, Boutros M, Camp JG, Clarke L, Clevers H, Knoblich JA, et al. The organoid cell atlas. *Nat Biotechnol.* 2021;39(1):13–17.

123. Gao D, Vela I, Sboner A, Iaquinta PJ, Karthaus WR, Gopalan A, et al. Organoid cultures derived from patients with advanced prostate cancer. *Cell.* 2014;159(1):176–187.
124. Sachs N, De Ligt J, Kopper O, Gogola E, Bounova G, Weeber F, et al. A living biobank of breast cancer organoids captures disease heterogeneity. *Cell.* 2018;172(1):373–86. e10.
125. Van de Wetering M, Francies HE, Francis JM, Bounova G, Iorio F, Pronk A, et al. Prospective derivation of a living organoid biobank of colorectal cancer patients. *Cell.* 2015;161(4):933–945.

https://doi.org/10.1142/9789819806782_bmatter

Index